The Handbook of Environmental Chemistry

Volume 133

Founding Editor: Otto Hutzinger

Series Editors: Damià Barceló • Andrey G. Kostianoy

In over four decades, *The Handbook of Environmental Chemistry* has established itself as the premier reference source, providing sound and solid knowledge about environmental topics from a chemical perspective. Written by leading experts with practical experience in the field, the series continues to be essential reading for environmental scientists, environmental managers, and decision-makers in industry, government, agencies and public-interest groups. The series also supports and advances the UN's sustainable development goals, mainly SDG14 (Life below water), SDG6 (Clean Water and Sanitation), SDG13 (Climate Change) and SDG15 (Life on Land).

Two distinguished Series Editors, internationally renowned volume editors as well as a prestigious Editorial Board safeguard the publication of volumes according to high scientific standards.

Presenting a wide spectrum of viewpoints and approaches in topical volumes, the scope of the series covers topics such as

- local and global changes of natural environment and climate
- anthropogenic impact on the environment
- water, air and soil pollution
- remediation and waste characterization
- environmental contaminants
- biogeochemistry and geoecology
- chemical reactions and processes
- chemical and biological transformations as well as physical transport of chemicals in the environment
- environmental modeling

A particular focus of the series lies in methodological advances in environmental analytical chemistry.

The Handbook of Envir onmental Chemistry is available both in print and online via http://link.springer.com/bookseries/698. Articles are published online as soon as they have been reviewed and approved for publication.

Meeting the needs of the scientific community, publication of volumes in subseries has been discontinued to achieve a broader scope for the series as a whole.

Sustainable Strategies for Air Pollution Mitigation

Development, Economics, and Technologies

Volume Editors: Matthew Chidozie Ogwu ·
Sylvester Chibeuze Izah

With contributions by

A. O. Aigberua · M. Akram · H. G. Al-Thani · T. Amabie ·
B. R. Barnes · R. S. Breitenstein · N. G. Etim · R. Glory ·
B. Ikhajiagbe · O. Imarhiagbe · R. J. Isaifan · S. C. Izah ·
E. A. Kosoe · T. P. L. Nguyen · M. U. Nwodo ·
E. G. Obahiagbon · T. C. Odubo · M. C. Ogwu ·
C. A. Onwudiegwu · M. E. Osawaru · N. A. Poggioli ·
G. Richard · B. Ro · A. Sarfraz · M. Sarfraz · Z. Sarfraz ·
W. E. Sawyer · A. Sharipov · L. Sylva · B. Thapa ·
O. P. Thompson · J. Xu

Springer

Editors
Matthew Chidozie Ogwu
Appalachian State University
Boone, NC, USA

Sylvester Chibeuze Izah
Bayelsa Medical University
Yenagoa, Nigeria

ISSN 1867-979X ISSN 1616-864X (electronic)
The Handbook of Environmental Chemistry
ISBN 978-3-031-74156-2 ISBN 978-3-031-74157-9 (eBook)
https://doi.org/10.1007/978-3-031-74157-9

This Springer imprint is published by the registered company Springer Nature Switzerland AG
The registered company address is: Gewerbestrasse 11, 6330 Cham, Switzerland

This book is dedicated to Almighty God in recognition of His infinite wisdom and knowledge.

Preface

Air pollution represents one of the most critical environmental challenges of our time, affecting not only human health but also the delicate balance of ecosystems worldwide. The detrimental impacts of air pollutants extend from urban centers to rural landscapes, contributing to climate change, degrading air quality, and impairing the natural systems upon which all life depends. In light of these escalating challenges, this edited book, *Sustainable Strategies for Air Pollution Mitigation: Development, Economics, and Technologies*, emerges as a timely and essential resource, aiming to bridge the gap between academic research, policy development, and practical applications in the field of air pollution control by providing a thorough and integrative examination of the diverse strategies and solutions necessary to address air pollution sustainably.

The book provides a holistic understanding of sustainable approaches to air pollution mitigation by integrating perspectives from development, economics, and technological innovation. It illuminates the intricate links between air pollution and broader societal and economic contexts, while also presenting advanced technological solutions and sustainable strategies. This approach underscores the importance of a holistic view in understanding and tackling the air pollution crisis. It is organized into four main sections, each delving into critical aspects of air pollution mitigation. The introductory section provides a foundational understanding of the environmental and health impacts of air pollution and sets the stage for the discussions in subsequent chapters by detailing the scope and scale of air pollution as a global challenge. The second part focuses on the *Development and Economic Impacts of Air Pollution*. Chapters in this section explore the direct and indirect environmental consequences of air pollution, including ecosystem damage and climate change implications as well as the effects of air pollution on social structures and economic productivity, highlighting how air quality influences public health, labor markets, and economic growth. The section also discusses cost–benefit analyses, economic losses, and the financial implications of various mitigation strategies and examines how air pollution affects businesses and industries, considering both the challenges it poses and the opportunities it creates for innovation and sustainable practices. It is known that air pollution has a disproportionately higher impact on marginalized and vulnerable communities, and we emphasized the importance of equity in air quality management. The final chapter in this section connects air pollution with the broader agenda of the United Nations Sustainable Development Goals (SDGs), illustrating how mitigating air pollution is integral to achieving these goals. The third part is on *Technological and Sustainable Strategies for Air Pollution Mitigation.* Chapters in this section outline sustainable practices and policies that can effectively reduce air pollution while promoting long-term environmental health, discuss cutting-edge technologies that offer innovative solutions for air quality improvement, from emission reduction technologies to advanced monitoring systems, and the roles of urban planning and green infrastructure in mitigating air pollution are explored, highlighting how cities can be designed to support cleaner air and healthier environments. An overview of policy frameworks and regulatory approaches that support sustainable air quality improvements is also presented along with case studies from the Middle East and North Africa region, showcasing regional actions and strategies for reducing air pollution and its health impacts. The concluding section issues a call to action for all stakeholders to commit to sustainable practices and policies that will lead to a world free from the adverse impacts of air pollution.

The book is intended for a broad audience including researchers, policymakers, practitioners, and students. It aims to serve as a reference for those engaged in the study and practice of air pollution mitigation and environmental sustainability. Information presented in this book can not only catalyze further research and policy development but also inspire innovative solutions to one of the most pressing environmental challenges of our time. There is a need to integrate development, economics, and technological perspectives from an air pollution standpoint, and this book provides a robust framework for understanding and mitigating air pollution sustainably. By addressing the intricate interplay between development, economics, and technology, this book will equip readers with the knowledge and tools to tackle one of the most pressing environmental issues of our time.

Boone, NC, USA Matthew Chidozie Ogwu
Bayelsa, Nigeria Sylvester Chibueze Izah
June 2024

Acknowledgments

Numerous individuals made significant contributions to ensure the completion of this book. In particular, we thank the editors of The Handbook of Environmental Chemistry for their invaluable guidance and advice. We are equally thankful to the dedicated editorial team at Springer, including Ms. Lydia Mueller, Dr. Sofia Costa, Ms. Neelofar Yasmeen, and others, for their unwavering commitment to the success of this book. Our heartfelt thanks also go to all the contributors who played a vital role in different work sections. We sincerely appreciate your shared knowledge, enriching humanity through this book.

Contents

About the Editors

Matthew Chidozie Ogwu is an Assistant Professor of Integrated Ecology and Sustainable Development in the Goodnight Family Sustainable Development Department at Appalachian State University, USA. He obtained his Ph.D. in Biological Sciences with a research focus in the Molecular Biology of Soil Microbial and Geographical Ecology from Seoul National University. He holds an M.Sc. (Distinction) and B.Sc. (First Class) in Plant Diversity and Conservation and Plant Biology and Biotechnology, respectively, from the University of Benin, Nigeria. After receiving his Ph.D., he worked as a Researcher at the Centre for Floristic Research of the Apennines, Italy, where he contributed to numerous European Union-funded research including the famous NATURA 2000 project for sustainable ecosystem and biodiversity management. Before that, Dr. Ogwu was a Lecturer at the University of Benin, Nigeria, where he taught courses in plant biology and biotechnology, biometry, economic botany, cytogenetics, etc. and supervised and co-supervised honors and graduate students' theses, respectively. Dr. Ogwu is an interdisciplinary academic with transdisciplinary skills pertinent to the assessment of coupled human and natural as well as socio-ecological systems and has numerous awards, research grants, and scholarships to his name. His research and teaching revolve around sustainable biodiversity and ecosystem management by assessing the realm of coupled systems for sustainable environmental development as well as addressing sustainable development issues arising from various scales of biogeographical interactions. Some of his research interests include soil microbial and geographical ecology, molecular and genomic characterization of economic plant species for breeding, conservation, and sustainable utilization as well as biodiversity issues and policies, climate change science, gut and fecal microbiota, bioinformatics, biosystems and bioeconomy, sustainable agriculture and food production systems, ethnoscience (including traditional biotechnological practices), land use patterns, waste management, microbial diversity, environmental and social systems integrity, landscape preservation, and pollution ecology. He is spearheading some convergence works (One Health and Eco Health) in the Human Environmental and Agricultural Laboratory (HEAL lab) at Appalachian State University. Dr. Ogwu has published over 100 peer-reviewed works including several in high-impact journals and has presented his results at different international conferences. Dr. Ogwu has attended numerous professional courses and serves on the board of and as a reviewer for many peer-reviewed international journals. He continues to volunteer his time and skills to promote sustainable community development, especially in the Global South. Dr. Ogwu is fluent in several languages and loves to travel and meet new people.

Sylvester Chibueze Izah holds a Ph.D. in Public Health, Applied Microbiology and Environmental Health, an M.Sc. in Applied Microbiology and a B.Sc. in Biological Sciences from Niger Delta University, Nigeria. He is a licensed Environmental Health Specialist in Nigeria. He brings over ten years of experience from his work with an environmental consultancy firm specializing in soil, water, and atmospheric assessments. During the early stages of his career, Dr. Izah conducted studies on vegetation cover and biodiversity, focusing on species diversity, composition, and utilization. Presently, he serves as a lecturer at Bayelsa Medical University in Yenagoa, Nigeria, where he holds the position of Assistant Director of Academic Planning, Research, and Innovations. Dr. Izah's extensive research portfolio encompasses water, air, and soil quality, applied microbiology, biotechnology, public and environmental health, risk assessment, bioenergy, toxicology, medicinal plants, herbal medicine, and biodiversity. He has an impressive track record of over 250 peer-reviewed publications, including journal articles, book chapters, and edited books. Dr. Izah's Google Scholar total citation, h-index, and i10-Index are >6604, 44, and 179, respectively. His current research interests are centered around Sustainable Human-Environmental Health Interactions, covering areas such as environmental sustainability trends, toxicology, hygiene, sanitation, food science, public-environmental health assessment, medicinal plants, bioactive and natural compounds, vegetation, and wildlife. Dr. Izah has also collaborated on research projects with colleagues worldwide and actively contributes as an editorial and review board member for several esteemed journals.

Part I
Introduction

Toward Understanding the Impacts of Air Pollution

Matthew C. Ogwu, Odoligie Imarhiagbe, Beckley Ikhajiagbe, and Moses Edwin Osawaru

Contents

M. C. Ogwu (✉)
Goodnight Family Department of Sustainable Development, Appalachian State University, Boone, NC, USA
e-mail: ogwumc@appstate.edu

O. Imarhiagbe
Department of Health and Social Science, London School of Science and Technology, London, UK
e-mail: odoligie.imarhiagbe@lsst.ac

B. Ikhajiagbe and M. E. Osawaru
Department of Plant Biology and Biotechnology, University of Benin, Benin City, Edo State, Nigeria
e-mail: Beckley.ikhajiagbe@uniben.edu; Moses.osawaru@uniben.edu

Matthew Chidozie Ogwu and Sylvester Izah (eds.), *Sustainable Strategies for Air Pollution Mitigation: Development, Economics and Technologies*,
Hdb Env Chem (2024) 133: 3–44, DOI 10.1007/698_2024_1107,

Published online: 28 August 2024

Abstract This chapter aims to highlight the impacts of air pollution. Air pollution is a pervasive environmental problem with significant implications for human health, ecosystems, and global climate. Examining the impacts of poor air quality on society will help develop a comprehensive understanding of air pollution and work toward a cleaner and healthier environment for all. Addressing air quality issues is essential for safeguarding public health, protecting the environment, promoting economic prosperity, fostering social equity, and advancing global cooperation. By implementing effective pollution control measures, investing in clean technologies, and adopting sustainable practices, human societies can create a healthier, more resilient, and sustainable future for current and future generations. Air pollution originates from a wide range of sources, including industrial emissions, transportation activities, agricultural practices, natural phenomena, and indoor combustion. Each source contributes to the complex mixture of pollutants in the air, affecting air quality, human health, ecosystems, and the climate. Some sustainable strategies for addressing air pollution include regulatory measures, technological advancements, and sustainable practices; we can improve air quality, protect public health, and safeguard the environment for current and future generations. Air pollution dispersion is influenced by a multitude of mechanisms, including atmospheric stability, wind patterns, topography, and atmospheric reactions. There is a need to consider the complex interplay of these dispersion mechanisms to understand the impacts of air pollution as well as address the challenges of air pollution, protect public health, assess air quality, predict pollutant concentrations, develop effective pollution control measures, and promote sustainable air quality management practices. The diverse impacts of air pollution on human health, the environment, climate change, and socioeconomic factors underscore the urgent need to address this global challenge through comprehensive strategies and collaborative efforts. Understanding the multifaceted effects of air pollution and the interconnected nature of its impacts can help prioritize pollution prevention, emission reductions, sustainable development practices, and public health protection measures to safeguard the well-being of present and future generations. Addressing the impacts of air pollution requires a coordinated approach involving government policies, industry regulations, scientific research, public awareness campaigns, and community engagement to create a cleaner, healthier, and more sustainable environment for all.

Keywords Air quality, Community engagement, Government policies, Industrial emissions, Pollution control measures, Public awareness, Sustainable environment

Abbreviations

CO	Carbon monoxide
SO2	Sulfur dioxide
NOx	Nitrogen oxides
O3	Ozone
CO2	Carbon dioxide
CH4	Methane
N2O	Nitrous oxide
PM_{10}	Coarse particulate matter
$PM_{2.5}$	Fine particulate matter
PM	Particulate matter
COPD	Chronic obstructive pulmonary disease
EPA	U.S. Environmental Protection Agency
HFCs	Hydrofluorocarbons
PFCs	Perfluorocarbons
SF_6	Sulfur hexafluoride
IARC	International Agency for Research on Cancer
UV	Ultraviolet
CFCs	Chlorofluorocarbons
ROS	Reactive oxygen species
DNA	Deoxyribonucleic acid
PAHs	Polycyclic aromatic hydrocarbons
ADHD	Attention deficit hyperactivity disorder
WHO	World Health Organization
NGOs	Non-governmental organizations
pH	Hydrogen potential
H2SO4	Sulfuric acid
HNO3	Nitric acid
FGD	Flue gas desulfurization
SCR	Selective catalytic reduction
Pb	Lead
Hg	Mercury
Cd	Cadmium
As	Arsenic

1 Introduction

Air pollution is a pervasive and complex environmental issue that affects human health, ecosystems, and the global climate. Defined as the presence of harmful or excessive substances in the air, air pollution poses significant risks to public health and the environment [1, 2]. At its core, air pollution encompasses the release of various pollutants into the atmosphere, altering its chemical composition and leading to adverse effects on human health and the environment. These pollutants can originate from natural sources, such as volcanic eruptions, wildfires, and biological processes, or anthropogenic activities, including industrial processes, transportation, and energy production. While some pollutants are emitted directly into the air, others are formed through complex chemical reactions involving precursor compounds [3, 4]. The types of pollutants found in the air are diverse and can vary in composition, size, and toxicity [5]. Common categories of air pollutants include criteria air pollutants, which are regulated by environmental agencies due to their widespread prevalence and known health impacts. These include carbon monoxide (CO), sulfur dioxide (SO_2), nitrogen oxides (NOx), ozone (O_3), particulate matter (PM), and lead (Pb). Additionally, there are greenhouse gases, such as carbon dioxide (CO_2), methane (CH_4), and nitrous oxide (N_2O), which contribute to global warming and climate change. Particulate matter (PM) is of particular concern due to its varied sources and adverse health effects. PM refers to tiny particles suspended in the air, including dust, soot, and aerosols, which can penetrate deep into the respiratory system and cause respiratory and cardiovascular diseases [6, 7]. PM is categorized based on its size, with fine particles ($PM_{2.5}$) posing the greatest health risks due to their ability to penetrate the lungs and enter the bloodstream, while coarse particles (PM_{10}) primarily affect the upper respiratory tract.

Understanding the mechanisms of air pollution dispersion is essential for assessing its impacts on human health and the environment. Pollutants released into the atmosphere undergo complex processes of transport, transformation, and deposition, influenced by meteorological conditions, atmospheric chemistry, and topographical features [8, 9]. Wind patterns, temperature inversions, and atmospheric stability can affect the dispersion and accumulation of pollutants in specific geographic regions, leading to localized air quality issues and health concerns [10]. The scope of air pollution extends beyond localized sources of contamination, with pollutants often traveling long distances through atmospheric transport mechanisms [11, 12]. This phenomenon, known as transboundary pollution, can result in the cross-border spread of pollutants, affecting regions far from their sources. For example, pollutants emitted from industrial facilities in one country can be carried by prevailing winds and deposited in neighboring countries, contributing to regional air quality degradation and diplomatic tensions [10, 13]. The implications of poor air quality are far-reaching and multifaceted, impacting human health, ecosystems, and the global climate. Exposure to air pollution is associated with a wide range of health effects, including respiratory diseases, cardiovascular disorders, adverse birth outcomes, and premature mortality. Vulnerable populations, such as children, the elderly, and individuals with preexisting health conditions, are particularly susceptible to the adverse effects of air pollution, exacerbating existing health disparities.

In addition to its direct impacts on human health, air pollution also poses significant risks to ecosystems and biodiversity. Elevated levels of ozone and nitrogen oxides can damage vegetation, impairing photosynthesis and reducing crop yields [14]. Acid deposition, resulting from the deposition of sulfur and nitrogen compounds, can acidify soils and surface waters, threatening aquatic ecosystems and biodiversity [15, 16]. Furthermore, air pollution contributes to climate change through the emission of greenhouse gases, leading to rising global temperatures, sea level rise, and extreme weather events. Poor air quality is a leading environmental health risk, contributing to a wide range of acute and chronic illnesses. Exposure to air pollution has been linked to respiratory diseases such as asthma, chronic obstructive pulmonary disease (COPD), and lung cancer, as well as cardiovascular diseases, strokes, and adverse birth outcomes [17]. By improving air quality, we can prevent illness, reduce healthcare costs, and enhance overall well-being. Air pollution adversely impacts ecosystems, biodiversity, and natural resources. According to the reports of Rowland et al. [18] and Emberson [19], ground-level ozone and nitrogen oxides can harm vegetation, impairing photosynthesis and reducing crop yields. Acid deposition damages forests, soils, and aquatic ecosystems, threatening biodiversity, and ecosystem services. Therefore, by mitigating air pollution, we can protect natural habitats, preserve biodiversity, and promote ecosystem resilience. Many air pollutants, such as CO2, methane CH4, and N2O, are also greenhouse gases that contribute to global warming and climate change. Reducing the emission of these pollutants can contribute to mitigating climate change impacts, such as rising temperatures, sea level rise, and extreme weather events. Transitioning to clean energy sources and sustainable transportation systems can help achieve greenhouse gas reduction targets and limit the severity of climate change impacts [20, 21]. Addressing air quality issues can yield significant economic benefits, both in terms of healthcare savings and increased productivity. By preventing illness and premature deaths associated with air pollution, healthcare costs can be reduced, relieving the burden on healthcare systems, and improving public finances. Additionally, investments in pollution control technologies, renewable energy infrastructure, and energy efficiency measures can create jobs, stimulate economic growth, and enhance competitiveness in the global market [22]. Air pollution disproportionately affects vulnerable populations, including children, the elderly, low-income communities, and marginalized groups [23]. These communities often bear the brunt of pollution from industrial facilities, transportation corridors, and other sources of contamination, leading to environmental injustice and health disparities. Therefore, social inequity, environmental injustice, and exclusive development are often associated with air pollution. However, air pollution knows no borders and requires coordinated action at the local, national, and international levels. Transboundary pollution from industrial activities, transportation networks, and deforestation can affect air quality in neighboring countries and regions, necessitating collaboration and cooperation among nations [13, 24]. International agreements, such as the Paris Agreement and the Montreal Protocol, provide frameworks for collective action on air pollution and climate change, fostering global solidarity and shared responsibility.

This chapter aims to highlight the impacts of air pollution. Air pollution is a pervasive environmental problem with significant implications for human health, ecosystems, and the global climate. Examining the impacts of poor air quality on society will help develop a comprehensive understanding of air pollution and work toward a cleaner and healthier environment for all. Addressing air quality issues is essential for safeguarding public health, protecting the environment, promoting economic prosperity, fostering social equity, and advancing global cooperation. By implementing effective pollution control measures, investing in clean technologies, and adopting sustainable practices, we can create a healthier, more resilient, and sustainable future for current and future generations.

2 Understanding Air Pollution

2.1 Sources of Air Pollution

Air pollution is a complex issue that arises from a variety of sources, both natural and anthropogenic. The different sources of air pollution range from industrial emissions to vehicular exhaust, agricultural practices, and natural phenomena. Industrial emissions constitute a significant source of air pollution, releasing a variety of pollutants into the atmosphere during manufacturing processes, energy production, and other industrial activities [25, 26]. Factories, power plants, refineries, and other industrial facilities emit large quantities of pollutants such as sulfur dioxide, nitrogen oxides, particulate matter, volatile organic compounds (VOCs), and heavy metals. These pollutants can contribute to smog formation, acid rain, respiratory diseases, and environmental degradation, affecting both local air quality and global climate change [27].

Transportation is another major source of air pollution, with vehicles emitting pollutants such as carbon monoxide, nitrogen oxides, particulate matter, and volatile organic compounds through tailpipe emissions [28, 29]. Cars, trucks, buses, motorcycles, and other modes of transportation powered by fossil fuels release large amounts of pollutants into the air, particularly in urban areas with high traffic congestion. Traffic-related air pollution is associated with respiratory diseases, cardiovascular issues, and reduced quality of life, highlighting the need for cleaner, more sustainable transportation options.

Agricultural practices also contribute to air pollution through the release of ammonia, methane, and nitrous oxide from livestock operations, fertilizers, manure management, and crop burning [30]. Livestock farming, in particular, is a significant source of ammonia emissions, which can react with other pollutants in the atmosphere to form fine particulate matter and exacerbate smog formation [31]. Additionally, the application of synthetic fertilizers and manure to agricultural fields can release nitrogen oxides and contribute to air pollution in rural and suburban areas.

Aside from anthropogenic sources, natural phenomena such as wildfires, volcanic eruptions, and dust storms can also introduce pollutants into the atmosphere and impact air quality on a regional or global scale [32]. Wildfires, for instance, release large quantities of particulate matter, carbon monoxide, and volatile organic compounds into the air, leading to hazy conditions, health advisories, and respiratory issues in affected areas. Volcanic eruptions can emit sulfur dioxide, ash, and other pollutants that can spread over vast distances and impact air quality, climate patterns, and aviation operations. Dust storms, fueled by wind-blown soil, can transport particulate matter, allergens, and pathogens over long distances, affecting air quality, visibility, and public health in their path.

Another source of air pollution that often goes unnoticed is indoor air pollution, which arises from the combustion of biomass fuels, coal, kerosene, and other traditional cooking fuels in poorly ventilated households. Indoor air pollution is a significant health concern in many developing countries, where households rely on biomass fuels for cooking and heating. The smoke and fumes produced by these traditional fuels contain pollutants such as carbon monoxide, particulate matter, volatile organic compounds, and carcinogens, posing serious risks to human health, particularly for women and children who spend extended periods indoors [33, 34].

2.2 *Types of Air Pollutants*

Air pollution is a major environmental issue that poses significant risks to human health, the environment, and the economy. Different types of air pollutants are released into the atmosphere from various sources, contributing to poor air quality and a range of adverse impacts including:

Criteria Air Pollutants Criteria air pollutants are a group of six common air pollutants regulated by the US Environmental Protection Agency (EPA) due to their adverse impacts on human health and the environment [35, 36]. The criteria pollutants include:

- Particulate matter: PM consists of tiny particles suspended in the air, varying in size from coarse dust to fine particles [37]. PM can originate from natural sources such as dust storms and wildfires, as well as anthropogenic sources like vehicle emissions, industrial processes, and combustion of fossil fuels. PM can irritate the respiratory tract, trigger asthma attacks, and increase the risk of cardiovascular and respiratory diseases [38].
- Sulfur dioxide: SO_2 is a colorless gas produced primarily by burning fossil fuels containing sulfur, such as coal and oil. SO_2 emissions can cause respiratory issues, exacerbate asthma, and contribute to acid rain formation when combined with water vapor in the atmosphere [39].
- Nitrogen dioxide: NO_2 is a reddish-brown gas emitted from vehicles, power plants, and industrial facilities burning fossil fuels. NO2 can react with other

pollutants to form ground-level ozone, a key component of smog that can lead to respiratory problems, lung damage, and reduced lung function [40].

- Carbon monoxide: CO is a colorless, odorless gas produced by incomplete combustion processes, such as vehicle exhaust and industrial emissions. CO can interfere with the blood's ability to carry oxygen, leading to headaches, dizziness, fatigue, and even death in high concentrations [41].
- Ozone: Ground-level ozone is a secondary pollutant formed by the reaction of NO_2 and VOCs in the presence of sunlight. Ozone can trigger respiratory issues, worsen asthma symptoms, and damage lung tissue, particularly in vulnerable populations.
- Lead: Lead is a toxic metal that can be released into the air from sources such as leaded gasoline, smelters, and industrial processes. Lead exposure can harm the nervous system, cardiovascular system, and kidneys, especially in children and pregnant women [42–44].

Greenhouse Gases Greenhouse gases are compounds that trap heat in the Earth's atmosphere, leading to global warming and climate change. Although greenhouse gases are essential for maintaining the planet's temperature, human activities have significantly increased their concentrations, causing the greenhouse effect to intensify [45, 46]. The main greenhouse gases include:

- Carbon dioxide: CO_2 is the most abundant greenhouse gas emitted from burning fossil fuels, deforestation, and industrial processes. CO_2 contributes to the buildup of heat-trapping gases in the atmosphere, leading to rising global temperatures, sea level rise, and changes in weather patterns.
- Methane: Methane is a potent greenhouse gas released from natural sources like wetlands and landfills, as well as human activities such as livestock farming, rice cultivation, and fossil fuel extraction. Methane has a greater warming potential than CO_2 but remains in the atmosphere for a shorter period.
- Nitrous oxide: Nitrous oxide is a greenhouse gas produced by agricultural activities, sewage treatment, and combustion processes. N_2O has a long atmospheric lifetime and contributes to both global warming and ozone depletion.
- Fluorinated gases: Fluorinated gases, including hydrofluorocarbons (HFCs), perfluorocarbons (PFCs), and sulfur hexafluoride (SF6), are synthetic greenhouse gases used in refrigeration, air conditioning, and industrial applications [47, 48]. These gases have high global warming potential and contribute to climate change.

Particulate Matter PM refers to a mixture of solid particles and liquid droplets suspended in the air, classified based on size into PM_{10} (particles with a diameter of 10 μm or smaller) and $PM_{2.5}$ (particles with a diameter of 2.5 μm or smaller). PM is generated from various sources, including vehicle emissions, industrial processes, construction activities, agricultural practices, and wildfires [6, 7]. Fine particles ($PM_{2.5}$) are particularly concerning due to their ability to penetrate deep into the respiratory system and enter the bloodstream, leading to a wide range of health effects, including respiratory diseases, cardiovascular issues, and adverse birth outcomes [49]. PM can also reduce visibility, contribute to haze, and impact air quality both locally and regionally.

Volatile Organic Compounds According to David and Niculescu [50], VOCs are a group of carbon-based chemicals that easily evaporate into the air at room temperature, contributing to air pollution and the formation of ground-level ozone (smog). VOCs are released from a variety of sources, such as vehicle exhaust, industrial emissions, solvents, paints, and household products [51]. Some common VOCs include benzene, toluene, formaldehyde, and xylene. Exposure to VOCs can cause eye, nose, and throat irritation, headaches, dizziness, and long-term health effects such as cancer and organ damage [52]. VOCs also play a role in the formation of fine particulate matter and ground-level ozone, which can exacerbate respiratory issues and degrade air quality.

Toxic Air Pollutants Toxic air pollutants, also known as hazardous air pollutants, are substances that pose serious health risks due to their toxic, carcinogenic, or mutagenic properties (Fig. 1). These pollutants can originate from industrial processes, vehicle emissions, waste incineration, and other sources. Some examples of toxic air pollutants include:

- **Benzene**: Benzene is a known carcinogen found in emissions from vehicle exhaust, industrial sources, and tobacco smoke [53]. Inhalation of benzene can lead to leukemia and other forms of cancer, as well as damage to the nervous system and immune system.
- **Mercury**: Mercury is a toxic metal released from coal-fired power plants, waste incineration, and gold mining activities. Mercury exposure can harm the brain,

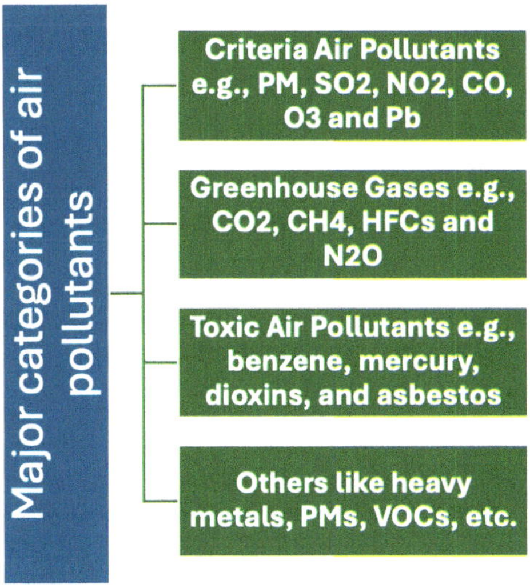

Fig. 1 Major categories of air pollutants

nervous system, kidneys, and cardiovascular system, particularly in developing fetuses and young children [54].

- **Dioxins**: Dioxins are a group of highly toxic chemicals produced by burning industrial waste, chlorine bleaching processes, and forest fires [55]. Dioxin exposure can cause cancer, reproductive disorders, immune system damage, and other health effects.
- **Asbestos**: Asbestos is a naturally occurring mineral used in building materials, insulation, and fireproofing products. Inhalation of asbestos fibers can lead to lung cancer, mesothelioma, and other respiratory diseases.

2.3 Mechanisms of Air Pollution Dispersion

Air pollution dispersion refers to the movement and spread of pollutants in the atmosphere, influenced by a variety of physical, chemical, and meteorological factors. Understanding the mechanisms of air pollution dispersion is essential for predicting pollutant concentrations, assessing potential exposure risks, and designing effective mitigation strategies. Some of these mechanisms include (Fig. 2):

Atmospheric Stability Atmospheric stability plays a crucial role in air pollution dispersion by affecting the vertical movement of air masses and pollutants. Stability refers to the tendency of air parcels to rise or sink in the atmosphere, influenced by temperature variations and the presence of inversions [56, 57]. According to Nugent et al. [58], three types of atmospheric stability conditions can impact pollutant dispersion (Fig. 3):

- **Unstable atmosphere**: In an unstable atmosphere, air parcels tend to rise due to buoyancy forces, leading to vertical mixing and dispersion of pollutants. Warm air near the surface rises, carrying pollutants upward and diluting their concentrations. Unstable conditions are favorable for pollutant dispersion, as pollutants are less likely to accumulate near the ground.
- **Stable atmosphere**: In a stable atmosphere, air parcels resist vertical movement and form layers with little mixing, trapping pollutants near the surface. Cold air near the ground may create temperature inversions, preventing pollutants from dispersing and leading to the accumulation of pollutants in urban areas. Stable conditions can result in poor air quality and elevated pollutant concentrations.
- **Neutral atmosphere**: A neutral atmosphere exhibits limited vertical movement of air parcels, with minimal buoyancy forces and mixing. Neutral conditions are

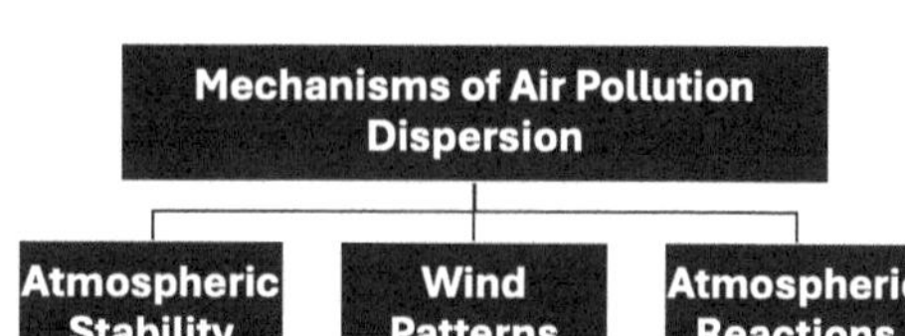

Fig. 2 Some key mechanisms of air pollution dispersion

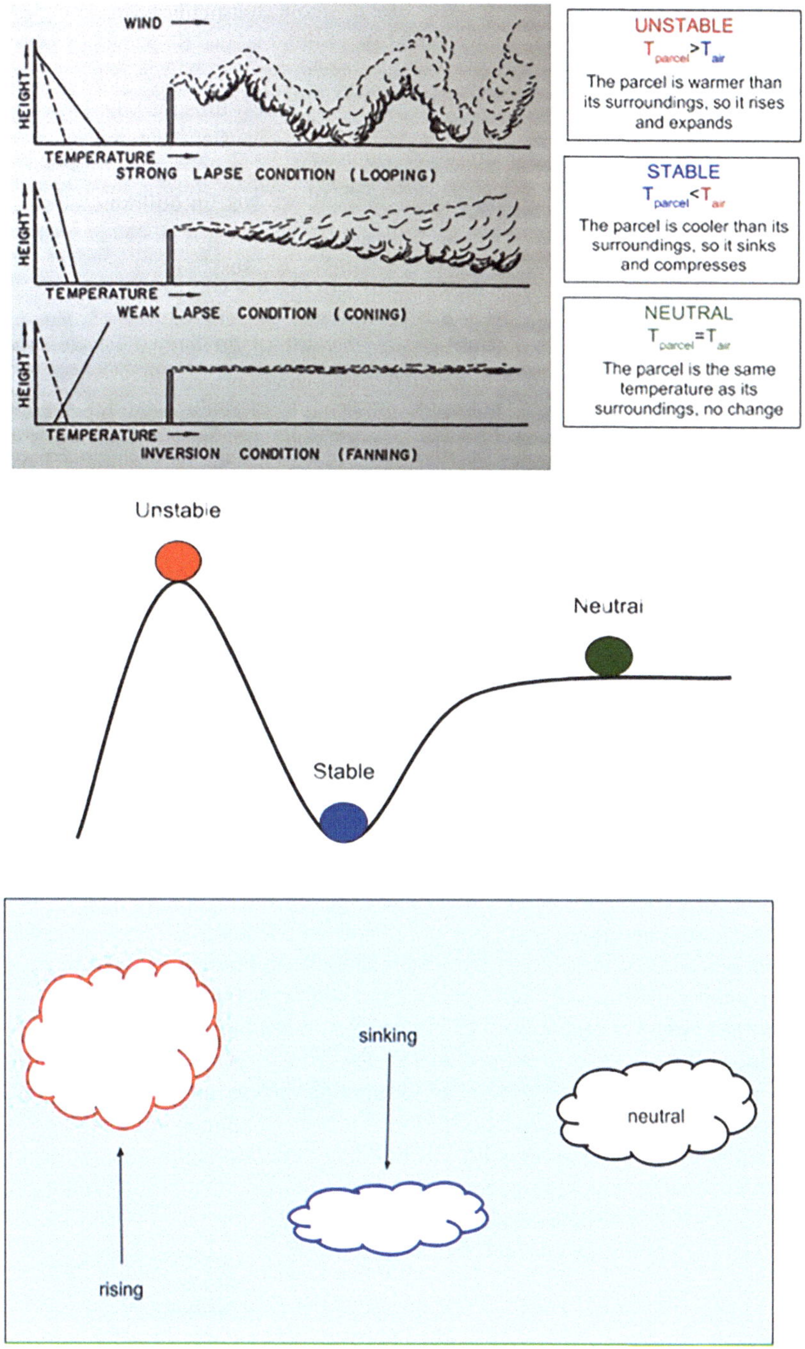

Fig. 3 Atmospheric stability conditions concerning air pollution dispersion. Source: Adapted from Repace [59]; Nugent et al. [58]

considered transitional between stable and unstable states, allowing for moderate pollutant dispersion. Under neutral stability, pollutants may disperse horizontally rather than vertically, depending on wind patterns and other factors.

Wind Patterns Wind patterns play a critical role in air pollution dispersion by transporting pollutants over long distances and influencing their distribution in the atmosphere (Fig. 2). Surface winds, as well as upper-level winds, can carry pollutants away from their source and disperse them across different regions [60]. The following wind patterns affect air pollution dispersion:

- **Local winds**: Local winds, such as sea breezes, mountain winds, and valley winds, can impact air pollution dispersion in specific geographical areas. These winds arise from local temperature differences, topographic features, and geographic settings, influencing the movement of pollutants near the surface [61].
- **Prevailing winds**: Prevailing winds, which blow consistently from a particular direction over a large area, play a key role in transporting pollutants over regional and continental scales [62]. Pollutants emitted in one location can be carried by prevailing winds to distant areas, affecting air quality in downwind regions.
- **Jet streams**: High-altitude jet streams are fast-flowing bands of air in the upper atmosphere that can influence the movement of pollutants over long distances [63]. Jet streams can act as atmospheric highways, transporting pollutants rapidly across continents and oceans, impacting air quality on a global scale.

Topography Topographic features such as mountains, valleys, and coastlines can affect air pollution dispersion by altering wind patterns, creating local microclimates, and influencing atmospheric stability [64–66]. Topography can lead to the following effects on air pollution dispersion:

- **Mountain valleys**: Mountain valleys can experience temperature inversions and stagnant air conditions, trapping pollutants and leading to poor air quality. Pollutants emitted in valleys may accumulate due to limited vertical mixing and the blocking of winds by surrounding mountain ranges [61].
- **Coastal areas**: Coastal regions can be subject to sea breezes, which bring clean marine air inland and disperse pollutants away from the coast. Coastal topography can enhance pollutant dispersion by promoting airflow and reducing pollutant concentrations near shorelines [67].
- **Urban areas**: Urban areas with tall buildings and structures can create local turbulence, affecting wind patterns and pollutant dispersion. Tall buildings can block or redirect airflow, leading to complex dispersion patterns and localized areas of high pollution concentrations [68].

Atmospheric Reactions Chemical reactions occurring in the atmosphere can affect air pollution dispersion by altering the composition and behavior of pollutants. Reactions between pollutants, sunlight, and atmospheric components can lead to the formation of secondary pollutants, such as ozone, nitrate aerosols, and volatile organic compounds [69, 70]. Atmospheric reactions influence air pollution dispersion in the following ways:

- **Photochemical smog**: Photochemical smog is a type of air pollution formed by the reaction of nitrogen oxides, volatile organic compounds, and sunlight. These reactions produce ground-level ozone and fine particulate matter, contributing to smog formation and respiratory issues [71]. Photochemical smog affects air quality in urban areas with high levels of vehicle emissions and industrial activities.
- **Acid deposition**: Acid deposition, including acid rain, snow, and fog, results from the reaction of sulfur dioxide and nitrogen oxides with atmospheric moisture. These pollutants form sulfuric acid and nitric acid, which can be deposited on vegetation, water bodies, and soil [72]. Acid deposition impacts ecosystems, aquatic life, and infrastructure, influencing air pollution dispersion and regional air quality.

3 Some Impacts of Air Pollution

Air pollution has become a pressing global issue with significant impacts on human health, the environment, and the economy. Air pollution is a significant environmental and public health issue that poses a wide range of adverse impacts on human health, the environment, and the economy. The pollutants present in the air, such as particulate matter, nitrogen oxides, sulfur dioxide, ozone, and volatile organic compounds, as well as their sources and mechanism of action, contribute to the wide-ranging effects on both physical and mental health [27, 73]. The presence of these harmful substances in the atmosphere can lead to a variety of detrimental consequences that affect individuals, communities, and ecosystems.

First and foremost, air pollution is closely linked to respiratory diseases. Particulate matter, which consists of fine particles suspended in the air, can penetrate deep into the lungs and cause or exacerbate conditions such as asthma, COPD, and bronchitis [74]. Long-term exposure to high levels of particulate matter has been associated with reduced lung function, increased risk of respiratory infections, and higher mortality rates among individuals with preexisting respiratory conditions [75, 76]. Nitrogen oxides and sulfur dioxide, which are produced primarily by vehicles, industrial processes, and power plants, can also have detrimental effects on respiratory health. These pollutants can irritate the respiratory tract, trigger asthma attacks, and increase susceptibility to respiratory infections. Children, the elderly, and individuals with compromised respiratory function are particularly vulnerable to the respiratory effects of nitrogen oxides and sulfur dioxide [77]. In addition to respiratory diseases, air pollution has been implicated in cardiovascular issues such as heart disease, stroke, and hypertension [78]. Fine particulate matter and nitrogen oxides can enter the bloodstream, causing inflammation, oxidative stress, and damage to blood vessels [79]. Prolonged exposure to these pollutants has been linked to an increased risk of coronary artery disease, heart attacks, and stroke. Furthermore, air pollution has been shown to exacerbate existing cardiovascular conditions and contribute to premature mortality [79]. Air pollution is a risk factor for cardiovascular diseases, including heart attacks, strokes, hypertension, and atherosclerosis. Pollutants such as nitrogen dioxide, sulfur dioxide, and carbon monoxide can trigger systemic inflammation, oxidative stress, and vascular dysfunction, contributing to the onset and progression of cardiovascular conditions.

Furthermore, emerging research suggests that air pollution can have adverse effects on fetal development and birth outcomes [80, 81]. Pregnant women exposed to high levels of air pollutants, such as particulate matter and nitrogen oxides, may face an increased risk of preterm birth, low birth weight, and developmental abnormalities in their offspring. These pollutants can cross the placental barrier and interfere with fetal growth and development, leading to long-term health consequences for the child [82]. Exposure to air pollutants during pregnancy can increase the risk of preterm birth, low birth weight, birth defects, and developmental delays in children, impacting the health and well-being of future generations. In the same way, emerging research suggests that air pollution may have adverse effects on the brain and nervous system, leading to cognitive decline, neurodevelopmental disorders in children, and an increased risk of neurodegenerative diseases like Alzheimer's and Parkinson's [83, 84]. Fine particulate matter and toxic air pollutants can penetrate the blood–brain barrier and cause neuroinflammation, oxidative damage, and neuronal dysfunction. Long-term exposure to carcinogenic air pollutants, such as benzene, formaldehyde, and diesel exhaust particles, can increase the risk of developing cancer, including lung cancer, bladder cancer, and leukemia [85, 86]. These pollutants have been classified as Group 1 human carcinogens by the International Agency for Research on Cancer (IARC), highlighting their cancer-causing potential. Air pollution has been linked to adverse effects on reproductive health, including infertility, pregnancy complications, and adverse birth outcomes [87]. Moreover, recent studies [47, 48, 88, 89] have indicated a potential link between air pollution and cognitive impairment. Fine particulate matter and other pollutants have been shown to access the brain through the olfactory nerve, triggering neuroinflammation, oxidative stress, and neuronal damage. Chronic exposure to air pollution has been associated with cognitive decline, impaired memory, and an increased risk of neurodegenerative diseases such as Alzheimer's and Parkinson's [90]. Children and the elderly are particularly susceptible to the cognitive effects of air pollution, as their developing and aging brains may be more vulnerable to environmental insults.

Beyond its profound impacts on human health, air pollution also carries significant economic consequences. The healthcare costs associated with treating air pollution-related illnesses, such as respiratory diseases, cardiovascular issues, and cognitive impairment, place a substantial burden on healthcare systems worldwide [89]. Additionally, lost productivity due to illness, absenteeism, and premature mortality resulting from air pollution contributes to economic losses at both the individual and societal levels. Furthermore, the environmental damage caused by air pollution, such as acid rain, smog, and ecosystem degradation, can have far-reaching economic implications, affecting agriculture, tourism, and other industries. To address the multifaceted impacts of air pollution, concerted efforts are needed at the local, national, and global levels to reduce emissions, enhance air quality monitoring, and promote sustainable practices [20, 35]. By safeguarding air quality and protecting public health, we can mitigate the far-reaching consequences of air pollution and create a healthier, more prosperous future for generations to come. Air pollution has profound effects on the environment, leading to degradation of ecosystems, biodiversity loss, and harm to wildlife populations. For instance, the deposition of acidic compounds, such as sulfuric acid and nitric acid, in the atmosphere can result in acid rain, which has detrimental effects on forests, lakes, soils, and aquatic organisms. Acid rain can leach nutrients from the soil, damage vegetation, sterilize water bodies, and disrupt aquatic ecosystems, leading to declines in plant and animal populations. Certain air pollutants, such as chlorofluorocarbons (CFCs) and other halogenated compounds, can contribute to the depletion of the ozone layer in the stratosphere [91]. Ozone depletion allows harmful ultraviolet (UV) radiation to reach the Earth's surface, increasing the risk of skin cancer, cataracts, and immune suppression in humans and damaging marine ecosystems. Pollutants like sulfur dioxide and nitrogen oxides can damage vegetation, disrupt food chains, and alter habitats, leading to declines in biodiversity and ecosystem services [46]. Endangered species and sensitive ecosystems are particularly vulnerable to the impacts of air pollution. Certain air pollutants, such as greenhouse gases like carbon dioxide, methane, and nitrous oxide, contribute to global warming and climate change. The accumulation of greenhouse gases in the atmosphere traps heat, leading to rising temperatures, melting ice caps, sea level rise, changes in precipitation patterns, and extreme weather events. Climate change has far-reaching consequences for ecosystems, agriculture, water resources, and human societies.

Air pollution has significant socioeconomic implications, affecting productivity, healthcare costs, quality of life, and environmental management. The burden of air pollution on public health results in substantial healthcare costs related to hospital admissions, emergency room visits, medication expenses, and treatment of pollution-related illnesses [92]. Individuals with respiratory and cardiovascular conditions may require ongoing medical care and face reduced quality of life due to air pollution exposure. Air pollution can impair cognitive function, reduce work performance, and lead to absenteeism and decreased productivity in the workforce [93]. Workers exposed to high levels of air pollution may experience fatigue, respiratory symptoms, and mental health issues, impacting their ability to perform job duties effectively and contributing to economic losses for businesses [73]. The remediation of air pollution and its environmental impacts requires costly cleanup efforts, pollution control technologies, and regulatory measures to reduce emissions from industrial sources, transportation, and energy production [94, 95]. Governments, businesses, and communities invest resources in pollution abatement, monitoring programs, and environmental compliance to mitigate the effects of air pollution on ecosystems and public health. Air pollution can diminish the quality of life for individuals living in polluted areas, affecting their well-being, sense of community, and social interactions. Poor air quality can limit outdoor activities, exacerbate respiratory symptoms, and create stress and anxiety related to environmental health risks. Residents of urban areas with high pollution levels may experience reduced quality of life and diminished environmental amenities.

4 Health Impacts of Air Pollution

Air pollution, a silent but pervasive threat, has emerged as a major public health concern that affects billions of people worldwide. Exposure to air pollutants can have a range of detrimental effects on human health, leading to respiratory, cardiovascular, neurological, and other health conditions. From bustling urban centers to remote rural areas, the air we breathe contains pollutants that pose significant risks to human health. The health effects of air pollution are mediated through complex mechanisms involving inflammation, oxidative stress, and genetic alterations. Inhalation of pollutants can trigger inflammatory responses in the respiratory system, leading to tissue damage and compromised lung function. Oxidative stress, induced by reactive oxygen species (ROS) generated by pollutants, can damage cellular structures and DNA, contributing to the development of chronic diseases. Additionally, genetic susceptibility plays a role in determining individual responses to air pollution exposure, highlighting the interplay between genetic and environmental factors in health outcomes.

4.1 Mechanisms of Harm

Air pollutants can harm human health through various mechanisms, including inflammation, oxidative stress, DNA damage, and disruption of cellular functions. The following are the key mechanisms of harm associated with exposure to air pollution:

- **Inflammation**: Air pollutants such as particulate matter, ozone, and nitrogen dioxide can trigger inflammation in the respiratory tract, leading to irritation, coughing, and shortness of breath. Chronic exposure to inflammatory stimuli can damage lung tissues, impair immune responses, and increase the risk of respiratory infections and chronic lung diseases [96, 97].
- **Oxidative stress**: Air pollutants generate ROS and free radicals in the body, causing oxidative stress and cellular damage. Oxidative stress can disrupt antioxidant defenses, promote inflammation, and contribute to the development of cardiovascular diseases, neurological disorders, and cancer [98].
- **DNA damage**: Certain air pollutants, including polycyclic aromatic hydrocarbons (PAHs) and heavy metals, have genotoxic properties that can interact with DNA and cause mutations, chromosomal aberrations, and DNA strand breaks. DNA damage induced by air pollution exposure may increase the risk of cancer and genetic disorders [99, 100].

- **Endothelial dysfunction**: Air pollutants can impair the function of endothelial cells lining blood vessels, leading to endothelial dysfunction and vascular inflammation. Endothelial dysfunction is a key early event in the development of atherosclerosis, hypertension, and cardiovascular diseases, increasing the risk of heart attacks and strokes [101].
- **Neuroinflammation**: Emerging evidence suggests that air pollution exposure can induce neuroinflammation in the brain, affecting neuronal function, synaptic transmission, and cognitive abilities [102]. Neuroinflammation is associated with neurodegenerative diseases like Alzheimer's and Parkinson's, as well as neurodevelopmental disorders in children.

4.2 Respiratory Health Impacts

Air pollution significantly contributes to respiratory diseases, with asthma and COPD being among the most prevalent. Exposure to pollutants such as PM, NO_2, and O_3 can exacerbate respiratory symptoms, trigger asthma attacks, and worsen COPD symptoms. Studies have shown a clear association between air pollution exposure and increased hospital admissions for respiratory conditions, highlighting the direct impact of polluted air on respiratory health [77, 103]. Air pollution is a common trigger for asthma symptoms, causing airway inflammation, bronchoconstriction, and increased production of mucus. Children, in particular, are vulnerable to the effects of air pollution on asthma, with exposure to high levels of pollutants linked to asthma onset and worsening of symptoms. Long-term exposure to air pollutants, especially fine particulate matter and nitrogen dioxide, is a risk factor for COPD, a progressive lung disease characterized by airflow limitation, cough, and breathlessness. COPD patients are more likely to experience exacerbations and hospitalizations due to air pollution exposure. Air pollution can irritate the bronchial tubes, leading to bronchitis, inflammation, and excess mucus production in the airways [35]. Chronic bronchitis is a common respiratory condition associated with exposure to particulate matter, sulfur dioxide, and other pollutants, resulting in coughing, phlegm, and respiratory discomfort.

4.3 Cardiovascular Health Impacts

Beyond respiratory effects, air pollution also poses grave risks to cardiovascular health. Fine particulate matter ($PM_{2.5}$) and other pollutants can penetrate deep into the lungs and enter the bloodstream, triggering inflammation, oxidative stress, and endothelial dysfunction [38, 74]. These mechanisms contribute to the development of cardiovascular diseases such as heart attacks, strokes, and hypertension. Epidemiological studies [78, 104] consistently demonstrated a link between long-term exposure to air pollution and an elevated risk of cardiovascular morbidity and mortality. Short-term exposure to air pollutants, particularly fine particulate matter, and nitrogen dioxide, can trigger heart attacks and exacerbate existing cardiovascular conditions [105]. Air pollution-induced inflammation, oxidative stress, and endothelial dysfunction can destabilize atherosclerotic plaques, leading to acute coronary events [79, 106]. Long-term exposure to air pollution is associated with an increased risk of stroke, including ischemic strokes caused by blockages in blood vessels and hemorrhagic strokes due to bleeding in the brain. Air pollutants like particulate matter and ozone can promote blood clot formation, blood vessel constriction, and systemic inflammation, contributing to strokes [107]. Air pollution exposure is a risk factor for hypertension, elevated blood pressure, and cardiovascular disease. Pollutants such as nitrogen dioxide and fine particulate matter can increase sympathetic nervous system activity, vascular resistance, and oxidative stress, leading to hypertension and arterial stiffness [108]. Air pollution promotes the development of atherosclerosis, a chronic inflammatory condition characterized by the buildup of plaque in the arteries. Particulate matter, nitrogen dioxide, and other pollutants contribute to the progression of atherosclerosis by inducing endothelial dysfunction, oxidative stress, and lipid oxidation, leading to cardiovascular events [79, 109].

4.4 Neurological and Cognitive Health Effects

Emerging research has highlighted the neurological and cognitive effects of air pollution exposure, revealing its impact on brain health, cognitive function, and mental well-being [84, 102]. Prenatal and early-life exposure to air pollutants has been linked to neurodevelopmental disorders in children, including autism spectrum disorder, attention deficit hyperactivity disorder (ADHD), and cognitive impairments [110]. Air pollution-induced neuroinflammation, oxidative stress, and neurotoxicity can disrupt brain development and function, affecting learning, behavior, and social skills. Long-term exposure to air pollution is associated with cognitive decline, memory impairment, and an increased risk of Alzheimer's disease and other forms of dementia in older adults. Air pollution exposure has been linked to mental health conditions, including depression, anxiety, and mood disorders. Chronic exposure to air pollutants can induce systemic inflammation, alter neurotransmitter levels, and impact the brain's stress response systems, contributing to the development of mental health disorders.

4.5 Cancer and Other Health Effects

In addition to respiratory and cardiovascular diseases, air pollution is implicated in the development of various types of cancer. Carcinogenic pollutants such as benzene, formaldehyde, and PAHs are present in polluted air and have been associated with increased cancer risk [4, 86]. Furthermore, air pollution has been linked to neurologic disorders, adverse birth outcomes, and developmental disorders in children. Prenatal exposure to air pollutants has been shown to increase the risk of low birth weight, preterm birth, and developmental delays, underscoring the importance of addressing air pollution for maternal and child health [81].

4.6 Vulnerable Populations

Certain populations are more vulnerable to the health impacts of air pollution, including children, the elderly, and individuals with preexisting health conditions. Children are particularly susceptible due to their developing respiratory and immune systems, while the elderly may experience exacerbated symptoms of chronic diseases [111]. Socioeconomically disadvantaged communities often bear the brunt of air pollution, residing in areas with higher pollution levels and facing barriers to accessing healthcare and resources for mitigation [112, 113]. Individuals with preexisting respiratory conditions, such as asthma, COPD, and allergies, are more susceptible to the effects of air pollution on lung function and respiratory health. Air pollutants can trigger asthma attacks, exacerbate COPD symptoms, and increase the frequency of respiratory exacerbations in vulnerable individuals. Socioeconomically disadvantaged communities, including low-income earners, minorities, and marginalized populations, are disproportionately affected by air pollution exposure due to environmental injustice, residential proximity to pollution sources, and lack of access to healthcare services. Environmental racism and inequity exacerbate the health disparities associated with air pollution in vulnerable communities.

4.7 Long-Term Health Impacts of Air Pollution

The long-term health impacts of air pollution extend beyond immediate symptoms and acute health effects, contributing to chronic diseases, disability, premature mortality, and societal burden. The cumulative effects of air pollution on public health include:

- **Chronic respiratory diseases**: Long-term exposure to air pollution is a risk factor for chronic respiratory diseases, such as COPD, asthma, bronchitis, and lung cancer [114]. Individuals exposed to high levels of air pollutants over extended periods are at increased risk of developing persistent respiratory symptoms, lung function decline, and respiratory impairments [5].
- **Cardiovascular diseases:** Prolonged exposure to air pollution contributes to the development of cardiovascular diseases, including heart attacks, strokes, hypertension, atherosclerosis, and heart failure [7]. Chronic inflammation, oxidative stress, and endothelial dysfunction caused by air pollutants can lead to progressive cardiovascular damage and increased risk of cardiovascular events [28].

- **Cognitive decline and neurological disorders**: Long-term exposure to air pollution is associated with cognitive decline, memory loss, and neurodegenerative disorders like Alzheimer's and Parkinson's [83, 84]. Aging populations exposed to high levels of air pollutants are at greater risk of developing neurological impairments, cognitive deficits, and dementia, affecting their quality of life and independence.
- **Premature mortality:** Air pollution is a leading environmental risk factor for premature mortality, contributing to millions of deaths worldwide each year. The cumulative effects of air pollution on respiratory, cardiovascular, and other health conditions increase the risk of premature death, particularly in vulnerable populations and regions with high pollution levels [105].

4.8 Mitigation and Prevention Strategies: Environmental and Policy Implications of the Health Impacts of Air Pollution

Addressing the health impacts of air pollution requires a multifaceted approach encompassing public health interventions, policy responses, and technological innovations. Air quality monitoring and surveillance are essential for assessing pollution levels and identifying high-risk areas [115]. Public health interventions such as health impact assessments and risk communication help raise awareness and empower communities to protect themselves from air pollution [116, 117]. Policy responses, including regulatory standards, urban planning measures, and investments in clean technologies, are crucial for reducing emissions and improving air quality [95]. Technological innovations such as pollution control technologies and clean energy solutions offer promising avenues for mitigating air pollution and protecting public health [26]. Air pollution is a global issue that transcends geographical boundaries and requires collaborative efforts at the local, national, and international levels. International agreements and initiatives, such as the World Health Organization (WHO) guidelines and the Paris Agreement on climate change, provide frameworks for collective action and cooperation. Cross-sectoral partnerships involving governments, non-governmental organizations (NGOs), academia, and industry are essential for implementing effective strategies and mobilizing resources to address air pollution and its health impacts.

Reducing the health impacts of air pollution requires a multifaceted approach involving regulatory policies, emission controls, urban planning, public health interventions, and community engagement [26]. Setting and enforcing ambient air quality standards for key pollutants like particulate matter, ozone, nitrogen dioxide, sulfur dioxide, and carbon monoxide are essential for protecting public health and reducing exposure levels [118]. Governments and regulatory agencies play a critical role in monitoring air quality, conducting emissions testing, and implementing pollution control measures. Implementing pollution control technologies, such as catalytic converters, particulate filters, and scrubbers, in industrial processes, vehicles, and power plants can reduce emissions of harmful pollutants and improve air quality [119]. Investing in clean energy, fuel-efficient technologies, alternative transportation, and renewable energy sources can help reduce air pollution emissions and mitigate health impacts. Designing cities with green spaces, pedestrian zones, bike lanes, and public transportation systems can reduce vehicle emissions, traffic congestion, and air pollution in urban areas. Urban planners, architects and policymakers can prioritize sustainable development, green infrastructure, and eco-friendly transportation options to create healthier and more livable cities [120]. Raising public awareness about the health risks of air pollution, promoting clean air practices, and encouraging behavior changes, such as reducing vehicle idling, carpooling, and using energy-efficient appliances, can empower individuals to protect their health and minimize exposure to air pollutants. Health education campaigns, community outreach programs, and environmental initiatives can enhance public understanding of air quality issues and promote sustainable lifestyle choices. Implementing health monitoring systems, air quality indices, and surveillance programs can track the health impacts of air pollution, identify vulnerable populations, and inform public health interventions. Monitoring air quality data, hospital admissions, emergency room visits, and mortality rates can help public health authorities assess the burden of disease associated with air pollution and prioritize prevention strategies [115].

5 Environmental Impacts of Air Pollution

Ecosystem degradation, alteration of climate patterns, acid rains, eutrophication, and biodiversity loss are some of the environmental impacts of air pollution. Therefore, pollution caused by various air pollutants released from natural sources and human activities, such as transportation, industrial processes, agriculture, and energy production, can harm the air, water, soil, and wildlife, leading to environmental degradation and ecosystem disruption. Air pollutants can have profound implications for natural resources, including water, soil, and biodiversity. Acid deposition can contaminate groundwater and surface waters, affecting drinking water supplies and aquatic ecosystems. Soil acidification can reduce soil fertility and crop productivity, leading to economic losses and food security concerns. Furthermore, air pollution can degrade natural habitats and disrupt ecosystem services such as pollination, nutrient cycling, and flood regulation, further compromising the resilience of ecosystems and their ability to support human well-being.

5.1 Ecosystem Degradation

Air pollution poses a significant threat to terrestrial and aquatic ecosystems, disrupting ecological processes and impairing biodiversity. Acid deposition, resulting from the deposition of SO_2 and NOx, can acidify soils and surface waters, leading to the loss of plant and animal species [15, 35]. Additionally, pollutants such as O3 and PMs can damage vegetation, impairing photosynthesis and reducing crop yields. Forests, wetlands, and other natural habitats are particularly vulnerable to air pollution, with long-term implications for ecosystem resilience and ecosystem services [16]. Acid rain is often a result of air pollutants like sulfur dioxide and nitrogen oxides combining with water vapor in the atmosphere, which can damage forest ecosystems by leaching nutrients from the soil, weakening tree roots, and inhibiting plant growth [72]. Acid deposition can lead to forest decline, tree dieback, and loss of biodiversity in sensitive forested areas. Acid rain can acidify freshwater bodies, such as lakes, rivers, and streams, affecting aquatic ecosystems and aquatic species. Acidification of water bodies can harm fish, amphibians, and aquatic plants by altering water pH, reducing food sources, and disrupting reproductive cycles, leading to declines in population and biodiversity. Air pollutants like heavy metals, ozone, and volatile organic compounds can settle onto the soil surface, contaminating soil quality, reducing soil fertility, and impacting soil microorganisms [14]. Soil pollution from air pollution can affect plant growth, crop yields, and ecosystem functioning, leading to environmental degradation and soil degradation. Air pollution can contribute to biodiversity loss by damaging habitats, disrupting ecosystems, and impairing the health of plant and animal species. Pollution-induced changes in air quality, water quality, and soil quality can threaten biodiversity hotspots, endangered species, and ecosystem services, reducing resilience and adaptability in natural ecosystems [46].

5.2 Biodiversity Loss

The degradation of ecosystems due to air pollution can have cascading effects on biodiversity, leading to the loss of species and disruption of ecological networks. Pollutants such as nitrogen compounds and heavy metals can accumulate in soils and water bodies, poisoning plants and animals and disrupting food chains. Endangered species and sensitive habitats are particularly at risk, with potential consequences for ecosystem stability and biodiversity conservation efforts. Furthermore, air pollution can exacerbate existing threats to biodiversity, such as habitat destruction and climate change, further increasing the risk of species extinction [121, 122]. Air pollutants such as SO_2, NOx, O_3, heavy metals, and VOCs can directly harm organisms by entering their respiratory systems, damaging tissues, and interfering with physiological functions. For example, sulfur dioxide and nitrogen oxides emitted from industrial processes and combustion activities can cause leaf damage, reduced growth rates, and even death in plants. Similarly, heavy metals like lead, mercury, and cadmium deposited from the atmosphere can accumulate in soils and water bodies, posing toxic risks to plants, animals, and microorganisms. Airborne pollutants can deposit onto soils and water bodies, leading to soil acidification, nutrient imbalances, and changes in soil microbial communities. Acid deposition resulting from sulfur dioxide and nitrogen oxide emissions can lower soil pH levels, affecting soil fertility and nutrient availability for plants [14]. Additionally, the deposition of heavy metals and other contaminants can contaminate soils and groundwater, impacting soil biota and aquatic organisms. These changes in soil and water quality can disrupt ecosystem processes and reduce habitat suitability for diverse plant and animal species. Air pollution can indirectly impact biodiversity by altering food availability and quality within ecosystems. For instance, pollutants can impair photosynthesis in plants, reducing primary productivity and limiting the energy available to herbivores and higher trophic levels. Furthermore, air pollution can disrupt the symbiotic relationships between plants and pollinators, leading to declines in pollinator populations and affecting the reproductive success of flowering plants. Changes in food availability and pollination services can cascade through food webs, affecting the abundance, distribution, and diversity of species within ecosystems. Air pollution can contribute to habitat fragmentation and degradation, further exacerbating biodiversity loss. For example, pollutants emitted from urban and industrial areas can degrade natural habitats by altering vegetation structure, reducing habitat complexity, and creating barriers to species movement. Additionally, pollutants like nitrogen oxides can promote the growth of nitrogen-tolerant plant species, leading to shifts in plant communities and changes in ecosystem composition. Habitat fragmentation and degradation can isolate populations, reduce genetic diversity, and increase the vulnerability of species to extinction. Air pollution can interact synergistically with other environmental stressors, such as climate change, habitat loss, invasive species, and overexploitation, amplifying the impacts on biodiversity [35]. For instance, increased temperatures and altered precipitation patterns resulting from climate change can exacerbate the effects of air pollution on ecosystems, leading to more severe droughts, heat stress, and wildfires. Similarly, habitat loss and fragmentation can exacerbate the effects of air pollution by reducing the resilience of ecosystems and limiting the ability of species to adapt to changing environmental conditions.

5.3 Acid Deposition

One of the most significant environmental impacts of air pollution is acid deposition, which occurs when SO_2 and NOx react with water vapor and other chemicals in the atmosphere to form sulfuric acid and nitric acid. Acid deposition can acidify soils, surface waters, and freshwater ecosystems, leading to the decline of aquatic species and the disruption of aquatic ecosystems [15, 72]. Acid rain, a form of acid deposition, can damage buildings, monuments, and infrastructure, further exacerbating environmental degradation and economic costs. This phenomenon occurs when SO_2 and NOx, primarily emitted from human activities such as industrial processes, transportation, and fossil fuel combustion, react with water vapor, oxygen, and other chemicals in the atmosphere to form sulfuric acid (H2SO4) and nitric acid (HNO3) [123]. These acidic compounds then fall to the ground in the form of rain, snow, fog, or dry deposition. Acid deposition can lower soil pH levels, leading to soil acidification. This can leach essential nutrients such as calcium, magnesium, and potassium from the soil, making them less available to plants. Soil acidification can also release aluminum and other toxic metals into the soil, posing risks to plant health. Acidic deposition can acidify surface waters, including lakes, rivers, and streams [124]. This can harm aquatic organisms such as fish, amphibians, and invertebrates, which may be sensitive to changes in pH levels. Acidic water can also mobilize aluminum from soils, leading to toxic effects on aquatic life. Acid rain can directly damage vegetation by causing leaf injury, stunted growth, and reduced photosynthesis. It can also weaken plants, making them more susceptible to diseases, pests, and harsh environmental conditions. Acid deposition can corrode buildings, monuments, and infrastructure made of limestone, marble, concrete, and metal. This can lead to structural damage, aesthetic degradation, and economic costs associated with repair and maintenance [125].

Implementing technologies such as flue gas desulfurization (FGD) and selective catalytic reduction (SCR) can reduce sulfur dioxide and nitrogen oxide emissions from industrial sources and power plants [126]. Transitioning to renewable energy sources such as solar, wind, and hydropower can help reduce reliance on fossil fuels and mitigate acid deposition. Enforcing regulations such as the Clean Air Act in the USA and the Gothenburg Protocol in Europe can help limit emissions of sulfur dioxide, nitrogen oxides, and other pollutants responsible for acid deposition. Planting trees and restoring natural habitats can help mitigate the impacts of acid deposition by stabilizing soils, improving water retention, and enhancing biodiversity. Adding lime or other soil amendments can also help neutralize acidic soils and restore fertility.

5.4 Water and Soil Contamination

Air pollution can lead to water and soil contamination through the deposition of pollutants, atmospheric deposition, and runoff from polluted surfaces. Acid rain and acid deposition can acidify surface waters and soils, affecting water quality, aquatic life, and terrestrial ecosystems. Acidified soils can release toxic metals, aluminum, and nutrients into water bodies, impacting aquatic organisms and disrupting nutrient cycles in ecosystems. Airborne heavy metals, such as lead, mercury, cadmium, and arsenic, can deposit onto soils and water bodies, causing soil contamination, groundwater pollution, and toxic effects on plants, animals, and humans [127]. Heavy metal contamination from air pollution can bioaccumulate the food chain, leading to health risks and environmental hazards. Air pollutants like nitrogen oxides and ammonia can deposit onto soils, altering soil chemistry, nutrient cycling, and microbial populations. Soil degradation from air pollution can reduce soil productivity, impair plant growth, and compromise ecosystem resilience, contributing to desertification, erosion, and loss of arable land [128]. Various pollutants emitted into the atmosphere can deposit onto land and water surfaces through processes such as dry deposition and precipitation, leading to contamination of soils and water bodies. Metals such as lead (Pb), mercury (Hg), cadmium (Cd), and arsenic (As) are emitted into the atmosphere from industrial processes, mining activities, fossil fuel combustion, and waste incineration [129]. These metals can then deposit onto soils and water bodies, where they accumulate and pose risks to ecosystems and human health. Also, VOCs, PAHs, and other organic pollutants are released from industrial activities, vehicle emissions, and agricultural operations. These compounds can undergo atmospheric reactions and deposit onto soils and water surfaces, where they can persist and bioaccumulate in the environment [130].

Sulfur dioxide (SO_2) and NOx emitted from industrial sources and combustion processes can contribute to soil acidification and water acidification, leading to leaching of metals and other contaminants from soils and release of aluminum into water bodies [14]. Particulate matter and gaseous pollutants can settle directly onto soil and water surfaces through dry deposition, where they can accumulate over time and impact soil and water quality. Acidic pollutants such as H_2SO_4 and HNO_3 can dissolve in precipitation and deposit onto soils and water bodies as acid rain. This can lead to soil acidification and leaching of metals and other contaminants into groundwater and surface waters. Pollutants can be transported over long distances by atmospheric currents before being deposited onto soils and water bodies, resulting in widespread contamination of terrestrial and aquatic ecosystems [127]. Heavy metals, organic pollutants, and acid deposition can contaminate soils, leading to reduced fertility, impaired plant growth, and risks to human health through the consumption of contaminated crops and groundwater [131, 132]. Contaminants deposited from the atmosphere can pollute surface waters and groundwater, leading to risks for aquatic ecosystems and human populations. Heavy metals and organic compounds can bioaccumulate in aquatic organisms, leading to toxicity and biomagnification through the food chain. Acid deposition can acidify surface waters, leading to declines in fish populations, loss of biodiversity, and degradation of aquatic habitats.

5.5 *Climate Change*

Air pollution plays a significant role in climate change, contributing to the warming of the Earth's atmosphere and the disruption of global climate patterns. Greenhouse gases such as CO2, CH4, and N2O trap heat in the atmosphere, leading to rising temperatures, melting ice caps, and sea level rise. Additionally, aerosols and black carbon particles emitted by combustion processes can absorb and scatter sunlight, further exacerbating warming trends and altering regional climate patterns [47, 48, 133]. Climate change impacts ecosystems, biodiversity, and natural resources, posing additional challenges for adaptation and mitigation efforts. Air pollutants like carbon dioxide, methane, and nitrous oxide are greenhouse gases that trap heat in the atmosphere, leading to global warming and climate change [45, 46]. Emissions from fossil fuel combustion, transportation, deforestation, and industrial activities contribute to the enhanced greenhouse effect, causing temperature rise and climate extremes [134]. Certain air pollutants, such as CFCs and halons, can deplete the ozone layer in the stratosphere, leading to ozone holes and increased UV radiation exposure on Earth's surface [135]. Ozone depletion from air pollution can harm human health, damage ecosystems, and disrupt climate patterns, affecting weather, sea levels, and biodiversity. Air pollution influences climate processes, such as cloud formation, precipitation patterns, and temperature regulation, affecting regional and global climate systems [136, 137]. Pollution-induced changes in atmospheric composition, aerosols, and radiation balance can exacerbate climate change impacts, including heat waves, droughts, storms, and sea level rise.

5.6 Mitigation and Prevention Strategies: Environmental and Policy Implications of the Environmental Impacts of Air Pollution

Addressing the environmental impacts of air pollution requires coordinated efforts to reduce emissions, promote sustainable practices, and protect ecosystems and natural resources. Transitioning to clean, renewable energy sources, such as solar, wind, hydropower, and geothermal energy, can reduce air pollution emissions, lower greenhouse gas levels, and promote sustainable development. Investing in clean energy technologies, energy efficiency, and green infrastructure can help mitigate environmental impacts and combat climate change. Implementing pollution control technologies, such as scrubbers, filters, catalytic converters, and emission reduction systems, in industrial facilities, vehicles, and power plants can reduce air pollutant emissions and improve air quality [138]. Enforcing emission standards, regulatory measures, and pollution monitoring programs can help minimize environmental impacts and protect ecosystems. Planting trees, restoring degraded landscapes, and implementing ecosystem-based approaches can help mitigate the environmental impacts of air pollution on forests, wetlands, and habitats. Reforestation projects, habitat restoration initiatives, and conservation efforts can enhance biodiversity, improve soil quality, and sequester carbon dioxide, promoting ecosystem resilience. Adopting sustainable agriculture practices, such as precision farming, organic farming, and agroecological methods, can reduce air pollution emissions from agricultural activities, minimize soil contamination, and protect water quality. Sustainable agriculture practices, crop rotation, and soil conservation techniques can enhance soil health, biodiversity, and ecosystem services [139]. Promoting international cooperation, multilateral agreements, and global partnerships to address transboundary air pollution, cross-border emissions, and environmental challenges can facilitate coordinated action to protect ecosystems, mitigate climate change impacts, and safeguard natural resources [140]. Collaborative efforts, information sharing, and capacity-building initiatives can advance sustainable development goals and environmental protection objectives.

6 Social and Economic Impacts of Air Pollution

The social and economic impacts of air pollution are multifaceted and wide-ranging, affecting individuals, communities, and economies at local, regional, and global scales. For instance, exposure to air pollution is associated with increased rates of hospital admissions, emergency room visits, and premature deaths, leading to a substantial public health burden and reduced quality of life for affected individuals and communities [141, 142]. The healthcare costs associated with treating air pollution-related illnesses can place strain on healthcare systems and budgets, diverting resources away from other essential services and treatments [143]. Certain groups, including children, the elderly, low-income communities, and marginalized populations, are disproportionately affected by air pollution due to factors such as limited access to healthcare, inadequate housing, and exposure to environmental injustices. Children are particularly vulnerable to the health impacts of air pollution, as their developing respiratory and immune systems make them more susceptible to respiratory infections, asthma, and other adverse health effects [144].

Also, air pollution can reduce productivity and economic output by causing illness, absenteeism, and impaired cognitive function among workers. Workers exposed to high levels of air pollution may experience respiratory symptoms, fatigue, and decreased cognitive performance, leading to decreased efficiency and productivity in the workplace [35]. Work absenteeism due to air pollution-related illnesses can result in lost wages, decreased economic productivity, and increased healthcare costs for employers and governments [145]. Air pollution disproportionately affects low-income communities and communities of color, exacerbating existing social inequalities and environmental injustices. These communities are more likely to live near industrial facilities, highways, and other sources of pollution, leading to higher exposure levels and increased health risks. Environmental justice initiatives seek to address these disparities by advocating for equitable access to clean air, water, and land for all communities, regardless of socioeconomic status or race. Air pollution can have negative impacts on tourism and recreational activities by degrading air quality, diminishing scenic views, and deterring visitors from outdoor destinations. High levels of air pollution in popular tourist destinations can lead to cancellations of travel plans, loss of revenue for local businesses, and damage to the tourism industry [146]. Investments in pollution control measures and sustainable tourism practices can help mitigate the negative effects of air pollution on tourism and support local economies.

Air pollution contributes to environmental degradation by harming ecosystems, contaminating soil and water, and disrupting biodiversity. Acid deposition, ozone depletion, and eutrophication are among the environmental impacts of air pollution that can lead to loss of habitat, decline in species diversity, and degradation of natural resources [16, 147]. The loss of ecosystem services, such as pollination, nutrient cycling, and water purification, can have cascading effects on human well-being, including reduced agricultural productivity, food security concerns, and increased vulnerability to natural disasters [72]. Air pollution contributes to increased healthcare expenditures due to the treatment of respiratory illnesses, cardiovascular diseases, and other health conditions linked to exposure to pollutants. The costs of hospitalizations, medications, doctor visits, and emergency care for individuals affected by air pollution-related illnesses place a significant burden on healthcare systems and public health budgets. Air pollution can lead to lost productivity in the workforce due to absenteeism, reduced work performance, and increased sick days associated with respiratory symptoms and health complications caused by exposure to pollutants. Reduced productivity levels can impact individual employees, businesses, and the economy, leading to economic losses and decreased competitiveness [148]. Air pollution can cause damage to infrastructure, including buildings, roads, bridges, and monuments, through the corrosion of materials, soiling of surfaces, and degradation of structural integrity [149]. The costs of repairing, maintaining, and replacing infrastructure affected by air pollution can strain public budgets, increase maintenance expenses, and hinder sustainable infrastructure development [143].

Air pollution exacerbates environmental justice issues by disproportionately affecting low-income communities, minority populations, and marginalized groups living in areas with high pollution levels and environmental hazards [113, 150]. These communities often face unequal exposure to pollutants, limited access to green spaces, inadequate healthcare services, and higher health risks associated with poor air quality. Vulnerable populations, such as children, the elderly, individuals with preexisting health conditions, and socioeconomically disadvantaged groups, are disproportionately impacted by the social effects of air pollution. These populations may experience higher rates of respiratory illnesses, reduced quality of life, increased healthcare costs, and limited access to resources to mitigate the impacts of pollution on their well-being. Air pollution influences urban planning decisions related to land use, transportation systems, building design, and environmental regulations to create healthier, sustainable urban environments [151]. Implementing green infrastructure, sustainable building practices, and public transportation options can help reduce emissions, improve air quality, and enhance the livability of urban spaces. Addressing air pollution requires sustainable transportation solutions, such as promoting public transit, cycling infrastructure, electric vehicles, and alternative fuels to reduce emissions from transportation sources. Investing in sustainable transportation options can decrease air pollution, improve public health, and support sustainable development goals for accessible, environmentally friendly transportation systems. Transitioning to renewable energy sources, such as solar, wind, hydropower, and geothermal energy, is crucial for reducing air pollution emissions, mitigating climate change impacts, and promoting sustainable development [152]. Investing in clean energy technologies, energy efficiency measures, and renewable energy infrastructure can help transition to a low-carbon economy, create green jobs, and support environmental sustainability goals.

7 Sustainable Air Pollution Impact Mitigation Strategies

7.1 Regulatory Frameworks

National Air Quality Standards Establishing and enforcing national air quality standards is a key regulatory approach to controlling air pollution. Governments set limits on pollutant concentrations in the air to protect public health and the environment. Monitoring air quality, implementing pollution control measures, and enforcing compliance with emission standards are essential components of national air quality regulations [153].

International Agreements and Treaties International agreements and treaties play a crucial role in addressing transboundary air pollution, global emissions reduction, and environmental protection [154]. Agreements such as the Paris Agreement, the Kyoto Protocol, and the Montreal Protocol set targets for reducing greenhouse gas emissions, controlling ozone-depleting substances, and promoting international cooperation to mitigate climate change and air pollution impacts.

7.2 Technological Solutions

Emission Control Technologies Adopting emission control technologies, such as scrubbers, catalytic converters, particulate filters, and low-emission engines, can reduce air pollutant emissions from industrial sources, vehicles, power plants, and other facilities [35, 155]. Investing in pollution control equipment, retrofitting existing infrastructure, and implementing the best available technologies can help improve air quality, mitigate environmental impacts, and comply with regulatory requirements.

Clean Energy Alternatives Transitioning to clean energy alternatives, such as renewable energy sources (solar, wind, hydropower), nuclear energy, and energy-efficient technologies, can reduce greenhouse gas emissions, decrease air pollution, and promote sustainable development [20]. Investing in clean energy infrastructure, implementing energy efficiency measures, and phasing out fossil fuels can support a low-carbon economy and combat climate change.

7.3 *Behavioral Changes*

Transportation Choices Promoting sustainable transportation choices, such as public transit, cycling, walking, carpooling, and electric vehicles, can help reduce air pollution emissions from transportation sources [156, 157]. Encouraging individuals to use alternative modes of transportation, reduce vehicle miles traveled, and adopt eco-friendly transportation habits can improve air quality, reduce traffic congestion, and enhance public health.

Energy Consumption Patterns Encouraging energy conservation, energy efficiency, and sustainable consumption practices can help reduce air pollution emissions associated with energy production and consumption [158]. Adopting energy-saving measures, using appliances with high energy efficiency ratings, and promoting resource conservation can lower energy demand, decrease emissions, and support environmental sustainability goals.

7.4 *Community Engagement and Advocacy*

Grassroots Movements Engaging local communities, grassroots organizations, and environmental activists in air pollution mitigation efforts can raise awareness, mobilize public support, and advocate for policy changes. Grassroots movements play a vital role in promoting environmental justice, holding polluters accountable, and empowering communities to address air pollution challenges through collective action [112, 159].

Public Awareness Campaigns Raising public awareness about the impacts of air pollution on human health, ecosystems, and the environment is essential for fostering behavioral changes, promoting sustainable practices, and advocating for policy responses. Public awareness campaigns, educational initiatives, and community outreach activities can inform and empower individuals to act, support clean air initiatives, and protect the environment for future generations.

8 Conclusion

The impacts of air pollution are profound and far-reaching, affecting human health, ecosystems, economies, and social well-being. From respiratory illnesses and cardiovascular diseases to environmental degradation and lost productivity, air pollution takes a heavy toll on individuals, communities, and societies worldwide. The burden of air pollution falls disproportionately on vulnerable populations, including children, the elderly, and low-income communities, exacerbating existing social inequalities and environmental injustices. Air pollution takes a heavy toll on human health, contributing to respiratory diseases, cardiovascular diseases, cancers, and other adverse health outcomes. Vulnerable populations, including children, the elderly, and socioeconomically disadvantaged communities, bear the brunt of these health impacts, exacerbating existing social inequalities and health disparities. The burden of air pollution-related illnesses extends beyond individual suffering to encompass economic costs, including healthcare expenditures, lost productivity, and premature mortality.

The interconnected nature of air pollution and its impacts underscores the need for comprehensive and concerted action to address this pressing issue. By reducing emissions of harmful pollutants, promoting clean energy technologies, and implementing sustainable practices, we can mitigate the health and environmental impacts of air pollution and create healthier, more resilient communities. Additionally, addressing air pollution requires collaboration and cooperation at the local, national, and international levels, involving governments, businesses, civil society organizations, and individuals. Beyond its effects on human health, air pollution also degrades ecosystems, contaminates soil and water, and disrupts biodiversity. Acid deposition, ozone depletion, and eutrophication are among the environmental consequences of air pollution that threaten ecosystem resilience and compromise ecosystem services. Biodiversity loss, habitat degradation, and declines in ecosystem productivity have far-reaching implications for food security, water quality, and natural resource management.

The social and economic impacts of air pollution are profound, affecting communities, economies, and societies worldwide. Lost productivity, work absenteeism, and healthcare expenditures associated with air pollution-related illnesses impose significant economic costs on individuals, businesses, and governments. Furthermore, air pollution disproportionately affects marginalized populations, exacerbating social inequalities and environmental injustices. Tourism, recreation, and outdoor activities also suffer as a result of poor air quality, leading to economic losses for local economies and businesses. Furthermore, investing in pollution control measures and environmental protection initiatives not only improves public health and well-being but also generates economic benefits, including job creation, healthcare savings, and increased productivity. By recognizing the social, economic, and environmental costs of air pollution and taking proactive steps to address them, we can build a more sustainable and equitable future for all.

The impacts of air pollution are vast and multifaceted, affecting every aspect of human life and the environment. Addressing this complex challenge requires coordinated efforts at the local, national, and international levels, involving governments, businesses, civil society organizations, and individuals.

References

1. Genc S, Zadeoglulari Z, Fuss SH, Genc K (2012) The adverse effects of air pollution on the nervous system. J Toxicol 2012:782462. https://doi.org/10.1155/2012/782462

2. Ghorani-Azam A, Riahi-Zanjani B, Balali-Mood M (2016) Effects of air pollution on human health and practical measures for prevention in Iran. J Res Med Sci 21:65. https://doi.org/10.4103/1735-1995.189646
3. Bălă GP, Râjnoveanu RM, Tudorache E, Motişan R, Oancea C (2021) Air pollution exposure-the (in)visible risk factor for respiratory diseases. Environ Sci Pollut Res Int 28(16):19615–19628. https://doi.org/10.1007/s11356-021-13208-x
4. IARC (2016) IARC monographs on the evaluation of carcinogenic risks to humans, no. 109. 1.2, sources of air pollutants. IARC working group on the evaluation of carcinogenic risks to humans. Outdoor air pollution. Lyon (FR): International Agency for Research on Cancer; 2016. Available from: https://www.ncbi.nlm.nih.gov/books/NBK368029/
5. Lee YG, Lee PH, Choi SM, An MH, Jang AS (2021) Effects of air pollutants on airway diseases. Int J Environ Res Public Health 18(18):9905. https://doi.org/10.3390/ijerph18189905
6. Comunian S, Dongo D, Milani C, Palestini P (2020) Air pollution and Covid-19: the role of particulate matter in the spread and increase of Covid-19's morbidity and mortality. Int J Environ Res Public Health 17(12):4487. https://doi.org/10.3390/ijerph17124487
7. Du Y, Xu X, Chu M, Guo Y, Wang J (2016) Air particulate matter and cardiovascular disease: the epidemiological, biomedical and clinical evidence. J Thorac Dis 8(1):E8–E19. https://doi.org/10.3978/j.issn.2072-1439.2015.11.37
8. Bai L, Wang J, Ma X, Lu H (2018) Air pollution forecasts: an overview. Int J Environ Res Public Health 15(4):780. https://doi.org/10.3390/ijerph15040780
9. Lee HJ, Jo YJ, Kim S, Kim D, Kim JM, Choi D, Jo HY, Bak J, Park SY, Jeon W, Kim CH (2022) Transboundary aerosol transport process and its impact on aerosol-radiation-cloud feedbacks in springtime over Northeast Asia. Sci Rep 12(1):4870. https://doi.org/10.1038/s41598-022-08854-1
10. Richard G, Izah SC, Sawyer EW, Ogwu MC (2023) Seasonal variations in the concentration of some atmospheric pollutants along a major highway in Southern Nigeria. J Environ Stud 30 (1):10–25. https://doi.org/10.21608/JESJ.2023.205673.1045
11. Dias D, Tchepel O (2018) Spatial and temporal dynamics in air pollution exposure assessment. Int J Environ Res Public Health 15(3):558. https://doi.org/10.3390/ijerph15030558
12. Zhang Z, Nie T, Wu Y, Ling J, Huang D (2022) The temporal and spatial distributions and influencing factors of transboundary pollution in China. Int J Environ Res Public Health 19 (8):4643. https://doi.org/10.3390/ijerph19084643
13. Kim MJ (2019) The effects of transboundary air pollution from China on ambient air quality in South Korea. Heliyon 5(12):e02953. https://doi.org/10.1016/j.heliyon.2019.e02953
14. Prakash J, Agrawal SB, Agrawal M (2023) Global trends of acidity in rainfall and its impact on plants and soil. J Soil Sci Plant Nutr 23(1):398–419. https://doi.org/10.1007/s42729-022-01051-z
15. Grennfelt P, Engleryd A, Forsius M, Hov Ø, Rodhe H, Cowling E (2020) Acid rain and air pollution: 50 years of progress in environmental science and policy. Ambio 49(4):849–864. https://doi.org/10.1007/s13280-019-01244-4
16. Irvine IC, Greaver T, Phelan J, Sabo RD, Houtven VAN, G. (2017) Terrestrial acidification and ecosystem services: effects of acid rain on bunnies, baseball, and Christmas trees. Ecosphere 8(6). https://doi.org/10.1002/ecs2.1857
17. Tiotiu AI, Novakova P, Nedeva D, Chong-Neto HJ, Novakova S, Steiropoulos P, Kowal K (2020) Impact of air pollution on asthma outcomes. Int J Environ Res Public Health 17 (17):6212. https://doi.org/10.3390/ijerph17176212
18. Rowland A, Murray AJ, Wellburn AR (1985) Oxides of nitrogen and their impact upon vegetation. Rev Environ Health 5(4):295–342
19. Emberson L (2020) Effects of ozone on agriculture, forests and grasslands. Philos Transact A Math Phys Eng Sci 378(2183):20190327. https://doi.org/10.1098/rsta.2019.0327

20. Ma R, Abid N, Yang S, Ahmad F (2023) From crisis to resilience: strengthening climate action in OECD countries through environmental policy and energy transition. Environ Sci Pollut Res Int 30(54):115480–115495. https://doi.org/10.1007/s11356-023-29970-z
21. Wang H, Luo J, Zhang M, Ling Y (2022) The impact of transportation restructuring on the intensity of greenhouse gas emissions: empirical data from China. Int J Environ Res Public Health 19(19):12960. https://doi.org/10.3390/ijerph191912960
22. Ali S, Khan KA, Gyamfi BA, Ofori EK, Tetteh D, Shamansurova Z (2024) Can clean energy and technology address environmental sustainability in G7 under the pre-set of human development? Environ Sci Pollut Res Int 31(9):13800–13814. https://doi.org/10.1007/s11356-024-32011-y
23. Wodtke GT, Ard K, Bullock C, White K, Priem B (2022) Concentrated poverty, ambient air pollution, and child cognitive development. Sci Adv 8(48):eadd0285. https://doi.org/10.1126/sciadv.add0285
24. Molina LT, Velasco E, Retama A, Zavala M (2019) Experience from integrated air quality management in the Mexico City Metropolitan Area and Singapore. Atmos 10:512. https://doi.org/10.3390/atmos10090512
25. Gao G, Wang S, Xue R, Liu D, Ren H, Zhang R (2021) Uncovering the characteristics of air pollutants emission in industrial parks and analyzing emission reduction potential: case studies in Henan, China. Sci Rep 11(1):23709. https://doi.org/10.1038/s41598-021-03193-z
26. Mailloux NA, Abel DW, Holloway T, Patz JA (2022) Nationwide and regional $PM_{2.5}$-related air quality health benefits from the removal of energy-related emissions in the United States. GeoHealth 6(5):e2022GH000603. https://doi.org/10.1029/2022GH000603
27. Breitner-Busch S, Mücke HG, Schneider A, Hertig E (2023) Impact of climate change on non-communicable diseases due to increased ambient air pollution. J Health Monit 8(Suppl 4):103–121. https://doi.org/10.25646/11655
28. Brugge D, Durant JL, Rioux C (2007) Near-highway pollutants in motor vehicle exhaust: a review of epidemiologic evidence of cardiac and pulmonary health risks. Environ Health 6:23. https://doi.org/10.1186/1476-069X-6-23
29. Zhang K, Batterman S (2013) Air pollution and health risks due to vehicle traffic. Sci Total Environ 450–451:307–316. https://doi.org/10.1016/j.scitotenv.2013.01.074
30. Xu RT, Pan SF, Chen J, Chen GS, Yang J, Dangal SRS, Shepard JP, Tian HQ (2018) Half-century ammonia emissions from agricultural systems in Southern Asia: magnitude, spatio-temporal patterns, and implications for human health. GeoHealth 2(1):40–53. https://doi.org/10.1002/2017GH000098
31. Grossi G, Goglio P, Vitali A, Williams AG (2018) Livestock and climate change: impact of livestock on climate and mitigation strategies. Anim Front 9(1):69–76. https://doi.org/10.1093/af/vfy034
32. Chandrappa R, Chandra Kulshrestha U (2015) Air pollution and disasters. Sustainable air pollution management: theory and practice, pp 325–343. https://doi.org/10.1007/978-3-319-21596-9_8
33. Idowu OS, De Azevedo LB, Zohoori FV, Kanmodi K, Pak T (2023) Health risks associated with the production and usage of charcoal: a systematic review. BMJ Open 13(7):e065914. https://doi.org/10.1136/bmjopen-2022-065914
34. Lin TC, Krishnaswamy G, Chi DS (2008) Incense smoke: clinical, structural and molecular effects on airway disease. Clin Mol Allergy 6:3. https://doi.org/10.1186/1476-7961-6-3
35. Manisalidis I, Stavropoulou E, Stavropoulos A, Bezirtzoglou E (2020) Environmental and health impacts of air pollution: a review. Front Public Health 8:14. https://doi.org/10.3389/fpubh.2020.00014
36. Suh HH, Bahadori T, Vallarino J, Spengler JD (2000) Criteria air pollutants and toxic air pollutants. Environ Health Perspect 108(Suppl 4):625–633
37. Kan H, London SJ, Chen G, Zhang Y, Song G, Zhao N, Jiang L, Chen B (2007) Differentiating the effects of fine and coarse particles on daily mortality in Shanghai, China. Environ Int 33(3):376–384. https://doi.org/10.1016/j.envint.2006.12.001

38. Basith S, Manavalan B, Shin TH, Park CB, Lee WS, Kim J, Lee G (2022) The impact of fine particulate matter 2.5 on the cardiovascular system: a review of the invisible killer. Nanomaterials 12(15):2656. https://doi.org/10.3390/nano12152656
39. Reno AL, Brooks EG, Ameredes BT (2015) Mechanisms of heightened airway sensitivity and responses to inhaled SO2 in asthmatics. Environ Health Insights 9(Suppl 1):13–25. https://doi.org/10.4137/EHI.S15671
40. Zoran MA, Savastru RS, Savastru DM, Tautan MN (2020) Assessing the relationship between ground levels of ozone (O_3) and nitrogen dioxide (NO_2) with coronavirus (COVID-19) in Milan, Italy. Sci Total Environ 740:140005. https://doi.org/10.1016/j.scitotenv.2020.140005
41. Kinoshita H, Türkan H, Vucinic S, Naqvi S, Bedair R, Rezaee R, Tsatsakis A (2020) Carbon monoxide poisoning. Toxicol Rep 7:169–173. https://doi.org/10.1016/j.toxrep.2020.01.005
42. Ikhajiagbe B, Ogwu MC (2020) Hazard quotient, microbial diversity and plant composition of spent crude oil polluted-soil. Beni-Suef Univ J Basic Appl Sci 9(26). https://doi.org/10.1186/s43088-020-00052-0
43. Ogwu MC (2018) Effects of indole-3-acetic acid on growth parameters of Citrullus lanatus (Thunberg) Matsum and Nakai. Momona Ethiop J Sci 10(1):109–125. https://doi.org/10.4314/mejs.v10i1.8
44. Ogwu MC, Chime AO, Osawaru ME, Emoekpere R (2014) Germination of watermelon [Citrullus lanatus (Thunberg) Matsumura and Nakai] in lead polluted petri dish. Italian J Occup Environ Hyg 5(3-4):126–130
45. Cassia R, Nocioni M, Correa-Aragunde N, Lamattina L (2018) Climate change and the impact of greenhouse gasses: CO_2 and NO, friends and foes of plant oxidative stress. Front Plant Sci 9:273. https://doi.org/10.3389/fpls.2018.00273
46. Shivanna KR (2022) Climate change and its impact on biodiversity and human welfare. Proc Indian Nat Sci Acad A Phys Sci 88(2):160–171. https://doi.org/10.1007/s43538-022-00073-6
47. Wang P, Yang Y, Xue D, Ren L, Tang J, Leung LR, Liao H (2023) Aerosols overtake greenhouse gases causing a warmer climate and more weather extremes toward carbon neutrality. Nat Commun 14(1):7257. https://doi.org/10.1038/s41467-023-42891-2
48. Wang M, Zhou XA, Curl C, Fitzpatrick A, Vedal S, Kaufman J (2023) Long-term exposure to ambient air pollution and cognitive function in older US adults: the multi-ethnic study of atherosclerosis. Environ Epidemiol 7(1):e242. https://doi.org/10.1097/EE9.0000000000000242
49. Xing YF, Xu YH, Shi MH, Lian YX (2016) The impact of PM2.5 on the human respiratory system. J Thorac Dis 8(1):E69–E74. https://doi.org/10.3978/j.issn.2072-1439.2016.01.19
50. David E, Niculescu VC (2021) Volatile organic compounds (VOCs) as environmental pollutants: occurrence and mitigation using nanomaterials. Int J Environ Res Public Health 18(24):13147. https://doi.org/10.3390/ijerph182413147
51. Montero-Montoya R, López-Vargas R, Arellano-Aguilar O (2018) Volatile organic compounds in air: sources, distribution, exposure and associated illnesses in children. Ann Glob Health 84(2):225–238. https://doi.org/10.29024/aogh.910
52. Shuai J, Kim S, Ryu H, Park J, Lee CK, Kim GB, Ultra Jr VU, Yang W (2018) Health risk assessment of volatile organic compounds exposure near Daegu dyeing industrial complex in South Korea. BMC Public Health 18(1):528. https://doi.org/10.1186/s12889-018-5454-1
53. Steiling K, Kathuria H (2024) Occupational benzene exposure: An unrecognized threat for lung cancer development. Am J Respir Crit Care Med 209(2):128–130. https://doi.org/10.1164/rccm.202311-2023ED
54. Fernandes Azevedo B, Barros Furieri L, Peçanha FM, Wiggers GA, Frizera Vassallo P, Ronacher Simões M, Fiorim J, Rossi de Batista P, Fioresi M, Rossoni L, Stefanon I, Alonso MJ, Salaices M, Valentim Vassallo D (2012) Toxic effects of mercury on the cardiovascular and central nervous systems. J Biomed Biotechnol 2012:949048. https://doi.org/10.1155/2012/949048

55. Nhung NTH, Nguyen XT, Long VD, Wei Y, Fujita T (2022) A review of soil contaminated with dioxins and biodegradation technologies: current status and future prospects. Toxics 10 (6):278. https://doi.org/10.3390/toxics10060278
56. Lakra K, Avishek K (2022) A review on factors influencing fog formation, classification, forecasting, detection and impacts. Rendiconti Lincei Scienze fisiche e naturali 33(2):319–353. https://doi.org/10.1007/s12210-022-01060-1
57. Rafkin S, Lora JM, Soto A, Battalio J (2022) The interaction of deep convection with the general circulation in Titan's atmosphere. Part 1: cloud resolving simulations. Icarus 373:114623. https://doi.org/10.1016/j.icarus.2021.114755
58. Nugent A, DeCou D, Russell S, Karamperidou C (2019) Atmospheric science: ATMO 200 companion text. Available online: https://pressbooks-dev.oer.hawaii.edu/atmo/front-matter/disclaimer/
59. Repace J (2015) Analysis of the air quality impact of hoyvik heating plant parts I, II, & III. https://doi.org/10.13140/RG.2.1.3322.6407
60. Keshavarzian E, Jin R, Dong K, Kwok KCS, Zhang Y, Zhao M (2020) Effect of pollutant source location on air pollutant dispersion around a high-rise building. App Math Model 81:582–602. https://doi.org/10.1016/j.apm.2020.01.019
61. Pérez IA, García MÁ, Sánchez ML, Pardo N, Fernández-Duque B (2020) Key points in air pollution meteorology. Int J Environ Res Public Health 17(22):8349. https://doi.org/10.3390/ijerph17228349
62. Zabrocki L, Alari A, Benmarhnia T (2022) Improving the design stage of air pollution studies based on wind patterns. Sci Rep 12(1):7917. https://doi.org/10.1038/s41598-022-11939-6
63. Isaacs JD (1963) Atmospheric jet streams. Science 141(3585):1045–1046. https://doi.org/10.1126/science.141.3585.1045
64. Giovannini L, Ferrero E, Karl T, Rotach MW, Staquet C, Trini Castelli S, Zardi D (2020) Atmospheric pollutant dispersion over complex terrain: challenges and needs for improving air quality measurements and modeling. Atmos 11:646. https://doi.org/10.3390/atmos11060646
65. Ulpiani G (2021) On the linkage between urban heat Island and urban pollution Island: three-decade literature review towards a conceptual framework. Sci Total Environ 751:141727. https://doi.org/10.1016/j.scitotenv.2020.141727
66. Wu M, Zhang G, Wang L, Liu X, Wu Z (2022) Influencing factors on airflow and pollutant dispersion around buildings under the combined effect of wind and buoyancy – a review. Int J Environ Res Public Health 19(19):12895. https://doi.org/10.3390/ijerph191912895
67. Geddes JA, Wang B, Li D (2021) Ozone and nitrogen dioxide pollution in a coastal urban environment: the role of sea breezes, and implications of their representation for remote sensing of local air quality. J Geophys Res Atmos 126(18):e2021JD035314. https://doi.org/10.1029/2021JD035314
68. Lim HD, Hertwig D, Grylls T, Gough H, Reeuwijk MV, Grimmond S, Vanderwel C (2022) Pollutant dispersion by tall buildings: laboratory experiments and large-Eddy simulation. Exp Fluids 63(6):92. https://doi.org/10.1007/s00348-022-03439-0
69. Hassen W, Hnaien N, Ben Said L, Albati FM, Ayadi B, Rajhi W, Kolsi L (2023) Air pollution dispersion in Hail city: climate and urban topography impact. Heliyon 9(10):e20608. https://doi.org/10.1016/j.heliyon.2023.e20608
70. Sbai SE, Bentayeb F, Yin H (2022) Atmospheric pollutants response to the emission reduction and meteorology during the COVID-19 lockdown in the north of Africa (Morocco). Stoch Environ Res Risk Assess 36(11):3769–3784. https://doi.org/10.1007/s00477-022-02224-z
71. Ebi KL, McGregor G (2008) Climate change, tropospheric ozone and particulate matter, and health impacts. Environ Health Perspect 116(11):1449–1455. https://doi.org/10.1289/ehp.11463
72. Driscoll CT, Driscoll KM, Mitchell MJ, Raynal DJ (2003) Effects of acidic deposition on forest and aquatic ecosystems in New York State. Environ Pollut 123(3):327–336. https://doi.org/10.1016/s0269-7491(03)00019-8

73. Bhui K, Newbury JB, Latham RM, Ucci M, Nasir ZA, Turner B, O'Leary C, Fisher HL, Marczylo E, Douglas P, Stansfeld S, Jackson SK, Tyrrel S, Rzhetsky A, Kinnersley R, Kumar P, Duchaine C, Coulon F (2023) Air quality and mental health: evidence, challenges and future directions. BJPsych Open 9(4):e120. https://doi.org/10.1192/bjo.2023.507
74. Thangavel P, Park D, Lee YC (2022) Recent insights into particulate matter ($PM_{2.5}$)-mediated toxicity in humans: an overview. Int J Environ Res Public Health 19(12):7511. https://doi.org/10.3390/ijerph19127511
75. Kyung SY, Jeong SH (2020) Particulate-matter related respiratory diseases. Tuberc Respir Dis 83(2):116–121. https://doi.org/10.4046/trd.2019.0025
76. Larson PS, Espira L, Glenn BE, Larson MC, Crowe CS, Jang S, O'Neill MS (2022) Long-term $PM_{2.5}$ exposure is associated with symptoms of acute respiratory infections among children under five years of age in Kenya, 2014. Int J Environ Res Public Health 19(5):2525. https://doi.org/10.3390/ijerph19052525
77. Domingo JL, Rovira J (2020) Effects of air pollutants on the transmission and severity of respiratory viral infections. Environ Res 187:109650. https://doi.org/10.1016/j.envres.2020.109650
78. Lee BJ, Kim B, Lee K (2014) Air pollution exposure and cardiovascular disease. Toxicol Res 30(2):71–75. https://doi.org/10.5487/TR.2014.30.2.071
79. Miller MR (2020) Oxidative stress and the cardiovascular effects of air pollution. Free Radic Biol Med 151:69–87. https://doi.org/10.1016/j.freeradbiomed.2020.01.004
80. Fussell JC, Jauniaux E, Smith RB, Burton GJ (2024) Ambient air pollution and adverse birth outcomes: a review of underlying mechanisms. BJOG 131(5):538–550. https://doi.org/10.1111/1471-0528.17727
81. Röllin HB (2022) Introduction to the special issue of *IJERPH* entitled "Prenatal Exposure to Environmental Pollutants and Other Stressors: Impacts on Fetal Development, Birth Outcomes, Children's Health and Beyond". Int J Environ Res Public Health 19(14):8816. https://doi.org/10.3390/ijerph19148816
82. Rani P, Dhok A (2023) Effects of pollution on pregnancy and infants. Cureus 15(1):e33906. https://doi.org/10.7759/cureus.33906
83. Kim H, Kim WH, Kim YY, Park HY (2020) Air pollution and central nervous system disease: a review of the impact of fine particulate matter on neurological disorders. Front Public Health 8:575330. https://doi.org/10.3389/fpubh.2020.575330
84. Roy R, D'Angiulli A (2024) Air pollution and neurological diseases, current state highlights. Front Neurosci 18:1351721. https://doi.org/10.3389/fnins.2024.1351721
85. Cani M, Turco F, Butticè S, Vogl UM, Buttigliero C, Novello S, Capelletto E (2023) How does environmental and occupational exposure contribute to carcinogenesis in genitourinary and lung cancers? Cancer 15(10):2836. https://doi.org/10.3390/cancers15102836
86. Turner MC, Andersen ZJ, Baccarelli A, Diver WR, Gapstur SM, Pope 3rd CA, Prada D, Samet J, Thurston G, Cohen A (2020) Outdoor air pollution and cancer: An overview of the current evidence and public health recommendations. CA Cancer J Clin 70:460. https://doi.org/10.3322/caac.21632
87. Loomis D, Huang W, Chen G (2014) The International Agency for Research on Cancer (IARC) evaluation of the carcinogenicity of outdoor air pollution: focus on China. Chin J Cancer 33(4):189–196. https://doi.org/10.5732/cjc.014.10028
88. Lee YG, Yoon SJ, Yoon SH, Kang SW, Jeon S, Kim M, Shin DA, Nam CM, Ye BS (2023) Air pollution is associated with faster cognitive decline in Alzheimer's disease. Ann Clin Transl Neurol 10(6):964–973. https://doi.org/10.1002/acn3.51779
89. Liu H, Hu T (2023) Correlation between air pollution and cognitive impairment among older individuals: empirical evidence from China. BMC Geriatr 23(1):366. https://doi.org/10.1186/s12877-023-03932-z
90. Kilian J, Kitazawa M (2018) The emerging risk of exposure to air pollution on cognitive decline and Alzheimer's disease – evidence from epidemiological and animal studies. Biom J 41(3):141–162. https://doi.org/10.1016/j.bj.2018.06.001

91. Hayman GD (1997) CFCS and the ozone layer. Br J Clin Pract Suppl 89:2–9
92. Romley JA, Hackbarth A, Goldman DP (2012) The impact of air quality on hospital spending. Rand Health Q 2(3):6
93. Bari MW, Saleem S, Bashir M, Ahmad B (2022) Impact of ambient air pollution on outdoor employees' performance: mediating role of anxiety. Front Psychol 13:926534. https://doi.org/10.3389/fpsyg.2022.926534
94. Asif Z, Chen Z, Wang H, Zhu Y (2022) Update on air pollution control strategies for coal-fired power plants. Clean Techn Environ Policy 24(8):2329–2347. https://doi.org/10.1007/s10098-022-02328-8
95. Jonidi Jafari A, Charkhloo E, Pasalari H (2021) Urban air pollution control policies and strategies: a systematic review. J Environ Health Sci Eng 19(2):1911–1940. https://doi.org/10.1007/s40201-021-00744-4
96. Bhat TA, Panzica L, Kalathil SG, Thanavala Y (2015) Immune dysfunction in patients with chronic obstructive pulmonary disease. Ann Am Thorac Soc 12(Suppl 2):S169–S175. https://doi.org/10.1513/AnnalsATS.201503-126AW
97. Chen L, Deng H, Cui H, Fang J, Zuo Z, Deng J, Li Y, Wang X, Zhao L (2017) Inflammatory responses and inflammation-associated diseases in organs. Oncotarget 9(6):7204–7218. https://doi.org/10.18632/oncotarget.23208
98. Pizzino G, Irrera N, Cucinotta M, Pallio G, Mannino F, Arcoraci V, Squadrito F, Altavilla D, Bitto A (2017) Oxidative stress: harms and benefits for human health. Oxid Med Cell Longev 2017:8416763. https://doi.org/10.1155/2017/8416763
99. Sopian NA, Jalaludin J, Abu Bakar S, Hamedon TR, Latif MT (2021) Exposure to particulate PAHs on potential genotoxicity and cancer risk among school children living near the petrochemical industry. Int J Environ Res Public Health 18(5):2575. https://doi.org/10.3390/ijerph18052575
100. Tavares AM, Viegas S, Louro H, Göen T, Santonen T, Luijten M, Kortenkamp A, Silva MJ (2022) Occupational exposure to hexavalent chromium, nickel and PAHs: a mixtures risk assessment approach based on literature exposure data from European countries. Toxics 10(8):431. https://doi.org/10.3390/toxics10080431
101. Hadi HA, Carr CS, Al Suwaidi J (2005) Endothelial dysfunction: cardiovascular risk factors, therapy, and outcome. Vasc Health Risk Manag 1(3):183–198
102. Brockmeyer S, D'Angiulli A (2016) How air pollution alters brain development: the role of neuroinflammation. Transl Neurosci 7(1):24–30. https://doi.org/10.1515/tnsci-2016-0005
103. Martins A, Scotto M, Deus R, Monteiro A, Gouveia S (2021) Association between respiratory hospital admissions and air quality in Portugal: a count time series approach. PLoS One 16(7):e0253455. https://doi.org/10.1371/journal.pone.0253455
104. Jalali S, Karbakhsh M, Momeni M, Taheri M, Amini S, Mansourian M, Sarrafzadegan N (2021) Long-term exposure to $PM_{2.5}$ and cardiovascular disease incidence and mortality in an eastern Mediterranean country: findings based on a 15-year cohort study. Environ Health 20(1):112. https://doi.org/10.1186/s12940-021-00797-w
105. de Bont J, Jaganathan S, Dahlquist M, Persson Å, Stafoggia M, Ljungman P (2022) Ambient air pollution and cardiovascular diseases: An umbrella review of systematic reviews and meta-analyses. J Intern Med 291(6):779–800. https://doi.org/10.1111/joim.13467
106. Araujo JA (2010) Particulate air pollution, systemic oxidative stress, inflammation, and atherosclerosis. Air Qual Atmos Health 4(1):79–93. https://doi.org/10.1007/s11869-010-0101-8
107. Lamorie-Foote K, Ge B, Shkirkova K, Liu Q, Mack W (2023) Effect of air pollution particulate matter on ischemic and hemorrhagic stroke: a scoping review. Cureus 15(10):e46694. https://doi.org/10.7759/cureus.46694
108. Bhatnagar A (2022) Cardiovascular effects of particulate air pollution. Annu Rev Med 73:393–406. https://doi.org/10.1146/annurev-med-042220-011549
109. Hamanaka RB, Mutlu GM (2018) Particulate matter air pollution: effects on the cardiovascular system. Front Endocrinol 9:680. https://doi.org/10.3389/fendo.2018.00680

110. Kaur S, Morales-Hidalgo P, Arija V, Canals J (2023) Prenatal exposure to air pollutants and attentional deficit hyperactivity disorder development in children: a systematic review. Int J Environ Res Public Health 20(8):5443. https://doi.org/10.3390/ijerph20085443
111. Dondi A, Carbone C, Manieri E, Zama D, Del Bono C, Betti L, Biagi C, Lanari M (2023) Outdoor air pollution and childhood respiratory disease: the role of oxidative stress. Int J Mol Sci 24(5):4345. https://doi.org/10.3390/ijms24054345
112. Mihaylov NL, Perkins DD (2015) Local environmental grassroots activism: contributions from environmental psychology, sociology and politics. Behav Sci 5(1):121–153. https://doi.org/10.3390/bs5010121
113. Hajat A, Hsia C, O'Neill MS (2015) Socioeconomic disparities and air pollution exposure: a global review. Curr Envir Health Rpt 2:440–450. https://doi.org/10.1007/s40572-015-0069-5
114. Jiang XQ, Mei XD, Feng D (2016) Air pollution and chronic airway diseases: what should people know and do? J Thorac Dis 8(1):E31–E40. https://doi.org/10.3978/j.issn.2072-1439.2015.11.50
115. Neo EX, Hasikin K, Mokhtar MI, Lai KW, Azizan MM, Razak SA, Hizaddin HF (2022) Towards integrated air pollution monitoring and health impact assessment using federated learning: a systematic review. Front Public Health 10:851553. https://doi.org/10.3389/fpubh.2022.851553
116. Ogwu MC (2023) Plants as monitors and managers of pollution. In: Srivastav AL, Grewal AS, Tiwari M, Pham TD (eds) Role of green chemistry in ecosystem restoration to achieve environmental sustainability. Elsevier, Amsterdam, Netherlands, pp 51–60. ISBN: 978-0-443-15291-7. https://doi.org/10.1016/B978-0-443-15291-7.00022-5
117. Ogwu MC (2019) Towards sustainable development in Africa: the challenge of urbanization and climate change adaptation. In: Cobbinah PB, Addaney M (eds) The geography of climate change adaptation in urban Africa. Springer Nature, Cham, pp 29–55. https://doi.org/10.1007/978-3-030-04873-0_2
118. Ross K, Chmiel JF, Ferkol T (2012) The impact of the Clean Air Act. J Pediatr 161(5):781–786. https://doi.org/10.1016/j.jpeds.2012.06.064
119. Filippelli G, Anenberg S, Taylor M, van Geen A, Khreis H (2020) New approaches to identifying and reducing the global burden of disease from pollution. GeoHealth 4(4): e2018GH000167. https://doi.org/10.1029/2018GH000167
120. Alderton A, Davern M, Nitvimol K, Butterworth I, Higgs C, Ryan E, Badland H (2019) What is the meaning of urban liveability for a city in a low-to-middle-income country? Contextualizing liveability for Bangkok, Thailand. Global Health 15(1):51. https://doi.org/10.1186/s12992-019-0484-8
121. Ogwu MC, Osawaru ME, Chime AO (2014) Comparative assessment of plant diversity and utilization patterns of tropical home gardens in Edo State, Nigeria. Scientia Africana 13 (2):146–162
122. Osawaru ME, Ogwu MC (2014) Ethnobotany and germplasm collection of two genera of cocoyam (Colocasia [Schott] and Xanthosoma [Schott], Araceae) in Edo State Nigeria. Sci Technol Arts Res J 3(3):23–28. https://doi.org/10.4314/star.v3i3.4
123. Moroń W, Ferens W, Wach J (2021) Emission of typical pollutants (NO_X, SO_2) in the oxygen combustion process with air in-leakages. Environ Sci Pollut Res Int 28(36):50683–50695. https://doi.org/10.1007/s11356-021-14292-9
124. Rahman MA, Lee SH, Ji HC, Kabir AH, Jones CS, Lee KW (2018) Importance of mineral nutrition for mitigating aluminum toxicity in plants on acidic soils: current status and opportunities. Int J Mol Sci 19(10):3073. https://doi.org/10.3390/ijms19103073
125. Millán Ramírez GP, Byliński H, Niedostatkiewicz M (2021) Deterioration and protection of concrete elements embedded in contaminated soil: a review. Materials 14(12):3253. https://doi.org/10.3390/ma14123253
126. Li Z, Jiang J, Ma Z, Fajardo OA, Deng J, Duan L (2017) Influence of flue gas desulfurization (FGD) installations on emission characteristics of $PM_{2.5}$ from coal-fired power plants

equipped with selective catalytic reduction (SCR). Environ Pollut 230:655–662. https://doi.org/10.1016/j.envpol.2017.06.103
127. Briffa J, Sinagra E, Blundell R (2020) Heavy metal pollution in the environment and their toxicological effects on humans. Heliyon 6(9):e04691. https://doi.org/10.1016/j.heliyon.2020.e04691
128. Borrelli P, Robinson DA, Panagos P, Lugato E, Yang JE, Alewell C, Wuepper D, Montanarella L, Ballabio C (2020) Land use and climate change impacts on global soil erosion by water (2015-2070). Proc Natl Acad Sci U S A 117(36):21994–22001. https://doi.org/10.1073/pnas.2001403117
129. Ikhajiagbe B, Ogwu MC (2021) Growth performance and ferulic acid composition of Amaranthus [L.] species and Mitracarpus villosus (Sw.) DC in competition with weed species. Environ Sustain. https://doi.org/10.1007/s42398-021-00163-z
130. Priya AK, Muruganandam M, Ali SS, Kornaros M (2023) Clean-up of heavy metals from contaminated soil by phytoremediation: a multidisciplinary and eco-friendly approach. Toxics 11(5):422. https://doi.org/10.3390/toxics11050422
131. Ikhajiagbe B, Ogwu MC, Lato NF (2021) Growth and yield response of soybean (Glycine max [L.] Merr.) accessions after exposure to cadmium. VEGETOS – Int J Plant Res Biotechnol. https://doi.org/10.1007/s.42535-021-00189-y
132. Ikhajiagbe B, Ogwu MC, Ekhator PO, Omoayena II (2022) Implications of soil nitrogen enhancement on the yield performance of soybean (Glycine max) in cadmium-polluted soil. J Pollut Effects Control 10(2):1–12. 1000334. https://doi.org/10.35248/23754-4397.22.10.334
133. Lei W, Li X, Yin Z, Zhang L, Zhao W (2024) Pollution characteristics and source apportionment of black carbon aerosols during spring in Beijing. Toxics 12(3):202. https://doi.org/10.3390/toxics12030202
134. Perera FP (2017) Multiple threats to child health from fossil fuel combustion: impacts of air pollution and climate change. Environ Health Perspect 125(2):141–148. https://doi.org/10.1289/EHP299
135. Neale RE, Barnes PW, Robson TM, Neale PJ, Williamson CE, Zepp RG, Wilson SR, Madronich S, Andrady AL, Heikkilä AM, Bernhard GH, Bais AF, Aucamp PJ, Banaszak AT, Bornman JF, Bruckman LS, Byrne SN, Foereid B, Häder DP, Hollestein LM et al (2021) Environmental effects of stratospheric ozone depletion, UV radiation, and interactions with climate change: UNEP Environmental Effects Assessment Panel, update 2020. Photochem Photobiol Sci 20(1):1–67. https://doi.org/10.1007/s43630-020-00001-x
136. Chang JH, Lee YL, Chang LT, Chang TY, Hsiao TC, Chung KF, Ho KF, Kuo HP, Lee KY, Chuang KJ, Chuang HC (2023) Climate change, air quality, and respiratory health: a focus on particle deposition in the lungs. Ann Med 55(2):2264881. https://doi.org/10.1080/07853890.2023.2264881
137. D'Amato G, Bergmann KC, Cecchi L, Annesi-Maesano I, Sanduzzi A, Liccardi G, Vitale C, Stanziola A, D'Amato M (2014) Climate change and air pollution: effects on pollen allergy and other allergic respiratory diseases. Allergo J Int 23(1):17–23. https://doi.org/10.1007/s40629-014-0003-7
138. Kalair AR, Seyedmahmoudian M, Stojcevski A, Abas N, Khan N (2021) Waste to energy conversion for a sustainable future. Heliyon 7(10):e08155. https://doi.org/10.1016/j.heliyon.2021.e08155
139. Wittwer RA, Bender SF, Hartman K, Hydbom S, Lima RAA, Loaiza V, Nemecek T, Oehl F, Olsson PA, Petchey O, Prechsl UE, Schlaeppi K, Scholten T, Seitz S, Six J, van der Heijden MGA (2021) Organic and conservation agriculture promote ecosystem multifunctionality. Sci Adv 7(34):eabg6995. https://doi.org/10.1126/sciadv.abg6995
140. King JS, Manning J, Woodward A (2023) In this together: international collaborations for environmental and human health. J Law Med Ethics 51(2):271–286. https://doi.org/10.1017/jme.2023.82

141. Teyton A, Baer RJ, Benmarhnia T, Bandoli G (2023) Exposure to air pollution and emergency department visits during the first year of life among preterm and full-term infants. JAMA Netw Open 6(2):e230262. https://doi.org/10.1001/jamanetworkopen.2023.0262
142. Ward-Caviness CK, Danesh Yazdi M, Moyer J, Weaver AM, Cascio WE, Di Q, Schwartz JD, Diaz-Sanchez D (2021) Long-term exposure to particulate air pollution is associated with 30-day readmissions and hospital visits among patients with heart failure. J Am Heart Assoc 10 (10):e019430. https://doi.org/10.1161/JAHA.120.019430
143. Li L, Du T, Zhang C (2020) The impact of air pollution on healthcare expenditure for respiratory diseases: evidence from the People's Republic of China. Risk Manag Healthc Policy 13:1723–1738. https://doi.org/10.2147/RMHP.S270587
144. Mathiarasan S, Hüls A (2021) Impact of environmental injustice on children's health-interaction between air pollution and socioeconomic status. Int J Environ Res Public Health 18 (2):795. https://doi.org/10.3390/ijerph18020795
145. Ganbat M, Erdenebileg N, Batbold C, Nergui S, Anderson R, Wigfall C, Amarsanaa N, Heikens A, Sarantuya M, Warburton D (2022) Integrating quantitative and qualitative approaches to assess wintertime illness-related absenteeism and its direct and indirect costs among the private sector in Ulaanbaatar. PLoS One 17(2):e0263220. https://doi.org/10.1371/journal.pone.0263220
146. Bogalecka M, Grobelna A (2023) Air pollution and its potential consequences for tourism and career development from students' perspective: a case study of the Gdańsk agglomeration in Poland. Int J Environ Res Public Health 20(3):2651. https://doi.org/10.3390/ijerph20032651
147. Stevens CJ, Bell JNB, Brimblecombe P, Clark CM, Dise NB, Fowler D, Lovett GM, Wolseley PA (2020) The impact of air pollution on terrestrial managed and natural vegetation. Philos Transact A Math Phys Eng Sci 378(2183):20190317. https://doi.org/10.1098/rsta.2019.0317
148. de Vries K, Erumban A, van Ark B (2021) Productivity and the pandemic: short-term disruptions and long-term implications: the impact of the COVID-19 pandemic on productivity dynamics by industry. IEEP 18(3):541–570. https://doi.org/10.1007/s10368-021-00515-4
149. Kelly FJ, Fussell JC (2011) Air pollution and airway disease. Clin Exp Allergy 41(8):1059–1071. https://doi.org/10.1111/j.1365-2222.2011.03776.x
150. Rentschler J, Leonova N (2023) Global air pollution exposure and poverty. Nat Commun 14 (1):4432. https://doi.org/10.1038/s41467-023-39797-4
151. Dannenberg AL, Jackson RJ, Frumkin H, Schieber RA, Pratt M, Kochtitzky C, Tilson HH (2003) The impact of community design and land-use choices on public health: a scientific research agenda. Am J Public Health 93(9):1500–1508. https://doi.org/10.2105/ajph.93.9.1500
152. Dong F, Qin C, Zhang X, Zhao X, Pan Y, Gao Y, Zhu J, Li Y (2021) Towards carbon neutrality: the impact of renewable energy development on carbon emission efficiency. Int J Environ Res Public Health 18(24):13284. https://doi.org/10.3390/ijerph182413284
153. WHO (2021) WHO global air quality guidelines: particulate matter (PM2.5 and PM10), ozone, nitrogen dioxide, sulfur dioxide and carbon monoxide [Internet]. World Health Organization, Geneva. 6, Implementation of the guidelines. Available from: https://www.ncbi.nlm.nih.gov/books/NBK574587/
154. Yamineva Y, Romppanen S (2017) Is law failing to address air pollution? Reflections on international and EU developments. Rev Eur Comp Int Environ Law 26(3):189–200. https://doi.org/10.1111/reel.12223
155. Jayaram V, Nigam A, Welch WA, Miller JW, Cocker 3rd DR (2011) Effectiveness of emission control technologies for auxiliary engines on ocean-going vessels. J Air Waste Manag Assoc 61(1):14–21. https://doi.org/10.3155/1047-3289.61.1.14
156. Mundorf N, Redding CA, Bao S (2018) Sustainable transportation and health. Int J Environ Res Public Health 15(3):542. https://doi.org/10.3390/ijerph15030542
157. Pietrzak K, Pietrzak O (2020) Environmental effects of electromobility in a sustainable urban public transport. Sustain For 12:1052. https://doi.org/10.3390/su12031052

158. Li C, Chen Z, Hu Y, Cai C, Zuo X, Shang G, Lin H (2023) The energy conservation and emission reduction co-benefits of China's emission trading system. Sci Rep 13(1):13758. https://doi.org/10.1038/s41598-023-40811-4
159. Chowkwanyun M (2023) Environmental justice: where it has been, and where it might be going. Annu Rev Public Health 44:93–111. https://doi.org/10.1146/annurev-publhealth-071621-064925

Part II
Development and Economic Impacts of Air Pollution

Environmental Impacts of Air Pollution

Glory Richard, Wisdom Ebiye Sawyer, and Avez Sharipov

Contents

Abstract Air pollution is a major environmental issue due to its substantial effects on ecosystem dynamics and conditions. This chapter discusses the consequences of air pollution on ecosystems, focusing on the causes, chemical composition, and resulting disruptions to ecology and the atmosphere. Air pollution stems from natural and anthropogenic factors, primarily the burning of fossil fuels, industrial

G. Richard (✉)
School of Community Health and Policy, Morgan State University, Baltimore, MD, USA

Department of Community Medicine, Faculty of Clinical Sciences, Niger Delta University, Wilberforce Island, Bayelsa, Nigeria

W. E. Sawyer
Department of Community Medicine, Faculty of Clinical Sciences, Niger Delta University, Wilberforce Island, Bayelsa, Nigeria

A. Sharipov
Department of Inorganic, Physics and Colloid Chemistry, Faculty of Pharmacy, Tashkent Pharmaceutical Institute, Tashkent, Uzbekistan

Scientific Laboratory of Innovative Pharmaceutical Compounds, Tashkent Pharmaceutical Institute, Tashkent, Uzbekistan

Matthew Chidozie Ogwu and Sylvester Izah (eds.), *Sustainable Strategies for Air Pollution Mitigation: Development, Economics and Technologies*,
Hdb Env Chem (2024) 133: 47–76, DOI 10.1007/698_2024_1114,

Published online: 13 September 2024

pollutants, farming methods, and automobile emissions. Secondary pollutants like smog and acid rain are created when air pollutants interact in the atmosphere, further harming ecosystems. Air pollution significantly impacts atmospheric processes. Pollutants influence Earth's radiative balance by absorbing and scattering sunlight, thus affecting weather patterns and contributing to climate change. Greenhouse gases trap heat in the atmosphere, leading to altered precipitation patterns, more frequent extreme weather events, and rising sea levels, all of which exacerbate global warming. These changes threaten biodiversity, agriculture, and water supply. Air pollution negatively affects terrestrial ecosystems through processes like acid deposition and nutrient imbalances. Acid rain increases soil acidity, releasing toxic metals and depleting essential minerals, which harms plant health, reduces agricultural output, and disturbs forest ecosystems. Ground-level ozone, created by photochemical reactions involving NOx and VOCs, can inhibit photosynthesis, reducing agricultural yields and hindering forest growth. Aquatic habitats are similarly affected by air pollution. Acid rain increases the acidity of water bodies, altering their chemical composition and affecting marine life viability. Elevated nitrogen levels from atmospheric deposition and agricultural runoff lead to eutrophication, characterized by excessive algal growth, which creates ocean dead zones devoid of marine life. This imbalance threatens biodiversity, jeopardizes fisheries, and impacts human health. Targeted strategies like emission-reducing legislation are essential to mitigating the harmful effects of air pollution on the environment. International cooperation is crucial, as air pollution is a transboundary issue requiring global solutions. Public awareness and involvement are also vital to support pollution-reducing legislation and encourage behavioral changes.

Keywords Greenhouse gases, Nitrogen dioxide, Particulate matter, Persistent organic pollutants, Sulfur dioxide, Trace metals, Volatile organic compounds

Abbreviations

$PM_{2.5}$	Particulate matter 2.5
PM_{10}	Particulate matter 10
SO_2	Sulfur dioxide
O_3	Ozone
BC	Black carbon
NO_2	Nitrogen dioxide
ROS	Reactive oxygen species
BBB	Blood-brain barrier
iNOS	Inducible nitric oxide synthase
TNFα	Tumor necrosis factor alpha
IL-1β	Interleukin 1 beta
COX-2	Cyclooxygenase 2
NFκB	Nuclear factor "kappa-light-chain-enhancer" of activated B-cells

NOX-2	NADPH oxidase isoform 2
oxLDL	Oxidized low-density lipoprotein
Aβ	Amyloid beta peptide
NOx	Oxides of nitrogen
COPD	Chronic obstructive pulmonary disease
EC	Elemental carbon
POPs	Persistent organic pollutants
DDT	Dichloro-diphenyl-trichloroethane
PCBs	Polychlorinated biphenyls
CO2	Carbon dioxide
HFCs	Hydrofluorocarbons
PFCs	Perfluorocarbons
CH4	Methane
CFCs	Chlorofluorocarbons
ODS	Ozone-depleting substances
4-HNE	4-hydroxynonenal
8-oxoG	8-hydroxyguanosine

1 Introduction

The presence of harmful compounds and particles in the atmosphere poses a significant threat to both human health and the overall well-being of the ecosystem [1]. This atmospheric contamination is commonly referred to as air pollution, and it arises directly from human activities, such as vehicular emissions, the burning of fossil fuels, industrial production, and agricultural practices. The impact of air pollution is severe, affecting millions of people globally and making it one of the most pressing challenges for our planet. Shockingly, outdoor air pollution is identified as one of the major risk factors for premature mortality worldwide, causing approximately millions of deaths each year [2, 3].

Air quality holds immense significance for both ecosystems and human populations. Poor air quality devastates the environment but also profoundly affects human health and well-being [4–6]. Air pollution is responsible for various health issues, including cancer, cardiovascular diseases, and respiratory ailments [7, 8]. In the most severe cases, these conditions can lead to premature death. Children, the elderly, and people with pre-existing medical disorders are among the vulnerable groups that are most vulnerable.

Furthermore, air pollution exerts a significant adverse impact on the natural environment. Its long-term consequences may include the depletion of the ozone layer and alterations in global climate patterns. Beyond disrupting ecosystems and affecting agriculture and vegetation, ozone layer depletion heightens the risk of skin cancer due to increased exposure to harmful ultraviolet rays from the sun [9, 10]. Climate change, driven by the accumulation of greenhouse gases in the atmosphere, manifests in rising sea levels, shifts in weather patterns, and the loss of habitats for numerous species [11, 12].

Ensuring and maintaining high air quality positively affects public health and the environment and plays a crucial role in fostering economic growth. Pollution and poor air quality can be detrimental to economic progress due to the increased healthcare costs and decreased worker productivity they incur [13, 14]. Moreover, air pollution can negatively impact ecosystems, leading to substantial cleanup or remediation expenses.

Consequently, it is imperative to prioritize actions that enhance air quality by enacting regulations that promote environmentally sustainable practices and technologies. These initiatives involve a variety of tactics, such as the development of renewable energy sources, a reduction in the consumption of fossil fuels, and the promotion of ecologically friendly modes of transportation like walking, cycling, and public transportation.

Individuals can significantly improve air quality through their daily activities, such as minimizing single-use plastics, reducing overall energy consumption, and responsibly disposing of hazardous waste [15, 16]. Instead of driving alone, people can opt for carpooling, switch to electric vehicles, or choose public transportation options.

Air quality influences human health, ecosystem health, and economic prosperity. Enhancing air quality can significantly benefit both the global environment and economic development, alongside improvements in public health [14, 17, 18]. Therefore, formulating policies that preserve and enhance air quality requires collaborative efforts from corporations, governments, and citizens. This chapter focuses on the diverse environmental effects of air pollution, showing how different pollutants harm both human well-being and the integrity of the ecosystem. The chapter provides a succinct summary of the main effects caused by air pollutants, including ozone (O3), sulfur oxides (SO2), nitrogen oxides (NOx), particulate matter, trace metals, and persistent organic pollutants. In addition, the chapter provides measures to reduce emissions and enhance air quality standards.

This chapter primarily discusses the environmental impacts of air pollution, focusing on various aspects. It provides an overview of how major air pollutants affect both human health and the environment. The environmental impacts explored include acid rain, ozone depletion, ecosystem damage, loss of biodiversity, human health impacts, and the economic costs associated with air pollution. Additionally, the chapter discusses solutions aimed at reducing air pollution.

2 Impacts of Air Pollutants on Environmental Matrices, Processes, and Status

Air pollution occurs when hazardous or excessive concentrations of a particular contaminant are present in the atmosphere, potentially harming human, animal, and ecological health. These air pollutants can either be physical, chemical, or biological.

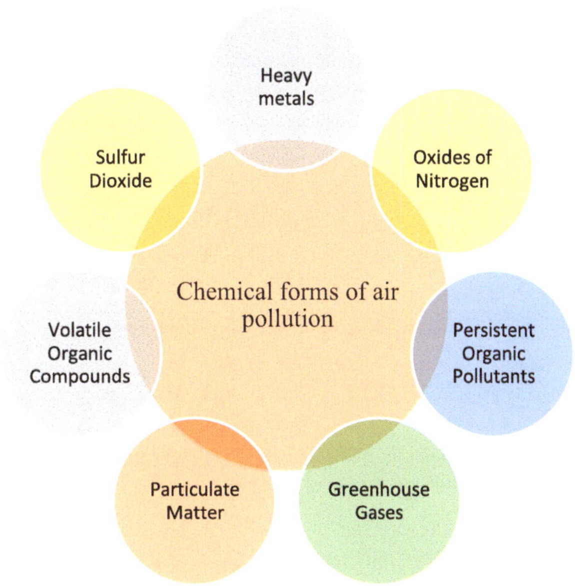

Fig. 1 Various forms of air pollution

Of these chemical air pollutants are the most studied. Chemical pollution refers to the presence of a wide range of chemical compounds and substances in the atmosphere, known for their potential to affect human health, ecosystems, and the environment adversely. These compounds can be emitted from natural sources and human activities and exist in gaseous, liquid, or solid states. Figure 1 shows the most frequently encountered types of chemical pollutants that can be identified in the atmosphere.

Air pollution consists of various pollutants, each with distinct sources and impacts (Fig. 2). Ozone (O3) is formed when certain oxides of nitrogen oxides react with volatile organic compounds (VOCs) under sunlight [20, 21]. Chemical solvents, industrial processes, and vehicular emissions are significant sources. The primary sources of sulfur oxides (SO and SO2) include industrial operations like smelting and the burning of fossil fuels like coal and oil [19]. Nitrogen oxides (NO and NO2) are emitted from vehicles, power plants, and agricultural activities [22, 23]. Particulate matter (PM), including PM2.5 and PM10, originates from combustion processes, construction activities, and natural sources like dust storms and wildfires. Black carbon (BC) and elemental carbon (EC) are products of incomplete combustion, commonly from diesel engines and biomass burning [19]. Ultrafine particles (UFPs), smaller than 0.1 μm, are produced by vehicle emissions, industrial processes, and secondary formation from gas to particle conversion [24, 25].

Organic substances resistant to environmental degradation, linger for extended periods, and bioaccumulate in food chains are known as persistent organic pollutants (POPs). Pesticides like dichloro-diphenyl-trichloroethane(DDT), industrial compounds like polychlorinated biphenyls (PCBs), and accidental by-products of

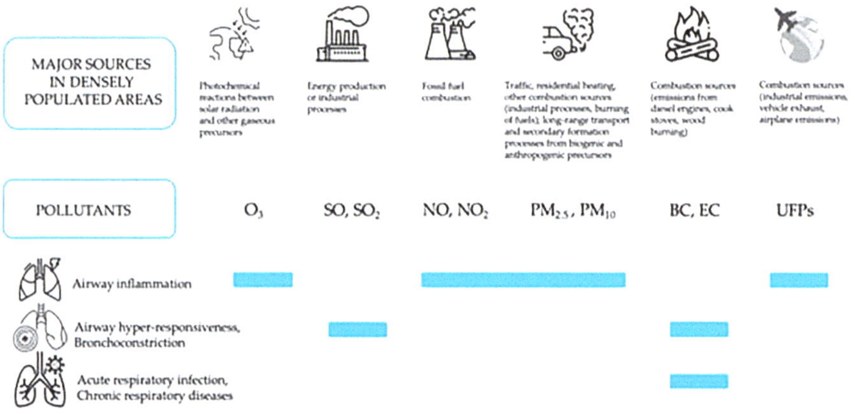

Fig. 2 Some major air pollutants and their major sources and health concerns. Source: [19]. https://creativecommons.org/licenses/by/4.0/

combustion and industrial operations are a few examples of these pollutants [26–28]. POPs threaten the environment and human health because they can travel great distances via air currents and deposit far from their sources. Numerous negative consequences are known to be caused by them, such as immune system suppression, developmental and reproductive problems, and carcinogenicity in both humans and wildlife [29, 30].

Toxic pollutants such as trace metals are released into the atmosphere by volcanic eruptions, automobile emissions, and industrial operations [31, 32]. These metals (such as lead, mercury, cadmium, and arsenic) build up in soils, waterways (including sediment and water), and living things as they linger in the environment.

Air pollutants can be absorbed into sediments after they reach water bodies [33–38]. Because sedimentary environments are generally stable, air pollutants can stay in sediments for extended periods. Pollutants as POPs and trace metals can attach themselves to sediment particles, forming a persistent source of contamination that can be rereleased by a variety of physical and chemical processes, such as redox or pH changes. Pollutants may be reintroduced into the water column due to this remobilization, affecting water quality and continuously endangering aquatic life. Thus, sediments affect the general health of marine ecosystems by serving as both sinks and possible secondary sources of pollutants [39, 40].

Air pollution has the potential to severely deteriorate water quality in bodies of water, impacting both human use and ecological integrity [41, 42]. Water's chemical balance can be changed by contaminants such as nitrates, sulfates, and trace metals. This can result in problems including eutrophication, acidification, and toxicity to marine life. Algal blooms, which lower the oxygen content of the water and endanger fish and other marine life, can be brought on by high nitrogen levels. Acidification can interfere with aquatic species' ability to reproduce and maintain their metabolic processes. Toxic compounds can also bioaccumulate in the food

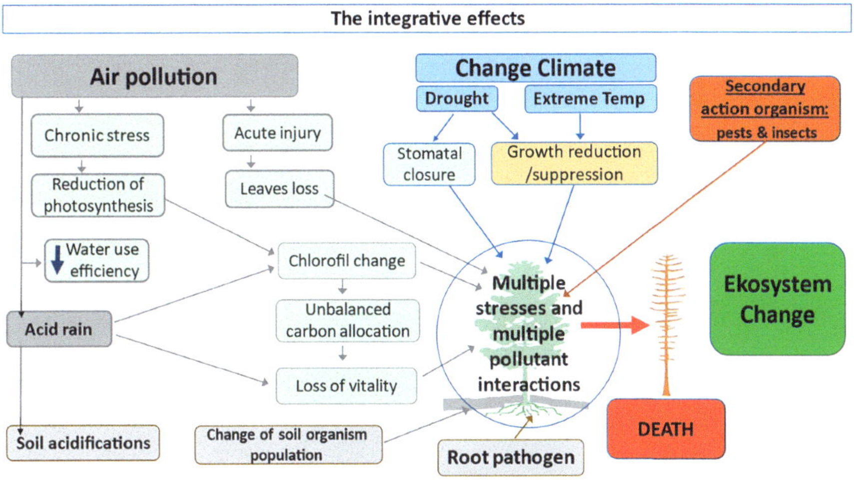

Fig. 3 An integrated effect on the environmental impacts of air pollution. Source: [46]; https://creativecommons.org/licenses/by/4.0/

chain, endangering predators and, eventually, people who drink contaminated water or seafood [33, 36].

Therefore, exposure to airborne trace metals can result in serious health consequences, such as cancer, kidney damage, neurological issues, and respiratory ailments [43]. Pregnant women and children are more susceptible to these consequences.

Air pollution significantly impacts human health, biodiversity, and environmental health, with various pollutants causing direct and indirect effects. Studies by Richard et al. [44, 45], Uzoekwe et al. [32], and Izah et al. [31] highlight the diverse impacts of air pollutants. These pollutants contribute to acid rain, climate change, ozone depletion, soil acidification, habitat destruction, and a decline in photosynthesis rates (Fig. 3). These effects lead to severe consequences such as respiratory and cardiovascular diseases in humans, the degradation of ecosystems, and the disruption of natural processes that sustain life on Earth.

3 Environmental Impacts of Air Pollution

3.1 Anthropogenic Sources and Environmental Impacts of Greenhouse Gas Emissions

Greenhouse gas emissions originate from various anthropogenic activities, each playing a distinct role in the ongoing climate crisis (Fig. 4). The combustion of fossil fuels stands as a cornerstone in the genesis of greenhouse gases. When coal,

Fig. 4 Sources of greenhouse gas

oil, and natural gas undergo combustion, carbon dioxide (CO2) is released into the atmosphere, acting as a significant contributor to the intensified greenhouse effect. This process, intrinsic to energy production, industrial operations, and transportation, is a primary driver of the elevated atmospheric concentrations of CO2. The widespread clearing of forests for agricultural expansion, logging, and other human activities leads to the release of stored carbon dioxide [47–49]. As trees are felled and subsequently decompose, the carbon they had sequestered is reintroduced into the atmosphere. Furthermore, the loss of forests diminishes Earth's capacity to absorb CO2 through photosynthesis, creating a dual impact on the greenhouse gas balance [49, 50]. Some industrial processes unleash potent greenhouse gases into the atmosphere, exacerbating the climate crisis. Notably, cement production releases substantial amounts of CO2, contributing significantly to overall emissions. Moreover, various manufacturing processes emit synthetic greenhouse gases such as hydrofluorocarbons (HFCs) and perfluorocarbons (PFCs), which possess heightened global warming potentials compared to CO2 [51–53]. Agricultural practices are multifaceted contributors to greenhouse gas emissions. Livestock, particularly cattle, release methane (CH_4) during digestion, a process known as enteric fermentation. Additionally, methane is emitted from decomposing biomass under anaerobic processes with the right conditions. These emissions compound the environmental impact of agriculture, making it a notable source of greenhouse gases. Improper waste disposal practices, especially in landfills, contribute significantly to greenhouse gas emissions. Methane is emitted into the atmosphere significantly during the anaerobic breakdown of organic waste. The mismanagement of waste exacerbates the environmental consequences, marking waste disposal as a notable source of methane emissions. The transportation sector, involving vehicles, airplanes, and ships, represents a major source of greenhouse gas emissions. Burning fossil fuels in internal combustion engines releases CO_2, methane, and nitrous oxide (N_2O) into the atmosphere, contributing significantly to the overall greenhouse gas burden.

The environmental impacts of greenhouse gases are shown in Fig. 5. The paramount environmental consequence of is global warming, driven by the heat trapping in Earth's atmosphere [54, 55]. Because of the greenhouse effect, rising temperatures cause significant changes in climate patterns, which increase the

Fig. 5 Environmental impacts of greenhouse gases

frequency and severity of extreme weather events [56]. The effects are felt in both terrestrial and marine situations, affecting entire ecosystems. Sea levels increase due to the polar ice caps and glaciers melting due to global warming. This phenomenon directly threatens vulnerable ecosystems, low-lying islands, and coastal settlements [57, 58]. The encroachment of seawater jeopardizes infrastructure, displaces populations, and exacerbates the risk of coastal flooding and erosion [59]. Elevated global temperatures are intrinsically linked to an escalation in the frequency and severity of extreme weather events [60, 61]. Hurricanes, droughts, floods, and heatwaves become, more prevalent, devastatingly impacting communities, agriculture, and ecosystems [62]. Adaptive strategies are necessary in light of the growing unpredictability of weather patterns, which threatens societal resilience. Ocean acidification seriously threatens marine life, especially plankton and other species with calcium carbonate shells. The disruption of aquatic ecosystems by acidification impacts fisheries, biodiversity, and the general well-being of the oceans. Ecosystems and biodiversity suffer from variations in precipitation and temperature and a rise in extreme weather events [63]. Some plant and animal species may disseminate around the planet because they cannot adapt or migrate quickly enough to live. This ecosystem upheaval has cascading effects on food webs, ecological interactions, and the overall stability of ecosystems. Climate change, driven by greenhouse gas emissions, poses direct and indirect health risks to human populations [64, 65]. Heat-related illnesses become more prevalent as temperatures rise, infectious diseases spread to new regions, and changes in air quality impact respiratory health. Vulnerable populations face heightened risks, necessitating comprehensive public health strategies to mitigate these impacts [66].

Alterations in temperature and precipitation patterns directly affect agriculture, influencing crop yields and overall productivity. Shifts in growing seasons and the increased frequency of extreme weather events, such as droughts and floods, pose significant challenges to global food security [67]. Sustainable agricultural practices and adaptive strategies are crucial to address these challenges. The combined effects of ecosystem disruption and changing climatic conditions escalate the risk of biodiversity loss. Species that cannot adapt or migrate swiftly enough face increased threats of extinction. Bibi and Rahman [56] observed that climate change impacts agriculture by reducing total crop production, weakening soil nutritional content, slowing photosynthesis, and inhibiting seed development. They further explained that climate variations, such as changes in global rainfall patterns, rising carbon dioxide levels, and increasing average temperatures, have heightened the frequency of extreme events like floods and droughts, threatening cereal crop productivity worldwide. Crops face adverse effects from various biotic and abiotic stresses due to these temperature and rainfall fluctuations, directly affecting their growth and maturity dates (Fig. 6).

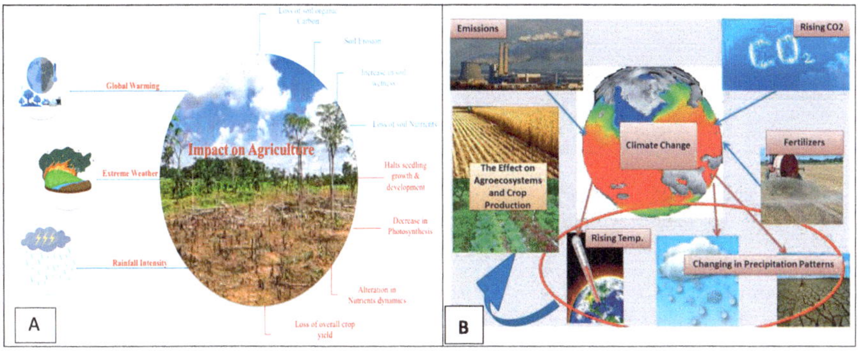

Fig. 6 The consequences of climate change on agriculture; (**a**) How crops and soil are harmed by extreme weather events [56], (**b**) How agroecosystems are affected by rising temperatures and climate change [68]. https://creativecommons.org/licenses/by/4.0/

3.2 *Acid Rain*

According to Zhang et al. [69], emissions of sulfur dioxide (SO2) and nitrogen oxides (NOx) are the primary causes of acid rain, a severe environmental problem. The primary human activities that release these pollutants include power plant and vehicle combustion of fossil fuels, and industrial processes. Sulfuric and nitric acids are created after reaching the atmosphere by a series of chemical reactions involving SO2 and NOx, facilitated by moisture and sunlight. Acid rain is created when these acids react with atmospheric moisture. The phenomenon is not localized; wind can transport it across long distances and affect areas far from the sources of pollution.

The impacts of acid rain are profound and multifaceted. When acid rain falls, it leads to the acidification of soils, which disrupts nutrient availability and negatively affects plant health [70]. Essential nutrients like calcium and magnesium are leached from the soil, while toxic metals like aluminum become more soluble and harmful to plant roots. Water bodies, such as lakes, rivers, and streams, also suffer from lowered pH levels due to acid rain, which harms aquatic life, including fish and other organisms, which experience physiological stress and decreased reproductive success. Additionally, acid rain damages vegetation directly by harming leaves and bark, thereby reducing photosynthesis and making plants more vulnerable to diseases and harsh weather (Fig. 7) [71, 72]. The built environment is also at risk, as acid rain accelerates the deterioration of buildings and monuments, particularly those made of limestone, marble, and concrete [73, 74], resulting in increased economic costs for maintenance and restoration.

Acid rain significantly alters soil composition by leaching essential nutrients such as calcium and magnesium from the soil [71]. These nutrients are crucial for maintaining soil health and supporting plant growth. When acid rain infiltrates the soil, it increases the solubility of these nutrients, causing them to be washed away into deeper soil layers or nearby water bodies. This depletion of vital nutrients

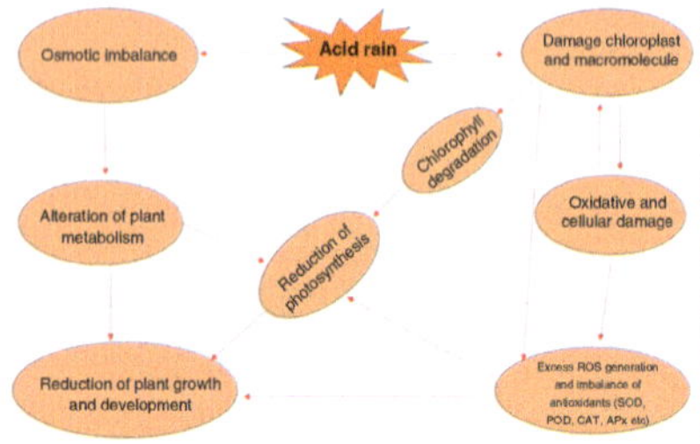

Significant damage to foliage due to acid rain simulation (pH 2.5). Liquidambar styraciflua (a–d). Fraxinus uhdei (e–h). Arrows indicate specific areas where the sprayed acidic solution has damaged the intercostal leaf zones and margins of both species. Source: [71]	Acid rain significantly affects photosynthesis, antioxidants, and overall plant growth. The primary enzymes involved in mitigating oxidative stress in plants are ascorbate peroxidase (APx), catalase (CAT), peroxidase (POD), and superoxide dismutase (SOD). These enzymes help manage the levels of reactive oxygen species (ROS), which are harmful byproducts of cellular metabolism that increase under acidic conditions. Elevated ROS levels can damage cellular structures, hinder photosynthesis, and impair plant growth. The role of APx, CAT, POD, and SOD is crucial in neutralizing ROS, thus protecting plants from the adverse effects of acid rain. Source: [72]

Fig. 7 Impacts of acid rain on vegetation. Source: [71, 72]. https://creativecommons.org/licenses/by/4.0/

disrupts nutrient cycles and weakens plant health [71], making vegetation more susceptible to diseases, pests, and environmental stresses. Furthermore, acid rain can increase the concentration of toxic metals like aluminum in the soil, which can be detrimental to plant root systems and further impair their ability to absorb water and nutrients.

Aquatic environments, particularly lakes and rivers, are profoundly affected by acid rain, which lowers the pH of these water bodies and makes them more acidic [75]. This acidification severely threatens aquatic life, including fish, amphibians, and various invertebrates. The lower pH levels can interfere with the reproductive cycles of these organisms, impair their development, and lead to increased mortality rates. For instance, fish eggs and larvae are susceptible to pH changes and may not survive in highly acidic conditions. Acid rain also disrupts the balance of aquatic ecosystems by affecting the availability of nutrients and altering the food web dynamics. As a result, these ecosystem's biodiversity and overall health can be significantly compromised, leading to long-term environmental consequences.

Acid rain has significant ecological consequences. Figure 8 depicts the impacts of acid rain on some parts of China. Zhang et al. [69] reported severe acid deposition in China, leading to economic losses, material corrosion, and reduced lifespan of

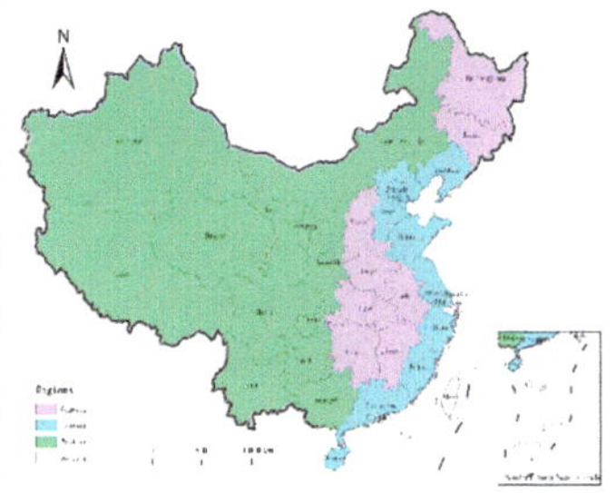

Eastern, Central, and Western regions in China

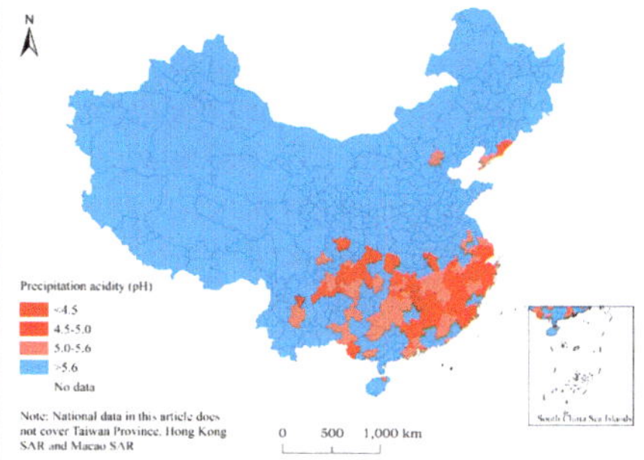

Distribution of acid deposition in China (Eastern, Central, and Western regions) in 2013.

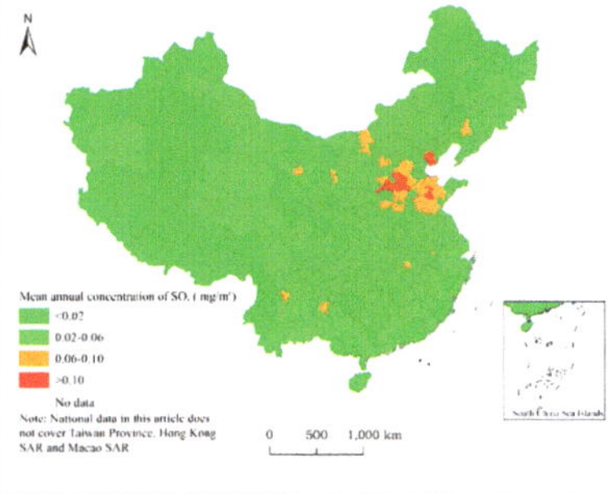

Distribution of mean annual SO_2 level in China (Eastern, Central, and Western regions) in 2013.

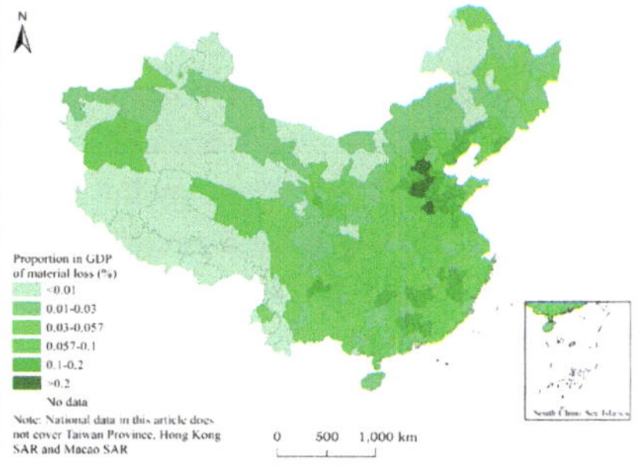

Distribution of proportion in cities' GDP of material loss caused by acid rain in China (Eastern, Central, and Western regions) in 2013.

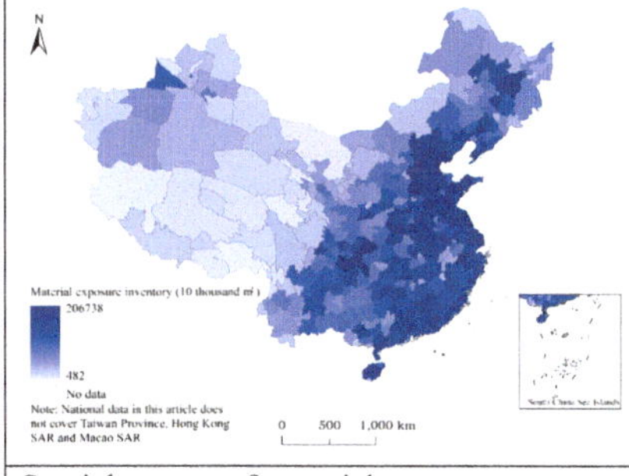

Spatial pattern of material exposure inventory in China (Eastern, Central, and Western regions) in 2013.

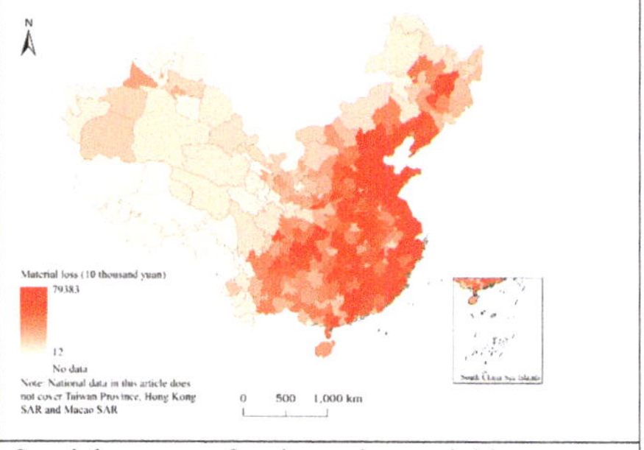

Spatial pattern of estimated material loss caused by acid deposition in China (Eastern, Central, and Western regions) in 2013.

Fig. 8 Acid deposition in Eastern, Central, and Western regions of China leading to economic losses, material corrosion, and reduced lifespan of materials. Source: [69]. https://creativecommons.org/licenses/by/4.0/

materials. Using data from the China National Environmental Monitoring Network and based on domestic and international research findings, the authors developed a material exposure inventory per capita for urban and rural areas in Eastern, Central, and Western China. To calculate the losses of building materials due to acid deposition, they used exposure-response functions. The study found that the material losses resulting from acid deposition in 2013 amounted to 32.165 billion yuan, which is 3.4% of the entire amount spent on controlling environmental pollution that year and 0.057% of China's GDP.

3.3 *Ozone Depletion*

Ozone serves dual roles in Earth's atmosphere: in the stratosphere, it acts protectively by shielding the planet from excessive ultraviolet radiation, whereas in the troposphere, it functions as an atmospheric pollutant [76] (Fig. 9a). According to Juráň et al. [77], UV solar energy initiates a complex photochemical process leading to tropospheric ozone (O3) formation (Fig. 9b). Initially, NO2 undergoes photolysis to produce NO and an electronically excited oxygen atom (O), which reacts with atmospheric molecular oxygen (O2) to form O3 [69]. In a photo-stationary state, O3 can regenerate NO2 in the presence of NO, sustaining the cycle. Net O3 production occurs when precursor gases like carbon monoxide (CO), methane (CH4), and volatile organic compounds (VOCs) reach sufficient concentrations [77]. Catalyzing these reactions are hydroxide anions (OH−), hydroperoxyl radicals (HO2), NO, and NO2 [77]. Enhanced O3 synthesis occurs under critical precursor levels and increased UV radiation. Urban and industrial emissions transport CO, NOx, and

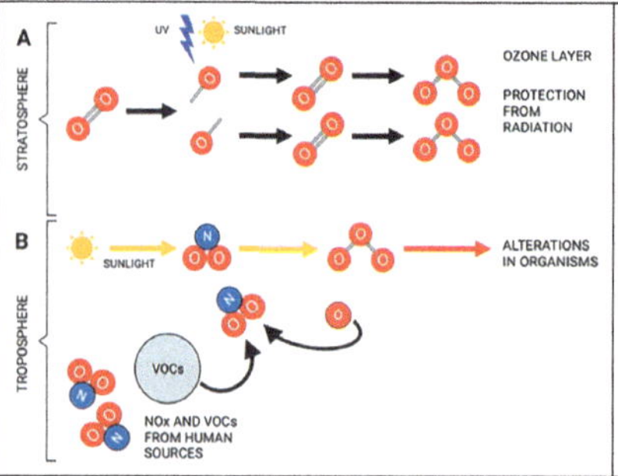

Ozone functions: yellow arrows show how solar radiation and nitrogen species combine to produce ozone. Source: [76]

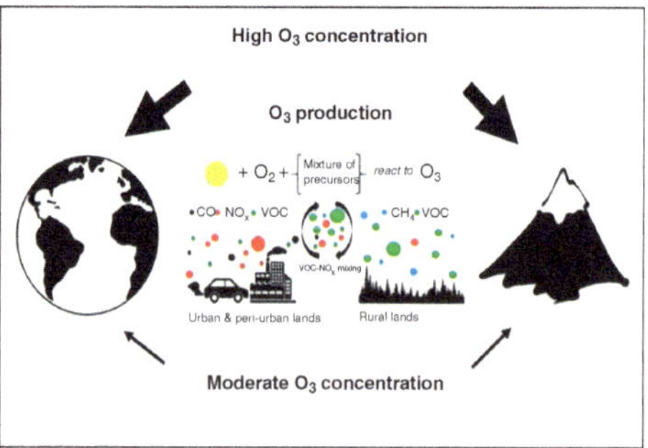

Tropospheric ozone (O_3) formation. (Source: Juráň et al., 77)

Fig. 9 Functions of ozone and formation of tropospheric ozone. Source: [76, 77]. https://creativecommons.org/licenses/by/4.0/

VOCs to rural areas, that mix with CH4 and clean air, creating a favorable VOC/NOx ratio for O3 generation [77]. Variations in UV radiation at different altitudes and the distribution of precursor sources linked to industrialization contribute to variations in O3 concentrations across latitudes and elevations [77].

Ozone depletion, primarily caused by air pollution, significantly negatively influences the environment, human health, and the economy. The depletion of the ozone layer, a protective shield in Earth's stratosphere, results from the release of certain chemicals, such as chlorofluorocarbons (CFCs), halons, and other ozone-depleting substances (ODS). These substances break down ozone (O3) molecules, reducing the layer's ability to absorb ultraviolet (UV) light from the sun. Increased UV exposure impairs immune systems, raises the risk of skin cancer and cataracts in humans, and harms aquatic and terrestrial ecosystems by decreasing phytoplankton populations and altering plant development [78, 79].

Ozone depletion harms the ecosystem because it increases the amount of UV radiation that reaches Earth's surface. Terrestrial ecosystems are harmed by increasing UV-B radiation (280–320 nm) because it interferes with photosynthesis, plant development, and nutrient cycling. Food web disruption and lower agricultural output are the outcomes of this. This affects crops and the whole food chain, sometimes with unanticipated effects for higher trophic levels and herbivores. Aquatic ecosystems are equally vulnerable since the basis of the marine food chain, phytoplankton, is extremely sensitive to UV-B radiation. An increase in UV exposure may cause phytoplankton levels to fall, affecting fish and other marine animals that depend on it.

Ozone depletion has a substantial impact on climate change. Stratospheric cooling is a significant consequence of ozone due to its ability to absorb UV rays and warm the stratosphere. Depletion of the ozone layer leads to less absorption of UV light, resulting in a decline in stratospheric temperatures. This cooling has the potential to alter atmospheric circulation patterns, which could impact global weather and climate systems. Numerous substances that weaken the ozone layer (ODS) are also potent greenhouse gases. Because of their high potential for global warming, they have a more significant impact on global warming even, if their atmospheric concentrations are lower than those of CO2. This means that ozone depletion affects the environment immediately due to increased UV radiation, but it also has long-term repercussions on global climate patterns. For this reason, continuous efforts to reduce ODS emissions and protect the ozone layer are crucial.

Ozone depletion contributes to agricultural losses and growing healthcare costs, and marine fisheries are all significantly impacted economically. Specifically, Juráň et al. [77] reported the harmful effects of O3 on cellular and leaf functions to impact carbon allocation at both the individual tree and ecosystem levels (Fig. 10a). This suggests that O3 can reduce the rate of photosynthesis, which in turn negatively affect crop yield. O3 can also affect human health. For instance, Rivas-Arancibia et al. [80] reported that chronic environmental ozone pollution can disrupt inflammatory response regulation by reaching the brain and lungs, promoting ROS

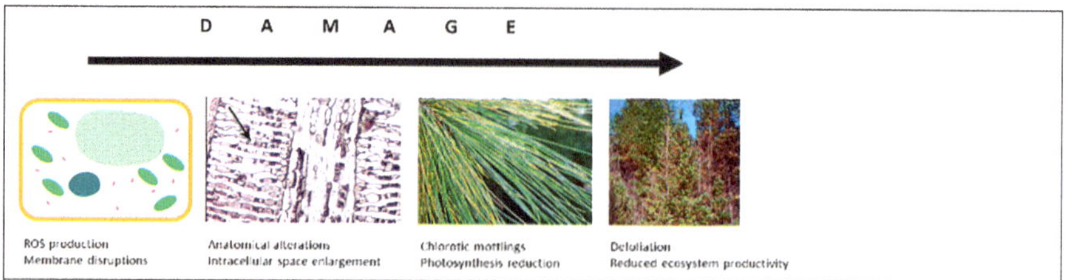

The distribution of carbon at the tree and ecosystem levels affects O3 effects at the cellular and leaf levels. Source: [77]

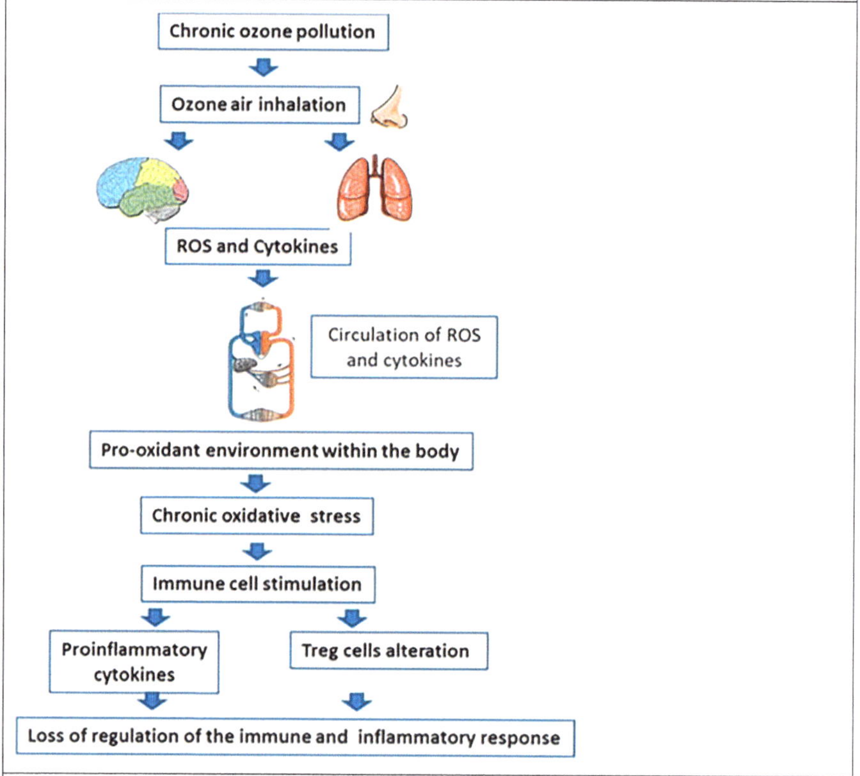

The effects of ozone pollution on the environment and how they affect the regulation of the inflammatory response. Source: [80]

Fig. 10 Impacts of ozone on plants and humans. Source: [77, 80]. https://creativecommons.org/licenses/by/4.0/

formation. The authors further reported that this increases prooxidants and cytokines, leading to heightened levels of proinflammatory cytokines and compromising immune response regulation in the body (Fig. 10b). In addition, the loss of ozone raises the concentrations of potent greenhouse gases and cools the stratosphere, both of which exacerbate climate change.

3.4 Ecosystem Damage

The chemical makeup of the soil is changed when pollutants including; sulfur dioxide, nitrogen oxides, trace metals, and particulate matter, settle on the ground. The introduction of hazardous compounds, unbalanced nutrient intake, and pH changes brought on by this pollution can all harm the health and growth of plants. Plants exposed to contaminated soils may grow more slowly, produce less photosynthesis, and become more vulnerable to pests and illnesses. Given that the health of the soil and vegetation is essential to the larger ecological balance, these changes can potentially affect not only individual plants but also entire ecosystems. For example, soil pollution can upset microbial communities that are critical to the cycling of nutrients, thus deteriorating the quality of the soil and influencing plant growth.

Another significant effect of air pollution on ecosystems is water pollution [81]. Through precipitation or direct deposition, airborne contaminants can end up in rivers, lakes, and seas. This procedure may result in eutrophication, a condition in which an overabundance of nutrients especially nitrogen and phosphorus encourages the proliferation of algae [82, 83]. Oxygen depletion in the water due to harmful algal blooms can result in dead zones where aquatic life cannot live. In addition, these blooms have the potential to release chemicals that are harmful to fish (shell and fin) and even people who drink contaminated water or eat seafood [84, 85]. Furthermore, contaminants like lead and mercury can build up in aquatic environments, endangering the health of marine organisms and perhaps making their way into the human food chain [33, 36, 86–89]. Overall, the effects on aquatic ecosystems can be catastrophic, resulting in a decline in water quality, food web disturbance, and biodiversity loss [86]. These effects can have a significant impact on human populations who depend on these water supplies as well as the environment.

Air pollution harms ecosystems by interacting with terrestrial and aquatic environments in complex ways that affect soil, land, and water quality; therefore, understanding these mechanisms is essential for developing air pollution-reduction measures.

3.4.1 Air Pollutant Deposition and Transformation

Sulfur dioxide (SO_2), nitrogen oxides (NO_x), ammonia, trace metals, and persistent organic pollutants (POPs) are emitted from agricultural activities, industrial processes, and automobiles. These pollutants can travel long distances in the atmosphere due to chemical changes. For instance, the oxidation of SO_2 and NO_x produces sulfuric and nitric acids, leading to acid rain. Ammonia can react with acidic pollutants to form fine particulate matter ($PM_{2.5}$) that settles on soil and water surfaces.

3.4.2 Soil Pollution

Air pollution alters soil chemistry and structure. Acid rain, deposits nitric and sulfuric acids, and can lower soil pH and acidify the soil. This acidity removes essential nutrients like calcium and magnesium, reducing soil fertility and plant growth. Acidic conditions can also release toxic metals such as aluminum [90], which harm soil microorganisms and plant roots.

Airborne trace metals like lead, mercury, and cadmium can persist in the soil for long periods due to their poor solubility and strong binding with soil particles. Soil pH, organic matter, and other factors influence the absorption of these metals. Overexposure to trace metals can injure plants and soil organisms, disrupt nitrogen cycling, and decrease microbial activity [91, 92]. These metals can enter the food chain through plant absorption, posing risks to humans and animals.

3.4.3 Land Deterioration

Air pollution can accelerate land degradation in multiple ways. Acid rain releases hazardous metals and leaches nutrients, reducing plant and soil productivity. Ammonia and nitrogen oxide deposition in intensively farmed areas can cause nitrogen saturation, exceeding the soil's capacity to process nitrogen [93, 94]. Overexposure to nitrogen can acidify soil, reduce biodiversity, and alter species composition, favoring nitrogen-loving plants.

Furthermore, POPs such as dioxins and PCBs can accumulate in the soil and biota due to their resistance to decomposition. These chemicals can disrupt endocrine systems, reduce reproduction rates, and increase wildlife mortality, leading to long-term ecological damage.

3.4.4 Water Quality Decline

Air pollution can directly and indirectly harm water quality. Acid rain lowers the pH of surface waters, making lakes, rivers, and streams more acidic [41]. Dissolved metals like aluminum, released from sediments and soils into acidic waters, can be toxic to aquatic life. These changes in pH and metal toxicity can harm fish and other aquatic organisms, reducing biodiversity and altering ecosystem dynamics [95, 96].

Nitrogen and sulfur deposition can lead to the eutrophication of water bodies, causing harmful algal blooms that produce toxins dangerous to humans and aquatic life [97, 98]. When these algal blooms decompose, they deplete oxygen levels in the water, causing hypoxia and killing fish and other sensitive organisms.

Trace metals and POPs can persist in sediments, leading to long-term pollution. These toxins can re-enter the water column during flooding, dredging, or chemical changes, affecting water quality. They bioaccumulate in aquatic organisms, posing risks to top predators, including humans, and biomagnifying up the food chain.

3.4.5 Integrated Impacts on Ecosystems

Soil contamination, land deterioration, and water quality decline can severely harm ecosystems. These impacts include ecological changes, disturbances in the nitrogen cycle, biodiversity loss, and alterations in species composition. Degradation in one element can cause cascading effects in aquatic and terrestrial ecosystems due to their interdependence. For example, low soil quality can increase runoff and sedimentation in water bodies, harming aquatic habitats and water quality.

3.5 Loss of Biodiversity

Air pollution plays a critical role in habitat degradation, significantly impacting biodiversity. Pollutants such as SO2, NOx, and particulate matter can deposit onto ecosystems, leading to acid rain and soil acidification. Acid rain alters the pH of soil and water bodies, making these environments inhospitable for many plants and aquatic species. Forests, wetlands, and other sensitive ecosystems suffer as their physical and chemical environments are disrupted, reducing habitat quality and availability. This degradation can lead to the decline of plant species that cannot to adapt to the altered conditions, consequently affecting the animals that depend on them for food and shelter. For example, acid rain can leach essential nutrients from the soil, stunting tree growth and leading to forest dieback [99, 100]. This in turn, reduces the habitat available for forest-dwelling species.

Species decline is another significant impact of air pollution on biodiversity. Many pollutants, such as ozone (O_3) and fine particulate matter ($PM_{2.5}$), can directly harm wildlife, particularly affecting their respiratory systems. Chronic exposure to these pollutants can cause respiratory diseases, reduce lung function, and increase mortality rates among wildlife. Additionally, pollutants can weaken animals' immune systems, making them more susceptible to diseases [101, 102] and reducing their ability to survive and reproduce.

For instance, fish and amphibians in polluted waters may experience decreased reproductive success due to the toxic effects of heavy metals and other contaminants [103, 104]. Trace metals like mercury and cadmium can interfere with reproductive hormones and lead to developmental abnormalities in their offspring. Amphibians, such as frogs and salamanders, are particularly sensitive to these pollutants due to their permeable skin, which easily absorbs contaminants from the water.

Birds and mammals exposed to high levels of air pollutants may also suffer from reduced fertility and developmental problems in their offspring. For example, studies have shown that birds living in areas with high air pollution levels exhibit thinner eggshells, reduced hatching success, and increased chick deformities [105, 106]. Mammals, including species like bats and small rodents, may experience lung tissue damage, reduced reproductive rates, and compromised immune systems, making them more vulnerable to predation and disease.

One notable example of biodiversity affected by air pollution is the decline in populations of the European starling (*Sturnus vulgaris*) [107]. This bird species, which is common in urban and industrial areas, has shown a decrease in population size due to air pollution. Research has indicated that exposure to some air pollutants can lead to respiratory issues and weakened immune systems in starlings [107, 108], resulting in higher mortality rates and lower reproductive success.

Another example is the decline in amphibian populations in the United States and Europe. Species such as the northern leopard frog (*Rana pipiens*) and the common toad (*Bufo bufo*) have experienced significant population declines due to pollution in their aquatic habitats [109, 110]. The presence of trace metals and acid rain in their breeding grounds can lower the survival rates of eggs and larvae, contributing to the overall decline of these species.

This direct harm to wildlife and habitat degradation, leads to significant declines in population sizes and the overall biodiversity of affected ecosystems [111–114]. The loss of species and the disruption of ecological interactions can have cascading effects on ecosystem function and resilience. For example, the decline of insect populations due to air pollution can affect the food availability for insectivorous birds and mammals, further impacting their populations.

3.6 Human Health Impacts

Air pollution profoundly impacts on human health, particularly concerning respiratory and cardiovascular diseases [4, 5]. Exposure to ground-level ozone, PM2.5, and NO2 can cause inflammation, reduced lung function, and deep lung penetration. These pollutants have been related to several health problems, including lung cancer, COPD, bronchitis, and asthma [4, 5, 7]. These toxins also affect the cardiovascular system since they can enter the bloodstream and lead to diseases like high blood pressure, heart attacks, strokes, etc. Higher rates of death and morbidity are the outcome of these health issues together, which significantly strain healthcare systems and reduce the quality of life for individuals impacted. Children, the elderly, and people with underlying medical disorders are among the vulnerable groups that are most impacted.

In addition to respiratory and cardiovascular health, air pollution can affect cognitive and developmental health [115–117]. Long-term exposure to air pollution has been shown to adversely affect children's brain development, which may lead to behavioral problems, IQ declines, and developmental delays. For instance, research indicates that elevated levels of NO2 and PM2.5 during pregnancy and the early years of life are associated with a decreased ability to develop cognitively as well as a higher chance of developing neurodevelopmental disorders, such as autism spectrum disorder and attention deficit hyperactivity disorder [115–117]. Due to its detrimental effects on cognitive function, long-term adult exposure to air pollution increases the risk of acquiring neurodegenerative diseases such as Parkinson's and Alzheimer's [118, 119]. It is thought that oxidative stress and neuroinflammatory

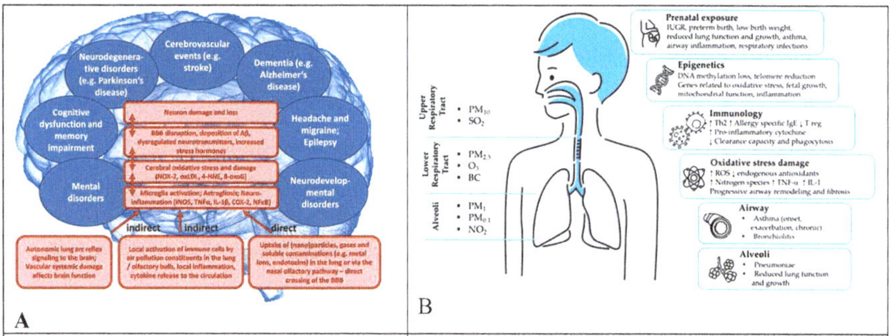

The impact of air pollution components on mental and neurological health conditions. Source: [120]	The main direct and indirect causes of damage to the respiratory system and the major air pollutants that are deposited. Source: [19]

Fig. 11 Human health impacts of air pollution as it concerns respiratory¸ neurological, and mental disorders. Source: [19, 120]. https://creativecommons.org/licenses/by/4.0/

pathways triggered by air pollution are mostly responsible for these detrimental effects.

Hahad et al. [120] elucidated the relationship between air pollution and neurological and mental disorders (Fig. 11a), demonstrating that fine particulate matter, reactive gases, secondary environmental toxins, heavy metals, endotoxins, and products of photoreaction or atmospheric chemistry, such as nitrated VOCs and reactive aldehydes, can enter the body through three primary pathways, both direct and indirect. They asserted that microglia activation causes cerebral oxidative stress and neuroinflammation, which obstruct essential brain functions and initiate pathological processes, including amyloid buildup, neuronal loss, and damage. They highlighted significant elements involved, such as amyloid beta peptide (Aβ), 4-hydroxynonenal (4-HNE), 8-hydroxyguanosine (8-oxoG), nuclear factor "kappa-light-chain-enhancer" of activated B-cells (NFκB), cyclooxygenase 2 (COX-2), iNOS, TNFα, and oxidized low-density lipoprotein (oxLDL).

Dondi et al. [19] further stated that air pollution impacts immunology, oxidative stress damage, epigenetics, prenatal exposure, and respiratory health, affecting both those directly and indirectly exposed. They also noted that air pollution may influence the respiratory system, including the upper and lower airways and the alveoli (Fig. 11b).

Air pollution has a substantial financial impact on societies, primarily through increased healthcare expenses [121]. When allergies, respiratory, cardiovascular, and other health issues are made worse by air pollution, there is a sudden and significant surge in need for medical services. PM2.5 and PM10, SO2, NOx, and O3 are examples of pollutants closely linked to diseases such as lung cancer, asthma, heart disease, and COPD. Healthcare systems are financially strained by the costs associated with hospital admissions, prescription medication costs, long-term care, and preventive care. Initiatives to mitigate the consequences of air pollution and public health campaigns also require a substantial financial commitment. In countries where there is universal healthcare, this can be highly taxing on public resources; in countries without such programs, individuals and families might be forced to pay uncontrollably high medical bills, which exacerbate economic disparities and diminish general well-being.

The detrimental health impacts of air pollution also extend to employees at work, which has a significant detrimental impact on output and productivity [122]. Workers who suffer from illnesses associated with air pollution may be less productive and miss more work. Employees who cannot perform at their peak levels may become less effective and productive. Chronic illnesses can cause people to take extended leaves of absence or even quit their occupations, leading to a loss of skilled labor. Furthermore, as pollution may raise health insurance premiums and necessitate funding additional staff safety procedures, it may deter business investment and economic activity in affected areas. Competitiveness and general economic growth may be hampered by economic setbacks, particularly in densely populated or heavily industrialized urban regions where pollution levels are highest.

One of the main economic repercussions of air pollution is the deterioration of infrastructure [123]. Pollutants like SO2 and NOx cause acid rain, which erodes essential infrastructure such as buildings, monuments, bridges, and other structures. Due to corrosion, these structures require costly maintenance, repairs, and replacements from time to time, which affects businesses, governments, and private property owners. Historical monuments and buildings with cultural value are particularly vulnerable since damage to them can permanently destroy cultural assets.

4 Solutions to Reduce the Environmental Impacts of Air Pollution

Clean Energy Sources: Reducing air pollution requires moving to renewable energy sources like solar, wind, and hydropower. When fossil fuels are replaced with these clean energy sources, there is also a notable decrease in the emissions of sulfur dioxide, nitrogen oxides, and particulate matter [124]. Furthermore, advances in grid and energy storage technologies further enhance the integration and reliability of renewable energy in existing infrastructures [125, 126], thereby supporting sustainable development.

Public Transportation: Investing in and promoting public transportation systems is vital for reducing automobile emissions, a significant contributor to air pollution. Efficient and well-connected public transit encourages people to use buses, trains, and subways instead of personal vehicles. Furthermore, expanding bike-sharing schemes and developing electric or hybrid buses also enhance sustainable urban mobility, reducing emissions and improving urban air quality [127, 128] while alleviating traffic congestion.

Regulations: Strict environmental regulations are essential to combat air pollution. Governments and regulatory bodies must establish and enforce emission standards for vehicles, factories, and other pollution sources. Implementing emission controls, cleaner technology requirements and monitoring systems can significantly reduce the release of hazardous substances into the atmosphere [129]. Effective laws promote environmentally conscious habits and cleaner practices among individuals and industries.

Personal Solutions: A person's lifestyle decisions and choices can have a big impact on lowering air pollution. Pollution can be reduced overall by adopting sustainable practices including cutting back on energy consumption, upgrading to energy-efficient appliances, and disposing of waste properly [17, 130]. Opting for eco-friendly transportation methods like walking or cycling instead of driving can decrease reliance on personal vehicles. Additionally, promoting green space projects and tree planting in communities [131] enhances ecosystems' natural ability to filter and purify the air. Important individual tactics include teaching communities about the effects of human activity and cultivating a feeling of environmental responsibility.

5 Conclusion

Air pollution affects numerous ecosystems and poses major hazards to human health and the delicate balance of the natural world. Its effects on the environment are extensive. Air pollution causes air, water, and soil contamination in addition to the detrimental impacts on flora, fauna, and climate, all of which call for prompt attention and concerted mitigation efforts. Particulate matter, SOx, NOx, VOC, and other pollutants combine to produce a complex web of environmental deterioration. These pollutants are released through human activities and have natural effects. Therefore, addressing air pollution requires a multimodal approach that combines individual and societal solutions. Cutting emissions entails moving to renewable energy sources such as solar, wind, and hydropower via a sustainable apptoach. Additionally, investing in public transportation and promoting bike-sharing schemes and electric buses can significantly reduce automobile emissions and improve urban air quality. Moreover, strict environmental regulations and personal lifestyle adjustments, such as using energy-efficient appliances and opting for eco-friendly transportation, are essential for effectively combating air pollution.

References

1. Sawyer WE, Izah SC, Richard G, Ogwu MC (2023) Microbial air contaminants: diversity and health effects. Ann Public Health Res 10(1):1120
2. Khomenko S, Cirach M, Pereira-Barboza E, Mueller N, Barrera-Gómez J, Rojas-Rueda D, de Hoogh K, Hoek G, Nieuwenhuijsen M (2021) Premature mortality due to air pollution in European cities: a health impact assessment. Lancet Planet Health 5(3):e121–e134

3. Wu Z, Zhang Y, Zhang L, Huang M, Zhong L, Chen D, Wang X (2019) Trends of outdoor air pollution and the impact on premature mortality in the Pearl River Delta region of southern China during 2006–2015. Sci Total Environ 690:248–260
4. Richard G, Izah SC, Morufu OR, Austin-Asomeji I (2023) Public and environmental health implications of artisanal petroleum refining and risk reduction strategies in The Niger Delta region of Nigeria. Bio-Research 21(1):1896–1910
5. Richard G, Izah SC, Ibrahim M (2023) Air pollution in The Niger Delta region of Nigeria: sources, health effects, and strategies for mitigation. J Environ Sci 29(1):1–15
6. Richard G, Izah SC, Sawyer EW, Ogwu MC (2023) Seasonal variations in the concentration of some atmospheric pollutants along a major highway in Southern Nigeria. J Environ Stud 30(1):10–25
7. Seiyaboh EI, Izah SC (2019) Impacts of soil pollution on air quality under Nigerian setting. J Soil Water Sci 3(1):45–53
8. Ohimain EI, Izah SC, Abah SO (2013) Air quality impacts of smallholder oil palm processing in Nigeria. J Environ Prot 4:83–98
9. de Gruijl F, Leun J (2000) Environment and health: 3. Ozone depletion and ultraviolet radiation. CMAJ 163(7):851–855
10. Diffey B (2003) Climate change, ozone depletion and the impact on ultraviolet exposure of human skin. Phys Med Biol 49(1):R1
11. Chakraborty S, Pattanayak A, Mandal S, Das M, Roychowdhury R (2014) An overview of climate change: causes, trends and implications. In: Crop improvement in the era of climate change. IK International Publishing House, New Delhi, pp 1–29
12. Singh V (2024) Global warming and climate change. In: Textbook of environment and ecology. Springer Nature, Singapore, pp 283–295
13. Quah E, Boon TL (2003) The economic cost of particulate air pollution on health in Singapore. J Asian Econ 14(1):73–90
14. Shafik N (1994) Economic development and environmental quality: an econometric analysis. Oxf Econ Pap 46(Supplement_1):757–773
15. Singh N, Tang Y, Ogunseitan OA (2020) Environmentally sustainable management of used personal protective equipment. Environ Sci Technol 54(14):8500–8502
16. Mallick SK, Pramanik M, Maity B, Das P, Sahana M (2021) Plastic waste footprint in the context of COVID-19: reduction challenges and policy recommendations towards sustainable development goals. Sci Total Environ 796:148951
17. Omer AM (2008) Energy, environment and sustainable development. Renew Sust Energ Rev 12(9):2265–2300
18. McMichael AJ (2000) The urban environment and health in a world of increasing globalization: issues for developing countries. Bull World Health Organ 78:1117–1126
19. Dondi A, Carbone C, Manieri E, Zama D, Del Bono C, Betti L, Biagi C, Lanari M (2023) Outdoor air pollution and childhood respiratory disease: the role of oxidative stress. Int J Mol Sci 24(5):4345
20. Jenkin ME, Clemitshaw KC (2000) Ozone and other secondary photochemical pollutants: chemical processes governing their formation in the planetary boundary layer. Atmos Environ 34(16):2499–2527
21. Sillman S (1999) The relation between ozone, NOx and hydrocarbons in urban and polluted rural environments. Atmos Environ 33(12):1821–1845
22. Rai R, Rajput M, Agrawal M, Agrawal SB (2011) Gaseous air pollutants: a review on current and future trends of emissions and impact on agriculture. J Sci Res 55(771):1
23. McDonald BC, Dallmann TR, Martin EW, Harley RA (2012) Long-term trends in nitrogen oxide emissions from motor vehicles at national, state, and air basin scales. J Geophys Res Atmos 117(D21)
24. Sioutas C, Delfino RJ, Singh M (2005) Exposure assessment for atmospheric ultrafine particles (UFPs) and implications in epidemiologic research. Environ Health Perspect 113(8):947–955

25. Kumar P, Wiedensohler A, Birmili W, Quincey P, Hallquist M (2016) Ultrafine particles pollution and measurements. In: Comprehensive analytical chemistry, vol 73. Elsevier, pp 369–390
26. Mohapatra P (2021) A critical review of effectiveness of the regulations in India on designated persistent organic pollutants (POPs) in the Stockholm convention. In: Mohapatra P (ed) Persistent organic pollutants. CRC Press, pp 181–200
27. Kumar R, Mukherji S (2018) Threat posed by persistent organochlorine pesticides and their mobility in the environment. Curr Org Chem 22(10):954–972
28. Link T (2006) Country situation on persistent organic pollutants (POPs) in India. Toxics Link, New Delhi
29. Thakur M, Pathania D (2020) Environmental fate of organic pollutants and effect on human health. In: Abatement of environmental pollutants. Elsevier, pp 245–262
30. Corsini E, Luebke RW, Germolec DR, DeWitt JC (2014) Perfluorinated compounds: emerging POPs with potential immunotoxicity. Toxicol Lett 230(2):263–270
31. Izah SC, Uzoekwe S, Aigberua A (2021) Source, geochemical spreading and risks of trace metals in particulate matter 2.5 within a gas flaring area in Bayelsa State, Nigeria. Adv Environ Technol 7(2):101–118
32. Uzoekwe SA, Izah SC, Aigberua AO (2021) Environmental and human health risk of heavy metals in atmospheric particulate matter (PM 10) around gas flaring vicinity in Bayelsa State, Nigeria. Toxicol Environ Heal Sci 13(4):323–335
33. Ogamba EN, Ebere N, Izah SC (2017) Levels of lead and cadmium in the bone and muscle tissues of Oreochromis niloticus and Clarias camerunensis. EC Nutrition 7(3):117–123
34. Seiyaboh EI, Inyang IR, Izah SC (2016) Seasonal variation of physico-chemical quality of sediment from Ikoli Creek, Niger Delta. Int J Innov Environ Stud Res 4(4):29–34
35. Seiyaboh EI, Izah SC, Oweibi S (2017) Physico-chemical characteristics of sediment from Sagbama Creek, Nigeria. Biotechnol Res 3(1):25–28
36. Ogamba EN, Ebere N, Izah SC (2017) Heavy metal concentration in water, sediment and tissues of Eichhornia crassipes from Kolo Creek, Niger Delta. Greener J Environ Manag Public Safety 6(1):001–005
37. Aghoghovwia OA, Izah SC, Miri FA (2018) Environmental risk assessment of heavy metals in sediment of Nun River around Gbarantoru and Tombia Towns, Bayelsa State, Nigeria. Biol Evid 8(3):21–31
38. Aigberua AO, Ogbuta AA, Izah SC (2020) Selected heavy metals in sediment of Taylor creek due to anthropogenic activities in The Niger Delta region of Nigeria: geochemical spreading and evaluation of environmental risk. Biodiversity Int J 4(2):67–80. https://doi.org/10.15406/bij.2020.04.00166
39. Adams WJ, Kimerle RA, Barnett Jr JW (1992) Sediment quality and aquatic life assessment. Environ Sci Technol 26(10):1864–1875
40. Chapman PM, Wang F, Caeiro SS (2013) Assessing and managing sediment contamination in transitional waters. Environ Int 55:71–91
41. Seiyaboh EI, Izah SC (2017) A review of impacts of gas flaring on vegetation and water resources in The Niger Delta region of Nigeria. Int J Econ Energy Environ 2(4):48–55
42. Seiyaboh EI, Izah SC (2017) Review of impact of anthropogenic activities in surface water resources in The Niger Delta region of Nigeria: a case of Bayelsa State. Int J Ecotoxicol Ecobiol 2(2):61–73
43. Izah SC, Chakrabarty N, Srivastav AL (2016) A review on heavy metal concentration in potable water sources in Nigeria: human health effects and mitigating measures. Expo Health 8:285–304
44. Richard G, Sawyer WE, Izah SC (2021) Health implications of air quality index of fine and coarse particulates during outdoor combustion of biomass in The Niger Delta, Nigeria. Int J World Policy Develop Stud 7(2):21–26
45. Richard G, Sawyer WE, Izah SC (2021) Outdoor air quality index of biomass combustion in The Niger Delta, Nigeria: a health impact perspective. J Adv Res Med Sci Technol 8(1):19–28

46. Chojnacka-Ożga L, Ożga W (2021) The impact of air pollution on the growth of Scots pine stands in poland on the basis of dendrochronological analyses. Forests 12(10):1421
47. Kupfer D, Karimanzira R (1991) Agriculture, forestry, and other human activities. In: Climate change: the IPCC response strategies. Island Press, Washington, pp 73–127
48. Houghton RA (1996) Converting terrestrial ecosystems from sources to sinks of carbon. Ambio 25(4):267–272
49. Malhi Y, Meir P, Brown S (2002) Forests, carbon and global climate. Philos Trans R Soc London, Ser A 360(1797):1567–1591
50. Musselman RC, Fox DG (1991) A review of the role of temperate forests in the global CO2 balance. J Air Waste Manage Assoc 41(6):798–807
51. Schwarz W, Leisewitz A (1999) Emissions and reduction potentials of hydrofluorocarbons, perfluorocarbons and sulphur hexafluoride in Germany. In: Report for the German Federal Environmental Agency. Umweltbundesamt, UBA, Berlin
52. Sovacool BK, Griffiths S, Kim J, Bazilian M (2021) Climate change and industrial F-gases: a critical and systematic review of developments, sociotechnical systems and policy options for reducing synthetic greenhouse gas emissions. Renew Sust Energ Rev 141:110759
53. Demirbas A (2006) Hazardous emissions, global climate change and environmental precautions. Ener Sour B Econ Plan Policy 1(1):75–84
54. Leggett J (1992) Global warming: the scientific evidence and its implications. Transnat'l L & Contemp Probs 2:1
55. Tuckett RP (2016) The role of atmospheric gases. In: Climate change. Elsevier, pp 375–397
56. Bibi F, Rahman A (2023) An overview of climate change impacts on agriculture and their mitigation strategies. Agriculture 13(8):1508
57. Khan AA (2024) An analytical study on the impact of global warming: effects on environment. In: Harale GD, Bhave AV, Pawar GG (eds) Recent trends in commerce, management, accountancy and business economics. Rayat Shikshan Sanstha's Abasaheb Marathe Arts and New Commerce Science College, Rajapur, p 54
58. Chen Y, Mai R (2024) Rising tides: long term impact of sea level rise on marine ecosystems. Int J Mar Sci 14(2):102–110
59. Kasperson RE, Kasperson JX (2012) Climate change, vulnerability and social justice. In: Social contours of risk. Routledge, pp 301–321
60. Ahmadalipour A, Moradkhani H (2018) Escalating heat-stress mortality risk due to global warming in the Middle East and North Africa (MENA). Environ Int 117:215–225
61. Hartmann H, Bastos A, Das AJ, Esquivel-Muelbert A, Hammond WM, Martínez-Vilalta J, McDowell NG, Powers JS, Pugh TA, Ruthrof KX, Allen CD (2022) Climate change risks to global forest health: emergence of unexpected events of elevated tree mortality worldwide. Annu Rev Plant Biol 73:673–702
62. Loucks DP (2021) Impacts of climate change on economies, ecosystems, energy, environments, and human equity: a systems perspective. In: The impacts of climate change. Elsevier, pp 19–50
63. Handmer J, Honda Y, Kundzewicz ZW, Arnell N, Benito G, Hatfield J, Mohamed IF, Peduzzi P, Wu S, Sherstyukov B, Yamano H (2012) Changes in impacts of climate extremes: human systems and ecosystems. Managing the risks of extreme events and disasters to advance climate change adaptation special report of the intergovernmental panel on climate change, pp 231–290
64. Balbus JM, Boxall AB, Fenske RA, McKone TE, Zeise L (2013) Implications of global climate change for the assessment and management of human health risks of chemicals in the natural environment. Environ Toxicol Chem 32(1):62–78
65. Franchini M, Mannucci PM (2015) Impact on human health of climate changes. Eur J Intern Med 26(1):1–5
66. Izah SC, Iyiola AO, Yarkwan B, Richard G (2023) Impact of air quality as a component of climate change on biodiversity-based ecosystem services. In: Srivastav AL, Dubey AK,

Kumar A, Narang SK, Khan MA (eds) Visualization techniques for climate change with machine learning and artificial intelligence. Elsevier, pp 123–148. ISBN: 978-0-323-99714-0
67. Haile M (2005) Weather patterns, food security and humanitarian response in sub-Saharan Africa. Philos Trans R Soc Lond B Biol Sci 360(1463):2169–2182
68. Elbasiouny H, El-Ramady H, Elbehiry F, Rajput VD, Minkina T, Mandzhieva S (2022) Plant nutrition under climate change and soil carbon sequestration. Sustainability 14(2):914
69. Zhang Y, Li Q, Zhang F, Xie G (2017) Estimates of economic loss of materials caused by acid deposition in China. Sustainability 9(4):488
70. Prakash J, Agrawal SB, Agrawal M (2023) Global trends of acidity in rainfall and its impact on plants and soil. J Soil Sci Plant Nutr 23(1):398–419
71. Rodríguez-Sánchez VM, Rosas U, Calva-Vásquez G, Sandoval-Zapotitla E (2020) Does acid rain alter the leaf anatomy and photosynthetic pigments in urban trees? Plan Theory 9(7):862
72. Debnath B, Sikdar A, Islam S, Hasan K, Li M, Qiu D (2021) Physiological and molecular responses to acid rain stress in plants and the impact of melatonin, glutathione and silicon in the amendment of plant acid rain stress. Molecules 26(4):862
73. Basu S, Orr SA, Aktas YD (2020) A geological perspective on climate change and building stone deterioration in London: implications for urban stone-built heritage research and management. Atmosphere 11(8):788
74. Brimblecombe P (2014) Environment and architectural stone. In: Stone in architecture: properties, durability. Springer, Berlin, pp 317–347
75. Haines TA (1981) Acidic precipitation and its consequences for aquatic ecosystems: a review. Trans Am Fish Soc 110(6):669–707
76. Rivas-Arancibia S, Miranda-Martínez A, Rodríguez-Martínez E, Hernández-Orozco E, Valdés-Fuentes M, De la Rosa-Sierra R (2023) Ozone environmental pollution: relationship between the intestine and neurodegenerative diseases. Antioxidants 12(7):1323
77. Juráň S, Grace J, Urban O (2021) Temporal changes in ozone concentrations and their impact on vegetation. Atmos 12(1):82
78. Andrady A, Aucamp PJ, Austin AT, Bais AF, Ballare CL, Barnes PW, Bernhard GH, Bjoern LO, Bornman JF, Congdon N, Zepp RG (2017) Environmental effects of ozone depletion and its interactions with climate change: Progress report, 2016. Photochem Photobiol Sci 16(2): 107
79. El-Sayed SZ, Van Dijken GL, Gonzalez-Rodas G (1996) Effects of ultraviolet radiation on marine ecosystems. Int J Environ Stud 51(3):199–216
80. Rivas-Arancibia S, Hernández-Orozco E, Rodríguez-Martínez E, Valdés-Fuentes M, Cornejo-Trejo V, Pérez-Pacheco N, Dorado-Martínez C, Zequeida-Carmona D, Espinosa-Caleti I (2022) Ozone pollution, oxidative stress, regulatory T cells and antioxidants. Antioxidants 11(8):1553
81. Richard G, Izah SC, Ogwu MC (2022) Implications of artisanal crude oil refining on sustainable food production in The Niger Delta region of Nigeria. J Environ Bioremed Toxicol 5(2):69–77
82. Wurtsbaugh WA, Paerl HW, Dodds WK (2019) Nutrients, eutrophication and harmful algal blooms along the freshwater to marine continuum. Wiley Interdiscip Rev Water 6(5):e1373
83. Tiwari AK, Pal DB (2022) Nutrients contamination and eutrophication in the river ecosystem. In: Ecological significance of river ecosystems. Elsevier, pp 203–216
84. Brown AR, Lilley M, Shutler J, Lowe C, Artioli Y, Torres R, Berdalet E, Tyler CR (2020) Assessing risks and mitigating impacts of harmful algal blooms on mariculture and marine fisheries. Rev Aquac 12(3):1663–1688
85. Boesch DF (1997) Harmful algal blooms in coastal waters: options for prevention, control and mitigation (No. 10). US Department of Commerce, National Oceanic and Atmospheric Administration, Coastal Ocean Office
86. Izah SC, Richard G, Stanley HO, Sawyer WE, Ogwu MC, Uwaeme OR (2023) Integrating the one health approach and statistical analysis for sustainable aquatic ecosystem management and

trace metal contamination mitigation. ES Food Agrofor 14:1012. https://doi.org/10.30919/esfaf1012
87. Ogamba EN, Izah SC, Isimayemiema F (2016) Bioaccumulation of heavy metals in the gill and liver of a common Niger Delta wetland fish, Clarias gariepinus. Br J Appl Res 1(1):17–20
88. Ogamba EN, Izah SC, Ofoni-Ofoni AS (2016) Bioaccumulation of chromium, Lead and cadmium in the bones and tissues of Oreochromis niloticus and Clarias camerunensis from Ikoli creek, Niger Delta, Nigeria. Adv Sci J Zool 1(1):13–16
89. Ogamba EN, Izah SC, Omonibo E (2016) Bioaccumulation of hydrocarbon, heavy metals and minerals in Tympanotonus fuscatus from coastal region of Bayelsa state, Nigeria. Int J Hydrol Res 1:1–7
90. Foy CD (1984) Physiological effects of hydrogen, aluminum, and manganese toxicities in acid soil. Soil Acidity Liming 12:57–97
91. Narayanan M, Ma Y (2023) Mitigation of heavy metal stress in the soil through optimized interaction between plants and microbes. J Environ Manag 345:118732
92. Rahman MA, Hossain A, Islam MR, Azim A, Gaber A, Aftab T (2022) Metals and metalloids stress in plants: microorganisms and phytoremediation based mitigation strategies. In: Metals metalloids soil plant water systems. Academic Press, pp 445–484
93. Ghaly AE, Ramakrishnan VV (2015) Nitrogen sources and cycling in the ecosystem and its role in air, water and soil pollution: a critical review. J Pollut Effects Control 3(2):1–26
94. Follett RF, Delgado JA (2002) Nitrogen fate and transport in agricultural systems. J Soil Water Conserv 57(6):402–408
95. Chakraborty SK (2021) River pollution and perturbation: perspectives and processes. In: Riverine ecology, Volume 2: Biodiversity conservation, conflicts and resolution. Springer, Cham, pp 443–530
96. Sharma AK, Sharma M, Sharma S, Malik DS, Sharma M, Sharma AK (2024) A systematic review on assessment of heavy metals toxicity in freshwater fish species: current scenario and remedial approaches. J Geochem Explor 262:107472
97. Prakash A, Khanam S (2021) Nitrogen pollution threat to mariculture and other aquatic ecosystems: an overview. J Pharm Pharmacol 9:428–433
98. Watson SB, Whitton BA, Higgins SN, Paerl HW, Brooks BW, Wehr JD (2015) Harmful algal blooms. In: Freshwater algae of North America. Academic Press, pp 873–920
99. Tomlinson GH (2003) Acidic deposition, nutrient leaching and forest growth. Biogeochemistry 65:51–81
100. Tomlinson GH (1990) Effects of acid deposition on the forests of Europe and North America. CRC Press
101. Pandey G, Madhuri S (2014) Heavy metals causing toxicity in animals and fishes. Res J Anim Vet Fish Sci 2(2):17–23
102. Wobeser GA (2013) Essentials of disease in wild animals. Wiley
103. Carr JA, Patiño R (2011) The hypothalamus–pituitary–thyroid axis in teleosts and amphibians: endocrine disruption and its consequences to natural populations. Gen Comp Endocrinol 170(2):299–312
104. Slaby S, Marin M, Marchand G, Lemiere S (2019) Exposures to chemical contaminants: what can we learn from reproduction and development endpoints in the amphibian toxicology literature? Environ Pollut 248:478–495
105. Padate G (2009) Effect of air pollution on breeding performance of birds. In: Eco-degradation due to air pollution. Scientific Publishers, p 311
106. Fry DM (1995) Reproductive effects in birds exposed to pesticides and industrial chemicals. Environ Health Perspect 103(suppl 7):165–171
107. North MA (2017) European starlings as sentinels for health effects of urban air pollution, Doctoral thesis, University of Calgary, Calgary, Canada. Retrieved from https://prism.ucalgary.ca. https://doi.org/10.11575/PRISM/25964

108. Ziegler AK, Jensen JK, Jiménez-Gallardo L, Rissler J, Gudmundsson A, Nilsson JÅ, Isaksson C (2024) Dietary fatty acids modulate oxidative stress response to air pollution but not to infection. Front Physiol 15:1391806
109. Rouse JD, Bishop CA, Struger J (1999) Nitrogen pollution: an assessment of its threat to amphibian survival. Environ Health Perspect 107(10):799–803
110. Gardner T (2001) Declining amphibian populations: a global phenomenon in conservation biology. Anim Biodivers Conserv 24(2):25–44
111. Izah SC, Angaye CN, Aigberua AO, Nduka JO (2017) Uncontrolled bush burning in The Niger Delta region of Nigeria: potential causes and impacts on biodiversity. Int J Mol Ecol Conserv 7(1):1–15
112. Izah SC (2018) Ecosystem of The Niger Delta region of Nigeria: potentials and threats. Biodivers Int J 2(4):338–345
113. Izah SC, Seiyaboh EI (2018) Challenges of wildlife with therapeutic properties in Nigeria; a conservation perspective. Int J Avian Wildlife Biol 3(4):259–264
114. Izah SC, Seiyaboh EI (2018) Changes in the protected areas of Bayelsa state, Nigeria. Int J Mol Evol Biodivers 8(1):1–11
115. Lertxundi A, Andiarena A, Martínez MD, Ayerdi M, Murcia M, Estarlich M, Guxens M, Sunyer J, Julvez J, Ibarluzea J (2019) Prenatal exposure to PM2. 5 and NO2 and sex-dependent infant cognitive and motor development. Environ Res 174:114–121
116. Morgan ZE, Bailey MJ, Trifonova DI, Naik NC, Patterson WB, Lurmann FW, Chang HH, Peterson BS, Goran MI, Alderete TL (2023) Prenatal exposure to ambient air pollution is associated with neurodevelopmental outcomes at 2 years of age. Environ Health 22(1):11
117. Castagna A, Mascheroni E, Fustinoni S, Montirosso R (2022) Air pollution and neurodevelopmental skills in preschool-and school-aged children: a systematic review. Neurosci Biobehav Rev 136:104623
118. Tan L, Nakanishi E, Lee M (2022) Association between exposure to air pollution and late-life neurodegenerative disorders: an umbrella review. Environ Int 158:106956
119. Lin FC, Chen CY, Lin CW, Wu MT, Chen HY, Huang P (2022) Air pollution is associated with cognitive deterioration of Alzheimer's disease. Gerontology 68(1):53–61
120. Hahad O, Lelieveld J, Birklein F, Lieb K, Daiber A, Münzel T (2020) Ambient air pollution increases the risk of cerebrovascular and neuropsychiatric disorders through induction of inflammation and oxidative stress. Int J Mol Sci 21(12):4306
121. Siriopoulos C, Samitas A, Dimitropoulos V, Boura A, AlBlooshi DM (2021) Health economics of air pollution. In: Pollution assessment for sustainable practices in applied sciences and engineering. Butterworth-Heinemann, pp 639–679
122. Muzayyanah S, Hong CY, Adha R, Yang SF (2023) The non-linear relationship between air pollution, labor insurance and productivity: multivariate adaptive regression splines approach. Sustainability 15(12):9404
123. Silva LF, Pinto D, Neckel A, Oliveira ML (2020) An analysis of vehicular exhaust derived nanoparticles and historical Belgium fortress building interfaces. Geosci Front 11(6): 2053–2060
124. Shahsavari A, Akbari M (2018) Potential of solar energy in developing countries for reducing energy-related emissions. Renew Sust Energ Rev 90:275–291
125. Worku MY (2022) Recent advances in energy storage systems for renewable source grid integration: a comprehensive review. Sustainability 14(10):5985
126. Olajiga OK, Ani EC, Olatunde TM, Sikhakane ZQ (2024) Assessing the potential of energy storage solutions for grid efficiency: a review. Eng Sci Technol J 5(3):1112–1124
127. Gordon A (2023) The role of public transportation systems in promoting sustainable mobility in green cities. Adv Urban Resilience Sustain City Design 15(6):1–18
128. Shajal AK, Mohamed AH (2023) Evaluating the effectiveness of bike sharing programs in encouraging sustainable transportation in urban areas. J Sustain Urban Futures 13(6):1–10

129. Endres S, Maes F, Hopkins F, Houghton K, Mårtensson EM, Oeffner J, Quack B, Singh P, Turner D (2018) A new perspective at the ship-air-sea-interface: the environmental impacts of exhaust gas scrubber discharge. Front Mar Sci 5:361303
130. Omer AM (2009) Energy use and environmental impacts: a general review. J Renew Sustain Energy 1(5):053101
131. Semeraro T, Scarano A, Buccolieri R, Santino A, Aarrevaara E (2021) Planning of urban green spaces: an ecological perspective on human benefits. Land 10(2):105

Societal Burdens and Ramifications of Air Pollution

Enoch Akwasi Kosoe and Tamaraukepreye Catherine Odubo

Contents

Abstract Air pollution has profound socioeconomic impacts that extend beyond public health and environmental concerns. This chapter aims to discuss the socio-economic impacts of air pollution and highlights the multifaceted issues, ranging from economic costs to social inequalities and environmental justice concerns. Socially, air pollution diminishes the quality of life for affected individuals and communities by reducing outdoor recreational opportunities, limiting access to green spaces, and impairing the enjoyment of natural environments. Air pollution imposes economic burdens on individuals, families, businesses, and governments through increased healthcare costs, lost productivity, and decreased property values. The

E. A. Kosoe (✉)
Department of Environment and Resource Studies, Faculty of Integrated Development Studies, Simon Diedong Dombo University of Business and Integrated Development Studies, Wa, Upper West Region, Ghana
e-mail: ekosoe@ubids.edu.gh

T. C. Odubo
Department of Microbiology, Bayelsa Medical University, Yenagoa, Bayelsa, Nigeria

Matthew Chidozie Ogwu and Sylvester Izah (eds.), *Sustainable Strategies for Air Pollution Mitigation: Development, Economics and Technologies*,
Hdb Env Chem (2024) 133: 77–110, DOI 10.1007/698_2024_1115,

Published online: 10 September 2024

public health burden of air pollution imposes significant socioeconomic costs on individuals, communities, and societies. These costs include healthcare expenditures, lost productivity due to illness and absenteeism, and premature mortality. Air pollution-related illnesses result in increased healthcare utilization, hospital admissions, emergency room visits, and medication costs, straining healthcare systems and budgets. Environmental racism, socioeconomic disparities, and unequal access to clean air and healthy environments perpetuate social inequalities and contribute to disparities in health outcomes, life expectancy, and quality of life. There is an urgency to address the economic costs, social inequalities, and environmental injustices associated with air pollution. Lost productivity due to air pollution-related morbidity and mortality affects individuals, businesses, and economies, leading to decreased economic output and competitiveness. Implementing pollution control measures, promoting clean energy technologies, and fostering social and environmental justice can help mitigate the socioeconomic impacts of air pollution and build healthier, more equitable societies for all. Tourist destinations are often impacted by air pollution and pose economic challenges to local communities and businesses. For example, cities such as Beijing, Delhi, and Los Angeles have experienced declines in tourism due to concerns about air pollution, resulting in decreased spending by visitors and adverse effects on the hospitality industry. Additionally, the socioeconomic burden of air pollution disproportionately affects low-income communities and disadvantaged populations, exacerbating existing social inequalities in health outcomes and access to healthcare services.

Keywords Clean technologies, Economic costs, Environmental injustices, Public health, Social inequalities

Abbreviations

ADHD	Attention-deficit/hyperactivity disorder
BAT	Best available techniques
CO	Carbon monoxide
COPD	Chronic obstructive pulmonary disease
CSR	Corporate social responsibility
DNA	Deoxyribonucleic acid
PM2.5	Fine particulate matter
GHGs	Greenhouse gases
GDP	Gross domestic product
NO2	Nitrogen dioxide
NOx	Nitrogen oxides
O3	Ozone
PM	Particulate matter
PAHs	Polycyclic aromatic hydrocarbons
R&D	Research and development

SO2	Sulfur dioxide
UNFCCC	United Nations Framework Convention on Climate Change
VOCs	Volatile organic compounds

1 Introduction

Air pollution is a pressing environmental issue that not only poses threats to human health and the environment but also carries significant socioeconomic implications. Understanding and addressing these impacts is necessary for building healthier and more equitable societies [1]. Air pollution is a global issue that affects billions of people and ecosystems around the world [2]. One of the most significant socioeconomic impacts of air pollution is its adverse effects on public health. In recent years, there has been a growing recognition of the importance of understanding the multifaceted impacts of air pollution on public health, economies, and social well-being [3, 4]. Air pollution is a major risk factor for various respiratory and cardiovascular diseases, including asthma, bronchitis, lung cancer, and heart attacks [5, 6]. Exposure to air pollutants such as particulate matter (PM), nitrogen oxides (NOx), sulfur dioxide (SO2), and ozone (O3) is linked to various respiratory and cardiovascular diseases, including asthma, bronchitis, lung cancer, and heart attacks. The resulting health burden leads to increased healthcare expenditures, lost productivity due to illness and absenteeism, and premature mortality. Vulnerable populations, such as children, the elderly, and low-income communities, are disproportionately affected by air pollution-related illnesses, exacerbating existing social inequalities in health outcomes [7]. By understanding the health impacts of air pollution, policymakers, healthcare professionals, and communities can develop targeted interventions and public health campaigns to reduce exposure and mitigate the health risks associated with poor air quality [8, 9]. Moreover, understanding the link between air pollution and public health enables the implementation of preventive measures, such as improving air quality standards, promoting cleaner transportation, and reducing emissions from industrial sources [10].

Air pollution takes a heavy toll on economies through direct and indirect costs associated with healthcare expenditures, lost productivity, and environmental damage [11]. Healthcare costs related to treating air pollution-related illnesses strain healthcare systems and budgets, diverting resources away from other essential services and treatments [12, 13]. Lost productivity due to illness and absenteeism among workers further impacts economic output and competitiveness [14, 15]. Additionally, businesses face costs associated with healthcare expenses, decreased worker productivity, and potential regulatory fines for non-compliance with pollution standards [16, 17]. Governments incur costs for implementing pollution control measures, enforcing regulations, and addressing environmental cleanup and remediation efforts [18, 19]. Therefore, understanding the economic impacts of air pollution is essential for promoting sustainable economic development. Air pollution imposes significant costs on economies through healthcare expenditures, lost productivity, and environmental damage. By quantifying these economic costs, policymakers and businesses can make informed decisions about investing in pollution control measures, transitioning to cleaner energy sources, and implementing sustainable practices. Furthermore, understanding the economic impacts of air pollution can help policymakers prioritize resource allocation, develop cost-effective policies, and stimulate innovation in pollution control technologies [18–20]. Ultimately, addressing air pollution can contribute to economic growth, job creation, and improved living standards for communities globally.

Air pollution exacerbates social inequalities and environmental injustices by disproportionately affecting marginalized populations and communities [21, 22]. Low-income communities, communities of color, and indigenous peoples are often located near industrial facilities, highways, and other sources of pollution, leading to higher exposure levels and increased health risks [23]. These communities face barriers to accessing healthcare, environmental information, and legal resources to address environmental injustices. Environmental justice initiatives should aim at addressing these disparities by advocating for equitable access to clean air, water, and land for all communities, regardless of socioeconomic status or race/ethnicity. By understanding the social dimensions of air pollution, policymakers and community leaders can develop targeted interventions and policies to address the needs of marginalized populations and promote environmental justice. This may include measures such as improving access to healthcare, investing in green infrastructure in underserved communities, and empowering communities to participate in decision-making processes related to air quality [24]. Additionally, understanding the social impacts of air pollution can help raise awareness, foster solidarity, and mobilize collective action to address this shared challenge. Air pollution also impacts tourism and recreational activities by degrading air quality, diminishing scenic views, and deterring visitors from outdoor destinations. High levels of air pollution in popular tourist destinations can lead to cancellations of travel plans, loss of revenue for local businesses, and damage to the tourism industry [25]. Investments in pollution control measures and sustainable tourism practices can help mitigate the negative effects of air pollution on tourism and support local economies.

This chapter aims to discuss the socioeconomic impacts of air pollution. Air pollution has profound socioeconomic impacts that extend beyond public health and environmental concerns. The economic costs, social inequalities, and environmental injustices associated with air pollution highlight the urgent need for comprehensive strategies to address this pressing issue.

2 Public Health Burdens of Air Pollution

The public health burden of air pollution is substantial, with significant impacts on human health, well-being, and quality of life. Air pollution is a major environmental health risk factor that affects populations worldwide, contributing to a range of adverse health outcomes. Exposure to air pollution has been linked to a wide range of health problems, including respiratory diseases, cardiovascular diseases, cancers, and neurological disorders [26, 27]. Air pollution can irritate the respiratory system; exacerbate asthma, bronchitis, and chronic obstructive pulmonary disease (COPD); and increase the risk of respiratory infections such as pneumonia and bronchiolitis [28]. Long-term exposure to air pollution is associated with an increased risk of heart attacks, strokes, hypertension, and other cardiovascular disorders, including arrhythmias, heart failure, and atherosclerosis [29, 30]. Air pollutants such as benzene, formaldehyde, and diesel exhaust particles have been classified as carcinogens by international health agencies and may increase the risk of lung cancer and other cancers [31, 32]. According to Volk et al. [33], prenatal and childhood exposure to air pollution has been linked to cognitive impairment, developmental delays, and neurobehavioral disorders such as autism spectrum disorder and attention-deficit/hyperactivity disorder (ADHD). Air pollution exposure during pregnancy has been associated with adverse pregnancy outcomes, including preterm birth, low birth weight, and congenital anomalies. Certain groups are more vulnerable to the health impacts of air pollution, including children, the elderly, pregnant women, individuals with pre-existing health conditions, and socioeconomically disadvantaged populations [34, 35]. In results from the study by Aithal et al. [36], children are particularly susceptible to the effects of air pollution due to their developing respiratory and immune systems, higher breathing rates, and increased outdoor activity. The elderly and individuals with pre-existing health conditions, such as asthma, COPD, and heart disease, are also at higher risk of adverse health outcomes from air pollution exposure [37]. Socioeconomically disadvantaged populations may face higher exposure to air pollution due to living in areas with high pollution levels, limited access to healthcare, and socioeconomic disparities in housing, education, and employment [38].

The public health burden of air pollution imposes significant socioeconomic costs on individuals, communities, and societies. These costs include healthcare expenditures, lost productivity due to illness and absenteeism, and premature mortality. Air pollution-related illnesses result in increased healthcare utilization, hospital admissions, emergency room visits, and medication costs, straining healthcare systems and budgets [39, 40]. Lost productivity due to air pollution-related morbidity and mortality affects individuals, businesses, and economies, leading to decreased economic output and competitiveness [41]. Additionally, the socioeconomic burden of air pollution disproportionately affects low-income communities and disadvantaged populations, exacerbating existing social inequalities in health outcomes and access to healthcare services. The socioeconomic implications of the public health burden of air pollution are profound, affecting individuals, communities, and societies in various ways including (Fig. 1):

Healthcare Expenditures: Air pollution-related illnesses result in increased healthcare utilization, including hospital admissions, emergency room visits, doctor consultations, and medication costs [42]. Healthcare expenditures related to treating air pollution-related health conditions place a significant financial

Fig. 1 Some public health burdens of air pollution

burden on individuals, families, healthcare systems, and governments [43]. The allocation of resources toward treating air pollution-related illnesses diverts funds away from other essential healthcare services and public health initiatives.

Lost Productivity: Air pollution-related morbidity and mortality lead to lost productivity due to illness, absenteeism, and premature mortality. Workers exposed to high levels of air pollution may experience respiratory symptoms, fatigue, and decreased cognitive function, leading to decreased efficiency and productivity in the workplace [44, 45]. Lost productivity affects individuals, businesses, and economies, leading to decreased economic output, competitiveness, and economic growth.

Economic Disparities: The socioeconomic burden of air pollution disproportionately affects low-income communities, disadvantaged populations, and marginalized groups [38]. Socioeconomically disadvantaged populations may face higher exposure to air pollution due to living in areas with high pollution levels, limited access to healthcare, and socioeconomic disparities in housing, education, and employment [46]. The economic costs of air pollution-related illnesses further exacerbate existing social inequalities in health outcomes and access to healthcare services, perpetuating cycles of poverty and disadvantage.

Environmental Justice: Air pollution often disproportionately affects environmental justice communities, including low-income communities, communities of color, and indigenous peoples [47]. These communities may be located near industrial facilities, highways, and other sources of pollution, leading to higher exposure levels and increased health risks. Environmental justice initiatives seek to address these disparities by advocating for equitable access to clean air, water, and land for all communities, regardless of socioeconomic status or race/ethnicity.

Public Policy and Regulation: The socioeconomic implications of the public health burden of air pollution highlight the importance of effective public policies and regulations aimed at reducing emissions, improving air quality, and protecting public health. Governments play a critical role in implementing pollution control measures, enforcing environmental regulations, and promoting sustainable development practices to mitigate the socioeconomic impacts of air pollution [18, 19, 48]. Collaborative efforts between policymakers, healthcare professionals, researchers, and communities are needed to develop evidence-based policies and interventions that address the root causes of air pollution and promote health equity.

2.1 Healthcare Expenditures Resulting from Air Pollution

Air pollution-related health conditions result in increased healthcare utilization and expenditures. Individuals and governments incur costs for hospital admissions, emergency room visits, doctor consultations, diagnostic tests, medications, and treatments related to respiratory and cardiovascular diseases, cancers, and other health problems caused or exacerbated by air pollution. Healthcare expenditures related to air pollution place a financial burden on individuals, families, healthcare systems, and governments, diverting resources away from other essential healthcare services and public health initiatives [49–51]. Increased healthcare expenditures due to air pollution are a significant economic burden on individuals, families, healthcare systems, and governments [11]. Air pollution-related illnesses lead to higher healthcare utilization, including hospital admissions, emergency room visits, doctor consultations, diagnostic tests, medications, and treatments [52].

Air pollution is a major risk factor for respiratory diseases such as asthma, COPD, bronchitis, and pneumonia, as well as cardiovascular diseases such as heart attacks, strokes, hypertension, and heart failure [53, 54]. Individuals exposed to high levels of air pollution are more likely to develop these conditions, leading to increased healthcare utilization and expenditures for diagnosis, treatment, and management [55]. Short-term exposure to air pollution can exacerbate existing health conditions and lead to acute health effects such as respiratory symptoms, chest pain, coughing, wheezing, and difficulty breathing [14, 15]. Long-term exposure to air pollution is associated with chronic health effects such as reduced lung function, accelerated aging of the lungs, and increased risk of chronic diseases [56]. According to the findings by Li et al. [57], the treatment and management of acute and chronic health effects of air pollution contribute to higher healthcare expenditures over time. Air pollution-related illnesses often require medical intervention, leading to increased hospital admissions and emergency room visits [58]. Individuals experiencing severe respiratory or cardiovascular symptoms due to air pollution may require immediate medical attention and treatment in hospital settings [59]. Hospitalizations and emergency room visits incur costs for medical services, procedures, medications, and other healthcare resources. Healthcare providers may order diagnostic tests such as pulmonary function tests, chest X-rays, blood tests, and electrocardiograms to assess the impact of air pollution on patients' health and diagnose respiratory and cardiovascular conditions. Additionally, individuals with air pollution-related illnesses may require medications such as bronchodilators, corticosteroids, antihistamines, and cardiovascular medications to manage their symptoms and improve their health outcomes [60]. The costs of diagnostic tests and medications contribute to higher healthcare expenditures for individuals and healthcare systems. Individuals with chronic respiratory and cardiovascular conditions exacerbated by air pollution require long-term health monitoring and management to prevent disease progression, manage symptoms, and improve quality of life [8, 61]. This may include regular medical check-ups, medication adjustments, lifestyle modifications, and disease management programs. The ongoing costs of long-term health monitoring and management add to the overall healthcare expenditures associated with air pollution-related illnesses.

3 Economic Costs of Air Pollution

Some of the economic burdens associated with air pollution includes (Fig. 2).

Low Productivity: Air pollution-related illnesses lead to decreased productivity due to illness, absenteeism, and premature mortality. Workers exposed to high levels of air pollution may experience respiratory symptoms, fatigue, and decreased cognitive function, resulting in decreased efficiency and productivity in the workplace [62]. Lost productivity affects businesses and economies, leading to decreased economic output and competitiveness. Low productivity due to air pollution is a significant concern that affects individuals, businesses, and economies worldwide [63]. Air pollution can have direct and indirect effects on productivity in various sectors, leading to economic losses and reduced competitiveness.

Health Impacts: Exposure to air pollution can lead to a range of health problems, including respiratory diseases, cardiovascular diseases, and neurological disorders [64]. Individuals who are affected by air pollution-related illnesses may experience symptoms such as coughing, wheezing, shortness of breath, fatigue, headaches, and dizziness. These health issues can reduce productivity by causing discomfort,

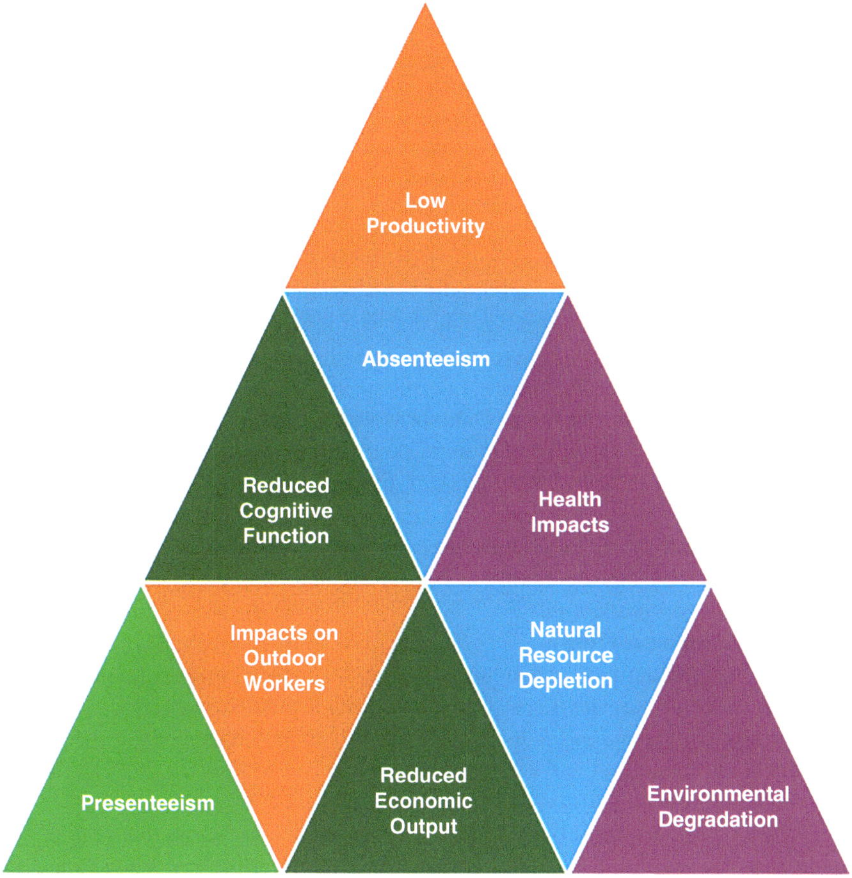

Fig. 2 Some economic burdens of air pollution

fatigue, and decreased concentration, leading to decreased efficiency and performance at work [14, 15]. Air pollution-related health issues lead to increased healthcare utilization and expenditures, including hospital admissions, emergency room visits, doctor consultations, diagnostic tests, medications, and treatments. The allocation of resources toward treating air pollution-related illnesses diverts funds away from other productive sectors of the economy, leading to decreased investment, innovation, and economic growth [65].

Absenteeism: Air pollution-related illnesses can result in increased absenteeism among workers. Individuals who experience respiratory symptoms or other health problems due to air pollution may need to take time off work to recover or seek medical treatment. Absenteeism disrupts workflow, reduces staffing levels, and can lead to delays in project completion and decreased productivity for businesses and organizations [66].

Presenteeism: Even when individuals continue to work despite experiencing health problems related to air pollution, their productivity may be affected by presenteeism. Presenteeism refers to the phenomenon where employees are physically present at work but are not fully productive due to health issues [66]. Unwell workers may have reduced concentration, motivation, and energy levels, leading to lower productivity and performance on tasks.

Reduced Cognitive Function: Air pollution has been linked to cognitive impairment and decreased cognitive function, particularly in children and older adults. Exposure to air pollutants such as fine particulate matter (PM2.5), nitrogen dioxide (NO2), and ozone (O3) can impair cognitive abilities such as memory, attention, and decision-making. Reduced cognitive function can affect performance in academic settings, workplaces, and daily activities, leading to lower productivity and efficiency [67].

Impacts on Outdoor Workers: Outdoor workers, such as construction workers, agricultural workers, and transportation workers, are particularly vulnerable to the effects of air pollution [68]. These workers may be exposed to high levels of air pollutants during their work activities, leading to increased health risks and decreased productivity. Extreme weather conditions, such as heatwaves and wildfires, exacerbated by climate change and air pollution, can further impact outdoor workers' health and productivity.

Reduced Economic Output: Air pollution hurts economic output and growth by affecting labor productivity, consumer spending, investment, and overall economic activity. Businesses may incur costs associated with healthcare expenses, decreased worker productivity, and regulatory compliance with pollution control measures. Reduced economic output and competitiveness can result in lower gross domestic product (GDP), decreased tax revenues, and diminished economic prosperity for communities and countries affected by air pollution. Reduced economic output due to air pollution is a significant consequence that affects businesses, industries, and economies at large. Air pollution can have direct and indirect impacts on economic productivity, leading to decreased efficiency, increased costs, and diminished competitiveness [69].

Decreased Labor Productivity: Air pollution-related health issues can lead to decreased labor productivity among workers. Individuals exposed to high levels of air pollution may experience respiratory symptoms, fatigue, and decreased cognitive function, which can impair their ability to perform tasks effectively. Reduced worker productivity results in lower output per worker, leading to decreased overall economic productivity and output.

Disruption of Supply Chains and Operations: Air pollution can disrupt supply chains and operations in various industries, particularly those that rely on outdoor activities and transportation [70]. Industries such as agriculture, construction, transportation, and manufacturing may experience delays, interruptions, and increased costs due to air pollution-related factors such as extreme weather events, transportation restrictions, and health and safety concerns for workers.

Environmental Degradation and Natural Resource Depletion: Air pollution contributes to environmental degradation and natural resource depletion, which can have long-term economic consequences. Pollution-related damage to ecosystems, soil, water, and biodiversity affects industries such as agriculture, fisheries, tourism, and forestry, leading to decreased productivity, revenue losses, and increased costs for environmental cleanup and remediation efforts [71].

3.1 Costs Incurred by Businesses, Governments, and Individuals

Businesses incur costs related to healthcare expenses, absenteeism decreased worker productivity, and potential regulatory fines for non-compliance with pollution standards [72]. Governments face costs for implementing pollution control measures, enforcing environmental regulations, and addressing environmental cleanup and remediation efforts [73]. Individuals bear the costs of healthcare expenditures, lost wages, decreased quality of life, and property damage caused by air pollution-related illnesses and environmental degradation [74]. The costs incurred by businesses, governments, and individuals due to air pollution are significant and multifaceted, encompassing various direct and indirect expenses related to healthcare, productivity losses, environmental damage, and regulatory compliance.

Businesses: Businesses may bear the costs associated with providing healthcare benefits to employees affected by air pollution-related illnesses. These costs include medical consultations, diagnostic tests, medications, and treatments for respiratory and cardiovascular conditions. Air pollution-related health issues can lead to decreased productivity among employees, resulting in lower output, efficiency, and profitability for businesses [75]. Productivity losses may arise from absenteeism, presenteeism, and decreased worker performance due to respiratory symptoms, fatigue, and decreased cognitive function. Businesses may incur costs for complying with environmental regulations and pollution control measures aimed at reducing emissions and improving air quality [76, 77]. These costs include investments in pollution control technologies, equipment upgrades, emissions monitoring, and regulatory compliance efforts.

Governments: Governments bear significant healthcare expenditures associated with air pollution-related illnesses, particularly through publicly funded healthcare systems. Governments allocate resources toward providing medical services, treatments, and medications for individuals affected by respiratory and cardiovascular diseases, cancers, and other health problems caused or exacerbated by air pollution [78]. Governments incur costs for addressing environmental damage and pollution-related impacts on ecosystems, water bodies, soil, and biodiversity. These costs include investments in environmental cleanup and remediation efforts, such as soil remediation, water treatment, habitat restoration, and pollution control initiatives. Governments invest resources in enforcing environmental regulations, monitoring air quality, and implementing pollution control measures to reduce emissions and protect public health [79]. These costs include funding for regulatory agencies, inspections, enforcement actions, and legal proceedings against polluting industries and businesses.

Individuals: Individuals bear the costs of seeking medical treatment for air pollution-related health conditions, including out-of-pocket expenses for doctor consultations, diagnostic tests, medications, and hospitalizations. These healthcare expenses can impose a financial burden on individuals and families, particularly those without adequate health insurance coverage. Individuals may experience lost income due to air pollution-related illnesses, absenteeism, and decreased productivity in the workplace [11, 80]. Lost income affects workers' earnings, financial stability, and quality of life, particularly for low-income individuals and families who may rely on wages from hourly or part-time employment [81, 82]. Individuals may incur costs for repairing or replacing property damaged by air pollution-related factors such as acid rain, industrial emissions, and vehicle exhaust. This property damage can include corrosion of buildings and infrastructure, deterioration of vehicles and equipment, and contamination of crops, gardens, and livestock.

4 Economic Benefits of Pollution Control Measures

Just like renewable energy technology, investing in pollution control measures and transitioning to cleaner energy sources can yield significant economic benefits [83, 84]. Pollution control technologies and policies aimed at reducing emissions of harmful pollutants can improve air quality, protect public health, and mitigate the socioeconomic impacts of air pollution. Transitioning to cleaner energy sources, such as renewable energy and energy efficiency measures, can create new job opportunities, stimulate economic growth, and reduce dependence on fossil fuels, which carry their own economic and environmental costs [85]. Pollution control measures offer various economic benefits that extend beyond environmental protection and public health improvement. These benefits include cost savings, job creation, technological innovation, enhanced competitiveness, and others (Fig. 3).

Cost Savings: Pollution control measures can lead to significant cost savings for businesses, governments, and individuals by reducing expenditures related to healthcare, environmental cleanup, and regulatory compliance [86]. By preventing pollution and mitigating its impacts, these measures help avoid the costs associated with treating air pollution-related illnesses, addressing environmental damage, and complying with pollution control regulations.

Improved Public Health: Pollution control measures contribute to improved public health outcomes by reducing exposure to harmful air pollutants and preventing associated health problems such as respiratory diseases, cardiovascular diseases, and cancers. Better public health leads to lower healthcare expenditures, increased productivity, and improved quality of life for individuals and communities, resulting in long-term economic benefits [87].

Job Creation and Economic Growth: According to the reports of Yang et al. [88] on China and globally from the findings of Peng et al. [89], investments in

Fig. 3 Some economic benefits of pollution control

pollution control measures stimulate economic activity and create job opportunities in various sectors, including environmental engineering, pollution monitoring, renewable energy, and sustainable infrastructure development. By promoting green technologies and clean energy sources, pollution control measures support the growth of green industries and contribute to job creation, economic diversification, and sustainable development.

Technological Innovation: Pollution control measures drive technological innovation and research and development (R&D) in environmental science, engineering, and clean technologies. Investments in pollution control technologies, emission reduction strategies, and sustainable practices spur innovation, foster technological advancements, and support the growth of the green economy [90, 91]. Innovations in pollution control contribute to the development of new products, services, and solutions that address environmental challenges while generating economic value.

Enhanced Competitiveness: Businesses that adopt pollution control measures and embrace sustainable practices gain a competitive edge in the marketplace [92]. By reducing environmental risks, improving resource efficiency, and enhancing corporate social responsibility (CSR), these businesses attract environmentally conscious consumers, investors, and stakeholders [93, 94]. Moreover, compliance with pollution control regulations and adherence to high environmental standards enhance the reputation and credibility of businesses, leading to increased market share, brand loyalty, and long-term profitability.

Environmental Protection and Resource Conservation: Pollution control measures promote environmental protection and resource conservation by reducing emissions of harmful pollutants, minimizing waste generation, and conserving natural resources [6, 31]. By preventing pollution and minimizing environmental degradation, these measures contribute to the preservation of ecosystems, biodiversity, and natural habitats, which are essential for supporting ecosystem services, such as clean air, water, and soil, that underpin economic activities and human well-being.

5 Social Impacts of Air Pollution

The social impacts of air pollution encompass a wide range of effects on communities, individuals, and society as a whole and manifest in various ways with significant implications for public health, well-being, social cohesion, and quality of life (Table 1). For instance, air pollution poses serious health risks to individuals, including respiratory and cardiovascular diseases, cancers, and other adverse health outcomes. Exposure to pollutants such as particulate matter (PM), nitrogen dioxide (NO2), sulfur dioxide (SO2), ozone (O3), and volatile organic compounds (VOCs) can lead to increased rates of respiratory infections, asthma exacerbations, heart attacks, strokes, and premature mortality [5, 70]. The health impacts of air pollution disproportionately affect vulnerable populations, including children, the elderly, pregnant women, and individuals with pre-existing health conditions, exacerbating health disparities and inequalities [115].

Air pollution diminishes the quality of life for affected individuals and communities by reducing outdoor recreational opportunities, limiting access to green spaces, and impairing the enjoyment of natural environments. Poor air quality can deter people from engaging in outdoor activities such as walking, jogging, cycling, and picnicking, leading to social isolation and reduced well-being [116]. Residents of highly polluted areas may experience discomfort, stress, and anxiety due to concerns about air pollution-related health risks, environmental degradation, and the impact on their families' health and safety [117, 118]. Air pollution imposes economic burdens on individuals, families, businesses, and governments through increased healthcare costs, lost productivity, and decreased property values [119]. Healthcare expenditures related to air pollution-related illnesses strain

Table 1 Social consequences of air pollution

Social impacts	Examples	References
Health effects (includes increased cost and health burdens)	Increased respiratory diseases such as asthma, bronchitis, and lung cancer, cardiovascular problems, premature deaths, and reduced life expectancy	[95, 96]
Economic impacts	Increased healthcare costs, decreased worker productivity, damage to crops and livestock, and loss of tourism revenue due to poor air quality	[97, 98]
Social inequalities	Greater impacts on low-income communities living in areas with higher pollution levels through disproportionate exposure to air pollution among marginalized populations	[99, 100]
Low social cohesion	Increased social tensions due to health disparities, community activism for environmental justice, impacts on cultural practices and traditions due to environmental degradation	[101, 102]
Educational impacts	Impaired cognitive development in children, increased absenteeism, reduced academic performance, and limited access to outdoor learning environments	[103, 104]
Mental health effects	Increased stress and anxiety due to air pollution-related health concerns and higher rates of depression in areas with poor air quality	[105, 106]
Higher rates of social disruptions	Displacement of communities due to environmental degradation and increased protests and conflicts over pollution-related issues	[107, 108]
Environmental injustice	Advocacy for equitable distribution of clean air resources and legal battles for pollution regulation and enforcement	[109, 110]
Community health advocacy	Grassroot movements for cleaner air and the formation of community organizations to address pollution issues	[111, 112]
Access to resources	Unequal access to information and resources for mitigating air pollution impacts and limited availability of healthcare services in affected communities	[113, 114]

healthcare systems and household budgets, leading to financial hardship and economic instability. Businesses may incur costs related to absenteeism, reduced productivity, and healthcare expenses for employees affected by air pollution-related health problems. Reduced consumer spending as well as decreased tourism revenues and increased insurance costs also contribute to economic losses in polluted areas (Table 1) [120]. Air pollution can negatively impact community well-being by eroding social cohesion, disrupting community networks, and undermining trust in institutions and authorities responsible for environmental protection and public health (Table 1). The works of Rimmler et al. [121] and van Horne et al. [122] suggest that concerns about air pollution-related health risks, environmental degradation, and the perceived lack of government action may lead to community distrust, activism, and advocacy for environmental justice. Community-based organizations, grassroots movements, and citizen science initiatives play a crucial role in raising awareness about air pollution issues, mobilizing communities, and advocating for policies and actions to address environmental concerns and protect public health.

5.1 Social Inequalities and Environmental Justice Cost of Air Pollutions

Social inequalities and environmental justice issues related to air pollution highlight disparities in exposure and health outcomes among different socioeconomic and racial/ethnic groups. Air pollution exacerbates social inequalities and environmental injustices by disproportionately affecting disadvantaged communities, low-income populations, minority groups, and marginalized individuals who bear a disproportionate burden of pollution-related health risks and environmental hazards [38, 21]. Environmental racism, socioeconomic disparities, and unequal access to clean air and healthy environments perpetuate social inequalities and contribute to disparities in health outcomes, life expectancy, and quality of life. Communities located near sources of pollution, such as highways, industrial facilities, and waste disposal sites, are often exposed to higher levels of air pollution and face greater health risks than more affluent areas with better environmental quality [123]. This disparity in exposure underscores the need for equitable policies and interventions to address environmental injustices and promote social equity. These disparities reflect systemic injustices in the distribution of environmental hazards and access to clean air, water, and other environmental resources.

Disparities in air pollution exposure refer to the unequal distribution of air pollution burdens among different socioeconomic, racial/ethnic, and geographic groups. These disparities are often driven by systemic factors such as environmental racism, socioeconomic status, urbanization patterns, and regulatory policies [124]. Low-income communities, communities of color, indigenous populations, and marginalized groups often bear a disproportionate burden of air pollution exposure due to the siting of polluting industries, highways, waste facilities, and other sources of pollution in their neighborhoods. Environmental racism and discrimination contribute to disparities in exposure to air pollution, as communities with higher proportions of racial/ethnic minorities and low-income residents are more likely to be located near sources of pollution and experience higher levels of pollution-related health risks. Factors such as residential segregation, historical disinvestment, and lack of political power exacerbate environmental injustices, perpetuating unequal exposure to air pollution and associated health disparities [125]. Low-income communities and marginalized populations often face higher levels of air pollution exposure compared to wealthier communities. Industries and transportation infrastructure, including highways, ports, and industrial facilities, are frequently located in or near low-income neighborhoods, exposing residents to emissions from vehicles, factories, and other sources of pollution. Lack of access to affordable housing and limited mobility options can also contribute to higher exposure among low-income populations, as individuals may be forced to live in areas with poor air quality due to financial constraints or limited housing choices.

Communities of color, including African American, Hispanic/Latinx, and Indigenous communities, are disproportionately impacted by air pollution due to historical patterns of environmental injustice and racial discrimination [126, 127]. Environmental racism plays a significant role in siting polluting facilities and toxic waste sites in minority neighborhoods, resulting in higher levels of pollution exposure and associated health risks. Environmental justice studies have consistently found that racial/ethnic minority communities bear a disproportionate burden of air pollution compared to predominantly white communities, highlighting systemic inequalities in environmental protection and public health. Urban areas, particularly densely populated cities and industrial zones, tend to experience higher levels of air pollution due to traffic congestion, industrial emissions, and other sources of pollution [128, 6]. Residents of urban neighborhoods, especially those located near highways, ports, and industrial facilities, face elevated risks of air pollution exposure and related health problems [123]. Rural communities may also be affected by air pollution, albeit to a lesser extent, due to agricultural activities, biomass burning, and transportation emissions. Rural populations, including farmworkers and residents of agricultural regions, may experience air pollution exposure from pesticides, dust, and agricultural burning practices.

Disparities in air pollution exposure often intersect with other environmental and social stressors, leading to cumulative impacts on vulnerable communities [129, 130]. For example, low-income communities of color may face multiple sources of pollution, as well as other environmental hazards such as lead contamination, hazardous waste sites, and a lack of access to green spaces and healthy food options. Cumulative exposure to air pollution, along with other environmental and social stressors, can exacerbate health disparities and contribute to higher rates of respiratory diseases, cardiovascular diseases, cancers, and other adverse health outcomes among disadvantaged populations [131, 132].

5.1.1 Health Impacts

Disadvantaged communities facing higher levels of air pollution exposure experience adverse health outcomes, including respiratory diseases, cardiovascular diseases, cancers, and other health problems [70]. Children, the elderly, pregnant women, and individuals with pre-existing health conditions are particularly vulnerable to the health effects of air pollution, exacerbating existing health disparities and inequities. Environmental injustice often results from a lack of access to quality healthcare services, health insurance coverage, and preventive care, further exacerbating health disparities and increasing the social and economic costs of air pollution-related illnesses [133].

Air pollution is a significant risk factor for respiratory diseases such as asthma, chronic obstructive pulmonary disease (COPD), bronchitis, and pneumonia. Inhalation of pollutants such as PM, NO2, SO2, O3, and VOCs can irritate the respiratory tract, exacerbate existing respiratory conditions, and impair lung function [14, 15]. Individuals exposed to high levels of air pollution may experience symptoms such as coughing, wheezing, shortness of breath, chest tightness, and respiratory infections. Long-term exposure to air pollution can lead to chronic respiratory conditions, decreased lung function, atherosclerosis, arterial stiffness, and increased risk of cardiovascular events, mortality, and increased susceptibility to respiratory infections [134, 135]. Air pollution is associated with an increased risk of cardiovascular diseases, including heart attacks, strokes, hypertension, arrhythmias, and heart failure. Fine particulate matter (PM2.5), in particular, can enter the bloodstream through the lungs, triggering systemic inflammation, oxidative stress, and endothelial dysfunction, which contribute to the development and progression of cardiovascular diseases. Individuals exposed to high levels of air pollution may experience symptoms such as chest pain, palpitations, fatigue, and dizziness. Certain air pollutants, such as benzene, formaldehyde, polycyclic aromatic hydrocarbons (PAHs), and diesel exhaust particles, are classified as carcinogens by international health agencies [136]. Long-term exposure to these carcinogens in air pollution is associated with an increased risk of lung cancer, bladder cancer, leukemia, and other cancers. Inhalation of carcinogenic air pollutants can lead to DNA damage, genetic mutations, and alterations in cell signaling pathways, which contribute to the initiation and progression of cancerous tumors. Prenatal and childhood exposure to air pollution has been linked to adverse neurodevelopmental outcomes, including cognitive impairment, developmental delays, behavioral problems, and neurodegenerative diseases such as Alzheimer's disease and Parkinson's disease [134, 137]. Air pollutants such as PM2.5, NO2, and PAHs can cross the blood-brain barrier, leading to neuroinflammation, oxidative stress, and neuronal damage, which affect cognitive function and behavior. Air pollution exposure during pregnancy is associated with adverse pregnancy outcomes, including preterm birth, low birth weight, and congenital anomalies. Pregnant women exposed to high levels of air pollution may experience complications such as preeclampsia, gestational diabetes, and placental dysfunction, which increase the risk of adverse birth outcomes [138]. Air pollutants such as PM, NO2, SO2, and CO can cross the placenta and affect fetal development, leading to growth restriction, organ damage, and developmental abnormalities.

Environmental justice advocates and policymakers work to address social inequalities and promote environmental justice through policy interventions such as environmental regulations, land use planning, zoning ordinances, and environmental impact assessments. Community-based organizations, grassroots activists, and environmental justice advocates engage in community organizing, advocacy campaigns, and public education efforts to raise awareness about environmental injustices, mobilize affected communities, and advocate for policy changes and solutions. Environmental justice organizations and legal advocacy groups pursue legal avenues, litigation, and civil rights complaints to challenge environmental injustices, hold polluters accountable, and seek justice for affected communities. Collaborative partnerships between government agencies, community organizations, academic institutions, and industry stakeholders facilitate dialogue, cooperation, and collective action to address environmental injustices, promote equitable development, and empower marginalized communities.

5.2 Impacts of Air Quality on Tourism and Recreation

Tourism and recreation are integral components of many economies, contributing to employment, revenue generation, and cultural exchange. However, air pollution can have detrimental effects on tourism and recreational activities, impacting air quality, scenic views, and visitor experiences [139]. Air pollution can degrade air quality in tourist destinations, affecting the health and well-being of visitors and residents alike. High levels of pollutants such as PM, NO2, SO2, and O3 can lead to respiratory problems, eye irritation, and discomfort among tourists. Poor air quality diminishes the attractiveness of tourist destinations, especially those known for their natural beauty and outdoor recreational opportunities. Visitors may be deterred from engaging in outdoor activities such as hiking, sightseeing, and wildlife viewing due to concerns about air pollution-related health risks [140].

Air pollution can obscure scenic views and diminish the aesthetic appeal of tourist attractions, national parks, and natural landmarks. Visibility impairment caused by haze, smog, and airborne particles reduces the visual quality of landscapes and undermines the overall visitor experience. Tourists often seek out destinations with pristine natural environments and unspoiled vistas. However, air pollution detracts from the scenic beauty of these destinations, leading to disappointment and dissatisfaction among visitors. According to the findings of Zhang et al. [49, 50] and Bogalecka and Grobelna [25], tourist destinations affected by air pollution may experience economic losses due to decreased visitor numbers, reduced tourism revenues, and negative publicity. Poor air quality and environmental degradation deter tourists from visiting affected destinations, resulting in lost income for businesses, hotels, restaurants, and recreational operators. Tourist destinations are often impacted by air pollution and pose economic challenges to local communities and businesses. For example, cities such as Beijing, Delhi, and Los Angeles have experienced declines in tourism due to concerns about air pollution, resulting in decreased spending by visitors and adverse effects on the hospitality industry [141, 142]. Air pollution poses significant challenges for tourism and recreational activities, affecting air quality, scenic views, and visitor experiences in destinations around the world [143].

6 Sustainable Policy and Governance Strategies for Mitigating the Socioeconomic Impacts of Air Pollution

Current air pollution policies and regulations vary significantly across countries and regions, reflecting differences in environmental priorities, regulatory frameworks, and institutional capacities [144, 145]. However, many jurisdictions have implemented a combination of regulatory measures, incentive-based approaches, and collaborative initiatives to address air pollution and protect public health, which includes (Fig. 4):

Emission Standards: Governments set emission standards and limits for various sources of air pollution, including vehicles, industrial facilities, power plants, and residential heating systems. These standards specify the maximum allowable levels of pollutants such as PM, NOx, SO2, VOCs, and GHGs [146]. Emission standards are typically enforced through regulations, permits, inspections, and compliance mechanisms, with penalties for non-compliance and incentives for meeting or exceeding standards [147].

Vehicle Regulations: Many countries have implemented stringent regulations to control vehicle emissions and promote cleaner transportation technologies. These regulations include emission standards for new vehicles, mandatory vehicle inspections and emissions testing programs, fuel quality standards, and incentives for electric and low-emission vehicles [148, 147]. Governments may also implement measures to reduce traffic congestion, promote public transit, encourage active transportation modes such as walking and cycling, and incentivize the use of clean fuels and alternative transportation options.

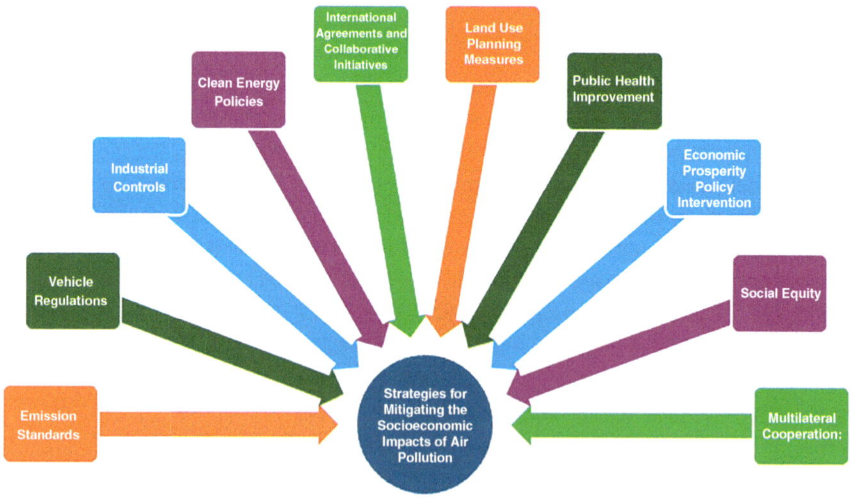

Fig. 4 Sustainable strategies for mitigating the socioeconomic impacts of air pollution

Industrial Controls: Industrial facilities are subject to regulations and permitting requirements aimed at controlling emissions of air pollutants and minimizing environmental impacts. These regulations may include emission limits, pollution control technologies, best available techniques (BAT), and pollution prevention measures [149]. These BATs are recognized as the basis for emission limit values. Governments conduct inspections, audits, and enforcement actions to ensure compliance with industrial emission standards and permit conditions, with penalties for violations and incentives for pollution prevention and environmental stewardship.

Clean Energy Policies: Many countries are adopting clean energy policies and transitioning toward renewable energy sources such as solar, wind, hydroelectric, and geothermal power to reduce air pollution emissions and combat climate change [150–152]. These policies include renewable energy targets, feed-in tariffs, tax incentives, subsidies, and carbon pricing mechanisms. Governments may also implement energy efficiency measures, building codes, and standards to reduce energy consumption, lower emissions, and improve air quality in residential, commercial, and industrial sectors. Assessing the effectiveness of policy interventions for mitigating the socioeconomic impacts of air pollution involves evaluating the outcomes and impacts of various regulatory measures, incentive-based approaches, and collaborative initiatives implemented by governments and stakeholders.

International Agreements and Collaborative Initiatives: Countries collaborate through international agreements, treaties, and protocols to address transboundary air pollution and global environmental challenges. Key agreements include the United Nations Framework Convention on Climate Change (UNFCCC), the Kyoto Protocol, the Paris Agreement, and regional air quality agreements. Governments participate in collaborative initiatives, knowledge-sharing platforms, and capacity-building programs to exchange best practices, share data and expertise, and support global efforts to improve air quality and protect public health.

Land Use Planning Measures: Land use planning measures involve zoning regulations, urban planning strategies, and development policies designed to minimize air pollution emissions and mitigate environmental impacts [153]. These measures include smart growth principles, mixed-use development, transit-oriented design, and green infrastructure. Land use planning measures aim to reduce vehicle miles travelled, promote compact and walkable communities, and limit urban sprawl, which contributes to air pollution and traffic congestion [154]. By integrating land use and transportation planning, governments can create more sustainable and livable cities with cleaner air and healthier environments.

Public Health Improvement: Policy interventions aimed at reducing air pollution emissions have been effective in improving public health outcomes by reducing exposure to harmful pollutants and mitigating the incidence of air pollution-related diseases [8]. Studies have shown that stricter air quality standards, emissions controls, and pollution abatement measures lead to reductions in respiratory and cardiovascular diseases, premature mortality, and healthcare expenditures [155–157]. Effective policy interventions should focus on setting and enforcing emissions standards, implementing vehicle regulations, controlling industrial emissions, and promoting clean energy technologies. Reducing ambient air pollution levels and limiting public exposure to pollutants will ultimately contribute to improved public health and well-being.

Economic Prosperity Policy Intervention: Interventions that aim to address air pollution should have positive economic impacts by reducing healthcare costs, improving productivity, and stimulating economic growth. Investments in pollution control measures, clean technologies, and sustainable infrastructure create jobs, spur innovation, and enhance competitiveness in green industries [49–51, 63]. Studies have shown that the economic benefits of pollution control outweigh the costs, with every dollar invested in air quality improvements yielding substantial returns in terms of healthcare savings, increased productivity, and environmental quality [11, 57]. By promoting cleaner air and healthier environments, policy interventions support economic prosperity and sustainable development.

Social Equity: Policy interventions play a crucial role in promoting social equity and reducing disparities in air pollution exposure and health outcomes among different socioeconomic and demographic groups. Measures such as emissions controls, vehicle regulations, and land use planning measures aim to minimize environmental injustices and ensure equitable access to clean air and healthy environments [158]. Environmental justice considerations are increasingly integrated into air quality management and policy-making processes, with efforts to address disproportionate burdens of pollution on vulnerable communities and promote inclusive decision-making. Collaborative initiatives, community engagement, and advocacy efforts contribute to advancing environmental justice goals and promoting social equity in air quality management.

Multilateral Cooperation: Multilateral cooperation and coordinated action are essential for addressing transboundary air pollution and global environmental challenges. International agreements, treaties, and protocols facilitate collaboration among countries to reduce emissions, mitigate climate change, and protect public health. Initiatives such as the United Nations Framework Convention on Climate Change (UNFCCC), the Paris Agreement, and regional air quality agreements promote information exchange, capacity-building, and joint efforts to address air pollution at the global, regional, and local levels. By working together, governments can achieve greater progress in addressing air pollution and its socioeconomic impacts.

7 Conclusion

The chapter has provided a comprehensive examination of the socioeconomic impacts of air pollution, highlighting the interconnectedness of environmental degradation, public health, economic prosperity, and social equity. From the public health burden imposed by air pollution-related diseases to the economic disruptions and social inequalities exacerbated by pollution, it is evident that air pollution poses significant challenges that require urgent attention and concerted action. The evidence presented underscores the critical importance of addressing air pollution not only as an environmental issue but also as a pressing public health and socioeconomic concern. Despite the daunting challenges posed by air pollution, there is cause for optimism. Through innovative policies, technological advancements, and community-driven initiatives, progress is being made in reducing emissions, protecting public health, and promoting sustainable development. However, much work remains to be done. The evidence is clear: reducing air pollution is essential for safeguarding public health, fostering economic prosperity, and promoting social justice.

Implementing pollution control measures and enacting stricter environmental regulations can help improve air quality and mitigate the negative effects of air pollution on tourism. Investments in clean technologies, emission reduction initiatives, and sustainable transportation solutions contribute to cleaner air and healthier environments in tourist destinations. Promoting sustainable tourism practices, such as eco-friendly transportation, waste management, and energy conservation, can reduce the environmental footprint of tourism activities and minimize air pollution impacts. Sustainable tourism certifications, eco-labeling programs, and responsible travel initiatives encourage tourists to support destinations committed to environmental stewardship. Raising public awareness about the link between air pollution and tourism can foster a greater appreciation for environmental conservation and encourage visitors to adopt environmentally friendly behaviors. Educational campaigns, interpretive signage, and guided tours highlighting local air quality issues and conservation efforts can empower tourists to make informed choices and minimize their environmental impact. By implementing pollution control measures, promoting sustainable tourism practices, and raising public awareness, stakeholders can work together to mitigate the negative effects of air pollution on tourism and ensure the long-term sustainability of tourist destinations for future generations.

Addressing air pollution requires collective action and collaboration across sectors and stakeholders. It requires innovative policies, investments in clean technologies, and community engagement to drive meaningful change. It also demands a commitment to environmental justice, ensuring that the benefits of clean air are equitably distributed among all communities, especially those most affected by pollution. It is crucial to continue advancing research, policy innovation, and community engagement efforts to tackle the complex challenges posed by air pollution. At the institutional level, there is a need to prioritize the development and implementation of evidence-based strategies to reduce emissions, protect vulnerable populations, and build resilient, sustainable societies. Moving forward, it is essential to prioritize collaborative approaches that engage stakeholders across sectors and foster meaningful dialogue and cooperation. There is a need to leverage collective expertise and resources, and commitment can develop and implement effective strategies to mitigate the socioeconomic impacts of air pollution and build healthier, more resilient communities. There is also a need to translate knowledge into action, mobilize resources, and empower individuals and communities to address the complex challenges posed by air pollution. Together, we can forge a path toward a cleaner, healthier, and more equitable future for all.

References

1. Gómez CA, Kleinman DV, Pronk N, Wrenn Gordon GL, Ochiai E, Blakey C, Johnson A, Brewer KH (2021) Addressing health equity and social determinants of health through healthy people 2030. J Public Health Manag Pract 27(Suppl 6):S249–S257. https://doi.org/10.1097/PHH.0000000000001297
2. Wolf MJ, Esty DC, Kim H, Bell ML, Brigham S, Nortonsmith Q, Zaharieva S, Wendling ZA, de Sherbinin A, Emerson JW (2022) New insights for tracking global and local trends in exposure to air pollutants. Environ Sci Technol 56(7):3984–3996. https://doi.org/10.1021/acs.est.1c08080
3. Bikis A (2023) Urban air pollution and greenness in relation to public health. J Environ Public Health 2023:8516622. https://doi.org/10.1155/2023/8516622
4. Fang Z, Wu PY, Lin YN, Chang TH, Chiu YH (2021) Air pollution's impact on the economic, social, medical, and industrial injury environments in China. Healthcare (Basel, Switzerland) 9(3):261. https://doi.org/10.3390/healthcare9030261
5. Bălă GP, Râjnoveanu RM, Tudorache E, Motișan R, Oancea C (2021) Air pollution exposure-the (in)visible risk factor for respiratory diseases. Environ Sci Pollut Res Int 28(16): 19615–19628. https://doi.org/10.1007/s11356-021-13208-x
6. Ogwu MC, Ogwu HI, Kosoe EA (2023) Plants used in the management and treatment of cardiovascular diseases: case study of the Benin people of southern Nigeria. In: Izah SC, Ogwu MC, Akram M (eds) Herbal medicine phytochemistry. Reference series in Phytochemistry. Springer, Cham. https://doi.org/10.1007/978-3-031-21973-3_4-1
7. Rentschler J, Leonova N (2023) Global air pollution exposure and poverty. Nat Commun 14(1):4432. https://doi.org/10.1038/s41467-023-39797-4
8. Giles LV, Barn P, Künzli N, Romieu I, Mittleman MA, van Eeden S, Allen R, Carlsten C, Stieb D, Noonan C, Smargiassi A, Kaufman JD, Hajat S, Kosatsky T, Brauer M (2011) From good intentions to proven interventions: effectiveness of actions to reduce the health impacts of air pollution. Environ Health Perspect 119(1):29–36. https://doi.org/10.1289/ehp.1002246
9. Hadley MB, Baumgartner J, Vedanthan R (2018) Developing a clinical approach to air pollution and cardiovascular health. Circulation 137(7):725–742. https://doi.org/10.1161/CIRCULATIONAHA.117.030377
10. Rives R, Elshorbany Y, Kaylor S (2023) The relationship between air quality, health outcomes, and socioeconomic impacts of the COVID-19 pandemic in the US. GeoHealth 7(5): e2022GH000735. https://doi.org/10.1029/2022GH000735
11. Li L, Du T, Zhang C (2020) The impact of air pollution on healthcare expenditure for respiratory diseases: evidence from the People's Republic of China. Risk Manag Healthc Policy 13:1723–1738. https://doi.org/10.2147/RMHP.S270587
12. Liu H, Hu T, Wang M (2021) Impact of air pollution on residents' medical expenses: a study based on the survey data of 122 cities in China. Front Public Health 9:743087. https://doi.org/10.3389/fpubh.2021.743087
13. Rice DP, Hodgson TA, Kopstein AN (1985) The economic costs of illness: a replication and update. Health Care Financ Rev 7(1):61–80
14. Lee DW, Lee J, Kim HR, Kang MY (2021) Health-related productivity loss according to health conditions among workers in South Korea. Int J Environ Res Public Health 18(14): 7589. https://doi.org/10.3390/ijerph18147589

15. Lee YG, Lee PH, Choi SM, An MH, Jang AS (2021) Effects of air pollutants on airway diseases. Int J Environ Res Public Health 18(18):9905. https://doi.org/10.3390/ijerph18189905
16. Anwar A, Hyder S, Bennett R, Younis M (2022) Impact of environmental quality on healthcare expenditures in developing countries: a panel data approach. Healthcare (Basel, Switzerland) 10(9):1608. https://doi.org/10.3390/healthcare10091608
17. Han A, Lee KH, Park J (2022) The impact of price transparency and competition on hospital costs: a research on all-payer claims databases. BMC Health Serv Res 22(1):1321. https://doi.org/10.1186/s12913-022-08711-x
18. Wang L, Wang H, Dong Z, Wang S, Cao Z (2021) The air pollution effect of government economic growth expectations: evidence from China's cities based on green technology. Environ Sci Pollut Res Int 28(22):27639–27654. https://doi.org/10.1007/s11356-021-12536-2
19. Wang M, Liao G, Li Y (2021) The relationship between environmental regulation, pollution and corporate environmental responsibility. Int J Environ Res Public Health 18(15):8018. https://doi.org/10.3390/ijerph18158018
20. Shen Y, Lyu M, Zhu J (2022) Air pollution and corporate green financial constraints: evidence from China's listed companies. Int J Environ Res Public Health 19(22):15034. https://doi.org/10.3390/ijerph192215034
21. Cheeseman MJ, Ford B, Anenberg SC, Cooper MJ, Fischer EV, Hammer MS, Magzamen S, Martin RV, van Donkelaar A, Volckens J, Pierce JR (2022) Disparities in air pollutants across racial, ethnic, and poverty groups at US public schools. GeoHealth 6(12):e2022GH000672. https://doi.org/10.1029/2022GH000672. PMID: 36467256; PMCID: PMC9714311
22. Goforth T, Nock D (2022) Air pollution disparities and equality assessments of US national decarbonization strategies. Nat Commun 13(1):7488. https://doi.org/10.1038/s41467-022-35098-4
23. Gochfeld M, Burger J (2011) Disproportionate exposures in environmental justice and other populations: the importance of outliers. American Journal of Public Health 101:S53_S63, https://doi.org/10.2105/AJPH.2011.300121
24. Bikomeye JC, Namin S, Anyanwu C, Rublee CS, Ferschinger J, Leinbach K, Lindquist P, Hoppe A, Hoffman L, Hegarty J, Sperber D, Beyer KMM (2021) Resilience and equity in a time of crises: investing in public urban greenspace is now more essential than ever in the US and beyond. Int J Environ Res Public Health 18(16):8420. https://doi.org/10.3390/ijerph18168420
25. Bogalecka M, Grobelna A (2023) Air pollution and its potential consequences for tourism and career development from students' perspective: a case study of the Gdańsk agglomeration in Poland. Int J Environ Res Public Health 20(3):2651. https://doi.org/10.3390/ijerph20032651
26. de Bont J, Jaganathan S, Dahlquist M, Persson Å, Stafoggia M, Ljungman P (2022) Ambient air pollution and cardiovascular diseases: An umbrella review of systematic reviews and meta-analyses. J Intern Med 291(6):779–800. https://doi.org/10.1111/joim.13467
27. Guo J, Chai G, Song X, Hui X, Li Z, Feng X, Yang K (2023) Long-term exposure to particulate matter on cardiovascular and respiratory diseases in low- and middle-income countries: a systematic review and meta-analysis. Front Public Health 11:1134341. https://doi.org/10.3389/fpubh.2023.1134341
28. Ko UW, Kyung SY (2022) Adverse effects of air pollution on pulmonary diseases. Tuberc Respir Dis 85(4):313–319. https://doi.org/10.4046/trd.2022.0116
29. Lee BJ, Kim B, Lee K (2014) Air pollution exposure and cardiovascular disease. Toxicol Res 30(2):71–75. https://doi.org/10.5487/TR.2014.30.2.071
30. Morris RD (2001) Airborne particulates and hospital admissions for cardiovascular disease: a quantitative review of the evidence. Environ Health Perspect 109(Suppl 4):495–500. https://doi.org/10.1289/ehp.01109s4495
31. Osawaru ME, Ogwu MC (2023) Plants used in the management and treatment of female reproductive health issues: case study from southern Nigeria. In: Izah SC, Ogwu MC, Akram

M (eds) Herbal medicine Phytochemistry. Reference series in Phytochemistry. Springer, Cham. https://doi.org/10.1007/978-3-031-21973-3_5-1
32. Turner MC, Andersen ZJ, Baccarelli A, Diver WR, Gapstur SM, Pope 3rd CA, Prada D, Samet J, Thurston G, Cohen A (2020) Outdoor air pollution and cancer: An overview of the current evidence and public health recommendations. CA Cancer J Clin. https://doi.org/10.3322/caac.21632. Advanced online publication
33. Volk HE, Perera F, Braun JM, Kingsley SL, Gray K, Buckley J, Clougherty JE, Croen LA, Eskenazi B, Herting M, Just AC, Kloog I, Margolis A, McClure LA, Miller R, Levine S, Wright R, Environmental influences on Child Health Outcomes (2021) Prenatal air pollution exposure and neurodevelopment: a review and blueprint for a harmonized approach within ECHO. Environ Res 196:110320. https://doi.org/10.1016/j.envres.2020.110320
34. Rani P, Dhok A (2023) Effects of pollution on pregnancy and infants. Cureus 15(1):e33906. https://doi.org/10.7759/cureus.33906
35. Sarizadeh R, Dastoorpoor M, Goudarzi G, Simbar M (2020) The association between air pollution and low birth weight and preterm labor in Ahvaz, Iran. Int J Women's Health 12: 313–325. https://doi.org/10.2147/IJWH.S227049
36. Aithal SS, Sachdeva I, Kurmi OP (2023) Air quality and respiratory health in children. Breathe (Sheff) 19(2):230040. https://doi.org/10.1183/20734735.0040-2023
37. Andrade A, D'Oliveira A, De Souza LC, Bastos ACRF, Dominski FH, Stabile L, Buonanno G (2023) Effects of air pollution on the health of older adults during physical activities: mapping review. Int J Environ Res Public Health 20(4):3506. https://doi.org/10.3390/ijerph20043506
38. Hajat A, Hsia C, O'Neill MS (2015) Socioeconomic disparities and air pollution exposure: a global review. Curr Environ Health Rep 2(4):440–450. https://doi.org/10.1007/s40572-015-0069-5
39. Ab Manan N, Noor Aizuddin A, Hod R (2018) Effect of air pollution and hospital admission: a systematic review. Ann Glob Health 84(4):670–678. https://doi.org/10.9204/aogh.2376
40. Slama A, Śliwczyński A, Woźnica J, Zdrolik M, Wiśnicki B, Kubajek J, Turżańska-Wieczorek O, Gozdowski D, Wierzba W, Franek E (2019) Impact of air pollution on hospital admissions with a focus on respiratory diseases: a time-series multi-city analysis. Environ Sci Pollut Res Int 26(17):16998–17009. https://doi.org/10.1007/s11356-019-04781-3
41. Le D, Li Y, Ren F (2022) Does air quality improvement promote enterprise productivity increase? Based on the spatial spillover effect of 242 cities in China. Front Public Health 10: 1050971. https://doi.org/10.3389/fpubh.2022.1050971
42. Pervin T, Gerdtham UG, Lyttkens CH (2008) Societal costs of air pollution-related health hazards: a review of methods and results. Cost Eff Resour Alloc C/E 6:19. https://doi.org/10.1186/1478-7547-6-19
43. Czechowski PO, Piksa K, Da Browiecki P, Oniszczuk-Jastrząbek AI, Czermański E, Owczarek T, Badyda AJ, Cirella GT (2022) Financing costs and health effects of air pollution in the tri-city agglomeration. Front Public Health 10:831312. https://doi.org/10.3389/fpubh.2022.831312
44. He Q, Ji XJ (2021) The labor productivity consequences of exposure to particulate matters: evidence from a Chinese National Panel Survey. Int J Environ Res Public Health 18(23): 12859. https://doi.org/10.3390/ijerph182312859
45. Zivin JG, Neidell M (2012) The impact of pollution on worker productivity. Am Econ Rev 102(7):3652–3673. https://doi.org/10.1257/aer.102.7.3652
46. Ferguson L, Taylor J, Davies M, Shrubsole C, Symonds P, Dimitroulopoulou S (2020) Exposure to indoor air pollution across socio-economic groups in high-income countries: a scoping review of the literature and a modelling methodology. Environ Int 143:105748. https://doi.org/10.1016/j.envint.2020.105748
47. Berberian AG, Gonzalez DJX, Cushing LJ (2022) Racial disparities in climate change-related health effects in the United States. Curr Environ Health Rep 9(3):451–464. https://doi.org/10.1007/s40572-022-00360-w

48. Liu N, Liu Y, Yu X (2023) The impact of public environmental concern on environmental pollution: the moderating effect of government environmental regulation. PLoS One 18(8): e0290255. https://doi.org/10.1371/journal.pone.0290255
49. Zhang P, Chen P, Xiao F, Sun Y, Ma S, Zhao Z (2022) The impact of information infrastructure on air pollution: empirical evidence from China. Int J Environ Res Public Health 19(21): 14351. https://doi.org/10.3390/ijerph192114351
50. Zhang X, Ma C, Li X, Xiong L, Nie S (2022) Assessing the impact of air pollution on inbound tourism along the Yangtze River across space and time. Int J Environ Res Public Health 19(17):10944. https://doi.org/10.3390/ijerph191710944
51. Zhang X, Xia Q, Lai Y, Wu B, Tian W, Miao W, Feng X, Xin L, Miao J, Wang N, Wu Q, Jiao M, Shan L, Du J, Li Y, Shi B (2022) Spatial effects of air pollution on the economic burden of disease: implications of health and environment crisis in a post-COVID-19 world. Int J Equity Health 21(1):161. https://doi.org/10.1186/s12939-022-01774-6
52. Taj T, Malmqvist E, Stroh E, Oudin Åström D, Jakobsson K, Oudin A (2017) Short-term associations between air pollution concentrations and respiratory health-comparing primary health care visits, hospital admissions, and emergency department visits in a multi-municipality study. Int J Environ Res Public Health 14(6):587. https://doi.org/10.3390/ijerph14060587
53. Jiang XQ, Mei XD, Feng D (2016) Air pollution and chronic airway diseases: what should people know and do? J Thorac Dis 8(1):E31–E40. https://doi.org/10.3978/j.issn.2072-1439.2015.11.50
54. Loaiza-Ceballos MC, Marin-Palma D, Zapata W, Hernandez JC (2022) Viral respiratory infections and air pollutants. Air Qual Atmos Health 15(1):105–114. https://doi.org/10.1007/s11869-021-01088-6
55. Raju S, Siddharthan T, McCormack MC (2020) Indoor air pollution and respiratory health. Clin Chest Med 41(4):825–843. https://doi.org/10.1016/j.ccm.2020.08.014
56. Eckhardt CM, Wu H (2021) Environmental exposures and lung aging: molecular mechanisms and implications for improving respiratory health. Curr Environ Health Rep 8(4):281–293. https://doi.org/10.1007/s40572-021-00328-2
57. Li D, Xiao H, Ma S, Zhang J (2022) Health benefits of air quality improvement: empirical research based on medical insurance reimbursement data. Front Public Health 10:855457. https://doi.org/10.3389/fpubh.2022.855457
58. Surit P, Wongtanasarasin W, Boonnag C, Wittayachamnankul B (2023) Association between air quality index and effects on emergency department visits for acute respiratory and cardiovascular diseases. PLoS One 18(11):e0294107. https://doi.org/10.1371/journal.pone.0294107
59. Shahi AM, Omraninava A, Goli M, Soheilarezoomand HR, Mirzaei N (2014) The effects of air pollution on cardiovascular and respiratory causes of emergency admission. Emergency (Tehran, Iran) 2(3):107–114
60. Naclerio R, Ansotegui IJ, Bousquet J, Canonica GW, D'Amato G, Rosario N, Pawankar R, Peden D, Bergmann KC, Bielory L, Caraballo L, Cecchi L, Cepeda SAM, Chong Neto HJ, Galán C, Gonzalez Diaz SN, Idriss S, Popov T, Ramon GD, Ridolo E et al (2020) International expert consensus on the management of allergic rhinitis (AR) aggravated by air pollutants: impact of air pollution on patients with AR: current knowledge and future strategies. World Allergy Organ J 13(3):100106. https://doi.org/10.1016/j.waojou.2020.100106
61. Laumbach R, Meng Q, Kipen H (2015) What can individuals do to reduce personal health risks from air pollution? J Thorac Dis 7(1):96–107. https://doi.org/10.3978/j.issn.2072-1439.2014.12.21
62. Orru H, Olstrup H, Hagenbjörk A, Nordin S, Orru K (2022) Exposures, symptoms and risk perception among office workers in relation to nanoparticles in the work environment. Int J Environ Res Public Health 19(10):5789. https://doi.org/10.3390/ijerph19105789

63. Mujtaba G, Shahzad SJH (2021) Air pollutants, economic growth and public health: implications for sustainable development in OECD countries. Environ Sci Pollut Res Int 28(10): 12686–12698. https://doi.org/10.1007/s11356-020-11212-1
64. Fu C, Kuang D, Zhang H, Ren J, Chen J (2022) Different components of air pollutants and neurological disorders. Front Public Health 10:959921. https://doi.org/10.3389/fpubh.2022.959921
65. Liao KJ, Hou X, Strickland MJ (2016) Resource allocation for mitigating regional air pollution-related mortality: a summertime case study for five cities in the United States. J Air Waste Manag Assoc 66(8):748–757. https://doi.org/10.1080/10962247.2016.1176085
66. Brouwer W, Verbooy K, Hoefman R, van Exel J (2023) Production losses due to absenteeism and presenteeism: the influence of compensation mechanisms and multiplier effects. Pharmaco Econ 41(9):1103–1115. https://doi.org/10.1007/s40273-023-01253-y
67. Shi Y, Qu S (2022) The effect of cognitive ability on academic achievement: the mediating role of self-discipline and the moderating role of planning. Front Psychol 13:1014655. https://doi.org/10.3389/fpsyg.2022.1014655
68. Moda HM, Filho WL, Minhas A (2019) Impacts of climate change on outdoor workers and their safety: some research priorities. Int J Environ Res Public Health 16(18):3458. https://doi.org/10.3390/ijerph16183458
69. Ren F, Zhu Y, Le D (2022) The spatial effect of air pollution governance on labor productivity: evidence from 262 Chinese cities. Int J Environ Res Public Health 19(20):13694. https://doi.org/10.3390/ijerph192013694
70. Manisalidis I, Stavropoulou E, Stavropoulos A, Bezirtzoglou E (2020) Environmental and health impacts of air pollution: a review. Front Public Health 8:14. https://doi.org/10.3389/fpubh.2020.00014
71. Kosoe EA, Achana GTW, Ogwu MC (2023) Meta-evaluation of the one health implication on food systems of agrochemical use. In: Ogwu MC, Chibueze Izah S (eds) One health implications of agrochemicals and their sustainable alternatives. Sustainable Development and Biodiversity 34, Springer, Singapore. https://doi.org/10.1007/978-981-99-3439-3_14
72. Nagata T, Mori K, Ohtani M, Nagata M, Kajiki S, Fujino Y, Matsuda S, Loeppke R (2018) Total health-related costs due to absenteeism, Presenteeism, and medical and pharmaceutical expenses in Japanese employers. J Occup Environ Med 60(5):e273–e280. https://doi.org/10.1097/JOM.0000000000001291
73. Ekins P, Zenghelis D (2021) The costs and benefits of environmental sustainability. Sustain Sci 16(3):949–965. https://doi.org/10.1007/s11625-021-00910-5
74. Zhou L, Zhong Q, Yang J (2022) Air pollution and household medical expenses: evidence from China. Front Public Health 9:798780. https://doi.org/10.3389/fpubh.2021.798780
75. Bari MW, Saleem S, Bashir M, Ahmad B (2022) Impact of ambient air pollution on outdoor employees' performance: mediating role of anxiety. Front Psychol 13:926534. https://doi.org/10.3389/fpsyg.2022.926534
76. Chai KC, Huang Y, Chang KC, Hu WJ (2020) Can environmental regulation reduce labor costs and improve business performance? Evidence from the air quality index. Front Public Health 7:398. https://doi.org/10.3389/fpubh.2019.00398
77. Li Y, Liu W (2023) Spatial effects of environmental regulation on high-quality economic development: from the perspective of industrial upgrading. Front Public Health 11:1099887. https://doi.org/10.3389/fpubh.2023.1099887
78. Marino E, Caruso M, Campagna D, Polosa R (2015) Impact of air quality on lung health: myth or reality? Ther Adv Chronic Dis 6(5):286–298. https://doi.org/10.1177/2040622315587256
79. Pei Y, Chen H, Liu Z, Hu H (2023) Does environmental regulation improve residents' health? Evidence from China. Front Public Health 11:973499. https://doi.org/10.3389/fpubh.2023.973499
80. Kankeu HT, Saksena P, Xu K, Evans DB (2013) The financial burden from non-communicable diseases in low- and middle-income countries: a literature review. Health Res Policy Syst 11:31. https://doi.org/10.1186/1478-4505-11-31

81. Gallagher S, Roy A, Domeracki SJ, Mohrmann T, Missar V, Jule J, Sharma S, DeWitt R (2021) The low-wage essential worker: occupational concerns and needs in the COVID-19 pandemic-a round table. Workplace Health Saf 69(4):154–160. https://doi.org/10.1177/2165079920988682
82. Pratap P, Dickson A, Love M, Zanoni J, Donato C, Flynn MA, Schulte PA (2021) Public health impacts of underemployment and unemployment in the United States: exploring perceptions, gaps and opportunities. Int J Environ Res Public Health 18(19):10021. https://doi.org/10.3390/ijerph181910021
83. Gyimah J, Fiati MK, Nwigwe UA, Vanessa AE, Yao X (2023) Exploring the impact of renewable energy on economic growth and carbon emissions: evidence from partial least squares structural equation modeling. PLoS One 18(12):e0295563. https://doi.org/10.1371/journal.pone.0295563
84. Xie JY, Suh DH, Joo SK (2021) A dynamic analysis of air pollution: implications of economic growth and renewable energy consumption. Int J Environ Res Public Health 18(18):9906. https://doi.org/10.3390/ijerph18189906
85. Bashir MF, Sadiq M, Talbi B, Shahzad L, Adnan Bashir M (2022) An outlook on the development of renewable energy, policy measures to reshape the current energy mix, and how to achieve sustainable economic growth in the post-COVID-19 era. Environ Sci Pollut Res Int 29(29):43636–43647. https://doi.org/10.1007/s11356-022-20010-w
86. Eckelman MJ, Sherman J (2016) Environmental impacts of the U.S. health care system and effects on public health. PLoS One 11(6):e0157014. https://doi.org/10.1371/journal.pone.0157014
87. Raghupathi V, Raghupathi W (2020) Healthcare expenditure and economic performance: insights from the United States data. Front Public Health 8:156. https://doi.org/10.3389/fpubh.2020.00156
88. Yang Z, Gao W, Li J (2022) Can economic growth and environmental protection achieve a "win-win" situation? Empirical evidence from China. Int J Environ Res Public Health 19(16):9851. https://doi.org/10.3390/ijerph19169851
89. Peng K, Feng K, Chen B, Shan Y, Zhang N, Wang P, Fang K, Bai Y, Zou X, Wei W, Geng X, Zhang Y, Li J (2023) The global power sector's low-carbon transition may enhance sustainable development goal achievement. Nat Commun 14(1):3144. https://doi.org/10.1038/s41467-023-38987-4
90. Ahmed F, Kousar S, Pervaiz A, Trinidad-Segovia JE, Del Pilar Casado-Belmonte M, Ahmed W (2022) Role of green innovation, trade and energy to promote green economic growth: a case of South Asian nations. Environ Sci Pollut Res Int 29(5):6871–6885. https://doi.org/10.1007/s11356-021-15881-4
91. Rauf F, Wanqiu W, Naveed K, Zhang Y (2024) Green R & D investment, ESG reporting, and corporate green innovation performance. PLoS One 19(3):e0299707. https://doi.org/10.1371/journal.pone.0299707
92. Sarango-Lalangui P, Castillo-Vergara M, Carrasco-Carvajal O, Durendez A (2023) Impact of environmental sustainability on open innovation in SMEs: An empirical study considering the moderating effect of gender. Heliyon 9(9):e20096. https://doi.org/10.1016/j.heliyon.2023.e20096
93. Chen CC, Khan A, Hongsuchon T, Ruangkanjanases A, Chen YT, Sivarak O, Chen SC (2021) The role of corporate social responsibility and corporate image in times of crisis: the mediating role of customer trust. Int J Environ Res Public Health 18(16):8275. https://doi.org/10.3390/ijerph18168275
94. Makhdoom ZH, Gao Y, Song X, Khoso WM, Baloch ZA (2023) Linking environmental corporate social responsibility to firm performance: the role of partnership restructure. Environ Sci Pollut Res Int 30(16):48323–48338. https://doi.org/10.1007/s11356-023-25776-1
95. Dockery DW, Pope III CA, Xu X, Spengler JD, Ware JH, Fay ME et al (1993) An association between air pollution and mortality in six US cities. N Engl J Med 329(24):1753–1759
96. WHO (2018) Ambient (outdoor) air quality and health. World Health Organization

97. Mullen K, Smith V, Loke M (2014) The social costs of ambient air pollution in the United States. J Environ Econ Manag 67(1):39–56
98. Organisation for Economic Co-operation and Development (OECD) (2016) The economic consequences of outdoor air pollution. OECD Publishing
99. Pastor M, Morello-Frosch R, Sadd J, Scoggins J (2002) Breathless: schools, air toxics, and environmental justice in California. Policy Stud J 30(2):379–402
100. Mohai P, Saha R (2015) Which came first, people or pollution? Assessing the disparate siting and post-siting demographic change hypotheses of environmental injustice. Environ Res Lett 10(11):115008
101. Agyeman J, Bullard RD, Evans B (eds) (2002) Just sustainabilities: development in an unequal world. MIT Press
102. Walker G (2012) Environmental justice: concepts, evidence and politics. Routledge
103. Guxens M, Sunyer J, Schembari A (2012) Air pollution and brain development. Curr Opin Pediatr 24(2):240–246
104. Zhang X, Chen X, Zhang X, Li Y (2018) The impact of exposure to air pollution on cognitive performance. Proc Natl Acad Sci 115(37):9193–9197
105. Clougherty JE (2010) A growing role for gender analysis in air pollution epidemiology. Environ Health Perspect 118(2):167–176
106. Lim YH, Kim H, Kim JH, Bae S, Park HY, Hong YC (2018) Air pollution and symptoms of depression in elderly adults. Environ Health Perspect 126(2):027009
107. Grossman GM, Krueger AB (1993) Environmental impacts of a North American free trade agreement. National Bureau of Economic Research
108. Sultana F (2013) Gendering climate change: geographical insights. Prof Geogr 65(1):126–133
109. Boyce JK (1994) Inequality as a cause of environmental degradation. Ecol Econ 11(3):169–178
110. Schlosberg D (2007) Defining environmental justice: theories, movements, and nature. Oxford University Press
111. Agyeman J (2005) Sustainable communities and the challenge of environmental justice. NYU Press
112. Pellow DN (2007) Resisting global toxics: transnational movements for environmental justice. MIT Press
113. Marshall JD, Wilson RD (1996) The association between environmental quality and personal income. J Environ Econ Manag 31(1):68–85
114. Taylor DE, Brody JG (2010) Biopolitics in the South Bronx: racial and ethnic disparities in asthma. Ann Assoc Am Geogr 100(4):908–917
115. Tiotiu AI, Novakova P, Nedeva D, Chong-Neto HJ, Novakova S, Steiropoulos P, Kowal K (2020) Impact of air pollution on asthma outcomes. Int J Environ Res Public Health 17(17):6212. https://doi.org/10.3390/ijerph17176212
116. Tainio M, Jovanovic Andersen Z, Nieuwenhuijsen MJ, Hu L, de Nazelle A, An R, Garcia LMT, Goenka S, Zapata-Diomedi B, Bull F, Sá TH (2021) Air pollution, physical activity and health: a mapping review of the evidence. Environ Int 147:105954. https://doi.org/10.1016/j.envint.2020.105954
117. Zhu J, Lu C (2023) Air quality, pollution perception, and Residents' health: evidence from China. Toxics 11(7):591. https://doi.org/10.3390/toxics11070591
118. Zundel CG, Ryan P, Brokamp C, Heeter A, Huang Y, Strawn JR, Marusak HA (2022) Air pollution, depressive and anxiety disorders, and brain effects: a systematic review. Neurotoxicology 93:272–300. https://doi.org/10.1016/j.neuro.2022.10.011
119. Syuhada G, Akbar A, Hardiawan D, Pun V, Darmawan A, Heryati SHA, Siregar AYM, Kusuma RR, Driejana R, Ingole V, Kass D, Mehta S (2023) Impacts of air pollution on health and cost of illness in Jakarta, Indonesia. Int J Environ Res Public Health 20(4):2916. https://doi.org/10.3390/ijerph20042916

120. Zhang J, Lu Y (2022) Exploring the effects of tourism development on air pollution: evidence from the panel smooth transition regression model. Int J Environ Res Public Health 19(14): 8442. https://doi.org/10.3390/ijerph19148442
121. Rimmler SM, Shaughnessy S, Tatum E, Muhammad N, Hawkins S, Lightfoot A, White-Williamson S, Woods CG (2023) Photovoice reveals Residents' concerns for air and water quality in industry-impacted rural community. Int J Environ Res Public Health 20(9):5656. https://doi.org/10.3390/ijerph20095656
122. Van Horne YO, Alcala CS, Peltier RE, Quintana PJE, Seto E, Gonzales M, Johnston JE, Montoya LD, Quirós-Alcalá L, Beamer PI (2023) An applied environmental justice framework for exposure science. J Expo Sci Environ Epidemiol 33(1):1–11. https://doi.org/10.1038/s41370-022-00422-z
123. Richard G, Izah SC, Sawyer EW, Ogwu MC (2023) Seasonal variations in the concentration of some atmospheric pollutants along a major highway in Southern Nigeria 30(1):10–25
124. Alvarez CH (2023) Structural racism as an environmental justice issue: a multilevel analysis of the state racism index and environmental health risk from air toxics. J Racial Ethn Health Disparities 10(1):244–258. https://doi.org/10.1007/s40615-021-01215-0
125. Woo B, Kravitz-Wirtz N, Sass V, Crowder K, Teixeira S, Takeuchi DT (2019) Residential segregation and racial/ethnic disparities in ambient air pollution. Race Soc Probl 11(1):60–67. https://doi.org/10.1007/s12552-018-9254-0
126. Kaufman JD, Hajat A (2021) Confronting environmental racism. Environ Health Perspect 129(5):51001. https://doi.org/10.1289/EHP9511
127. Roberts JD, Dickinson KL, Hendricks MD, Jennings V (2022) "I Can't breathe": examining the legacy of American racism on determinants of health and the ongoing pursuit of environmental justice. Curr Environ Health Rep 9(2):211–227. https://doi.org/10.1007/s40572-022-00343-x
128. Ogwu MC (2019) Towards sustainable development in Africa: the challenge of urbanization and climate change adaptation. In: Cobbinah PB, Addaney M (eds) The geography of climate change adaptation in urban Africa. Palgrave Macmillan, Cham. https://doi.org/10.1007/978-3-030-04873-0_2
129. Lee H, Kravitz-Wirtz N, Rao S, Crowder K (2023) Effects of prolonged exposure to air pollution and neighborhood disadvantage on self-rated health among adults in the United States: evidence from the panel study of income dynamics. Environ Health Perspect 131(8): 87001. https://doi.org/10.1289/EHP11268
130. Sadd JL, Pastor M, Morello-Frosch R, Scoggins J, Jesdale B (2011) Playing it safe: assessing cumulative impact and social vulnerability through an environmental justice screening method in the South Coast Air Basin, California. Int J Environ Res Public Health 8(5):1441–1459. https://doi.org/10.3390/ijerph8051441
131. Behinaein P, Hutchings H, Knapp T, Okereke IC (2023) The growing impact of air quality on lung-related illness: a narrative review. J Thorac Dis 15(9):5055–5063. https://doi.org/10.21037/jtd-23-544
132. Martenies SE, Milando CW, Williams GO, Batterman SA (2017) Disease and health inequalities attributable to air pollutant exposure in Detroit, Michigan. Int J Environ Res Public Health 14(10):1243. https://doi.org/10.3390/ijerph14101243
133. Landrigan PJ, Rauh VA, Galvez MP (2010) Environmental justice and the health of children. Mt Sinai J Med New York 77(2):178–187. https://doi.org/10.1002/msj.20173
134. Costa LG, Cole TB, Dao K, Chang YC, Coburn J, Garrick JM (2020) Effects of air pollution on the nervous system and its possible role in neurodevelopmental and neurodegenerative disorders. Pharmacol Ther 210:107523. https://doi.org/10.1016/j.pharmthera.2020.107523
135. Kurt OK, Zhang J, Pinkerton KE (2016) Pulmonary health effects of air pollution. Curr Opin Pulm Med 22(2):138–143. https://doi.org/10.1097/MCP.0000000000000248
136. Loh MM, Levy JI, Spengler JD, Houseman EA, Bennett DH (2007) Ranking cancer risks of organic hazardous air pollutants in the United States. Environ Health Perspect 115(8): 1160–1168. https://doi.org/10.1289/ehp.9884

137. Morgan ZEM, Bailey MJ, Trifonova DI, Naik NC, Patterson WB, Lurmann FW, Chang HH, Peterson BS, Goran MI, Alderete TL (2023) Prenatal exposure to ambient air pollution is associated with neurodevelopmental outcomes at 2 years of age. Environ Health Glob Access Sci Source 22(1):11. https://doi.org/10.1186/s12940-022-00951-y
138. Hao H, Yoo SR, Strickland MJ, Darrow LA, D'Souza RR, Warren JL, Moss S, Wang H, Zhang H, Chang HH (2023) Effects of air pollution on adverse birth outcomes and pregnancy complications in the U.S. state of Kansas (2000-2015). Sci Rep 13(1):21476. https://doi.org/10.1038/s41598-023-48329-5
139. Baloch QB, Shah SN, Iqbal N, Sheeraz M, Asadullah M, Mahar S, Khan AU (2023) Impact of tourism development upon environmental sustainability: a suggested framework for sustainable ecotourism. Environ Sci Pollut Res Int 30(3):5917–5930. https://doi.org/10.1007/s11356-022-22496-w
140. Odubo TC et al (2023) Socioeconomic values of herbal medicine. In: Izah SC, Ogwu MC, Akram M (eds) Herbal medicine Phytochemistry. Reference series in Phytochemistry. Springer, Cham. https://doi.org/10.1007/978-3-031-21973-3_3-1
141. Hemmati F, Dabbaghi F, Mahmoudi G (2020) Relationship between international tourism and concentrations of PM $_{2.5}$: an ecological study based on WHO data. J Environ Health Sci Eng 18(2):1029–1035. https://doi.org/10.1007/s40201-020-00524-6
142. Wu W, Xia X, Cui C, Qiu F (2023) Haze and inbound tourism: empirical evidence from China. Front Psychol 13:1056673. https://doi.org/10.3389/fpsyg.2022.1056673
143. Xu X, Dong D, Wang Y, Wang S (2019) The impacts of different air pollutants on domestic and inbound tourism in China. Int J Environ Res Public Health 16(24):5127. https://doi.org/10.3390/ijerph16245127
144. Fuller R, Landrigan PJ, Balakrishnan K, Bathan G, Bose-O'Reilly S, Brauer M, Caravanos J, Chiles T, Cohen A, Corra L, Cropper M, Ferraro G, Hanna J, Hanrahan D, Hu H, Hunter D, Janata G, Kupka R, Lanphear B, Lichtveld M et al (2022) Pollution and health: a progress update. Lancet Planet Health 6(6):e535–e547. https://doi.org/10.1016/S2542-5196(22)00090-0
145. WHO (2021) WHO global air quality guidelines: particulate matter (PM2.5 and PM10), ozone, nitrogen dioxide, sulfur dioxide and carbon monoxide [internet]. World Health Organization, Geneva. 6, Implementation of the guidelines. Available from: https://www.ncbi.nlm.nih.gov/books/NBK574587/
146. Chlebnikovas A, Paliulis D, Bradulienė J, Januševičius T (2023) Short-term field research on air pollution within the boundaries of the large city in the Baltic region. Environ Sci Pollut Res Int 30(34):81950–81965. https://doi.org/10.1007/s11356-022-23798-9
147. Singh S, Kulshrestha MJ, Rani N, Kumar K, Sharma C, Aswal DK (2023) An overview of vehicular emission standards. Mapan 38(1):241–263. https://doi.org/10.1007/s12647-022-00555-4
148. Jafari AJ, Delikhoon M, Rastani MJ, Baghani AN, Sorooshian A, Rohani-Rasaf M, Kermani M, Kalantary RR, Golbaz S, Golkhorshidi F (2021) Characteristics of gaseous and particulate air pollutants at four different urban hotspots in Tehran, Iran. Sustainable Cities and Society 70:102907. https://doi.org/10.1016/j.scs.2021.102907
149. Mahjouri M, Ishak MB, Torabian A, Manaf LA, Halimoon N (2018) An integrated methodology for establishing industrial effluent limits in developing countries: Iran as a case study. J Environ Health Sci Eng 16(2):181–192. https://doi.org/10.1007/s40201-018-0306-6
150. Arora NK, Mishra I (2022) Progress of sustainable development goal 7: clean and green energy for all as the biggest challenge to combat climate crisis. Environ Sustain (Singapore) 5(4):395–399. https://doi.org/10.1007/s42398-022-00257-2
151. Azam W, Khan I, Ali SA (2023) Alternative energy and natural resources in determining environmental sustainability: a look at the role of government final consumption expenditures in France. Environ Sci Pollut Res Int 30(1):1949–1965. https://doi.org/10.1007/s11356-022-22334-z

152. Dilanchiev A, Nuta F, Khan I, Khan H (2023) Urbanization, renewable energy production, and carbon dioxide emission in BSEC member states: implications for climate change mitigation and energy markets. Environ Sci Pollut Res Int 30(25):67338–67350. https://doi.org/10.1007/s11356-023-27221-9
153. Wang L (2024) Assessment of land use change and carbon emission: a Log Mean Divisa (LMDI) approach. Heliyon 10(3):e25669. https://doi.org/10.1016/j.heliyon.2024.e25669
154. Nieuwenhuijsen MJ (2021) New urban models for more sustainable, liveable and healthier cities post covid19; reducing air pollution, noise and heat Island effects and increasing green space and physical activity. Environ Int 157:106850. https://doi.org/10.1016/j.envint.2021.106850
155. Bouza E, Vargas F, Alcázar B, Álvarez T, Asensio A, Cruceta G, Gracia D, Guinea J, Gil MA, Linares C, Muñoz P, Pastor P, Pedro-Botet ML, Querol X, Tovar J, Urrutia I, Villar F, Palomo E (2022) Air pollution and health prevention: a document of reflection. Revista Espanola de Quimioterapia: Publicacion Oficial de la Sociedad Espanola de Quimioterapia 35(4):307–332. https://doi.org/10.37201/req/171.2021
156. Nie D, Shen F, Wang J, Ma X, Li Z, Ge P, Ou Y, Jiang Y, Chen M, Chen M, Wang T, Ge X (2021) Changes in air quality and its associated health and economic burden in 31 provincial capital cities in China during the COVID-19 pandemic. Atmos Res 249:105328. https://doi.org/10.1016/j.atmosres.2020.105328
157. Ross K, Chmiel JF, Ferkol T (2012) The impact of the clean air act. J Pediatr 161(5):781–786. https://doi.org/10.1016/j.jpeds.2012.06.064
158. Glazener A, Wylie J, van Waas W, Khreis H (2022) The impacts of car-free days and events on the environment and human health. Curr Environ Health Rep 9(2):165–182. https://doi.org/10.1007/s40572-022-00342-y

Economic Dimensions of Air Pollution: Cost Analysis, Valuation, and Policy Impacts

Ebiuwa G. Obahiagbon and Enoch Akwasi Kosoe

Contents

Abstract Air pollution is a pervasive environmental problem with far-reaching economic implications. This chapter provides an analysis of the economic impacts of air pollution, focusing on the costs incurred to illustrate the impacts, as well as policy implications, and recommendations for mitigation strategies. The economic costs of air pollution are substantial and multifaceted, encompassing direct healthcare expenditures, productivity losses, and environmental degradation. Studies

E. G. Obahiagbon
Department of Management and Human Resource Management, Glasgow School for Business and Society, Glasgow Caledonian University, London, UK

E. A. Kosoe (✉)
Department of Environment and Resource Studies, Faculty of Integrated Development Studies, Simon Diedong Dombo University of Business and Integrated Development Studies, Wa, Ghana
e-mail: ekosoe@ubids.edu.gh

Matthew Chidozie Ogwu and Sylvester Izah (eds.), *Sustainable Strategies for Air Pollution Mitigation: Development, Economics and Technologies*,
Hdb Env Chem (2024) 133: 111–140, DOI 10.1007/698_2024_1116,

Published online: 12 September 2024

have estimated the economic burden of air pollution at both the national and global levels, underscoring the urgency of implementing effective pollution control measures to mitigate economic losses and protect public health. For instance, economic losses because of the effects of ground-level ozone on wheat yields totaled EUR 1,418 million in Europe. The annual economic impact of air pollution in China is estimated at USD 900 billion a year, while in the United States, it is USD 600 billion, and India follows closely as the three nations bear the greatest economic brunt of air pollution. The economic framework for analyzing air pollution encompasses cost-benefit analysis, economic valuation methods, and market-based instruments. Cost-benefit analysis allows policymakers to compare the costs of pollution control measures with the benefits of reduced pollution-related damages, aiding in decision-making and resource allocation. Economic valuation assigns monetary values to the health and environmental impacts of pollution, facilitating the quantification of pollution-related costs and benefits. Market-based instruments, such as emissions trading systems and pollution taxes, provide economic incentives for pollution reduction by internalizing the external costs of pollution and stimulating innovation in pollution control technologies. However, there is a need to develop cost-effective pollution control strategies, implement market-based instruments, and reallocate resources to priority areas such as healthcare and environmental restoration. By investing in pollution control measures, governments can mitigate the economic costs of air pollution, promote sustainable development, and safeguard public health. In conclusion, the economic analysis of air pollution provides valuable insights into the costs and benefits of pollution control efforts, guiding policymakers in decision-making and resource allocation. Further research is needed to refine economic models, improve data quality, and enhance our understanding of the complex interactions between air pollution, economic development, and public health.

Keywords Air pollution control, Clean Air Act, Economic valuation, Emission trading, Health impact assessment, Pollution pricing, Regulatory impact analysis, Social cost of carbon

Abbreviations

PM	Particulate matter
O3	Ozone
NO2	Nitrogen dioxide
SO2	Sulfur dioxide
CO	Carbon monoxide
VOCs	Volatile organic compounds
COPD	Chronic obstructive pulmonary disease
CBA	Cost-benefit analysis
WHO	World Health Organization

WTP	Willingness to pay
CVM	Contingent valuation method
SEIA	Social or environmental impact assessment
EUR	Euros
USD	United States dollar

1 Introduction

Air pollution refers to the release of harmful substances into the atmosphere, which can have negative effects on human health, the environment, and ecosystems. Air pollutants are substances present in the atmosphere in concentrations that may harm human health, ecosystems, or property [1]. There are several types of air pollutants, including particulate matter (PM), ozone (O3), nitrogen dioxide (NO2), sulfur dioxide (SO2), carbon monoxide (CO), and volatile organic compounds (VOCs). Each type of pollutant has distinct sources, properties, and impacts on human health and the environment ([2, 3]). Air pollution can be caused by diverse sources, including the following: the burning of fossil fuels such as coal, oil, and gas for energy production is a major source of air pollutants, industrial processes such as manufacturing, mining, and construction can also release harmful pollutants into the atmosphere, transportation through the emission of pollutants such as PM, NO2, and CO into the atmosphere by vehicles and agricultural activities such as fertilization and pest control can release pollutants into the atmosphere, and improper waste disposal can also lead to air pollution [4]. According to Jiang et al. [5], air pollution can exacerbate respiratory problems such as asthma and chronic obstructive pulmonary disease (COPD). Some of the ailments associated with air pollution and air pollutants are treatable like COPD [6]. Exposure to air pollutants has been linked to an increased risk of cardiovascular disease. Some air pollutants have been linked to an increased risk of cancer [7]. Exposure to air pollutants has been linked to neurological damage [8]. Air pollution can damage ecosystems and harm biodiversity. Air pollutants can contribute to climate change by trapping heat in the atmosphere and can also contaminate water sources [9].

Air pollution is a significant environmental and public health challenge worldwide, with economic ramifications that extend far beyond its immediate health and environmental effects. The economic impacts of air pollution are far-reaching and multifaceted [10]. Air pollution can lead to increased healthcare costs, reduced productivity, and damage to crops and infrastructure, among other consequences. In addition, air pollution can also have significant social and environmental impacts, including increased mortality rates, decreased quality of life, and harm to ecosystems [4]. Therefore, understanding the economic impacts of air pollution is essential for developing effective policies and strategies to mitigate its effects.

The economic analysis of air pollution relies on several key principles and methodologies. One fundamental approach is cost-benefit analysis (CBA), which compares the costs of pollution control measures with the benefits of reduced pollution-related damages [11, 12]. CBA helps policymakers evaluate the efficiency and effectiveness of various pollution control strategies and prioritize interventions based on their net social benefits. Another essential concept is economic valuation, which assigns monetary values to the health and environmental impacts of air pollution [13]. This involves estimating the costs associated with pollution-related health outcomes, such as medical expenses and lost productivity, as well as the value of ecosystem services affected by pollution, such as clean water and biodiversity. Market-based instruments, such as emissions trading systems and pollution taxes, offer economic incentives for pollution reduction by internalizing the externality (i.e., among other things the external) costs of pollution [14, 15]. These mechanisms create economic efficiencies by allowing polluters to find the most cost-effective ways to reduce emissions while stimulating innovation in pollution control technologies. The economic costs of air pollution are substantial. Direct costs include healthcare expenditures for treating pollution-related illnesses, such as respiratory diseases and cardiovascular disorders. Indirect costs encompass productivity losses due to absenteeism, decreased worker productivity, and premature mortality. Moreover, air pollution imposes environmental costs by degrading ecosystems, harming wildlife, and diminishing natural resources. Quantifying these costs requires sophisticated economic modeling techniques and data analysis. Studies have estimated the economic burden of air pollution at both the national and global levels, highlighting the significant economic implications of inaction on pollution control [4, 16–18].

The economic analysis of air pollution has important implications for policy formulation and decision-making [19]. Policymakers can use economic insights to design cost-effective pollution control strategies that maximize societal welfare while minimizing economic distortions [20]. Market-based instruments, such as emissions trading schemes, offer flexible and efficient mechanisms for achieving pollution reduction targets. Furthermore, economic analysis can inform the allocation of resources to priority areas, such as healthcare infrastructure and environmental restoration efforts [21]. By investing in pollution prevention and mitigation measures, governments can mitigate the economic costs of air pollution and promote sustainable development. By quantifying the costs and benefits of these policies, researchers and policymakers can assess their overall impact on society and identify opportunities for optimization. The economic framework for analyzing air pollution provides valuable insights into the costs and benefits of pollution control efforts [22]. By employing tools such as cost-benefit analysis, economic valuation, and market-based instruments, policymakers can make informed decisions to address the economic and environmental challenges posed by air pollution [12, 23]. However, research is needed to refine economic models, improve data quality, and enhance our understanding of the complex interactions between air pollution, economic development, and public health.

This chapter aims to provide an in-depth analysis of the economic impacts of air pollution, focusing on both the microeconomic and macroeconomic aspects of this issue. The chapter is divided into several sections, each examining a specific aspect of the economic impacts of air pollution. The chapter addresses the economic costs associated with air pollution, presents case studies illustrating economic analyses of air pollution impacts, discusses policy implications and recommendations, and concludes with a synthesis of key contributions to the subject.

2 Economic Costs of Air Pollution

Air pollution is a pressing environmental and public health issue that affects communities worldwide. With the rapid industrialization and urbanization of modern society, the emission of pollutants into the atmosphere has reached alarming levels. From vehicle exhaust and industrial emissions to agricultural practices and household activities, various sources contribute to air pollution [24]. The consequences of air pollution extend beyond environmental degradation and public health concerns; they also have significant economic implications (Fig. 1; Table 1).

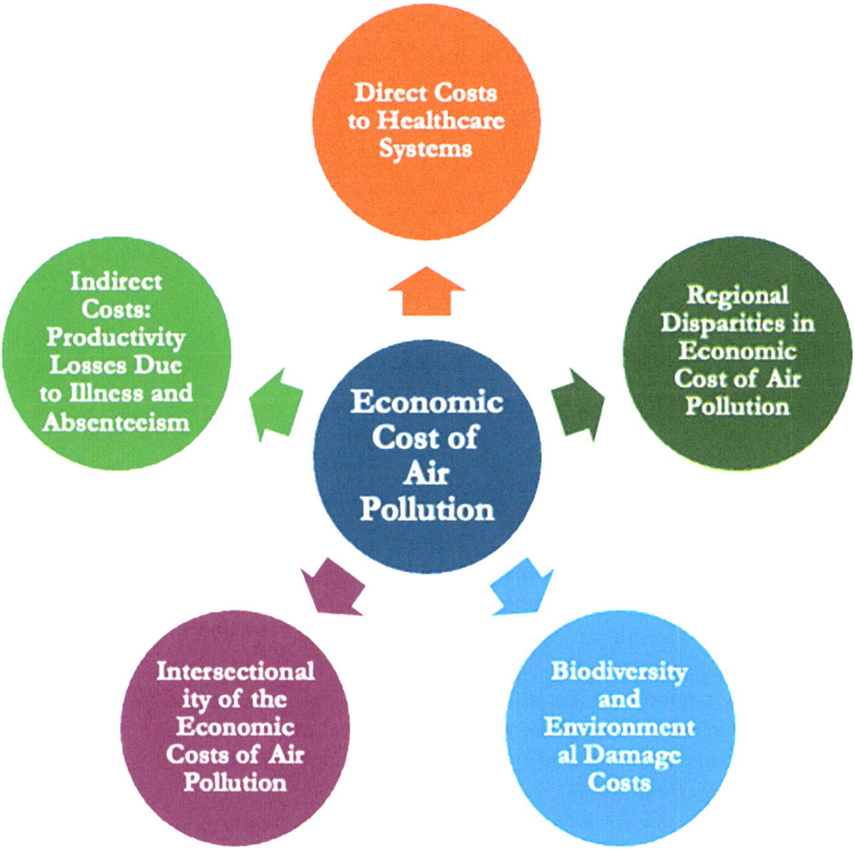

Fig. 1 Economic cost of air pollution

Table 1 Economic implications of air pollution

Economic dimension	Costs of air pollution	Examples	References
Healthcare costs	Medical treatment for respiratory diseases, cardiovascular diseases, and other health issues caused by air pollution	Treatment costs for asthma, lung cancer, and heart disease	Li et al. [25]
Lost productivity	Reduced productivity due to illness, absenteeism, and premature death caused by air pollution	Decreased work output, missed workdays, and lower economic output	Mitchell and Bates [26]; Le et al. [27]
Environmental damage	Costs associated with ecosystem degradation, biodiversity loss, and damage to natural resources caused by air pollution	Decline in crop yields, loss of biodiversity, and damage to ecosystems	Manisalidis et al. [4]
Climate change impacts	Costs resulting from air pollution contribute to climate change, including extreme weather events, sea-level rise, and changes in agricultural patterns	Increased frequency of hurricanes, flooding, droughts, and heat waves	D'Amato et al. [28], Bell et al. [29]
Infrastructure damage	Costs related to the deterioration of infrastructure such as buildings, roads, and bridges due to air pollution and acid rain	Corrosion of buildings, bridges, and monuments	Chandrappa and Chandra [30]

Understanding the economic costs associated with air pollution is crucial for policymakers, businesses, and society to make informed decisions and implement effective interventions.

2.1 *Direct Costs to Healthcare Systems*

Air pollution poses significant direct health risks to individuals, leading to a wide range of respiratory and cardiovascular illnesses. The healthcare costs associated with treating these illnesses represent a substantial economic burden on healthcare systems globally. From the report of Li et al. [25], it is no news that air pollution contributes to a myriad of health problems, ranging from minor respiratory irritations to severe chronic conditions and premature mortality. The direct healthcare costs incurred because of these illnesses encompass expenses related to diagnosis, treatment, hospitalization, medication, and rehabilitation [31]. These costs can be substantial, particularly in regions with high levels of air pollution and densely populated urban areas where exposure to pollutants is more prevalent as well as within developing economies. Numerous studies [25, 32–34] have investigated the relationship between air pollution and healthcare expenditure, providing insights into the economic impact of air pollution on healthcare systems. These studies analyze healthcare utilization patterns, hospital admission rates, emergency room visits, outpatient clinic visits, and medication prescriptions associated with air pollution-related illnesses and found correlations. By quantifying the financial burden on healthcare systems, policymakers gain a better understanding of the magnitude of the problem and the need for targeted interventions. For example, research conducted in urban areas with high levels of air pollution has revealed increased hospital admissions and healthcare expenditures due to respiratory diseases such as asthma, chronic bronchitis, and pneumonia [35]. Similarly, studies on the health effects of air pollution in vulnerable populations, such as children, the elderly, and individuals with pre-existing health conditions, highlight the disproportionate burden borne by these groups [14].

The allocation of resources for addressing air pollution-related health problems is a critical consideration for healthcare systems. Policymakers must balance competing priorities and allocate sufficient funding for preventive measures, medical treatment, public health programs, and infrastructure improvements to mitigate air pollution. However, budgetary constraints and resource limitations often pose challenges to implementing comprehensive strategies, particularly in low- and middle-income countries where healthcare resources are already stretched thin. Analyzing healthcare expenditure, examining case studies and statistics, and considering the allocation of resources are essential steps in understanding the direct economic burden of air pollution on healthcare systems [25]. By addressing these costs through targeted interventions and policy measures, healthcare systems can mitigate the adverse health effects of air pollution and alleviate the financial strain on individuals and society.

2.2 Indirect Costs: Productivity Losses Due to Illness and Absenteeism

Air pollution not only affects individual health but also has significant implications for workforce productivity due to illness and absenteeism. Productivity losses refer to the reduction in output or efficiency resulting from factors that impede the ability of workers to perform their tasks effectively [26]. In the context of air pollution, productivity losses occur when workers experience health problems, such as respiratory infections, allergies, or other illnesses, that diminish their capacity to work at optimal levels. These losses manifest as decreased work output, increased error rates, longer completion times, and higher rates of absenteeism, ultimately affecting overall productivity and economic performance [36, 37]. Air pollution can have a profound impact on worker productivity, particularly in industries or occupations that require outdoor labor or exposure to polluted environments. Workers exposed to high levels of air pollution are more likely to experience respiratory symptoms, fatigue, headaches, and other health issues that impair their ability to perform their job duties effectively [38]. Chronic exposure to air pollutants can exacerbate existing health conditions, leading to increased absenteeism, reduced work performance, and decreased job satisfaction among workers.

Quantifying productivity losses attributable to air pollution involves estimating the economic value of reduced work output and increased absenteeism resulting from pollution-related health problems. This can be achieved through various methods, including cost-of-illness studies, econometric analyses, and surveys of worker productivity and absenteeism. By assigning monetary values to the time lost due to illness and absenteeism, researchers can assess the economic impact of air pollution on productivity at the individual, organizational, and societal levels [17]. Certain industries or sectors are more susceptible to productivity losses due to air pollution, depending on the nature of their operations and the extent of exposure to pollutants. Industries such as construction, manufacturing, agriculture, transportation, and outdoor recreation are particularly vulnerable to the adverse effects of air pollution on worker health and productivity [36]. Workers in these sectors may face higher levels of exposure to pollutants, leading to increased rates of illness, absenteeism, and reduced on-the-job performance. However, an earlier report by Zivin and Neidell [39] opined that there is no clear connection between pollution and economic losses.

Mitigating productivity losses attributable to air pollution requires a multifaceted approach that addresses both the root causes of pollution and the underlying health problems affecting workers. Strategies may include implementing pollution control measures, improving workplace ventilation systems, providing personal protective equipment, offering health education and wellness programs, and promoting telecommuting or flexible work arrangements to reduce exposure to polluted environments [4, 40].

2.3 Environmental Damage Costs

Air pollution contributes to a wide range of environmental problems, including global warming, acid rain, ozone depletion, and the degradation of air, water, and soil quality. This can result from pollutants emitted into the atmosphere, such as particulate matter, sulfur dioxide, nitrogen oxides, and volatile organic compounds, which can undergo chemical reactions and deposition, leading to the formation of smog, haze, and secondary pollutants. Therefore, air pollutants have far-reaching effects on ecosystems, natural resources, and ecological processes, posing threats to biodiversity, habitat integrity, and ecosystem functioning [41].

The economic valuation of environmental degradation involves assessing the costs associated with the loss or impairment of ecosystem services and natural resources due to air pollution [42]. Ecosystem services, such as air and water purification, pollination, carbon sequestration, and climate regulation, provide essential benefits to society, including clean air and water, fertile soil, and climate stability. Air pollution has profound effects on ecosystems and biodiversity, disrupting ecological processes, altering species distributions, and compromising ecosystem resilience. Air pollutants can damage vegetation, impair photosynthesis, and reduce crop yields, leading to agricultural losses and food insecurity [43–45]. Additionally, air pollution contributes to forest decline, soil acidification, freshwater eutrophication, and marine pollution, threatening the health and vitality of terrestrial and aquatic ecosystems [46]. Biodiversity loss, habitat destruction, and species extinction are among the most severe consequences of air pollution, with cascading effects on ecosystem services and human well-being. The economic costs associated with environmental damage caused by air pollution include expenses related to environmental monitoring, pollution control measures, ecosystem restoration projects, and natural resource management efforts [47]. These costs encompass both direct expenditures, such as cleanup and remediation activities, and indirect costs, such as lost ecosystem services, reduced recreational opportunities, and decreased property values. Assessing the economic costs of environmental damage provides insights into the true cost of pollution and informs decision-making processes aimed at minimizing these costs and promoting sustainable resource management practices.

2.4 Intersectionality of the Economic Costs of Air Pollution

The economic costs of air pollution are multifaceted and interconnected, with various factors contributing to the overall burden on society. The intersectionality of economic costs associated with air pollution includes the interconnectedness between direct healthcare costs, productivity losses, and environmental damage, as well as how these costs compound and interact with each other to create complex challenges for policymakers, businesses, and individuals [10, 25]. Air pollution impacts human health, productivity, and the environment in interconnected ways, creating a web of economic costs that affect multiple sectors of society. For example, individuals exposed to high levels of air pollution are more likely to experience respiratory illnesses, leading to increased healthcare expenditures and productivity losses due to illness and absenteeism. Additionally, environmental damage caused by air pollution, such as crop failures, ecosystem degradation, and property damage, can further exacerbate economic costs by disrupting economic activities, increasing resource scarcity, and necessitating costly cleanup and restoration efforts. This can result in economic losses, as in 2019, the economic losses due to the impacts of ground-level ozone on wheat yields totaled EUR 1,418 million across 35 European countries [48]. According to Greenpeace research, the economic impact of air pollution in China is estimated at $900 billion a year, while in the USA, it is $600 billion, and India follows closely as the three nations bear the greatest economic brunt of air pollution [49].

The economic costs of air pollution compound and interact with each other, creating complex challenges for policymakers and stakeholders. For instance, the direct healthcare costs associated with treating air pollution-related illnesses can strain healthcare systems and lead to resource shortages, exacerbating disparities in access to healthcare services and health outcomes [50]. Similarly, productivity losses due to illness and absenteeism can hinder economic growth, reduce competitiveness, and undermine social cohesion, perpetuating cycles of poverty and inequality [51]. Moreover, environmental damage costs can impose long-term economic burdens on future generations, as the depletion of natural resources and degradation of ecosystems limit opportunities for sustainable development and economic prosperity.

The holistic economic impact of air pollution extends beyond individual sectors or regions, encompassing broader societal and economic dimensions. For example, air pollution-related health problems can diminish human capital and workforce productivity, reducing overall labor force participation rates and economic output [52]. Similarly, environmental damage caused by air pollution can undermine the resilience of ecosystems and natural systems, increasing the vulnerability of communities to environmental hazards, such as floods, droughts, and wildfires, and disrupting vital ecosystem services, such as water purification, pollination, and climate regulation. By considering the interconnectedness of economic costs, policymakers can develop more effective strategies for addressing air pollution and promoting sustainable development.

2.5 Regional Disparities in Economic Cost of Air Pollution and Vulnerable Populations

Air pollution does not affect all regions or populations equally, leading to disparities in both health outcomes and economic impacts [53]. Air pollution levels vary significantly across regions due to differences in industrial activity, transportation patterns, population density, and geographical features. Urban areas and industrial centers often experience higher levels of pollution compared to rural areas, leading to disparities in health outcomes and economic costs. For example, cities with heavy traffic congestion and industrial emissions may have higher rates of respiratory illnesses, hospital admissions, and healthcare expenditures related to air pollution [54]. Similarly, regions with limited access to healthcare services and environmental resources may face greater challenges in addressing the economic costs of pollution.

Certain demographic groups are more vulnerable to the health effects of air pollution, including children, the elderly, individuals with pre-existing health conditions, low-income communities, and minority populations [55, 56]. These groups often experience higher exposure to pollutants due to factors such as residential location, occupational hazards, and limited access to healthcare. Children are particularly susceptible to air pollution-related health problems due to their developing immune systems and higher breathing rates, while the elderly may face increased risks of cardiovascular and respiratory diseases [57]. Additionally, low-income communities and minority populations may bear a disproportionate burden of pollution-related health disparities due to environmental injustice and systemic inequalities. The regional disparities in economic costs of air pollution exacerbate existing social and economic inequalities, perpetuating cycles of poverty, ill health, and environmental degradation [53, 58]. Vulnerable populations, already marginalized and disadvantaged, are further marginalized by pollution-related health problems and economic burdens. For example, low-income communities may lack access to quality healthcare services, face barriers to employment and education, and suffer from inadequate housing conditions and environmental pollution. These disparities may also contribute to social unrest, political instability, and economic stagnation, hindering efforts to achieve sustainable development and equitable prosperity.

3 Economic Frameworks for Analyzing Air Pollution Impacts

Economic analysis plays a crucial role in understanding the costs and benefits of air pollution control measures, assessing the economic impact of pollution-related health problems, and informing policy decisions aimed at mitigating pollution and promoting sustainable development [23, 25, 59]. Economic analysis involves the systematic study of the allocation of scarce resources to meet society's unlimited wants and needs. In the context of air pollution, economic analysis aims to assess the economic costs and benefits associated with pollution control measures, quantify the monetary value of pollution-related health and environmental impacts, and analyze the trade-offs between environmental protection and economic development (Table 2).

3.1 Cost-Benefit Analysis

Cost-benefit analysis (CBA) is a widely used economic tool for evaluating the efficiency of policies, projects, and interventions by comparing the costs incurred with the benefits gained [80]. In the context of air pollution, CBA may be used to assess the costs of implementing pollution control measures, such as regulations, technological upgrades, and emission reduction programs, against the benefits of improved air quality, reduced health risks, and enhanced environmental quality [12]. Therefore, CBA is a systematic approach used to assess the economic efficiency of policies, projects, or interventions by comparing their costs with their benefits. When applied to air pollution impacts, CBA helps policymakers evaluate the costs of pollution control measures against the benefits of improved air quality, health outcomes, and environmental protection.

Step 1: Identification of Costs and Benefits The first step in conducting a CBA of air pollution impacts is to identify all relevant costs and benefits associated with pollution control measures and their impacts. Costs may include capital investments

Table 2 Some relevant economic analytical tools for assessing air pollution impacts

Analytical approach	Description	Application examples	References
Cost-benefit analysis (CBA) of air pollution	1. CBA of air pollution control measures	CBA of implementing emission reduction technologies in power plants	Boardman et al. [60], World Bank [61]
	2. Identification of costs and benefits associated with pollution control	Assessing the economic benefits of improving air quality in urban areas	Hanley et al. [62], Levinson [63], Smith [64]
	3. Estimation of monetary values for costs and benefits using appropriate valuation methods	Quantifying the economic value of avoided healthcare expenditures and productivity losses due to air pollution	Pearce and Turner [65], Freeman [66]
	4. Discounting future flows and conducting sensitivity analysis to assess the robustness of results	Conducting sensitivity analysis to evaluate the uncertainty in cost-benefit estimates for pollution control measures	Pearce et al. [67], Wang and Zhang [68]
Economic valuation methods for health and environmental impacts	1. Survey of economic valuation methods used to assess health and environmental impacts of air pollution	Economic valuation of the health benefits of reducing air pollution-related morbidity and mortality	Navrud and Ready [69], Johnson and Smith [70]
	2. Application of contingent valuation, stated preference methods, hedonic pricing, and travel cost methods	Hedonic pricing analysis to estimate the economic value of improvements in air quality on property values	Smith and Huang [71], Rahman and Khan [72], Hanemann [73]
	3. Explanation of how these methods quantify the economic value of pollution-related improvements and benefits	Stated preference methods to assess individuals' willingness to pay for pollution reduction programs	Hanemann [73], Garcia and Martinez [74]
Opportunity cost analysis of air pollution in developing economies	1. Analysis of the opportunity cost of air pollution on developing economies	Estimating the opportunity cost of pollution-related health problems on economic growth and human capital formation	Dasgupta et al. [75], Rahman and Khan [72], Nguyen and Tran [76]
	2. Examination of how air pollution affects human capital formation, economic development, and poverty alleviation efforts	Assessing the impact of air pollution on labor productivity and agricultural yields in low-income countries	Gennaioli and Tavoni [77], Brown and Green [78]
	3. Discussion on the impact of pollution-related health problems and environmental degradation on economic growth and sustainability	Examining the link between air pollution, environmental degradation, and poverty traps in developing economies	WHO [79], Navrud and Ready [69]

in pollution control technologies, operating and maintenance expenses, compliance costs, and administrative expenses [4, 81]. On the other hand, benefits may include avoided healthcare expenditures, improved worker productivity, enhanced environmental quality, and increased recreational opportunities. It is essential to consider both direct and indirect costs and benefits and account for any distributional impacts across different stakeholders.

Step 2: Estimation of Monetary Values Once costs and benefits have been identified, the next step is to estimate their monetary values using appropriate valuation methods. For example, the economic value of avoided health risks can be assessed using the human capital approach, which quantifies the monetary value of premature mortality and morbidity based on individuals' willingness to pay to avoid these outcomes. Similarly, the economic value of environmental improvements, education, and attitude can be estimated using contingent valuation surveys, which elicit individuals' preferences and willingness to pay for ecosystem services and environmental amenities [72]. It is essential to use reliable data and robust estimation techniques to ensure the accuracy and validity of monetary values.

Step 3: Discounting Future Flows In CBA, future costs and benefits are discounted to account for the time value of money and compare them consistently. Discounting adjusts future cash flows to their present value using a discount rate that reflects the opportunity cost of capital and the rate of time preference [82]. By discounting future flows, policymakers can assess the net present value of air pollution impacts and determine whether pollution control measures are economically justified over their lifecycle. However, choosing an appropriate discount rate is often challenging and requires careful consideration of factors such as risk, inflation, and social preferences.

Step 4: Sensitivity Analysis Sensitivity analysis is a critical component of CBA that assesses the robustness of results to changes in key assumptions, parameters, and input variables. By varying input values within plausible ranges and examining the sensitivity of outcomes to different scenarios, analysts can identify sources of uncertainty and potential biases in the analysis [83]. Sensitivity analysis helps policymakers make informed decisions under conditions of uncertainty and ensures that policy recommendations are robust and reliable. Sensitivity analysis can also inform risk management strategies and help prioritize research and data collection efforts to reduce uncertainty in future analyses [84].

3.2 Economic Valuation Methods

Economic valuation methods are used to assign monetary values to nonmarket goods and services, such as human health, environmental quality, and ecosystem services, which are not traded in conventional markets [85]. In the case of air pollution, economic valuation methods help quantify the economic value of pollution-related health improvements, reduced morbidity and mortality, and environmental benefits, such as improved water and soil quality, enhanced biodiversity, and climate change mitigation [86]. Common valuation techniques include contingent valuation, stated preference methods, hedonic pricing, and travel cost methods, each with its strengths and limitations depending on the context and application.

Economic valuation methods play a crucial role in analyzing the economic costs of air pollution by assigning monetary values to the health and environmental impacts associated with pollution. These methods help quantify the economic burden of pollution-related illnesses, environmental degradation, and loss of ecosystem services, providing policymakers with valuable information for decision-making and policy development [87]. Some key economic valuation methods commonly used for analyzing the economic costs of air pollution:

- **Contingent Valuation Method (CVM):** CVM is a survey-based approach used to elicit individuals' preferences and willingness to pay (WTP) for improvements in environmental quality [88, 89]). In the context of air pollution, CVM surveys can be used to estimate the economic value that individuals place on reducing pollution levels and improving air quality [90]. Respondents are presented with hypothetical scenarios describing changes in air quality and asked to state their WTP for these improvements. The aggregated responses are used to estimate demand curves and derive the economic value of pollution reductions in monetary terms.
- **Stated Preference Methods**: This includes methods, such as choice experiments and conjoint analysis, which are survey-based techniques used to assess individuals' preferences and trade-offs for different environmental attributes and policy alternatives [91]. In choice experiments, respondents are presented with a series of hypothetical scenarios and asked to choose between alternative options with varying levels of air quality. By analyzing choice data and estimating preference functions, analysts can quantify the economic value that individuals attach to improvements in air quality and pollution reductions.
- **Hedonic Pricing Method:** The hedonic pricing method is a revealed preference approach used to estimate the economic value of environmental amenities, such as clean air, through their impact on property values and market transactions [92]. In the context of air pollution, hedonic pricing models regress property prices on environmental quality indicators, such as pollution levels, controlling for other property characteristics and location factors. By estimating the marginal price of clean air, analysts can infer the implicit value that individuals place on pollution reductions and improvements in air quality.
- **Health-Based Valuation Methods**: Health-based valuation methods quantify the economic costs of air pollution-related health impacts, such as morbidity, mortality, and healthcare expenditures [93]. These methods typically use epidemiological studies and health impact assessments to estimate the incidence and severity of air pollution-related health problems. Economic costs are then calculated based on factors such as medical expenses, lost productivity, and reduced quality of life associated with air pollution-related illnesses.

- **Social or Environmental Impact Assessment (SEIA):** SEIA is a comprehensive process used to evaluate the social and environmental consequences of proposed projects, policies, or interventions, including air pollution [94]. SEIA considers the economic costs of environmental degradation caused by air pollution, such as damage to ecosystems, loss of biodiversity, and degradation of natural resources. Note that economic valuation techniques, such as contingent valuation and hedonic pricing, may be incorporated into SEIA to assess the economic value of ecosystem services and environmental amenities affected by air pollution.
- **Hedonic Pricing**: Hedonic pricing is a revealed preference method used to estimate the economic value of environmental amenities, such as clean air, through their impact on property values and market transactions [92]. In hedonic pricing models, analysts regress property prices on environmental quality indicators, such as air pollution levels, proximity to parks, and access to green spaces, controlling for other property characteristics and location factors [95]. By estimating the marginal price of environmental amenities, analysts can infer the implicit value that individuals place on pollution-related improvements and environmental quality enhancements. Hedonic pricing is widely used in real estate and urban economics but may be sensitive to model specification and data quality issues.
- **Travel Cost Methods**: Travel cost methods are used to estimate the economic value of recreational and tourism activities through their impact on travel behavior and expenditures. In travel cost studies, analysts assess individuals' willingness to pay for visiting natural parks, beaches, and other recreational sites by analyzing travel patterns, trip frequencies, and associated expenses [96, 97]. By modeling travel costs as a function of site characteristics, such as air quality, water quality, and scenic beauty, analysts can estimate the economic value of pollution-related improvements and recreational opportunities. Travel cost methods provide a direct measure of individuals' preferences and can inform policy decisions related to park management and environmental conservation.

3.3 Opportunity Cost of Air Pollution on Developing Economies

The opportunity cost of air pollution refers to the value of forgone opportunities and resources that could have been allocated to alternative uses in the absence of pollution-related health problems and environmental degradation [10, 34]. In developing economies, air pollution imposes significant economic costs in terms of reduced labor productivity, increased healthcare expenditures, and environmental damage, diverting scarce resources away from investments in education, infrastructure, and economic growth [10]. The opportunity cost of pollution is particularly acute in low-income countries, where limited resources and institutional capacities constrain efforts to address pollution-related challenges and promote sustainable development [16]. Simply put, the opportunity cost of air pollution refers to the value of forgone opportunities and resources that could have been allocated to alternative uses in the absence of pollution-related health problems and environmental degradation. In developing economies, air pollution imposes significant economic costs in terms of reduced labor productivity, increased healthcare expenditures, and environmental damage, diverting scarce resources away from investments in education, infrastructure, and economic growth [3, 25]. Some of the opportunity costs of air pollution in developing economies include:

- **Impact on Human Capital Formation**: Air pollution has adverse effects on human health, including respiratory diseases, cardiovascular disorders, and premature mortality, which can hinder human capital formation in developing economies [7]. Children exposed to air pollution may experience developmental delays, cognitive impairments, and lower academic achievement, limiting their future earning potential and economic productivity [98]. Poor health outcomes also increase absenteeism and reduce workforce participation, leading to a less skilled and less productive labor force. This, in turn, constrains economic growth and undermines efforts to improve living standards and reduce poverty.
- **Impact on Economic Development**: Air pollution poses significant challenges to economic development in developing economies by imposing direct costs on healthcare systems, reducing labor productivity, and undermining investments in physical and human capital [99]. The economic costs of pollution-related health problems drain scarce resources from productive sectors of the economy, diverting funds away from education, infrastructure, and poverty alleviation programs. Environmental degradation caused by air pollution further exacerbates economic challenges, affecting agriculture, tourism, and other key sectors vital for economic growth and sustainability [100]. As a result, pollution-related constraints on economic development perpetuate cycles of poverty and inequality, hindering progress toward achieving sustainable development goals.
- **Poverty Alleviation Efforts:** Air pollution exacerbates poverty in developing economies by imposing disproportionate burdens on low-income households and marginalized communities. Poor households are more likely to reside in areas with high levels of pollution, such as urban slums, industrial zones, and informal settlements, where exposure to pollutants is exacerbated by inadequate housing, limited access to healthcare services, and lack of environmental amenities [101]. Pollution-related health problems, combined with limited coping mechanisms and social safety nets, push vulnerable populations deeper into poverty, trapping them in a cycle of ill-health, indebtedness, and economic vulnerability [53]. This undermines poverty alleviation efforts and perpetuates social inequalities, making it more challenging to achieve inclusive and sustainable development.

Addressing the opportunity cost of air pollution in developing economies requires comprehensive policy interventions that promote environmental sustainability, public health, and inclusive economic growth. Policymakers should prioritize investments in pollution control measures, healthcare infrastructure, and social protection programs to mitigate the adverse effects of pollution on human capital formation, economic development, and poverty alleviation efforts. Additionally, policies aimed at promoting clean energy transitions, sustainable transportation, and green urban planning can help reduce pollution levels, enhance environmental quality, and create new opportunities for sustainable livelihoods in vulnerable communities. By integrating environmental considerations into development strategies and fostering cross-sectoral collaboration, policymakers can address the opportunity cost of air pollution and promote resilient, inclusive, and sustainable development pathways for all.

4 Economic Impacts of Carbon Tax on Air Pollution

A carbon tax is a fee imposed on the burning of carbon-based fuels (coal, oil, gas) and is designed to reduce greenhouse gas emissions and address climate change. The economic impacts of a carbon tax on air pollutant emission extend to air pollution reduction, yielding both direct and indirect effects (Table 3). For instance, the reduction of healthcare costs by discouraging the use of fossil fuels and a carbon tax leads to reduced emissions of pollutants like SO2, NOx, and PM especially PM2.5 [102]. Better air quality decreases respiratory and cardiovascular diseases, reducing healthcare expenditures [103]. The results of the study by Limaye et al. [104] found that the health benefits from reduced air pollution could be as significant as the climate benefits themselves. Improved air quality will likely result in fewer workdays lost due to illness, enhancing overall productivity. This can lead to economic gains as a healthier workforce is more efficient and incurs fewer medical-related absences [105]. A carbon tax makes fossil fuels more expensive, encouraging investment in renewable energy sources like wind, solar, and hydropower. This can drive technological innovation and create new industries and jobs in the clean energy sector [106]. Businesses and consumers may adopt energy-efficient technologies and practices to lower their energy costs, further reducing emissions and stimulating the economy through energy savings. Higher costs of carbon-intensive goods can shift consumer preferences toward greener alternatives, promoting sustainable consumption. This shift can spur market growth for eco-friendly products and services, contributing to economic diversification. Governments can use carbon tax revenues to fund public services, reduce other taxes, or invest in green infrastructure. This recycling of revenue can mitigate any regressive effects of the tax and support economic growth. For example, revenue can be used to subsidize renewable energy projects or provide rebates to low-income households.

Due to carbon tax, fossil fuel industries may experience economic downturns due to decreased demand. However, this is balanced by growth in renewable energy sectors, which benefit from increased investment and demand. Sectors heavily reliant on fossil fuels may face higher operating costs. This can lead to increased prices for goods and services, which could be passed on to consumers. To mitigate

Table 3 Some sectoral, direct, and indirect economic effects of carbon tax on air pollution

Category	Potential and actual impacts	Description	References
Direct economic impacts			
Reduction in healthcare costs and productivity gains	Improved air quality	Reduced emissions of SO2, NOx, and PM2.5 lead to fewer respiratory and cardiovascular diseases	Ambasta and Buonocore [102]
	Health benefits	Decreased healthcare expenditures and significant health benefits from reduced air pollution	Barwick et al. [103]
	Workforce health	Fewer workdays are lost due to illness, enhancing overall productivity and economic gains	Bialowolski et al. [105]
Indirect economic impacts			
Innovation and green technologies	Incentivizing clean energy	Higher fossil fuel costs encourage investment in renewable energy sources and drive technological innovation	Liu et al. [106]
	Energy efficiency	Adoption of energy-efficient technologies and practices, reducing emissions and stimulating the economy	Ahmadi et al. [107]
Shift in consumption patterns	Consumer behavior	Higher costs of carbon-intensive goods shift preferences toward greener alternatives, promoting sustainable consumption. Spurs market growth for eco-friendly products and services, contributing to economic diversification	Zhou et al. [111], Jiang et al. [112]
Revenue recycling	Economic redistribution	Carbon tax revenues fund public services, reduce other taxes, or invest in green infrastructure, supporting economic growth. Revenue can subsidize renewable energy projects or provide rebates to low-income households	Mujtaba and Shahzad [99]
Sectoral impacts			
Energy sector	Fossil fuel reduction	Fossil fuel industries may face downturns, but renewable energy sectors benefit from increased investment and demand	Andreoni et al. [113]
Manufacturing and transport	Cost increases	Higher operating costs for fossil fuel-reliant sectors, potentially leading to increased prices for goods and services	Andreoni et al. [113]
	Efficiency improvements	Investment in energy efficiency and cleaner technologies reduces carbon footprint and operational costs	Saqib et al. [114]

these costs, companies may invest in energy efficiency and cleaner technologies, ultimately reducing their carbon footprint and operational costs in the long term. British Columbia introduced a carbon tax in 2008, which led to a significant reduction in fuel consumption and greenhouse gas emissions [108, 102]). According to the work of Ahmadi et al. [109], the province saw a decrease in air pollutants and reported minimal adverse economic impacts, as the tax revenue was used to lower other taxes, maintaining economic stability and growth. Sweden's carbon tax, introduced in 1991, has significantly reduced emissions, while the economy has grown [110]. The tax encouraged energy efficiency and a shift toward renewable energy, improving air quality and public health. Therefore, a carbon tax not only helps mitigate climate change by reducing greenhouse gas emissions but also significantly impacts air pollution levels and the economy. The reduction in air pollution leads to substantial health and productivity benefits, while the economic impacts include shifts in consumption patterns, stimulation of green technologies, and potential redistribution of tax revenues [99]. Although there are challenges, such as potential cost increases in certain sectors, the overall benefits, particularly when revenue is effectively recycled, make the carbon tax a powerful tool for both environmental and economic policy.

5 Policy Implications and Recommendations

Policymakers play a crucial role in addressing the economic costs of air pollution through the implementation of effective regulations and incentives. Existing policies, such as emission standards, pollution taxes, and cap-and-trade systems, aim to reduce air pollution and its economic impacts [115]. However, there is a need for stronger enforcement mechanisms and greater international cooperation to address transboundary air pollution and global environmental challenges. Additionally, investments in clean energy technologies, sustainable transportation infrastructure, and green urban planning can help mitigate air pollution while promoting economic growth and public health [116]. Addressing environmental damage costs associated with air pollution requires coordinated policy interventions at the local, national, and international levels. Policies aimed at reducing emissions of air pollutants, such as emission standards, pollution taxes, and cap-and-trade systems, can help mitigate environmental degradation and promote cleaner production practices [117]. Additionally, investments in pollution control technologies, renewable energy infrastructure, and ecosystem restoration initiatives can help offset the economic costs of environmental damage and foster sustainable development. By integrating environmental considerations into policy frameworks and decision-making processes, policymakers can balance economic growth with environmental protection and ensure the long-term viability of ecosystems and natural resources.

Addressing regional disparities in the economic costs of air pollution requires targeted policy interventions that prioritize equity and social justice [118]. Policymakers should implement measures to reduce pollution hotspots, improve environmental monitoring and enforcement, and enhance access to healthcare services and resources in underserved communities. Additionally, investments in clean energy infrastructure, sustainable transportation, and green urban planning can help mitigate pollution-related health problems and promote economic development in disadvantaged regions [119]. Furthermore, community engagement, public participation, and stakeholder collaboration are essential for identifying local priorities, building resilience, and fostering inclusive decision-making processes.

Existing policy and regulatory instruments with effects on the economic costs of air pollution include:

- **Regulatory Frameworks**: Many countries have implemented regulatory frameworks to control air pollution, including emission standards for industrial facilities, vehicle emissions regulations, and ambient air quality standards [117]. These policies aim to limit pollutant emissions and protect public health and the environment.
- **Market-Based Instruments**: Market-based instruments, such as emissions trading schemes and pollution taxes, provide economic incentives for reducing air pollution by placing a price on emissions [120]. These policies encourage polluters to adopt cleaner technologies and practices while generating revenue for environmental protection initiatives.
- **Technology Standards**: Governments often set technology standards and performance requirements for pollution control equipment and processes to reduce emissions from stationary and mobile sources [23]. These standards drive innovation and investment in cleaner technologies while ensuring compliance with environmental regulations.
- **Public Awareness and Education**: Public awareness campaigns and educational programs play a crucial role in raising awareness about air pollution issues, promoting sustainable behaviors, and fostering community engagement in pollution prevention and mitigation efforts.

Some recommendations for policymakers:

- **Strengthen Enforcement and Compliance:** Policymakers should strengthen enforcement mechanisms and compliance monitoring to ensure that air pollution control policies are effectively implemented and enforced. This may involve investing in monitoring infrastructure, enhancing regulatory oversight, and imposing penalties for non-compliance.
- **Promote Innovation and Technology Adoption**: Policymakers should incentivize innovation and technology adoption in pollution control by providing financial incentives, research grants, and tax credits for developing and deploying cleaner technologies and practices. This can stimulate economic growth, create jobs, and drive sustainable development.
- **Enhance International Cooperation**: Air pollution is a transboundary issue that requires coordinated action at the international level. Policymakers should prioritize international cooperation and collaboration to address cross-border pollution sources, share best practices, and harmonize regulatory standards to achieve global air quality goals.
- **Invest in Sustainable Transportation**: Policymakers should prioritize investments in sustainable transportation infrastructure, including public transit, cycling

infrastructure, and electric vehicle charging stations, to reduce vehicle emissions and improve air quality in urban areas.
- **Address Environmental Justice Concerns**: Policymakers should pay particular attention to addressing environmental justice concerns and ensuring that air pollution control policies benefit all communities, especially those disproportionately affected by pollution, such as low-income and minority populations. This may involve conducting equity assessments, engaging with stakeholders, and prioritizing investments in environmental justice initiatives.

Some future directions for research and policy development to mitigate the economic impacts of air pollution include:

- **Air Quality Monitoring and Data Collection**: Continued investment in air quality monitoring infrastructure and data collection efforts is essential for accurately assessing pollution levels, identifying hotspots, and tracking progress toward air quality goals.
- **Health Impact Assessments**: Further research is needed to better understand the health impacts of air pollution, including the long-term effects of exposure to pollutants on vulnerable populations and the economic costs associated with pollution-related illnesses.
- **Climate Change and Air Quality:** Given the interconnectedness between air quality and climate change, future research should explore synergies and trade-offs between air pollution mitigation strategies and climate change mitigation and adaptation efforts.
- **Policy Evaluation and Effectiveness**: A rigorous evaluation of air pollution control policies is necessary to assess their effectiveness, cost-effectiveness, and societal impacts. This may involve conducting cost-benefit analyses, impact assessments, and stakeholder consultations to inform evidence-based policy decisions.
- **Community Engagement and Participation**: Engaging with local communities and stakeholders in the development and implementation of air pollution control policies is critical for fostering public trust, building resilience, and ensuring the success of pollution prevention and mitigation efforts.

6 Conclusion

Air pollution imposes significant economic costs on society, including direct healthcare expenses, productivity losses, and environmental damage. Addressing these economic costs requires coordinated efforts from policymakers, businesses, and individuals to reduce emissions, protect public health, and safeguard the environment. An understanding of the interconnectedness of these economic impacts and implementing evidence-based policies and interventions is necessary to mitigate the adverse effects of air pollution and build a more sustainable and resilient future for generations to come.

Regional disparities in economic costs of air pollution and vulnerabilities among certain populations highlight the need for comprehensive and equitable approaches to environmental management and public health. By addressing the root causes of pollution, promoting social inclusion, and empowering communities, policymakers can create healthier, more resilient societies and reduce the economic burdens of air pollution on the most vulnerable members of society. Addressing air pollution requires a multifaceted approach that involves a combination of regulatory measures, market-based instruments, technological innovation, and public engagement. By implementing effective policies, promoting sustainable practices, and investing in research and collaboration, policymakers can mitigate air pollution, protect public health, and promote environmental sustainability for current and future generations.

Future endeavors should focus on enhancing air quality monitoring, conducting health impact assessments, exploring synergies between air quality and climate change, evaluating policy effectiveness, and fostering community engagement to foster evidence-based policymaking and sustainable urban development. By adopting a comprehensive approach encompassing policy interventions, technological advancements, and community engagement, stakeholders can mitigate the economic burdens associated with air pollution and foster sustainable, inclusive urban development for present and future generations. The economic analysis of air pollution serves as a critical tool for informing policymaking, guiding resource allocation, and advancing collective efforts to combat pollution, protect public health, and safeguard the environment for present and future generations.

References

1. Perera F (2017) Pollution from fossil-fuel combustion is the leading environmental threat to global pediatric health and equity: solutions exist. Int J Environ Res Public Health 15(1):16. https://doi.org/10.3390/ijerph15010016
2. Shetty SS, Deepthi D, Harshitha S, Sonkusare S, Naik PB, Kumari NS, Madhyastha H (2023) Environmental pollutants and their effects on human health. Heliyon 9(9):e19496. https://doi.org/10.1016/j.heliyon.2023.e19496
3. Ogwu MC (2023) Plants as monitors and managers of pollution. In: Srivastav AL, Grewal AS, Tiwari M, Pham TD (eds) Role of green chemistry in ecosystem restoration to achieve environmental sustainability. Elsevier, Amsterdam, Netherlands. pp. 51–60. ISBN: 978-0-443-15291-7. https://doi.org/10.1016/B978-0-443-15291-7.00022-5
4. Manisalidis I, Stavropoulou E, Stavropoulos A, Bezirtzoglou E (2020) Environmental and health impacts of air pollution: a review. Front Public Health 8:14. https://doi.org/10.3389/fpubh.2020.00014
5. Jiang XQ, Mei XD, Feng D (2016) Air pollution and chronic airway diseases: what should people know and do? J Thorac Dis 8(1):E31–E40. https://doi.org/10.3978/j.issn.2072-1439.2015.11.50
6. Agarwal AK, Raja A, Brown BD (2023) Chronic obstructive pulmonary disease. [updated 2023 Aug 7]. In: StatPearls [Internet]. StatPearls Publishing, Treasure Island, 2024 Jan-. Available from: https://www.ncbi.nlm.nih.gov/books/NBK559281/

7. de Bont J, Jaganathan S, Dahlquist M, Persson Å, Stafoggia M, Ljungman P (2022) Ambient air pollution and cardiovascular diseases: an umbrella review of systematic reviews and meta-analyses. J Intern Med 291(6):779–800. https://doi.org/10.1111/joim.13467
8. Genc S, Zadeoglulari Z, Fuss SH, Genc K (2012) The adverse effects of air pollution on the nervous system. J Toxicol 2012:782462. https://doi.org/10.1155/2012/782462
9. Chang JH, Lee YL, Chang LT, Chang TY, Hsiao TC, Chung KF, Ho KF, Kuo HP, Lee KY, Chuang KJ, Chuang HC (2023) Climate change, air quality, and respiratory health: a focus on particle deposition in the lungs. Ann Med 55(2):2264881. https://doi.org/10.1080/07853890.2023.2264881
10. Usman M, Ma Z, Wasif Zafar M, Haseeb A, Ashraf RU (2019) Are air pollution, economic and non-economic factors associated with per capita health expenditures? Evidence from emerging economies. Int J Environ Res Public Health 16(11):1967. https://doi.org/10.3390/ijerph16111967
11. Gamarra AR, Lechón Y, Vivanco MG, Garrido JL, Martín F, Sánchez E, Theobald MR, Gil V, Santiago JL (2021) Benefit analysis of the 1st Spanish air pollution control programme on health impacts and associated externalities. Atmos 12:32. https://doi.org/10.3390/atmos12010032
12. Voorhees SS, Sakai R, Araki S, Sato H, Otsu A (2001) Cost-benefit analysis methods for assessing air pollution control programs in urban environments – a review. Environ Health Prev Med 6(2):63–73. https://doi.org/10.1007/BF02897948
13. Guijarro F, Tsinaslanidis P (2020) Analysis of the academic literature on environmental valuation. Int J Environ Res Public Health 17(7):2386. https://doi.org/10.3390/ijerph17072386
14. Deng J, Yang J, Liu Z, Tan Q (2023) Environmental protection tax and green innovation of heavily polluting enterprises: a quasi-natural experiment based on the implementation of China's environmental protection tax law. PLoS One 18(6):e0286253. https://doi.org/10.1371/journal.pone.0286253
15. Wang S, Chen G, Han X (2021) An analysis of the impact of the emissions trading system on the green total factor productivity based on the spatial difference-in-differences approach: the case of China. Int J Environ Res Public Health 18(17):9040. https://doi.org/10.3390/ijerph18179040
16. Marcantonio R, Javeline D, Field S, Fuentes A (2021) Global distribution and coincidence of pollution, climate impacts, and health risk in the Anthropocene. PLoS One 16(7):e0254060. https://doi.org/10.1371/journal.pone.0254060
17. Martinez GS, Spadaro JV, Chapizanis D, Kendrovski V, Kochubovski M, Mudu P (2018) Health impacts and economic costs of air pollution in the metropolitan area of Skopje. Int J Environ Res Public Health 15(4):626. https://doi.org/10.3390/ijerph15040626
18. Vilcins D, Christofferson RC, Yoon JH, Nazli SN, Sly PD, Cormier SA, Shen G (2024) Updates in air pollution: current research and future challenges. Ann Glob Health 90(1):9. https://doi.org/10.5334/aogh.4363
19. Persis J, Ben Amar A (2023) Predictive modeling and analysis of air quality - visualizing before and during COVID-19 scenarios. J Environ Manage 327:116911. https://doi.org/10.1016/j.jenvman.2022.116911
20. USEPA (2024) Economic and cost analysis of air pollution regulations. Available online: https://www.epa.gov/economic-and-cost-analysis-air-pollution-regulations
21. Bai J, Ma W, Wang Y, Jiang J (2024) Political incentives in market-based environmental regulation: evidence from China's carbon emissions trading scheme. Heliyon 10(4):e25730. https://doi.org/10.1016/j.heliyon.2024.e25730
22. Ou Y, West JJ, Smith SJ, Nolte CG, Loughlin DH (2020) Air pollution control strategies directly limiting national health damages in the US. Nat Commun 11(1):957. https://doi.org/10.1038/s41467-020-14783-2
23. Ekins P, Zenghelis D (2021) The costs and benefits of environmental sustainability. Sustain Sci 16(3):949–965. https://doi.org/10.1007/s11625-021-00910-5

24. Ullah S, Ullah N, Rajper SA, Ahmad I, Li Z (2021) Air pollution and associated self-reported effects on the exposed students at Malakand division, Pakistan. Environ Monit Assess 193(11):708. https://doi.org/10.1007/s10661-021-09484-2
25. Li L, Du T, Zhang C (2020) The impact of air pollution on healthcare expenditure for respiratory diseases: evidence from the People's Republic of China. Risk Manag Healthcare Policy 13:1723–1738. https://doi.org/10.2147/RMHP.S270587
26. Mitchell RJ, Bates P (2011) Measuring health-related productivity loss. Popul Health Manag 14(2):93–98. https://doi.org/10.1089/pop.2010.0014
27. Le D, Li Y, Ren F (2022) Does air quality improvement promote enterprise productivity increase? Based on the spatial spillover effect of 242 cities in China. Front Public Health 10: 1050971. https://doi.org/10.3389/fpubh.2022.1050971
28. D'Amato G, Bergmann KC, Cecchi L, Annesi-Maesano I, Sanduzzi A, Liccardi G, Vitale C, Stanziola A, D'Amato M (2014) Climate change and air pollution: effects on pollen allergy and other allergic respiratory diseases. Allergo J Int 23(1):17–23. https://doi.org/10.1007/s40629-014-0003-7
29. Bell JE, Brown CL, Conlon K, Herring S, Kunkel KE, Lawrimore J, Luber G, Schreck C, Smith A, Uejio C (2018) Changes in extreme events and the potential impacts on human health. J Air Waste Manag Assoc 68(4):265–287. https://doi.org/10.1080/10962247.2017.1401017
30. Chandrappa R, Chandra KU (2015) Major issues of air pollution. Sustain Air Pollut Manag:1–48. https://doi.org/10.1007/978-3-319-21596-9_1
31. Czechowski PO, Piksa K, Da Browiecki P, Oniszczuk-Jastrząbek AI, Czermański E, Owczarek T, Badyda AJ, Cirella GT (2022) Financing costs and health effects of air pollution in the Tri-City agglomeration. Front Public Health 10:831312. https://doi.org/10.3389/fpubh.2022.831312
32. An J, Heshmati A (2019) The relationship between air pollutants and healthcare expenditure: empirical evidence from South Korea. Environ Sci Pollut Res Int 26(31):31730–31751. https://doi.org/10.1007/s11356-019-06145-3
33. Shen JS, Wang Q, Shen HP (2021) Does industrial air pollution increase health care expenditure? Evidence from China. Front Public Health 9:695664. https://doi.org/10.3389/fpubh.2021.695664
34. Xu X, Yang H, Li C (2022) Theoretical model and actual characteristics of air pollution affecting health cost: a review. Int J Environ Res Public Health 19(6):3532. https://doi.org/10.3390/ijerph19063532
35. Gryech I, Ghogho M, Mahraoui C, Kobbane A (2022) An exploration of features impacting respiratory diseases in urban areas. Int J Environ Res Public Health 19(5):3095. https://doi.org/10.3390/ijerph19053095
36. Bari MW, Saleem S, Bashir M, Ahmad B (2022) Impact of ambient air pollution on outdoor employees' performance: mediating role of anxiety. Front Psychol 13:926534. https://doi.org/10.3389/fpsyg.2022.926534
37. Mendoza DL, Pirozzi CS, Crosman ET, Liou TG, Zhang Y, Cleeves JJ, Bannister SC, Anderegg WRL, Robert 3rd P (2020) Impact of low-level fine particulate matter and ozone exposure on absences in K-12 students and economic consequences. Environ Res Lett 15(11): 114052. https://doi.org/10.1088/1748-9326/abbf7a
38. Ahmad I, Balkhyour MA (2020) Occupational exposure and respiratory health of workers at small scale industries. Saudi J Biol Sci 27(3):985–990. https://doi.org/10.1016/j.sjbs.2020.01.019
39. Zivin JG, Neidell M (2012) The impact of pollution on worker productivity. Am Econ Rev 102(7):3652–3673. https://doi.org/10.1257/aer.102.7.3652
40. Zumrut IB, Kale OA, Tetik YO, Baradan S (2024) Mitigation strategies to reduce particulate matter concentrations in civil engineering laboratories. Environ Sci Pollut Res Int 31(8): 12340–12350. https://doi.org/10.1007/s11356-024-31926-w

41. Lovett GM, Tear TH, Evers DC, Findlay SE, Cosby BJ, Dunscomb JK, Driscoll CT, Weathers KC (2009) Effects of air pollution on ecosystems and biological diversity in the eastern United States. Ann N Y Acad Sci 1162:99–135. https://doi.org/10.1111/j.1749-6632.2009.04153.x
42. Bherwani H, Nair M, Musugu K, Gautam S, Gupta A, Kapley A, Kumar R (2020) Valuation of air pollution externalities: comparative assessment of economic damage and emission reduction under COVID-19 lockdown. Air Qual Atmos Health 13(6):683–694. https://doi.org/10.1007/s11869-020-00845-3
43. Vwioko DE, Okoekhian I, Ogwu MC (2018) Stress analysis of Amaranthus hybridus L. and Lycopersicon esculentum mill. exposed to sulphur and nitrogen dioxide. Pertanika Journal of Tropical Agricultural Sciences 41(3):1169–1191
44. Ikhajiagbe B, Ogwu MC (2020) Hazard quotient, microbial diversity and plant composition of spent crude oil polluted-soil. Beni-Suef University Journal of Basic and Applied Sciences 9 (26). https://doi.org/10.1186/s43088-020-00052-0
45. Lobell DB, Di Tommaso S, Burney JA (2022) Globally ubiquitous negative effects of nitrogen dioxide on crop growth. Sci Adv 8(22):eabm9909. https://doi.org/10.1126/sciadv.abm9909
46. De Marco A, Sicard P, Feng Z, Agathokleous E, Alonso R, Araminiene V, Augustatis A, Badea O, Beasley JC, Branquinho C, Bruckman VJ, Collalti A, David-Schwartz R, Domingos M, Du E, Garcia Gomez H, Hashimoto S, Hoshika Y, Jakovljevic T, McNulty S et al (2022) Strategic roadmap to assess forest vulnerability under air pollution and climate change. Glob Chang Biol 28(17):5062–5085. https://doi.org/10.1111/gcb.16278
47. Kjellstrom T, Lodh M, McMichael T, Ranmuthugala G, Shrestha R, Kingsland S (2006) Air and water pollution: burden and strategies for control. In: Jamison DT, Breman JG, Measham AR, Alleyne G, Claeson M, Evans DB, Jha P, Musgrive P (eds) Disease control priorities in developing countries2nd edn. The International Bank for Reconstruction and Development/ The World Bank, Washington. Chapter 43. Available from: https://www.ncbi.nlm.nih.gov/books/NBK11769/ Co-published by Oxford University Press, New York
48. EEA [European Environment Agency] (2022) Economic cost of wheat yields lost due to ground-level ozone exposure in some European countries in 2019 (EUR million). Available online: https://www.eea.europa.eu/data-and-maps/daviz/production-loss-by-country-in-1#tab-chart_1
49. Schleeter R (2020) New research: air pollution from fossil fuels costs the world $8 billion every day. Available online https://www.greenpeace.org/usa/news/new-research-air-pollution-from-fossil-fuels-costs-the-world-8-billion-every-day/
50. Liu H, Hu T, Wang M (2021) Impact of air pollution on Residents' medical expenses: a study based on the survey data of 122 cities in China. Front Public Health 9:743087. https://doi.org/10.3389/fpubh.2021.743087
51. Brouwer W, Verbooy K, Hoefman R, van Exel J (2023) Production losses due to absenteeism and Presenteeism: the influence of compensation mechanisms and multiplier effects. Pharmacoeconomics 41(9):1103–1115. https://doi.org/10.1007/s40273-023-01253-y
52. He Q, Ji XJ (2021) The labor productivity consequences of exposure to particulate matters: evidence from a Chinese National Panel Survey. Int J Environ Res Public Health 18(23): 12859. https://doi.org/10.3390/ijerph182312859
53. Hajat A, Hsia C, O'Neill MS (2015) Socioeconomic disparities and air pollution exposure: a global review. Curr Environ Health Rep 2(4):440–450. https://doi.org/10.1007/s40572-015-0069-5
54. Richard G, Izah SC, Sawyer EW, Ogwu MC (2023) Seasonal variations in the concentration of some atmospheric pollutants along a major highway in Southern Nigeria. J Environ Studies 30 (1):10–25. https://doi.org/10.21608/JESJ.2023.205673.1045
55. EEA [European Environment Agency] (2019) Unequal exposure and unequal impacts: social vulnerability to air pollution, noise and extreme temperatures in Europe. EEA Report No 22/2018
56. Ferguson L, Taylor J, Davies M, Shrubsole C, Symonds P, Dimitroulopoulou S (2020) Exposure to indoor air pollution across socio-economic groups in high-income countries: A

scoping review of the literature and a modelling methodology. Environ Int 143:105748. https://doi.org/10.1016/j.envint.2020.105748
57. Dondi A, Carbone C, Manieri E, Zama D, Del Bono C, Betti L, Biagi C, Lanari M (2023) Outdoor air pollution and childhood respiratory disease: the role of oxidative stress. Int J Mol Sci 24(5):4345. https://doi.org/10.3390/ijms24054345
58. Rentschler J, Leonova N (2023) Global air pollution exposure and poverty. Nat Commun 14(1):4432. https://doi.org/10.1038/s41467-023-39797-4
59. Remoundou K, Koundouri P (2009) Environmental effects on public health: an economic perspective. Int J Environ Res Public Health 6(8):2160–2178. https://doi.org/10.3390/ijerph6082160
60. Boardman AE, Greenberg DH, Vining AR, Weimer DL (2017) Cost-benefit analysis: concepts and practice.5th edn. Pearson
61. World Bank (2019) Environmental economics: a simple introduction. Available online: https://openknowledge.worldbank.org/handle/10986/31755
62. Hanley N, Shogren J, White B (2001) Introduction to environmental economics. Oxford University Press
63. Levinson A (2012) Valuing public goods using happiness data: the case of air quality. J Public Econ 96(9–10):869–880
64. Smith J (2018) Economic analysis of air pollution control measures. Environ Econ J 10(2): 45–62
65. Pearce D, Turner RK (1990) Economics of natural resources and the environment. Harvester Wheatsheaf
66. Freeman III AM (2003) The measurement of environmental and resource values: theory and methods. Resources for the future2nd edn
67. Pearce DW, Atkinson GD, Mourato S (2006) Cost-benefit analysis and the environment: recent developments. OECD Publishing
68. Wang L, Zhang S (2019) Cost-benefit analysis of reducing air pollution from power plants. Energy Econ 40(3):123–135
69. Navrud S, Ready RC (2007) Valuing cultural heritage: applying environmental valuation techniques to historic buildings, monuments, and artifacts. Edward Elgar Publishing
70. Johnson D, Smith P (2018) Economic valuation of health impacts from air pollution. J Health Econ 15(4):321–335
71. Smith VK, Huang JC (1995) Can markets value air quality? A meta-analysis of hedonic property value models. J Polit Econ 103(1):209–227
72. Khan I, Lei H, Ali G, Ali S, Zhao M (2019) Public attitudes, preferences and willingness to pay for river ecosystem services. Int J Environ Res Public Health 16(19):3707. https://doi.org/10.3390/ijerph16193707
73. Hanemann MW (1994) Valuing the environment through contingent valuation. J Econ Perspect 8(4):19–43
74. Garcia R, Martinez L (2020) Economic valuation of ecosystem services from air quality improvement. Ecol Econ 28(1):89–104
75. Dasgupta S, Laplante B, Mamingi N (2001) Pollution and Capital Markets in Developing Countries. J Environ Econ Manag 42(3):310–335
76. Nguyen H, Tran T (2018) Economic costs of air pollution on low-income communities. Dev Policy Rev 35(2):187–202
77. Gennaioli C, Tavoni M (2017) Clean and dirty international technology diffusion. Rev Econ Stud 84(4):1481–1507
78. Brown M, Green T (2017) Economic evaluation of vehicle emission standards. Transp Res Part D: Transp Environ 25(2):45–58
79. World Health Organization (2019) Air pollution. World Health Organization. https://www.who.int/health-topics/airpollution#tab=tab_1

80. Novikova T, Kaneva M, Zafarjonova M (2023) Cost-benefit analysis for health project evaluation (example of a Russian outpatient clinics' project in the Novosibirsk region). Front Public Health 11:1073964. https://doi.org/10.3389/fpubh.2023.1073964
81. Woollacott J, Alsufyani W, Beach RH, Morrison LTR, Bean de Hernández A, Rakic S, AlOmran M, Alsukait RF, Herbst CH, AlBalawi S (2022) Effective options for addressing air quality- related environmental public health burdens in Saudi Arabia. Heliyon 8(9):e10335. https://doi.org/10.1016/j.heliyon.2022.e10335
82. Torgerson DJ, Raftery J (1999) Economic notes discounting. BMJ 319(7214):914–915. https://doi.org/10.1136/bmj.319.7214.914
83. Chung PC, Lin IF (2023) Sensitivity analysis of selection bias: a graphical display by bias-correction index. PeerJ 11:e16411. https://doi.org/10.7717/peerj.16411
84. Zang X, Jalal H, Krebs E, Pandya A, Zhou H, Enns B, Nosyk B (2020) Prioritizing additional data collection to reduce decision uncertainty in the HIV/AIDS response in 6 US cities: a value of information analysis. Value Health 23(12):1534–1542. https://doi.org/10.1016/j.jval.2020.06.013
85. Schmidt S, Manceur AM, Seppelt R (2016) Uncertainty of monetary valued ecosystem services - value transfer functions for global mapping. PLoS One 11(3):e0148524. https://doi.org/10.1371/journal.pone.0148524
86. Im U, Brandt J, Geels C, Hansen KM, Christensen JH, Andersen MS, Solazzo E, Kioutsioukis I, Alyuz U, Balzarini A, Baro R, Bellasio R, Bianconi R, Bieser J, Colette A, Curci G, Farrow A, Flemming J, Fraser A, Jimenez-Guerrero P et al (2018) Assessment and economic valuation of air pollution impacts on human health over Europe and the United States as calculated by a multi-model ensemble in the framework of AQMEII3. Atmos Chem Phys 18(8):5967–5989. https://doi.org/10.5194/acp-18-5967-2018
87. Williams ML, Beevers S, Kitwiroon N (2018) Public health air pollution impacts of pathway options to meet the 2050 UK Climate Change Act target: a modelling study. NIHR Journals Library, Southampton, 2018 Jun. (Public Health Research, No. 6.7.) Chapter 9, Economic valuation of the air quality impact on mortality. Available from: https://www.ncbi.nlm.nih.gov/books/NBK507610/
88. Diener A, O'Brien B, Gafni A (1998) Health care contingent valuation studies: a review and classification of the literature. Health Econ 7(4):313–326. https://doi.org/10.1002/(sici)1099-1050(199806)7:4<313::aid-hec350>3.0.co;2-b
89. Steigenberger C, Flatscher-Thoeni M, Siebert U, Leiter AM (2022) Determinants of willingness to pay for health services: a systematic review of contingent valuation studies. Eur J Health Econ 23(9):1455–1482. https://doi.org/10.1007/s10198-022-01437-x
90. Morantes G, Rincon G, Chanaba A, Jones B (2024) Addressing air quality challenges: comparative analysis of Barcelona, Venezuela, and Guayaquil, Ecuador. Heliyon 10(8): e29211. https://doi.org/10.1016/j.heliyon.2024.e29211
91. Szinay D, Cameron R, Naughton F, Whitty JA, Brown J, Jones A (2021) Understanding uptake of digital health products: methodology tutorial for a discrete choice experiment using the Bayesian efficient design. J Med Internet Res 23(10):e32365. https://doi.org/10.2196/32365
92. Kim SG, Cho SH, Lambert DM, Roberts RK (2010) Measuring the value of air quality: application of the spatial hedonic model. Air Qual Atmos Health 3(1):41–51. https://doi.org/10.1007/s11869-009-0049-8
93. Safari Z, Fouladi-Fard R, Vahedian M, Mahmoudian MH, Rahbar A, Fiore M (2022) Health impact assessment and evaluation of economic costs attributed to $PM_{2.5}$ air pollution using BenMAP-CE. Int J Biometeorol 66(9):1891–1902. https://doi.org/10.1007/s00484-022-02330-1
94. Gulis G, Krishnankutty N, Boess ER, Lyhne I, Kørnøv L (2022) Environmental impact assessment, human health and the sustainable development goals. Int J Public Health 67: 1604420. https://doi.org/10.3389/ijph.2022.1604420

95. Czembrowski P, Łaszkiewicz E, Kronenberg J, Engström G, Andersson E (2019) Valuing individual characteristics and the multifunctionality of urban green spaces: the integration of sociotope mapping and hedonic pricing. PLoS One 14(3):e0212277. https://doi.org/10.1371/journal.pone.0212277
96. Chen HS (2020) The construction and validation of a sustainable tourism development evaluation model. Int J Environ Res Public Health 17(19):7306. https://doi.org/10.3390/ijerph17197306
97. Fitch A, Kuyer J, Kharadi N, Gower J, Roberts C, Dewey N, Hull S, Jones L (2022) Under the influence of nature: the contribution of natural capital to tourism spend. PLoS One 17(6):e0269790. https://doi.org/10.1371/journal.pone.0269790
98. Lu W, Hackman DA, Schwartz J (2021) Ambient air pollution associated with lower academic achievement among US children: A nationwide panel study of school districts. Environ Epidemiol 5(6):e174. https://doi.org/10.1097/EE9.0000000000000174
99. Mujtaba G, Shahzad SJH (2021) Air pollutants, economic growth and public health: implications for sustainable development in OECD countries. Environ Sci Pollut Res Int 28(10):12686–12698. https://doi.org/10.1007/s11356-020-11212-1
100. Ogwu MC (2019) Towards sustainable development in Africa: the challenge of urbanization and climate change adaptation. In: Cobbinah PB, Addaney M (eds) The Geography of climate change adaptation in urban Africa. Springer Nature, Switzerland. pp. 29–55. https://doi.org/10.1007/978-3-030-04873-0_2
101. Egondi T, Kyobutungi C, Ng N, Muindi K, Oti S, van de Vijver S, Ettarh R, Rocklöv J (2013) Community perceptions of air pollution and related health risks in Nairobi slums. Int J Environ Res Public Health 10(10):4851–4868. https://doi.org/10.3390/ijerph10104851
102. Ambasta A, Buonocore JJ (2018) Carbon pricing: a win-win environmental and public health policy. Can J Public Health 109(5–6):779–781. https://doi.org/10.17269/s41997-018-0099-5
103. Barwick PJ, Li S, Rao D, Zahur N (2018) The morbidity cost of air pollution: evidence from consumer spending in China. NBER Work Pap 5:1–63. https://doi.org/10.2139/ssrn.2999068
104. Limaye VS, Magal A, Joshi J, Maji S, Dutta P, Rajput P, Pingle S, Madan P, Mukerjee P, Bano S, Beig G, Mavalankar D, Jaiswal A, Knowlton K (2023) Air quality and health co-benefits of climate change mitigation and adaptation actions by 2030: an interdisciplinary modeling study in Ahmedabad, India. Environ Res Health 1(2):021003. https://doi.org/10.1088/2752-5309/aca7d8
105. Bialowolski P, McNeely E, VanderWeele TJ, Weziak-Bialowolska D (2020) Ill health and distraction at work: costs and drivers for productivity loss. PLoS One 15(3):e0230562. https://doi.org/10.1371/journal.pone.0230562
106. Liu W, Liu M, Liu T, Li Y, Hao Y (2022) Does a recycling carbon tax with technological progress in clean electricity drive the green economy? Int J Environ Res Public Health 19(3):1708. https://doi.org/10.3390/ijerph19031708
107. Ahmadi M, Sharifi A, Khalili S (2021) Presentation of a developed sub-epidemic model for estimation of the COVID-19 pandemic and assessment of travel-related risks in Iran. Environ Sci Pollut Res Int 28(12):14521–14529. https://doi.org/10.1007/s11356-020-11644-9
108. Elgie S (2015) British Columbia's carbon tax shift: an environmental and economic success. Available online: https://blogs.worldbank.org/en/climatechange/british-columbia-s-carbon-tax-shift-environmental-and-economicsuccess
109. Ahmadi Y, Yamazaki A, Kabore P (2022) How do carbon taxes affect emissions? Plant-level evidence from manufacturing. Environ Resource Econ 82(2):285–325. https://doi.org/10.1007/s10640-022-00678-x
110. GoS (n.d.) Sweden's carbon tax. Available online: https://www.government.se/government-policy/swedens-carbon-tax/swedens-carbon-tax/
111. Zhou Q, Wu J, Imran M, Nassani AA, Binsaeed RH, Zaman K (2023) Examining the trade-offs in clean energy provision: focusing on the relationship between technology transfer, renewable energy, industrial growth, and carbon footprint reduction. Heliyon 9(10):e20271. https://doi.org/10.1016/j.heliyon.2023.e20271

112. Jiang L, Niu H, Ru Y, Tong A, Wang Y (2023) Can carbon finance promote high quality economic development: evidence from China. Heliyon 9(12):e22698. https://doi.org/10.1016/j.heliyon.2023.e22698
113. Andreoni P, Aleluia Reis L, Drouet L, Dessens O, Fragkos P, Pietzcker R, Pye S, Rodrigues R, Tavoni M (2023) Fossil extraction bans and carbon taxes: assessing their interplay through multiple models. iScience 26(4, 106377). https://doi.org/10.1016/j.isci.2023.106377
114. Saqib N, Radulescu M, Usman M, Balsalobre-Lorente D, Cilan T (2023) Environmental technology, economic complexity, renewable electricity, environmental taxes and CO2 emissions: implications for low-carbon future in G-10 bloc. Heliyon 9(6):e16457. https://doi.org/10.1016/j.heliyon.2023.e16457
115. Tu Z, Liu B, Jin D, Wei W, Kong J (2022) The effect of carbon emission taxes on environmental and economic systems. Int J Environ Res Public Health 19(6):3706. https://doi.org/10.3390/ijerph19063706
116. Ma N, Liu P, Xiao Y, Tang H, Zhang J (2022) Can Green technological innovation reduce hazardous air pollutants?-An empirical test based on 283 cities in China. Int J Environ Res Public Health 19(3):1611. https://doi.org/10.3390/ijerph19031611
117. Jafari JA, Charkhloo E, Pasalari H (2021) Urban air pollution control policies and strategies: a systematic review. J Environ Health Sci Eng 19(2):1911–1940. https://doi.org/10.1007/s40201-021-00744-4
118. Picciano P, Qiu M, Eastham SD, Yuan M, Reilly J, Selin NE (2023) Air quality related equity implications of U.S. decarbonization policy. Nat Commun 14(1):5543. https://doi.org/10.1038/s41467-023-41131-x
119. Kruize H, van der Vliet N, Staatsen B, Bell R, Chiabai A, Muiños G, Higgins S, Quiroga S, Martinez-Juarez P, Aberg Yngwe M, Tsichlas F, Karnaki P, Lima ML, García de Jalón S, Khan M, Morris G, Stegeman I (2019) Urban green space: creating a triple win for environmental sustainability, health, and health equity through behavior change. Int J Environ Res Public Health 16(22):4403. https://doi.org/10.3390/ijerph16224403
120. Yu H, Zhang X (2024) Are carbon emissions trading and green financial instruments synergistic? -comprehensive quantitative research based on content analysis. PLoS One 19(3):e0298601. https://doi.org/10.1371/journal.pone.0298601

Corporate and Environmental Synergy for Air Pollution: Risks, Strategies, and Opportunities

Odoligie Imarhiagbe, Matthew C. Ogwu, Ebiuwa G. Obahiagbon, and Nicolas A. Poggioli

Contents

Abstract Air pollution is a critical environmental concern affecting various sectors worldwide, including industries and businesses. With rapid industrial growth and

O. Imarhiagbe (✉)
Department of Health and Social Science, London School of Science and Technology, London, UK
e-mail: odoligie.imarhiagbe@lsst.ac

M. C. Ogwu
Goodnight Family Department of Sustainable Development, Appalachian State University, Boone, NC, USA
e-mail: ogwumc@appstate.edu

E. G. Obahiagbon
Glasgow School for Business and Society, Department of Management and Human Resource Management, Glasgow Caledonian University, London, UK

N. A. Poggioli
Walker College of Business, Appalachian State University, Boone, NC, USA
e-mail: poggiolin@appstate.edu

Matthew Chidozie Ogwu and Sylvester Izah (eds.), *Sustainable Strategies for Air Pollution Mitigation: Development, Economics and Technologies*,
Hdb Env Chem (2024) 133: 141–172, DOI 10.1007/698_2024_1117,

Published online: 12 September 2024

urbanization, the release of pollutants into the atmosphere has intensified, leading to detrimental effects on human health, ecosystems, and the economy. The impacts of air pollution on industries and businesses are substantial and multifaceted, encompassing health, economic, environmental, and reputational dimensions. Addressing these challenges requires concerted efforts from governments, businesses, and civil society to reduce emissions, improve air quality, and promote sustainable development. By adopting cleaner technologies, engaging stakeholders, and adhering to stringent regulatory standards, industries can mitigate the adverse effects of air pollution and contribute to a healthier environment and a more resilient economy. This chapter aims to highlight the wide-ranging and significant business impacts of air pollution across various sectors. The chapter discusses how businesses also have opportunities to innovate, adapt, and collaborate to address these challenges effectively and contribute to a cleaner, healthier, and more sustainable future. This chapter explores the multifaceted impacts of air pollution on industries and businesses, analyzing the challenges they face and the strategies they employ to mitigate these impacts. By prioritizing environmental sustainability, investing in clean technologies, and engaging with stakeholders, businesses can mitigate the adverse effects of air pollution and build resilience in the face of environmental risks.

Keywords Air quality, Business operations, Economic costs, Environmental sustainability, Pollution control

1 Introduction

"Air pollution is contamination of the indoor or outdoor environment by any chemical, physical or biological agent that modifies the natural characteristics of the atmosphere" [1]. The multifaceted impacts of air pollution can be felt by industries, economies, and societies worldwide and can influence businesses across different sectors, posing both challenges and opportunities for innovation and adaptation [2–4]. Air pollutants of concern include but are not limited to black carbon (particulate matter), carbon monoxide, nitrogen dioxide, sulfur dioxide, methane, tropospheric ozone, and hydrofluorocarbons [1, 5]. These pollutants increase morbidity and mortality and impose significant economic costs on businesses, governments, and societies that manifest in various forms, including healthcare expenditures, productivity losses, property damage, and environmental remediation expenses [6]. For businesses, the economic burden of air pollution can be substantial, with direct and indirect costs impacting their bottom line [7, 8]. Industries that are reliant on outdoor operations, such as agriculture, construction, and tourism, may experience disruptions and financial losses during periods of poor air quality. Moreover, increased healthcare costs and absenteeism from air pollution-related illnesses can further strain businesses' finances and hinder productivity [6, 7].

Air pollution can disrupt supply chains by hampering transportation networks, delaying deliveries, and affecting the availability and quality of raw materials and goods [10]. Poor air quality can impede the movement of goods and people, leading to logistical challenges and increased shipping costs. Extreme weather events and environmental disasters exacerbated by air pollution, such as wildfires and hurricanes, can disrupt production facilities, distribution centers, and shipping routes, leading to supply shortages and increased costs for businesses [11]. Additionally, disruptions in the supply chain can affect the reliability and timeliness of product delivery, impacting customer satisfaction and brand reputation [12, 13].

Businesses may face reputational risks and public backlash if they are perceived as contributing to air pollution or failing to address environmental concerns adequately [14, 15]. Stakeholders, including customers, investors, and regulatory authorities, are increasingly scrutinizing businesses' environmental practices and holding them accountable for their impact on air quality and public health [3, 16, 17]. Negative publicity related to air pollution can tarnish a company's brand image, erode customer trust, and result in lost sales and market share [18–20]. Conversely, companies that prioritize environmental sustainability and demonstrate a commitment to reducing air pollution can enhance their reputation, attract environmentally conscious consumers, and differentiate themselves from competitors [21–23].

Stricter environmental regulations and emissions standards compel businesses to invest in pollution control technologies, adopt cleaner production processes, and adhere to compliance requirements to avoid fines, penalties, and legal liabilities. Noncompliance with environmental regulations can tarnish a company's reputation, lead to legal disputes, and result in financial penalties and operational restrictions [24–26]. Additionally, businesses operating in highly regulated industries, such as energy, manufacturing, and transportation, face increased scrutiny from regulatory authorities and may incur significant compliance costs to meet stringent environmental standards [27, 28]. Failure to comply with air quality regulations can also impact businesses' ability to secure permits, licenses, and financing for expansion projects [29]. The growing emphasis on environmental sustainability and clean technologies presents opportunities for businesses to innovate and develop environmentally friendly products and services [10]. Companies that proactively address air pollution issues and embrace sustainable practices can enhance their competitive advantage, attract environmentally conscious consumers, and access new markets and investment opportunities [30, 31]. For example, the renewable energy sector has experienced significant growth as businesses and consumers seek cleaner alternatives to fossil fuels. Similarly, advancements in pollution control technologies, such as catalytic converters, scrubbers, and emissions monitoring systems, have created opportunities for businesses to reduce their environmental footprint and comply with regulatory requirements [32–35]. Businesses can collaborate with government agencies, nonprofit organizations, and other stakeholders to address air pollution collectively through initiatives such as corporate social responsibility programs, industry partnerships, and community engagement efforts [36–39]. Collaborative platforms and industry alliances can facilitate knowledge sharing, technology transfer, and collective action to mitigate air pollution and promote sustainable development. By working together, businesses can leverage their collective resources and expertise to develop innovative solutions, advocate for policy changes, and drive positive environmental outcomes. Additionally, collaborative initiatives can enhance businesses' reputations, strengthen stakeholder relationships, and foster a culture of environmental stewardship within the corporate sector [40–42].

This chapter aims to highlight the wide-ranging and significant business impacts of air pollution across various sectors. The chapter discusses how businesses also have opportunities to innovate, adapt, and collaborate to address these challenges effectively and contribute to a cleaner, healthier, and more sustainable future. This chapter explores the multifaceted impacts of air pollution on industries and businesses, analyzing the challenges they face and the strategies they employ to mitigate these impacts. By prioritizing environmental sustainability, investing in clean technologies, and engaging with stakeholders, businesses can mitigate the adverse effects of air pollution and build resilience in the face of environmental risks [43–45].

2 Nexus of Business, Air Pollution, and Sustainable Environmental Development

Business is a broad term that encompasses a wide range of activities related to the production, buying, selling, and exchange of goods or services for profit [46, 47]. It involves various stakeholders, including entrepreneurs, investors, employees, customers, suppliers, and governments, who interact within a dynamic economic environment to create value and generate revenue. Businesses can operate in different sectors, such as manufacturing, retail, finance, technology, healthcare, and services, and can vary in size from small-scale enterprises to multinational corporations [48–50]. According to Yim [51], the key aspects of business include (Fig. 1):

- **Entrepreneurship**: It is the process of identifying opportunities, innovating, and taking risks to start and manage a business venture.
- **Management**: The practice of planning, organizing, leading, and controlling resources and activities within an organization to achieve its objectives effectively and efficiently.
- **Marketing**: The activities involved in identifying customer needs, promoting products or services, and building relationships with customers to drive sales and profitability.
- **Finance**: This is the management of financial resources, including budgeting, investment decisions, fundraising, and financial reporting, to support business operations and growth.
- **Operations**: The design, production, and delivery of goods or services, as well as the management of supply chains and logistics to meet customer demand.

Fig. 1 Aspects of a business enterprise

- **Human Resources**: The management of personnel, including recruitment, training, performance management, and employee relations, to support organizational goals and employee development.
- **Legal and Ethical Considerations**: Compliance with laws, regulations, and ethical standards governing business operations, including contracts, intellectual property, environmental sustainability, and corporate social responsibility.

Businesses play a vital role in driving economic growth, job creation, innovation, and social development [52, 53]. They contribute to wealth creation, poverty reduction, and improvement in living standards by producing goods and services that satisfy consumer needs and preferences. Additionally, businesses foster competition, drive efficiency, and spur technological advancements, leading to increased productivity and prosperity [54, 55]. In today's globalized and interconnected world, businesses face various challenges and opportunities, including technological disruptions, changing consumer preferences, market volatility, regulatory changes, and environmental sustainability. Successful businesses adapt to these dynamics by embracing innovation, fostering agility, and embracing responsible and sustainable business practices to thrive in a rapidly evolving landscape [56, 57].

Businesses and air pollution are closely interconnected, as industrial and commercial activities often contribute significantly to the emission of pollutants into the atmosphere [58–60]. The impact of businesses on air quality can vary depending on the sector, size, location, and operational practices of the enterprise [61–63]. Business activities from one or more of the following activities result in air pollution.

Industrial Processes and Emissions Many industries, such as manufacturing, energy production, transportation, and agriculture, release various pollutants into the air as by-products of their operations [64]. These pollutants include greenhouse gases (e.g., carbon dioxide, methane), particulate matter, nitrogen oxides, sulfur dioxide, volatile organic compounds, and hazardous air pollutants (e.g., benzene, formaldehyde) [65]. Industrial emissions can result from combustion processes, chemical reactions, and fugitive emissions from equipment and facilities [66]. Certain industrial processes, such as chemical manufacturing, metal smelting, and cement production, release air pollutants directly into the atmosphere [67]. These emissions can include greenhouse gases, particulate matter, heavy metals, and toxic chemicals, posing risks to human health and the environment.

Energy Generation Businesses that rely on fossil fuels for energy generation, such as coal-fired power plants, oil refineries, and natural gas processing facilities, are major sources of air pollution [68]. Burning fossil fuels releases carbon dioxide, sulfur dioxide, nitrogen oxides, and particulate matter into the atmosphere, contributing to smog formation, acid rain, and climate change [69].

Transportation Businesses involved in transportation activities, including shipping, aviation, and road transport, emit pollutants from vehicle exhausts, such as carbon monoxide, nitrogen oxides, volatile organic compounds, and particulate matter [70, 71]. Emissions from diesel trucks, ships, and airplanes can have significant local and regional air quality impacts, particularly in urban areas and near transportation hubs [72, 73].

Agricultural Activities Agricultural practices, such as livestock farming, crop burning, and fertilizer application, can also contribute to air pollution through the release of ammonia, methane, nitrous oxide, and dust particles [74, 75]. Livestock farming, in particular, is a significant source of methane emissions, a potent greenhouse gas that contributes to climate change [76].

Urban Development Urbanization and commercial development can lead to increased air pollution levels due to construction activities, demolition, land clearing, and increased traffic congestion. Urban areas often experience higher levels of air pollution from vehicular emissions, industrial activities, and residential heating and cooking [77, 78].

The nexus of business, air pollution, and sustainable environmental development highlights the interconnectedness of economic activities, environmental degradation, and the need for sustainable solutions. This relationship underscores the critical role that businesses play in both contributing to air pollution and mitigating its adverse effects on the environment and human health [65]. Businesses across various sectors contribute to air pollution through their operations, including industrial production, energy generation, transportation, and agriculture [79, 80]. Industrial activities often involve the combustion of fossil fuels, the release of chemical pollutants, and emissions from manufacturing processes, leading to the release of greenhouse gases, particulate matter, and other air pollutants into the atmosphere [81, 82]. Air pollution has significant environmental impacts, including climate change, biodiversity loss, ecosystem degradation, and habitat destruction. Pollutants emitted by businesses can lead to the acidification of water bodies, contamination of soil and water resources, and disruption of ecological balance, posing threats to ecosystems and wildlife. Air pollution is a major public health concern, contributing to respiratory diseases, cardiovascular ailments, cancer, and other adverse health outcomes [83, 84]. Businesses located in areas with poor air quality can face challenges related to employee health and productivity, as well as increased healthcare costs and liabilities associated with pollution-related illnesses [75]. Businesses are subject to environmental regulations and standards aimed at controlling air pollution and protecting public health and the environment. Compliance with these regulations often requires investments in pollution control technologies, emission reduction measures, and environmental management systems to minimize the environmental footprint of business operations [79, 84]. Adopting sustainable business practices is essential for mitigating the impact of business activities on air quality and promoting environmental sustainability. Businesses can implement measures to improve energy efficiency, reduce emissions, conserve resources, and minimize waste generation through pollution prevention, recycling, and waste management initiatives [82]. Addressing the nexus of business, air pollution, and sustainable development requires innovation, collaboration, and collective action from businesses, governments, civil society, and other stakeholders. Collaboration among businesses, research institutions, and government agencies can foster the development and adoption of clean technologies, green products, and sustainable business models to transition toward a low-carbon, resource-efficient economy. Businesses have a responsibility to minimize their environmental impact, uphold ethical business practices, and contribute to sustainable development goals. Corporate social responsibility initiatives, environmental stewardship programs, and sustainability reporting can enhance transparency, accountability, and stakeholder engagement, demonstrating a commitment to environmental protection and social responsibility.

3 Importance of Understanding the Business Impacts of Air Pollution

Understanding the business impacts of air pollution is of paramount importance in today's industrialized world. As pollution levels continue to rise due to industrial activities, vehicular emissions, and other human activities, the implications for businesses across various sectors are significant. From economic costs to operational disruptions and reputational damage, air pollution can have far-reaching effects on businesses, making it essential for organizations to understand and address these impacts effectively.

Economic Costs One of the most significant reasons for understanding the business impacts of air pollution is its economic costs. Air pollution imposes substantial financial burdens on businesses, including healthcare expenditures, lost productivity, property damage, and environmental cleanup costs [85]. For example, businesses may incur significant expenses in treating employees for pollution-related illnesses, repairing equipment damaged by pollutants, or cleaning up contaminated sites. By understanding the economic costs of air pollution, businesses can better assess their financial risks and liabilities and take proactive measures to mitigate them [86].

Operational Disruptions Air pollution can disrupt business operations in various ways. For instance, high levels of air pollution can pose health risks to employees, leading to increased absenteeism and reduced productivity [87]. Additionally, air pollution can damage equipment and infrastructure, leading to costly repairs and downtime. Moreover, transportation disruptions caused by poor air quality can affect supply chains, logistics, and distribution networks, further impacting business operations [10, 88].

Regulatory Compliance Governments worldwide are enacting stricter regulations and standards to control air pollution and reduce emissions [43]. Businesses operating in regulated industries must comply with these regulations to avoid penalties, fines, and legal liabilities [89]. Understanding the regulatory landscape and compliance requirements is essential for businesses to ensure adherence to environmental laws and regulations. Moreover, businesses that proactively address air pollution can gain a competitive advantage by demonstrating their commitment to sustainability and environmental responsibility [90].

Reputational Damage Air pollution can tarnish a company's reputation and brand image, leading to loss of customer trust and loyalty. Businesses that are perceived as contributing to air pollution or failing to address it adequately may face public backlash and negative publicity [3, 21]. On the other hand, companies that prioritize environmental sustainability and take proactive steps to reduce their environmental footprint can enhance their reputation and attract environmentally conscious consumers and investors [91, 92]. There is a need for businesses to understand the potential reputational risks associated with air pollution to implement sustainable strategies to protect and enhance their brand image.

Business Opportunities Despite the challenges posed by air pollution, it also creates opportunities for businesses to innovate and develop solutions to address environmental challenges [45]. For example, there is a growing demand for clean technologies, pollution control services, and sustainability consulting. By understanding the business impacts of air pollution, companies can identify emerging market trends and capitalize on opportunities for growth and innovation. Additionally, businesses that adopt environmentally friendly practices and products can gain a competitive advantage in the market and appeal to a growing segment of environmentally conscious consumers [3].

Health and Safety Concerns Air pollution not only affects the environment but also poses significant health and safety risks to employees and communities. Exposure to pollutants such as particulate matter, nitrogen oxides, sulfur dioxide, and volatile organic compounds can lead to respiratory problems, cardiovascular diseases, and other adverse health effects [5, 7]. In workplaces with poor indoor air quality, employees may experience symptoms such as headaches, fatigue, and respiratory irritation, leading to decreased productivity and increased absenteeism [93, 94]. Businesses need to prioritize measures to protect the well-being of their employees and surrounding communities, including implementing ventilation systems, providing personal protective equipment, and conducting regular air quality assessments.

Supply Chain Resilience Air pollution can disrupt supply chains and create vulnerabilities in global supply networks [95, 96]. For example, transportation disruptions caused by air quality issues can delay shipments, increase transportation costs, and impact inventory management. Moreover, businesses that rely on raw materials or components sourced from regions with high levels of air pollution may face challenges in ensuring the quality and availability of their supplies [97]. Businesses need to understand the potential risks associated with air pollution in their supply chains; businesses can diversify their supplier base, invest in alternative transportation methods, and implement contingency plans to mitigate disruptions and ensure supply chain resilience [10].

Long-Term Sustainability Addressing air pollution is essential for ensuring the long-term sustainability and viability of businesses. Environmental degradation caused by air pollution can have far-reaching consequences for ecosystems, natural resources, and biodiversity, ultimately affecting the long-term viability of businesses and their operations [11, 98–111]. By understanding the long-term environmental impacts of air pollution, businesses can adopt sustainable practices and technologies that minimize their environmental footprint and contribute to the preservation of natural resources. Moreover, businesses that prioritize environmental sustainability are better positioned to meet evolving consumer preferences, regulatory requirements, and investor expectations, enhancing their resilience and competitiveness in the long run.

4 Economic Costs of Air Pollution

The economic costs of air pollution encompass a broad range of impacts that affect individuals, businesses, governments, and society [7]. These costs arise from both direct and indirect consequences of air pollution, including healthcare expenditures, lost productivity, property damage, and environmental remediation efforts. Understanding the economic burden of air pollution is crucial for policymakers, businesses, and communities to develop effective strategies for pollution control and mitigation. Air pollution contributes to a variety of respiratory and cardiovascular diseases, including asthma, bronchitis, heart attacks, and lung cancer [101, 102]. As a result, healthcare systems incur significant costs in diagnosing, treating, and managing air pollution-related illnesses. These costs include hospitalization expenses, medication costs, doctor's visits, and long-term healthcare services. Individuals and families also bear financial burdens associated with out-of-pocket medical expenses and health insurance premiums, leading to increased healthcare costs at both individual and societal levels.

Air pollution can impair individuals' health and well-being, leading to decreased productivity in the workplace (Jishi 2008; [93]). Employees who suffer from respiratory symptoms or chronic illnesses due to air pollution may experience absenteeism, reduced work performance, and lower job satisfaction [103, 104]. Additionally, air pollution-related illnesses may result in disability and premature mortality, further reducing individuals' ability to participate in the labor force. Lost productivity due to air pollution not only affects individual workers but also has broader implications for businesses, industries, and national economies, leading to decreased economic output and competitiveness [104, 105]. Air pollution can cause damage to property and infrastructure, resulting in repair and maintenance costs for businesses, homeowners, and governments. Acid rain, for example, caused by air pollutants such as sulfur dioxide and nitrogen oxides, can corrode building materials, vehicles, and infrastructure, leading to deterioration and reduced lifespan. Additionally, air pollution can tarnish the aesthetic appeal of buildings and monuments, impacting tourism revenue and property values in affected areas. Businesses may incur expenses related to cleaning, repainting, and repairing property damaged by air pollution, further exacerbating the economic costs (Jishi 2008; [100]).

Efforts to mitigate and remediate the environmental impacts of air pollution require substantial financial resources from governments, businesses, and other stakeholders [106]. Cleanup and restoration activities to address contaminated air, water, and soil resources contaminated by air pollutants entail significant costs, including monitoring, testing, and remediation efforts [107]. Governments may allocate funds for environmental cleanup and restoration projects, while businesses may invest in pollution control technologies and infrastructure upgrades to comply with environmental regulations and reduce their environmental footprint [106]. These expenditures contribute to the overall economic burden of air pollution on society. The healthcare sector bears a significant portion of the economic costs of air pollution, as hospitals, clinics, and healthcare providers treat patients suffering from air pollution-related illnesses [29, 108]. Healthcare expenditures associated with air pollution include costs related to hospital admissions, emergency room visits, outpatient care, medications, and medical procedures. Individuals and families may incur out-of-pocket expenses for healthcare services not covered by insurance, further straining household budgets and financial resources [108]. Additionally, governments and taxpayers subsidize healthcare services and programs aimed at addressing air pollution-related health issues, placing additional financial burdens on public healthcare systems [110]. Air pollution disproportionately affects vulnerable populations, including low-income communities, minorities, and marginalized groups, exacerbating existing economic disparities. These populations often reside in areas with higher levels of air pollution due to proximity to industrial facilities, transportation corridors, and other sources of pollution. As a result, they bear a disproportionate burden of air pollution-related health problems and economic costs, including higher healthcare expenses, lost productivity, and reduced property values. Addressing economic disparities and promoting environmental justice is essential for mitigating the economic costs of air pollution and ensuring equitable access to clean air and healthy environments for all individuals and communities.

5 Impacts of Air Pollution on Industries and Businesses

Air pollution is a pressing environmental issue that affects various sectors of society, including industries and businesses. With the rapid industrialization and urbanization observed globally, the emission of pollutants into the atmosphere has escalated, leading to detrimental consequences for human health, ecosystems, and the economy [11, 111, 112]. Air pollution has profound impacts on business operations across various sectors, ranging from increased operational costs to diminished productivity and reputational damage [53, 93]. Air pollution poses significant challenges to business operations, ranging from increased costs and reduced productivity to supply chain disruptions and reputational damage.

For instance, in the manufacturing sector, air pollution can significantly disrupt manufacturing operations by causing equipment corrosion, reducing efficiency, and increasing maintenance requirements [81]. Pollutants such as particulate matter and sulfur dioxide can contaminate production facilities, leading to product defects and quality issues [80, 83]. Manufacturing industries often face substantial expenses associated with meeting environmental regulations and emission standards [22, 28]. Investments in pollution control technologies, such as scrubbers and filtration systems, are necessary to comply with regulatory requirements [23]. Compliance costs may include permit fees, monitoring expenses, and fines for noncompliance, impacting the profitability of businesses [113, 114]. In the tourism and hospitality industry, air pollution can deter tourists from visiting destinations known for poor air quality, resulting in decreased tourism revenue for affected regions. Smog and haze can obscure scenic views and natural landscapes, diminishing the attractiveness of tourist destinations [115]. Cultural and historical

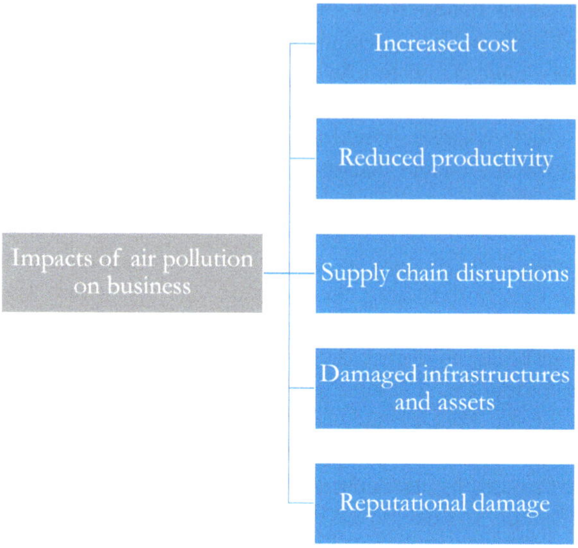

Fig. 2 Effects of air pollution on businesses

sites, such as monuments, temples, and archaeological sites, are susceptible to damage from air pollution [116]. Acid rain and airborne pollutants can corrode building materials and degrade the integrity of cultural heritage sites, leading to costly restoration efforts. Air pollution can harm agricultural productivity by damaging crops through leaf injury, reduced photosynthesis, and stunted growth [117, 118]. Ozone pollution negatively impacts crop yields by impairing plant development and decreasing overall harvests. Airborne pollutants can deposit onto soil and water surfaces, contaminating agricultural produce and water sources. Heavy metals, such as lead and mercury, can accumulate in crops and livestock, posing risks to human health through food consumption. Acid deposition from air pollution can acidify soil and water bodies, affecting nutrient availability and compromising ecosystem health. These impacts underscore the far-reaching consequences of air pollution on various industries and businesses, highlighting the urgent need for concerted efforts to address air quality issues and promote sustainable development practices [119]. Some key impacts of air pollution on business operations include (Fig. 2):

Increased Costs Businesses must invest in pollution control technologies and measures to adhere to environmental regulations and emission standards, leading to higher operational expenses. Poor air quality can result in adverse health effects among employees, leading to increased healthcare expenditures for businesses, including medical treatment, absenteeism, and disability claims [120]. Companies operating in areas with high levels of air pollution may face higher insurance premiums due to increased risks of property damage and liability claims associated with pollution-related incidents.

Reduced Productivity Air pollution can lead to respiratory problems, allergies, and other health issues among employees, reducing their productivity and overall well-being [121]. Employees may be absent from work more frequently due to illnesses caused or exacerbated by poor air quality, leading to decreased workforce productivity and increased staffing costs [122]. Exposure to air pollutants, such as particulate matter and volatile organic compounds, can impair cognitive function, concentration, and decision-making abilities, impacting job performance and efficiency [123, 124].

Supply Chain Disruptions Poor air quality can disrupt transportation networks, leading to delays in the delivery of raw materials, components, and finished goods, affecting production schedules and inventory management [125, 126]. Businesses reliant on suppliers located in areas with severe air pollution may face supply chain disruptions, shortages, and increased costs associated with sourcing alternative suppliers or mitigating risks.

Damage to Infrastructure and Assets Airborne pollutants, including sulfur dioxide and nitrogen oxides, can corrode building materials, machinery, and infrastructure, necessitating costly repairs and maintenance [81]. Acid rain, a consequence of air pollution, can damage buildings, vehicles, and outdoor equipment, leading to the depreciation of assets and increased capital expenditures for replacement.

Reputational Damage Businesses operating in regions with poor air quality may face negative consumer perception and backlash, impacting brand reputation, customer loyalty, and sales [127]. Investors and stakeholders may be reluctant to invest in companies perceived to have a negative environmental impact, leading to decreased shareholder value and funding opportunities [128].

6 Financial Risks and Liabilities from Air Pollution

Financial risks and liabilities are inherent aspects of business operations and refer to potential threats or obligations that may adversely affect a company's financial health, profitability, or ability to meet its financial obligations [129]. These risks can arise from various sources and may have significant implications for businesses if not properly managed. Financial risks and liabilities pose significant challenges for businesses of all sizes and industries, requiring proactive risk management, robust internal controls, and effective governance mechanisms to safeguard financial stability, protect shareholder value, and ensure long-term business sustainability [130, 131]. By identifying, assessing, and mitigating potential risks, businesses can enhance resilience, seize growth opportunities, and navigate uncertain economic conditions with greater confidence and agility. Some common types of financial risks and liabilities:

Market Risk Market risk refers to the potential loss resulting from changes in market prices or market conditions, including fluctuations in interest rates, exchange rates, commodity prices, and equity prices [132]. Businesses that are exposed to market risk may experience reduced revenues, increased costs, or losses on investments and assets due to adverse market movements.

Credit Risk Credit risk, also known as default risk, arises from the possibility that a counterparty may fail to fulfill its contractual obligations, leading to financial losses for the business [133]. This risk is particularly relevant for businesses that extend credit to customers, invest in debt securities, or engage in lending and borrowing activities [134]. Failure to assess and manage credit risk effectively can result in bad debts, impaired assets, and liquidity problems.

Operational Risk Operational risk stems from internal processes, systems, human error, or external events that may disrupt business operations or cause financial losses. This includes risks associated with technology failures, supply chain disruptions, regulatory compliance, fraud, legal disputes, and natural disasters [135, 136]. Failure to mitigate operational risk can lead to production delays, reputational damage, regulatory penalties, and financial liabilities [95].

Liquidity Risk Liquidity risk refers to the inability of a business to meet its short-term financial obligations or fund its operations due to insufficient cash flow or access to liquid assets. Businesses facing liquidity risk may struggle to pay suppliers, meet payroll, or cover operating expenses, leading to financial distress, insolvency, or bankruptcy [104]. Lack of liquidity can also hinder growth opportunities and limit the company's ability to seize strategic opportunities.

Legal and Regulatory Risk Legal and regulatory risks arise from noncompliance with laws, regulations, and industry standards governing business operations, transactions, and conduct [137]. Businesses operating in highly regulated industries or across multiple jurisdictions face legal and regulatory challenges related to taxation, environmental protection, labor practices, data privacy, product safety, and intellectual property rights [137]. Violations of laws or regulations can result in fines, penalties, lawsuits, reputational damage, and legal liabilities.

Environmental and Social Risk Environmental and social risks pertain to the impact of business activities on the environment, communities, and stakeholders. This includes risks associated with environmental pollution, resource depletion, climate change, social unrest, human rights violations, and ethical lapses. Businesses that fail to address environmental and social risks may face public scrutiny, protests, boycotts, and damage to brand reputation, leading to financial losses and long-term sustainability concerns.

Strategic Risk Strategic risk arises from the uncertainty and unpredictability of business strategy, market dynamics, competitive landscape, technological advancements, and changing consumer preferences [139, 140]. Businesses that fail to adapt to evolving market trends, anticipate competitive threats, or innovate strategically may lose market share, revenue opportunities, or competitive advantage, resulting in financial underperformance or decline.

Financial risks and liabilities stemming from air pollution can have profound implications for businesses across various industries. Although the direct financial impacts may vary depending on the nature of the business and its exposure to air pollution-related risks, several common financial risks and liabilities can arise including one of the following:

Operational Disruption Costs Air pollution, particularly from sources such as industrial emissions or wildfires, can disrupt business operations by causing facility closures, production delays, or supply chain disruptions [141]. These operational disruptions can lead to lost revenues, increased operating expenses, and potential penalties for failing to meet contractual obligations or service agreements.

Health and Safety Costs Exposure to air pollution poses risks to employee health and safety, potentially resulting in increased healthcare costs, workers' compensation claims, absenteeism, and productivity losses [142, 143]. Businesses may incur expenses related to providing medical treatment, implementing safety measures, or compensating workers for occupational illnesses or injuries caused or exacerbated by air pollution exposure.

Regulatory Compliance Costs Businesses operating in regions with stringent air quality regulations may face compliance costs associated with emissions monitoring, reporting, and mitigation measures [144]. Failure to comply with environmental regulations can lead to fines, penalties, legal expenses, and reputational damage. Additionally, regulatory changes or stricter enforcement of air quality standards may require businesses to invest in pollution control technologies or transition to cleaner production processes, resulting in capital expenditures and operational adjustments.

Legal Liabilities and Lawsuits Businesses emitting pollutants into the air may face legal liabilities and lawsuits from affected parties, including nearby residents, communities, or environmental advocacy groups [137]. Legal claims may allege harm or damages resulting from air pollution exposure, such as property damage, personal injury, adverse health effects, or loss of livelihood. Litigation costs, settlements, and judgments arising from air pollution-related lawsuits can be substantial and may impact the company's financial performance and reputation [138].

Reputation and Brand Damage Public awareness and concern about air pollution and its adverse effects on health and the environment have increased, leading to heightened scrutiny of businesses' environmental practices and emissions. Negative publicity, media coverage, or public outcry related to air pollution incidents or environmental violations can tarnish the company's reputation, erode consumer trust, and damage brand value [145]. Rebuilding trust and repairing reputational damage may require costly marketing campaigns, corporate social responsibility initiatives, or stakeholder engagement efforts.

Insurance Costs and Coverage Limitations Businesses may experience higher insurance premiums, coverage limitations, or exclusions for air pollution-related risks, particularly in high-risk industries or regions with poor air quality. Insurers may view air pollution exposures as unpredictable or costly liabilities and adjust insurance policies, accordingly, leading to increased insurance costs or difficulty obtaining coverage [146]. Inadequate insurance coverage for air pollution risks can leave businesses vulnerable to financial losses and liabilities in the event of pollution-related incidents or claims.

Market and Investor Risks Concerns about air pollution and environmental sustainability can influence investor perceptions, investment decisions, and market valuations of businesses. Companies with significant air pollution liabilities, regulatory compliance risks, or reputational issues may face higher borrowing costs, reduced access to capital, or stock price volatility [147]. Investors, shareholders, and stakeholders increasingly consider environmental factors, including air pollution, in assessing business risks, financial performance, and long-term value creation [148].

7 Business Opportunities in Addressing Air Pollution

Addressing air pollution presents not only challenges but also opportunities for businesses to innovate, develop new technologies, and contribute to environmental sustainability including through one or more of the following avenues:

Clean Energy Production Technologies Clean energy technologies produce energy in ways that cause less air pollution than existing energy technologies or no pollution. Developing, manufacturing, installing, and maintaining renewable energy sources like solar, wind, hydroelectric, and geothermal power presents significant opportunities for businesses to meet the growing demand for sustainable energy solutions [150]. The end-of-life management, decommissioning, and removal of polluting technologies like coal, gas, and oil power plants also offer business opportunities [151].

Emission Reduction Solutions Businesses that specialize in designing and implementing emission reduction technologies and strategies play a crucial role in helping industries and governments meet air quality standards [151]. Established business opportunities center on the installation and maintenance of pollutant monitoring and control systems. Opportunities are emerging around technologies to address materials more recently classified as pollutants. For example, a nascent industry is emerging around reducing emissions of carbon dioxide through technologies that capture it from air and use it as a feedstock for products like transportation fuel [152] or permanently store it in underground geological formations [153].

Green Transportation With the transportation sector being a major contributor to air pollution, there is a growing demand for cleaner and more sustainable transportation solutions [154]. Greening transportation involves two general paths: substituting cleaner technologies for polluting technologies and altering transportation patterns to reduce energy needs. Businesses involved in producing and maintaining electric vehicles, supporting public transportation systems, and producing and delivering alternative fuels such as hydrogen, sustainable aviation fuel, and biofuels have significant growth potential [155]. There are significant growth opportunities for logistics companies in reducing the distance goods must travel across value chains and making supply chains more resilient to disruption [156].

Air Quality Monitoring and Management There is an increasing need for accurate and reliable air quality monitoring systems to track pollution levels and assess the effectiveness of pollution control measures [157, 158]. Companies that develop sensors, monitoring devices, and software solutions for air quality management can capitalize on this growing market. There are significant innovation opportunities in developing and deploying novel forms of monitoring and management, including business opportunities that might arise from new satellite air pollution monitoring systems [159], in the same way significant business opportunities arose from satellite systems that monitor weather [160].

Green Building Technologies Buildings are responsible for a significant portion of urban air pollution, primarily through venting emissions from heating, cooling, ventilation, and industrial systems into surrounding communities. Businesses that provide energy-efficient HVAC systems and smart building technologies can help reduce pollution by lowering the amount of pollution emissions from buildings [161].

Waste Management and Recycling Waste disposal causes air pollution through landfill emissions and incineration processes [162]. Businesses that specialize in waste management, recycling, and resource recovery offer solutions to reduce pollution and promote a circular economy [163]. There is a significant business opportunity in capturing methane biogas emitted from landfills and selling it for power production and other uses [164]. Waste disposal services that substitute for on-site waste incineration can improve local air quality [165]. There are widespread business opportunities in creating more circular business models that could reduce air pollution by diverting waste stream materials back into production systems [166].

Carbon Offsetting and Climate Finance As businesses strive to reduce their carbon footprint, there is a growing demand for carbon offset projects and climate finance solutions. Companies that invest in renewable energy projects, reforestation initiatives, and other carbon offset activities can generate revenue while helping to mitigate air pollution [167]. Offsetting and climate finance require substantial governance implementation to ensure that offsetting claims are verified. There are significant business opportunities in offset auditing and verification [168].

Environmental Consulting and Advisory Services Businesses that offer environmental consulting, auditing, and advisory services play a vital role in helping organizations navigate environmental regulations, assess environmental risks, and develop sustainable business practices to reduce air pollution [169]. Sustainability reporting requirements have created strong demand for consulting and advisory services related to air pollution and other sustainability outcomes [170].

8 Policy and Regulatory Implications of Air Pollution on Businesses

The impacts of air pollution on businesses extend beyond operational challenges to encompass policy and regulatory implications that can significantly affect their operations and bottom line.

Environmental Regulations Governments worldwide have enacted environmental regulations aimed at controlling air pollution and reducing emissions of harmful pollutants such as particulate matter, sulfur dioxide, nitrogen oxides, and volatile organic compounds [3, 89]. These regulations often impose limits on emissions from industrial facilities, power plants, vehicles, and other sources of pollution. Businesses must comply with these regulations, which may require investing in pollution control technologies, implementing emission reduction measures, and reporting emissions data to regulatory authorities.

Permitting and Compliance Requirements Businesses that emit pollutants into the air are typically required to obtain permits from regulatory agencies authorizing their operations and specifying emission limits and control measures. Obtaining and maintaining these permits can be a complex and time-consuming process, requiring businesses to demonstrate compliance with environmental regulations through monitoring, reporting, and record-keeping requirements.

Emission Trading Programs Some jurisdictions have implemented emission trading programs, such as cap-and-trade systems, to reduce air pollution cost-effectively [171, 172]. Under these programs, businesses are allocated a certain number of emission allowances, which they can buy, sell, or trade with other entities. Participation in emissions trading programs can provide businesses with flexibility in meeting their emission reduction targets while potentially generating revenue from the sale of excess allowances.

Carbon Pricing Mechanisms Carbon pricing mechanisms, such as carbon taxes and emissions trading schemes, aim to internalize the costs of carbon pollution and incentivize businesses to reduce their carbon footprint [57]. By putting a price on carbon emissions, governments create economic incentives for businesses to invest in cleaner technologies, improve energy efficiency, and transition to low-carbon energy sources.

Public Disclosure and Transparency Requirements Increasingly, businesses are required to disclose information about their environmental performance, including air emissions, as part of corporate reporting and transparency initiatives [173]. Stakeholders, including investors, customers, and the public, are demanding greater transparency and accountability regarding businesses' environmental impacts, prompting companies to adopt more robust reporting practices and sustainability disclosure standards.

Litigation and Legal Liability Businesses may face legal challenges and liability claims related to air pollution, particularly if their activities result in harm to public health or the environment. Lawsuits alleging violations of environmental laws, negligence, nuisance, or other legal theories can result in costly litigation, fines, penalties, and reputational damage for businesses found liable for air pollution-related harm.

International Agreements and Treaties Businesses operating across borders may be subject to international agreements and treaties addressing air pollution and climate change, such as the Paris Agreement and the Kyoto Protocol. These agreements set global emission reduction targets and establish frameworks for international cooperation on climate mitigation and adaptation efforts, influencing businesses' strategies, investments, and operations worldwide.

The policy and regulatory landscape surrounding air pollution continues to evolve, with governments enacting stricter regulations, implementing market-based mechanisms, and increasing enforcement efforts to address environmental challenges [174]. Businesses must stay abreast of these developments, proactively manage their environmental risks, adopt sustainable practices to navigate the complex regulatory environment, and maintain their social license to operate. Examples of these policies and regulations are presented in Table 1, and they play crucial roles in shaping business practices, driving investments in pollution control technologies, promoting sustainable development, and protecting public health and the environment. Businesses must stay informed about regulatory requirements, monitor compliance obligations, and adapt their operations to meet evolving air quality standards and emission reduction targets.

9 Conclusion

The impacts of air pollution on businesses are multifaceted and far-reaching, affecting various sectors, operations, and financial outcomes. Businesses face significant challenges and opportunities in navigating the complex landscape of air quality regulation, environmental sustainability, and public health concerns. It is essential to recognize that air pollution poses substantial risks to businesses across diverse industries. From manufacturing plants to hospitality establishments, companies are increasingly vulnerable to regulatory scrutiny, reputational damage, and financial liabilities associated with their environmental performance. The manufacturing sector, for instance, must contend with stringent emissions standards, compliance costs, and operational disruptions resulting from regulatory enforcement actions. Similarly, businesses in the tourism and hospitality industry face challenges related to reduced tourist traffic, property damage from pollution-induced environmental degradation, and heightened consumer awareness of health and safety risks.

Furthermore, air pollution exerts a significant toll on agricultural and food production systems, threatening crop yields, food security, and supply chain

Table 1 Some relevant air pollution policies and regulations

Policy/regulation	Description	Impact on businesses	References
Clean Air Act (CAA)	Federal law in the United States aimed at controlling air pollution, including emissions from stationary sources (e.g., industrial facilities) and mobile sources (e.g., vehicles)	Businesses must comply with emission limits, install pollution control equipment, obtain permits, and report emissions data. Noncompliance can result in fines, penalties, and legal action. The CAA also promotes the development and adoption of cleaner technologies and fuels	Ross et al. [175], Nethery et al. [176]
European Union Emissions Trading System (EU ETS)	Cap-and-trade program established by the European Union to limit greenhouse gas emissions from industries such as power generation, manufacturing, and aviation	Businesses are allocated or required to purchase emissions allowances, which can be traded on the carbon market. Compliance costs depend on emission levels, market prices, and regulatory requirements. The EU ETS incentivizes emissions reduction efforts and investments in cleaner technologies	Bayer and Aklin [177]
National Ambient Air Quality Standards (NAAQS)	US federal standards set by the Environmental Protection Agency (EPA) to protect public health and the environment by limiting concentrations of criteria air pollutants, including ozone, particulate matter, nitrogen dioxide, sulfur dioxide, carbon monoxide, and lead	Businesses located in areas exceeding NAAQS must implement pollution control measures to reduce emissions and improve air quality. Non-attainment areas may face additional regulatory requirements and restrictions on new or expanding industrial activities	McClellan [178], Faridi et al. [179]
Kyoto Protocol	International treaty aimed at reducing greenhouse gas emissions to mitigate climate change. It established binding emission reduction targets for developed countries and established mechanisms for emissions trading and project-based emission reductions (clean development mechanism, joint implementation)	Businesses in countries subject to emission reduction targets must implement measures to reduce greenhouse gas emissions. Participation in emissions trading and offset projects can provide revenue opportunities or compliance flexibility	Kim et al. [180]
Corporate Environmental Responsibility (CER) Regulations	Various regulations and standards require businesses to assess and manage their environmental impacts, including air emissions, as	Businesses are increasingly expected to disclose and reduce their environmental footprint, including air pollution emissions, as part of their	Wang et al. [144], Zhou and Jin [181]

(continued)

Table 1 (continued)

Policy/regulation	Description	Impact on businesses	References
	part of their corporate social responsibility (CSR) efforts	sustainability initiatives and to meet stakeholder expectations. Compliance with CER regulations can enhance corporate reputation and competitiveness	
Vehicle Emission Standards	Regulations setting limits on exhaust emissions from vehicles, including cars, trucks, buses, and off-road equipment, to reduce air pollution and improve air quality	Automotive manufacturers must design and produce vehicles that meet emission standards, such as limits on nitrogen oxides (NOx), particulate matter (PM), carbon monoxide (CO), and hydrocarbons (HC). Compliance costs may include investments in emission control technologies and testing equipment	Singh et al. [182]

integrity. Farmers and food processors must grapple with the detrimental effects of air pollutants on soil health, water quality, and crop resilience, leading to reduced agricultural productivity, increased input costs, and compromised food safety. As consumers become more discerning about the origin and quality of their food products, businesses throughout the agricultural value chain must prioritize environmental stewardship and sustainable farming practices to maintain market competitiveness and consumer trust. Despite these challenges, it is essential to recognize that businesses also play a pivotal role in driving solutions to mitigate air pollution and promote environmental sustainability. By embracing cleaner production technologies, adopting energy-efficient practices, and investing in pollution control measures, companies can minimize their environmental footprint and enhance operational efficiency. Moreover, businesses have a unique opportunity to leverage innovation, collaboration, and strategic partnerships to develop scalable solutions that address air quality challenges while creating shared value for stakeholders. One promising avenue for businesses to mitigate air pollution is through the adoption of renewable energy sources and low-carbon technologies. By transitioning to cleaner energy sources such as solar, wind, and hydroelectric power, companies can reduce their reliance on fossil fuels, decrease greenhouse gas emissions, and contribute to the decarbonization of the economy. Additionally, investments in energy efficiency measures, such as building retrofits, industrial process optimization, and fleet electrification, can yield significant cost savings while reducing air pollutant emissions and enhancing corporate sustainability.

Businesses can play a critical role in advocating for policy reforms and regulatory frameworks that incentivize pollution abatement, promote sustainable development, and foster innovation. By engaging with policymakers, industry associations, and civil society organizations, companies can influence the development of evidence-based policies that strike a balance between environmental protection and economic growth. Additionally, businesses can demonstrate leadership by voluntarily committing to ambitious emission reduction targets, transparent reporting practices, and sustainable business practices that align with global best practices and industry standards. The impacts of air pollution on businesses are profound and multifaceted, spanning various sectors, operations, and financial outcomes. However, businesses also have a unique opportunity to drive positive change by embracing sustainable practices, investing in clean technologies, and advocating for policy reforms that prioritize environmental protection and public health. By adopting a proactive and collaborative approach to addressing air quality challenges, businesses can not only mitigate risks but also capitalize on emerging opportunities to create value, drive innovation, and contribute to a cleaner, healthier, and more sustainable future for all.

References

1. World Health Organization (2024) Air pollution. https://www.who.int/health-topics/air-pollution
2. Fang Z, Wu PY, Lin YN, Chang TH, Chiu YH (2021) Air pollution's impact on the economic, social, medical, and industrial injury environments in China. Healthcare 9(3):261. https://doi.org/10.3390/healthcare9030261
3. Manisalidis I, Stavropoulou E, Stavropoulos A, Bezirtzoglou E (2020) Environmental and health impacts of air pollution: a review. Front Public Health 8:14. https://doi.org/10.3389/fpubh.2020.00014
4. Vilcins D, Christofferson RC, Yoon JH, Nazli SN, Sly PD, Cormier SA, Shen G (2024) Updates in air pollution: current research and future challenges. Ann Glob Health 90(1):9. https://doi.org/10.5334/aogh.4363
5. Climate and Clean Air Coalition. (n.d.). Climate & Clean Air Coalition. https://www.ccacoalition.org/
6. Anwar A, Hyder S, Bennett R, Younis M (2022) Impact of environmental quality on healthcare expenditures in developing countries: a panel data approach. Healthcare 10(9):1608. https://doi.org/10.3390/healthcare10091608
7. Pervin T, Gerdtham UG, Lyttkens CH (2008) Societal costs of air pollution-related health hazards: a review of methods and results. Cost Eff Resour Alloc 6:19. https://doi.org/10.1186/1478-7547-6-19
8. Xu X, Yang H, Li C (2022) Theoretical model and actual characteristics of air pollution affecting health cost: a review. Int J Environ Res Public Health 19(6):3532. https://doi.org/10.3390/ijerph19063532
9. Liu H, Hu T, Wang M (2021) Impact of air pollution on residents' medical expenses: a study based on the survey data of 122 cities in China. Front Public Health 9:743087. https://doi.org/10.3389/fpubh.2021.743087
10. Obahiagbon EG, Ogwu MC (2023) Sustainable supply chain management in the herbal medicine industry. In: Izah SC, Ogwu MC, Akram M (eds) Herbal medicine phytochemistry. Reference series in phytochemistry. Springer, Cham. https://doi.org/10.1007/978-3-031-21973-3_64-1
11. Ogwu MC (2019) Towards sustainable development in Africa: the challenge of urbanization and climate change adaptation. In: Cobbinah PB, Addaney M (eds) The geography of climate

change adaptation in urban Africa. Springer Nature, pp 29–55. https://doi.org/10.1007/978-3-030-04873-0_2
12. Prior DD (2023) Customer relationship management. In: B2B customer engagement strategy. Palgrave Macmillan, Cham. https://doi.org/10.1007/978-3-031-23409-5_6
13. Tong X, Zhang H (2019) Perception, amplification and communication: a case study of food safety risks. China's Emerg Manag:71–115. https://doi.org/10.1007/978-981-13-9140-8_3
14. Shin M, Werner AK, Strosnider H, Hines LB, Balluz L, Yip FY (2019) Public perceptions of environmental public health risks in the United States. Int J Environ Res Public Health 16(6): 1045. https://doi.org/10.3390/ijerph16061045
15. Wang A, Zhang M, Zhou S (2022) Air pollution, environmental violation risk, and the cost of debt: evidence from China. Int J Environ Res Public Health 19(6):3584. https://doi.org/10.3390/ijerph19063584
16. Baum FE, Sanders DM, Fisher M, Anaf J, Freudenberg N, Friel S, Labonté R, London L, Monteiro C, Scott-Samuel A, Sen A (2016) Assessing the health impact of transnational corporations: its importance and a framework. Glob Health 12(1):27. https://doi.org/10.1186/s12992-016-0164-x
17. Kelly FJ, Fussell JC (2015) Air pollution and public health: emerging hazards and improved understanding of risk. Environ Geochem Health 37(4):631–649. https://doi.org/10.1007/s10653-015-9720-1
18. Allioui H, Mourdi Y (2023) Exploring the full potentials of IoT for better financial growth and stability: a comprehensive survey. Sensors 23(19):8015. https://doi.org/10.3390/s23198015
19. Kiihamäki SP, Korhonen M, Jaakkola JJK (2021) Ambient particulate air pollution and daily stock market returns and volatility in 47 cities worldwide. Sci Rep 11(1):8628. https://doi.org/10.1038/s41598-021-88041-w
20. Ramondt S, Ramírez AS (2020) Media reporting on air pollution: health risk and precautionary measures in national and regional newspapers. Int J Environ Res Public Health 17(18):6516. https://doi.org/10.3390/ijerph17186516
21. Moisescu OI, Gică OA (2020) The impact of environmental and social responsibility on customer loyalty: a multigroup analysis among generations X and Y. Int J Environ Res Public Health 17(18):6466. https://doi.org/10.3390/ijerph17186466
22. Wang M, Liu Z (2022) How do green innovation strategies contribute to firm performance under supply chain risk? Evidence from China's manufacturing sector. Front Psychol 13: 894766. https://doi.org/10.3389/fpsyg.2022.894766
23. Zhu Y, Zhang H, Siddik AB, Zheng Y, Sobhani FA (2023) Understanding corporate green competitive advantage through green technology adoption and green dynamic capabilities: does green product innovation matter? Systems 11:461. https://doi.org/10.3390/systems11090461
24. He B, Qi M, Wang N, Zhang Z (2022) Avoiding real harm but false good: the influence mechanism of political relations on the effectiveness of environmental regulation policies. Int J Environ Res Public Health 19(23):15953. https://doi.org/10.3390/ijerph192315953
25. Preble KA, Early BR (2023) Enforcing economic sanctions by tarnishing corporate reputations. Bus Polit:1–22. https://doi.org/10.1017/bap.2023.22
26. von Berlepsch D, Lemke F, Gorton M (2022) The importance of corporate reputation for sustainable supply chains: a systematic literature review, bibliometric mapping, and research agenda. J Bus Ethics:1–26. https://doi.org/10.1007/s10551-022-05268-x
27. Ali Q, Salman A, Parveen S (2022) Evaluating the effects of environmental management practices on environmental and financial performance of firms in Malaysia: the mediating role of ESG disclosure. Heliyon 8(12):e12486. https://doi.org/10.1016/j.heliyon.2022.e12486
28. Hendlin YH, Bialous SA (2020) The environmental externalities of tobacco manufacturing: a review of tobacco industry reporting. Ambio 49(1):17–34. https://doi.org/10.1007/s13280-019-01148-3

29. Shen Y, Lyu M, Zhu J (2022) Air pollution and corporate green financial constraints: evidence from China's listed companies. Int J Environ Res Public Health 19(22):15034. https://doi.org/10.3390/ijerph192215034
30. An Y, Madni GR (2023) Factors affecting the green investment and assessing sustainable performance of firms in China. PLoS One 18(12):e0296099. https://doi.org/10.1371/journal.pone.0296099
31. Kruize H, van der Vliet N, Staatsen B, Bell R, Chiabai A, Muiños G, Higgins S, Quiroga S, Martinez-Juarez P, Aberg Yngwe M, Tsichlas F, Karnaki P, Lima ML, García de Jalón S, Khan M, Morris G, Stegeman I (2019) Urban green space: creating a triple win for environmental sustainability, health, and health equity through behavior change. Int J Environ Res Public Health 16(22):4403. https://doi.org/10.3390/ijerph16224403
32. DuPont A (2018) Improving and monitoring air quality. Environ Sci Pollut Res Int 25(15): 15253–15263. https://doi.org/10.1007/s11356-018-1897-2
33. Filippelli G, Anenberg S, Taylor M, van Geen A, Khreis H (2020) New approaches to identifying and reducing the global burden of disease from pollution. GeoHealth 4(4): e2018GH000167. https://doi.org/10.1029/2018GH000167
34. Ramaswamy K, Jule LT, Nagaprasad N, Subramanian K, Shanmugam R, Priyanka Dwarampudi L, Seenivasan V (2022) Reduction of environmental chemicals, toxicity and particulate matter in wet scrubber device to achieve zero emissions. Sci Rep 12(1):9170. https://doi.org/10.1038/s41598-022-13369-w
35. Singh U, Algren M, Schoeneberger C, Lavallais C, O'Connell MG, Oke D, Liang C, Das S, Salas SD, Dunn JB (2022) Technological avenues and market mechanisms to accelerate methane and nitrous oxide emissions reductions. iScience 25(12):105661. https://doi.org/10.1016/j.isci.2022.105661
36. Parviainen T, Lehikoinen A, Kuikka S et al (2018) How can stakeholders promote environmental and social responsibility in the shipping industry? WMU J Marit Affairs 17:49–70. https://doi.org/10.1007/s13437-017-0134-z
37. Salvador Costa MJ, Leitão A, Silva R, Monteiro V, Melo P (2022) Climate change prevention through community actions and empowerment: a scoping review. Int J Environ Res Public Health 19(22):14645. https://doi.org/10.3390/ijerph192214645
38. Walker ET (2012) Putting a face on the issue: corporate stakeholder mobilization in professional grassroots lobbying campaigns. Bus Soc 51(4):561–601. https://doi.org/10.1177/0007650309350210
39. Werner WJ (2009) Corporate social responsibility initiatives addressing social exclusion in Bangladesh. J Health Popul Nutr 27(4):545–562. https://doi.org/10.3329/jhpn.v27i4.3401
40. Ali H, Yin J, Manzoor F, An M (2023) The impact of corporate social responsibility on firm reputation and organizational citizenship behavior: the mediation of organic organizational cultures. Front Psychol 13:1100448. https://doi.org/10.3389/fpsyg.2022.1100448
41. Chen CC, Khan A, Hongsuchon T, Ruangkanjanases A, Chen YT, Sivarak O, Chen SC (2021) The role of corporate social responsibility and corporate image in times of crisis: the mediating role of customer trust. Int J Environ Res Public Health 18(16):8275. https://doi.org/10.3390/ijerph18168275
42. Nguyen NTT, Nguyen NP, Thanh Hoai T (2021) Ethical leadership, corporate social responsibility, firm reputation, and firm performance: a serial mediation model. Heliyon 7(4):e06809. https://doi.org/10.1016/j.heliyon.2021.e06809
43. Abubakar IR, Maniruzzaman KM, Dano UL, AlShihri FS, AlShammari MS, Ahmed SMS, Al-Gehlani WAG, Alrawaf TI (2022) Environmental sustainability impacts of solid waste management practices in the global south. Int J Environ Res Public Health 19(19):12717. https://doi.org/10.3390/ijerph191912717
44. Baloch QB, Shah SN, Iqbal N, Sheeraz M, Asadullah M, Mahar S, Khan AU (2023) Impact of tourism development upon environmental sustainability: a suggested framework for sustainable ecotourism. Environ Sci Pollut Res Int 30(3):5917–5930. https://doi.org/10.1007/s11356-022-22496-w

45. Ekins P, Zenghelis D (2021) The costs and benefits of environmental sustainability. Sustain Sci 16(3):949–965. https://doi.org/10.1007/s11625-021-00910-5
46. Fletcher AC, Bourne PE (2012) Ten simple rules for starting a company. PLoS Comput Biol 8(3):e1002439. https://doi.org/10.1371/journal.pcbi.1002439
47. Wood B, Williams O, Baker P, Nagarajan V, Sacks G (2021) The influence of corporate market power on health: exploring the structure-conduct-performance model from a public health perspective. Glob Health 17(1):41. https://doi.org/10.1186/s12992-021-00688-2
48. Bohr A, Memarzadeh K (2020) The rise of artificial intelligence in healthcare applications. Artificial intelligence in healthcare, pp 25–60. https://doi.org/10.1016/B978-0-12-818438-7.00002-2
49. Farida I, Setiawan D (2022) Business strategies and competitive advantage: the role of performance and innovation. J Open Innov Technol Mark Complex 8:163. https://doi.org/10.3390/joitmc8030163
50. Shah SS, Asghar Z (2023) Dynamics of social influence on consumption choices: a social network representation. Heliyon 9(6):e17146. https://doi.org/10.1016/j.heliyon.2023.e17146
51. Yim EC (2015) How to innovate and make successful business. Springerplus 4(Suppl 2):K2. https://doi.org/10.1186/2193-1801-4-S2-K2
52. Wang W (2022) Toward economic growth and value creation through social entrepreneurship: modelling the mediating role of innovation. Front Psychol 13:914700. https://doi.org/10.3389/fpsyg.2022.914700
53. Zhang T, Wang W, Li C, Wang X (2023) The role of innovation and entrepreneurship in promoting common prosperity in China: empirical evidence from a two-way fixed effects model. PLoS One 18(12):e0295752. https://doi.org/10.1371/journal.pone.0295752
54. Kim HM, Sabri S, Kent A (2021) Smart cities as a platform for technological and social innovation in productivity, sustainability, and livability: a conceptual framework. Smart Cities Technol Soc Innov:9–28. https://doi.org/10.1016/B978-0-12-818886-6.00002-2
55. Solomon EM, van Klyton A (2020) The impact of digital technology usage on economic growth in Africa. Util Policy 67:101104. https://doi.org/10.1016/j.jup.2020.101104
56. Corrado L, Cropper M, Rao A (2023) Space exploration and economic growth: new issues and horizons. Proc Natl Acad Sci U S A 120(43):e2221341120. https://doi.org/10.1073/pnas.2221341120
57. Cui Y, Zhong C, Cao J, Guo M, Zhang M (2024) Spatial effect of carbon neutrality target on high-quality economic development-channel analysis based on total factor productivity. PLoS One 19(1):e0295426. https://doi.org/10.1371/journal.pone.0295426
58. Moreno-Cruz J (2019) Understanding the industrial contribution to pollution offers opportunities to further improve air quality in the United States. Proc Natl Acad Sci U S A 116(40): 19768–19770. https://doi.org/10.1073/pnas.1913999116
59. Mujtaba G, Shahzad SJH (2021) Air pollutants, economic growth and public health: implications for sustainable development in OECD countries. Environ Sci Pollut Res Int 28(10): 12686–12698. https://doi.org/10.1007/s11356-020-11212-1
60. Oncioiu I, Dănescu T, Popa MA (2020) Air-pollution control in an emergent market: does it work? Evidence from Romania. Int J Environ Res Public Health 17(8):2656. https://doi.org/10.3390/ijerph17082656
61. Bai L, Wang J, Ma X, Lu H (2018) Air pollution forecasts: an overview. Int J Environ Res Public Health 15(4):780. https://doi.org/10.3390/ijerph15040780
62. Bikis A (2023) Urban air pollution and greenness in relation to public health. J Environ Public Health 2023:8516622. https://doi.org/10.1155/2023/8516622
63. Nishida C, Yatera K (2022) The impact of ambient environmental and occupational pollution on respiratory diseases. Int J Environ Res Public Health 19(5):2788. https://doi.org/10.3390/ijerph19052788
64. von Schneidemesser E, Driscoll C, Rieder HE, Schiferl LD (2020) How will air quality effects on human health, crops and ecosystems change in the future? Philos Transact A Math Phys Eng Sci 378(2183):20190330. https://doi.org/10.1098/rsta.2019.0330

65. Cormier SA, Lomnicki S, Backes W, Dellinger B (2006) Origin and health impacts of emissions of toxic by-products and fine particles from combustion and thermal treatment of hazardous wastes and materials. Environ Health Perspect 114(6):810–817. https://doi.org/10.1289/ehp.8629
66. Turner MC, Andersen ZJ, Baccarelli A, Diver WR, Gapstur SM, Pope 3rd CA, Prada D, Samet J, Thurston G, Cohen A (2020) Outdoor air pollution and cancer: an overview of the current evidence and public health recommendations. CA Cancer J Clin. https://doi.org/10.3322/caac.21632
67. Tchounwou PB, Yedjou CG, Patlolla AK, Sutton DJ (2012) Heavy metal toxicity and the environment. Experientia Suppl 2012(101):133–164. https://doi.org/10.1007/978-3-7643-8340-4_6
68. Russo PN, Carpenter DO (2019) Air emissions from natural gas facilities in New York state. Int J Environ Res Public Health 16(9):1591. https://doi.org/10.3390/ijerph16091591
69. Zhang HW, Kok VC, Chuang SC, Tseng CH, Lin CT, Li TC, Sung FC, Wen CP, Hsiung CA, Hsu CY (2019) Long-term ambient hydrocarbons exposure and incidence of ischemic stroke. PLoS One 14(12):e0225363. https://doi.org/10.1371/journal.pone.0225363
70. Han S, Na S, Lim NG (2020) Evaluation of road transport pollutant emissions from transporting building materials to the construction site by replacing old vehicles. Int J Environ Res Public Health 17(24):9316. https://doi.org/10.3390/ijerph17249316
71. Jonidi Jafari A, Charkhloo E, Pasalari H (2021) Urban air pollution control policies and strategies: a systematic review. J Environ Health Sci Eng 19(2):1911–1940. https://doi.org/10.1007/s40201-021-00744-4
72. Masiol M, Harrison RM (2014) Aircraft engine exhaust emissions and other airport-related contributions to ambient air pollution: a review. Atmos Environ 95:409–455. https://doi.org/10.1016/j.atmosenv.2014.05.070
73. Fu X (2019) In-vehicle exposures at transportation and the health concerns. Indoor environmental quality and health risk toward healthier environment for all, pp 113–126. https://doi.org/10.1007/978-981-32-9182-9_6
74. Giltrap D, Cavanagh J, Stevenson B, Ausseil AG (2021) The role of soils in the regulation of air quality. Philos Trans R Soc Lond B Biol Sci 376(1834):20200172. https://doi.org/10.1098/rstb.2020.0172
75. Glibert PM (2020) From hogs to HABs: impacts of industrial farming in the US on nitrogen and phosphorus and greenhouse gas pollution. Biogeochemistry 150(2):139–180. https://doi.org/10.1007/s10533-020-00691-6
76. Dopelt K, Radon P, Davidovitch N (2019) Environmental effects of the livestock industry: the relationship between knowledge, attitudes, and behavior among students in Israel. Int J Environ Res Public Health 16(8):1359. https://doi.org/10.3390/ijerph16081359
77. Chandrappa R, Chandra Kulshrestha U (2015) Major issues of air pollution. Sustainable air pollution management: theory and practice, pp 1–48. https://doi.org/10.1007/978-3-319-21596-9_1
78. Wang Q (2018) Urbanization and global health: the role of air pollution. Iran J Public Health 47(11):1644–1652
79. Biró K, Szalmáné Csete M (2021) Corporate social responsibility in agribusiness: climate-related empirical findings from Hungary. Environ Dev Sustain 23(4):5674–5694. https://doi.org/10.1007/s10668-020-00838-3
80. Erickson LE, Jennings M (2017) Energy, transportation, air quality, climate change, health nexus: sustainable energy is good for our health. AIMS Public Health 4(1):47–61. https://doi.org/10.3934/publichealth.2017.1.47
81. Imran M, Khan S, Nassani AA, Haffar M, Khan HUR, Zaman K (2023) Access to sustainable healthcare infrastructure: a review of industrial emissions, coal fires, and particulate matter. Environ Sci Pollut Res Int 30(26):69080–69095. https://doi.org/10.1007/s11356-023-27218-4
82. Hanif S, Lateef M, Hussain K, Hyder S, Usman B, Zaman K, Asif M (2022) Controlling air pollution by lowering methane emissions, conserving natural resources, and slowing

urbanization in a panel of selected Asian economies. PLoS One 17(8):e0271387. https://doi.org/10.1371/journal.pone.0271387
83. Perera F (2017) Pollution from fossil-fuel combustion is the leading environmental threat to global pediatric health and equity: solutions exist. Int J Environ Res Public Health 15(1):16. https://doi.org/10.3390/ijerph15010016
84. Ahmad K, Irshad Younas Z, Manzoor W, Safdar N (2023) Greenhouse gas emissions and corporate social responsibility in USA: a comprehensive study using dynamic panel model. Heliyon 9(3):e13979. https://doi.org/10.1016/j.heliyon.2023.e13979
85. Syuhada G, Akbar A, Hardiawan D, Pun V, Darmawan A, Heryati SHA, Siregar AYM, Kusuma RR, Driejana R, Ingole V, Kass D, Mehta S (2023) Impacts of air pollution on health and cost of illness in Jakarta, Indonesia. Int J Environ Res Public Health 20(4):2916. https://doi.org/10.3390/ijerph20042916
86. Martinez GS, Spadaro JV, Chapizanis D, Kendrovski V, Kochubovski M, Mudu P (2018) Health impacts and economic costs of air pollution in the metropolitan area of Skopje. Int J Environ Res Public Health 15(4):626. https://doi.org/10.3390/ijerph15040626
87. Bari MW, Saleem S, Bashir M, Ahmad B (2022) Impact of ambient air pollution on outdoor employees' performance: mediating role of anxiety. Front Psychol 13:926534. https://doi.org/10.3389/fpsyg.2022.926534
88. Zivin JG, Neidell M (2012) The impact of pollution on worker productivity. Am Econ Rev 102(7):3652–3673. https://doi.org/10.1257/aer.102.7.3652
89. Jafari AJ, Charkhloo E, Pasalari H (2021) Urban air pollution control policies and strategies: a systematic review. J Environ Health Sci Eng 19(2):1911–1940. https://doi.org/10.1007/s40201-021-00744-4
90. Milner J, Hamilton I, Woodcock J, Williams M, Davies M, Wilkinson P, Haines A (2020) Health benefits of policies to reduce carbon emissions. BMJ 368
91. Ramírez AS, Ramondt S, Van Bogart K, Perez-Zuniga R (2019) Public awareness of air pollution and health threats: challenges and opportunities for communication strategies to improve environmental health literacy. J Health Commun 24(1):75–83. https://doi.org/10.1080/10810730.2019.1574320
92. Zhao Z, Liu L, Zhang F, Yang G (2022) Does air pollution prompt corporations to implement green management? Evidence from China. Environ Sci Pollut Res Int 29(6):8933–8946. https://doi.org/10.1007/s11356-021-16272-5
93. Surawattanasakul V, Sirikul W, Sapbamrer R, Wangsan K, Panumasvivat J, Assavanopakun P, Muangkaew S (2022) Respiratory symptoms and skin sick building syndrome among Office Workers at University Hospital, Chiang Mai, Thailand: associations with indoor air quality, AIRMED project. Int J Environ Res Public Health 19(17):10850. https://doi.org/10.3390/ijerph191710850
94. Joshi SM (2008) The sick building syndrome. Indian J Occup Environ Med 12(2):61–64. https://doi.org/10.4103/0019-5278.43262
95. Katsaliaki K, Galetsi P, Kumar S (2022) Supply chain disruptions and resilience: a major review and future research agenda. Ann Oper Res 319(1):965–1002. https://doi.org/10.1007/s10479-020-03912-1
96. Khan J, Ishizaka A, Mangla SK (2022) Assessing risk of supply chain disruption due to COVID-19 with fuzzy VIKORSort. Ann Oper Res:1–26. https://doi.org/10.1007/s10479-022-04940-9
97. Siddiqua A, Hahladakis JN, Al-Attiya WAKA (2022) An overview of the environmental pollution and health effects associated with waste landfilling and open dumping. Enviro Sci Pollut Res Int 29(39):58514–58536. https://doi.org/10.1007/s11356-022-21578-z
98. Ogwu MC (2019) Understanding the composition of food waste: an "-Omics" approach to food waste management. In: Gunjal AP, Waghmode MS, Patil NN, Bhatt P (eds) Global initiatives for waste reduction and cutting food loss. IGI Global, Pennsylvania, pp 212–236. https://doi.org/10.4018/978-1-5225-7706-5.ch011

99. Ogwu MC, Iyiola AO, Izah SC (2023) Overview of African biological resources and environment. In: Izah SC, Ogwu MC (eds) Sustainable utilization and conservation of Africa's biological resources and environment. Sustainable development and biodiversity, vol 888. Springer, Singapore, pp 1–34. https://doi.org/10.1007/978-981-19-6974-4_1
100. Ogwu MC, Izah SC, Iyiola AO (2022) An overview of the potentials, threats and conservation of biodiversity in Africa. In: Izah SC (ed) Biodiversity in Africa: potentials, threats and conservation. Sustainable development and biodiversity, vol 29. Springer, Singapore, pp 3–20. https://doi.org/10.1007/978-981-19-3326-4_1
101. Kurt OK, Zhang J, Pinkerton KE (2016) Pulmonary health effects of air pollution. Curr Opin Pulm Med 22(2):138–143. https://doi.org/10.1097/MCP.0000000000000248
102. Lee BJ, Kim B, Lee K (2014) Air pollution exposure and cardiovascular disease. Toxicol Res 30(2):71–75. https://doi.org/10.5487/TR.2014.30.2.071
103. Carrer P, Wolkoff P (2018) Assessment of indoor air quality problems in office-like environments: role of occupational health services. Int J Environ Res Public Health 15(4):741. https://doi.org/10.3390/ijerph15040741
104. Hou Y, Gao M, Huang L, Wang Q (2021) Air pollution reduces interpersonal trust: the roles of emotion and emotional susceptibility. Int J Environ Res Public Health 18(11):5631. https://doi.org/10.3390/ijerph18115631
105. Orru H, Olstrup H, Hagenbjörk A, Nordin S, Orru K (2022) Exposures, symptoms and risk perception among office workers in relation to nanoparticles in the work environment. Int J Environ Res Public Health 19(10):5789. https://doi.org/10.3390/ijerph19105789
106. Onofrei M, Gavriluţă Vatamanu AF, Bostan I, Filip BF, Popescu CL, Jitaru G (2020) Impacts of the allocation of governmental resources for improving the environment, an empirical analysis on developing European countries. Int J Environ Res Public Health 17(8):2783. https://doi.org/10.3390/ijerph17082783
107. Hevey D, Perko T, Martell M, Bradley G, Apers S, Rovenská KN (2023) A psycho-social-environmental lens on radon air pollutant: authorities', mitigation contractors', and residents' perceptions of barriers and facilitators to domestic radon mitigation. Front Public Health 11: 1252804. https://doi.org/10.3389/fpubh.2023.1252804
108. Bouza E, Vargas F, Alcázar B, Álvarez T, Asensio A, Cruceta G, Gracia D, Guinea J, Gil MA, Linares C, Muñoz P, Pastor P, Pedro-Botet ML, Querol X, Tovar J, Urrutia I, Villar F, Palomo E (2022) Air pollution and health prevention: a document of reflection. Rev Esp Quimioter 35(4):307–332. https://doi.org/10.37201/req/171.2021
109. Romley JA, Hackbarth A, Goldman DP (2012) The impact of air quality on hospital spending. Rand Health Quarterly 2(3):6
110. Mohammadzadeh Y, Sheikhmali A, Yahyavi Dizaj J, Mosadeghrad AM, Yusefzadeh H, Refah Kahriz A (2023) The impact of government subsidy programs on equity in health financing. Cost Eff Resour Alloc 21(1):54. https://doi.org/10.1186/s12962-023-00460-w
111. Ikhajiagbe B, Ogwu MC (2020) Hazard quotient, microbial diversity and plant composition of spent crude oil polluted-soil. Beni-Suef Univ J Basic Appl Sci 9(26). https://doi.org/10.1186/s43088-020-00052-0
112. Ogwu MC (2023) Plants as monitors and managers of pollution. In: Srivastav AL, Grewal AS, Tiwari M, Pham TD (eds) Role of green chemistry in ecosystem restoration to achieve environmental sustainability. Elsevier, Amsterdam, pp 51–60. ISBN: 978-0-443-15291-7. https://doi.org/10.1016/B978-0-443-15291-7.00022-5
113. Kong E, Ji Y (2023) Provision of hospital price information after increases in financial penalties for failure to comply with a US Federal Hospital Price Transparency Rule. JAMA Netw Open 6(6):e2320694. https://doi.org/10.1001/jamanetworkopen.2023.20694
114. Kuipers MAG, Nuyts PAW, Willemsen MC, Kunst AE (2022) Tobacco retail licencing systems in Europe. Tob Control 31(6):784–788. https://doi.org/10.1136/tobaccocontrol-2020-055910

115. Bogalecka M, Grobelna A (2023) Air pollution and its potential consequences for tourism and career development from students' perspective: a case study of the Gdańsk agglomeration in Poland. Int J Environ Res Public Health 20(3):2651. https://doi.org/10.3390/ijerph20032651
116. Imarhiagbe O, Ogwu MC (2022) Sacred groves in the global south: a panacea for sustainable biodiversity conservation. In: Izah SC (ed) Biodiversity in Africa: potentials, threats and conservation. Sustainable development and biodiversity, Vol 29. Springer, Singapore, pp. 525–546. https://doi.org/10.1007/978-981-19-3326-4_20
117. Richard G, Izah SC, Sawyer EW, Ogwu MC (2023) Seasonal variations in the concentration of some atmospheric pollutants along a major highway in southern Nigeria. J Environ Stud 30(1): 10–25. https://doi.org/10.21608/JESJ.2023.205673.1045
118. Richard G, Izah SC, Ogwu MC (2022) Implications of artisanal crude oil refining on sustainable food production in The Niger Delta region of Nigeria. J Environ Bioremediat Toxicol 5(2):69–77. https://doi.org/10.54987/jebat.v5i2.775
119. Ovuru KF, Izah SC, Ogidi OI, Imarhiagbe I, Ogwu MC (2023) Slaughterhouse facilities in developing nations: sanitation and hygiene practices, microbial contamination and sustainable management system. Food Sci Biotechnol. https://doi.org/10.1007/s10068-023-01406-x
120. Chai KC, Huang Y, Chang KC, Hu WJ (2020) Can environmental regulation reduce labor costs and improve business performance? Evidence from the air quality index. Front Public Health 7:398. https://doi.org/10.3389/fpubh.2019.00398
121. Santos UP, Arbex MA, Braga ALF, Mizutani RF, Cançado JED, Terra-Filho M, Chatkin JM (2021) Environmental air pollution: respiratory effects. J Bras Pneumol 47(1):e20200267
122. Schulte PA, Jacklitsch BL, Bhattacharya A, Chun H, Edwards N, Elliott KC, Flynn MA, Guerin R, Hodson L, Lincoln JM, MacMahon KL, Pendergrass S, Siven J, Vietas J (2023) Updated assessment of occupational safety and health hazards of climate change. J Occup Environ Hyg 20(5–6):183–206. https://doi.org/10.1080/15459624.2023.2205468
123. Chandra M, Rai CB, Kumari N, Sandhu VK, Chandra K, Krishna M, Kota SH, Anand KS, Oudin A (2022) Air pollution and cognitive impairment across the life course in humans: a systematic review with specific focus on income level of study area. Int J Environ Res Public Health 19(3):1405. https://doi.org/10.3390/ijerph19031405
124. Torres G, Mourad M, Leheste JR (2022) Indoor air pollution and decision-making behavior: an interdisciplinary review. Cureus 14(6):e26247. https://doi.org/10.7759/cureus.26247
125. Obahiagbon EG, Ogwu MC (2023) Consumer perception and demand for sustainable herbal medicine products and market. In: Izah SC, Ogwu MC, Akram M (eds) Herbal medicine phytochemistry. Reference series in phytochemistry. Springer, Cham. https://doi.org/10.1007/978-3-031-21973-3_65-1
126. Obahiagbon EG, Ogwu MC (2024) The nexus of business, sustainability, and herbal medicine. In: Izah SC, Ogwu MC, Akram M (eds) Herbal medicine phytochemistry. Reference series in phytochemistry. Springer, Cham. https://doi.org/10.1007/978-3-031-21973-3_67-1
127. Boz Z, Korhonen V, Koelsch Sand C (2020) Consumer considerations for the implementation of sustainable packaging: a review. Sustain For 12:2192. https://doi.org/10.3390/su12062192
128. Li C, Yan Y, Lu Y, Zeng G, Zhou L, Jin H, Xu Y, Chen Y (2022) Air pollution, investor sentiment and stock liquidity. Front Public Health 10:989457. https://doi.org/10.3389/fpubh.2022.989457
129. Kedarya T, Elalouf A, Cohen RS (2023) Calculating strategic risk in financial institutions. Glob J Flex Syst Manag:1–12. https://doi.org/10.1007/s40171-023-00342-3
130. Laghari F, Ahmed F, López García MLN (2023) Cash flow management and its effect on firm performance: empirical evidence on non-financial firms of China. PLoS One 18(6):e0287135. https://doi.org/10.1371/journal.pone.0287135
131. McGowan J, Wojahn A, Nicolini JR. Risk management event evaluation and responsibilities. [Updated 2023 Aug 23]. In: StatPearls [Internet]. StatPearls Publishing, Treasure Island, 2024 Jan-. Available from: https://www.ncbi.nlm.nih.gov/books/NBK559326/
132. Vo DH (2023) Market risk, financial distress and firm performance in Vietnam. PLoS One 18(7):e0288621. https://doi.org/10.1371/journal.pone.0288621

133. Naili M, Lahrichi Y (2022) Banks' credit risk, systematic determinants and specific factors: recent evidence from emerging markets. Heliyon 8(2):e08960. https://doi.org/10.1016/j.heliyon.2022.e08960
134. Banai Á, Berlinger E, Dömötör B (2022) Adjustable-rate mortgages in the era of global reflation: how to model additional default risk? PLoS One 17(3):e0263599. https://doi.org/10.1371/journal.pone.0263599
135. Gultekin B, Demir S, Gunduz MA, Cura F, Ozer L (2022) The logistics service providers during the COVID-19 pandemic: the prominence and the cause-effect structure of uncertainties and risks. Comput Ind Eng 165:107950. https://doi.org/10.1016/j.cie.2022.107950
136. Settembre-Blundo D, González-Sánchez R, Medina-Salgado S, García-Muiña FE (2021) Flexibility and resilience in corporate decision making: a new sustainability-based risk management system in uncertain times. Glob J Flex Syst Manag 22(Suppl 2):107–132. https://doi.org/10.1007/s40171-021-00277-7
137. Clarke R (2022) Research opportunities in the regulatory aspects of electronic markets. Electron Mark 32(1):179–200. https://doi.org/10.1007/s12525-021-00469-0
138. Ahi AA, Sinkovics N, Sinkovics RR (2023) E-commerce policy and the global economy: a path to more inclusive development? Manag Int Rev 63(1):27–56. https://doi.org/10.1007/s11575-022-00490-1
139. Matthews R, Rutherford BN, Edmondson D, Matthews L (2022) Uncertainty in industrial markets: the COVID-19 pandemic. Ind Mark Manag 102:364–376. https://doi.org/10.1016/j.indmarman.2022.02.006
140. Sanasi S, Ghezzi A (2022) Pivots as strategic responses to crises: evidence from Italian companies navigating Covid-19. Strateg Organ:14761270221122933. https://doi.org/10.1177/14761270221122933
141. Yu Z, Razzaq A, Rehman A, Shah A, Jameel K, Mor RS (2022) Disruption in global supply chain and socio-economic shocks: a lesson from COVID-19 for sustainable production and consumption. Oper Manag Res 15(1–2):233–248. https://doi.org/10.1007/s12063-021-00179-y
142. Masterson EA, Wurzelbacher SJ, Bushnell PT, Tseng CY (2023) Workers' compensation costs for occupational hearing loss claims in the United States, 2009-2013. Semin Hear 44(4):412–436. https://doi.org/10.1055/s-0043-1769587
143. Lawin H, Ayi Fanou L, Hinson AV, Stolbrink M, Houngbegnon P, Kedote NM, Fayomi B, Kagima J, Katoto P, Ouendo EMD, Mortimer K (2018) Health risks associated with occupational exposure to ambient air pollution in commercial drivers: a systematic review. Int J Environ Res Public Health 15(9):2039. https://doi.org/10.3390/ijerph15092039
144. Wang Y, Yang Y, Fu C, Fan Z, Zhou X (2021) Environmental regulation, environmental responsibility, and green technology innovation: empirical research from China. PLoS One 16(9):e0257670. https://doi.org/10.1371/journal.pone.0257670
145. Wu H, Feng T, Jiang W, Kong T (2022) Environmental penalties, investor attention and stock market reaction: moderating roles of air pollution and industry saliency. Int J Environ Res Public Health 19(5):2660. https://doi.org/10.3390/ijerph19052660
146. Zhao W (2020) Effect of air pollution on household insurance purchases. Evidence from China household finance survey data. PLoS One 15(11):e0242282. https://doi.org/10.1371/journal.pone.0242282
147. Zhang C, Jin S (2022) How does an environmental information disclosure of a Buyer's Enterprise affect green technological innovations of Sellers' Enterprise? Int J Environ Res Public Health 19(22):14715. https://doi.org/10.3390/ijerph192214715
148. Ao XY, Ong TS, Aprile R, Di Vaio A (2023) Environmental uncertainty and digital technologies corporate in shaping corporate green behavior and tax avoidance. Sci Rep 13(1):22170. https://doi.org/10.1038/s41598-023-49687-w
149. Abu-Rumman G, Khdair AI, Khdair SI (2020) Current status and future investment potential in renewable energy in Jordan: An overview. Heliyon 6(2):e03346. https://doi.org/10.1016/j.heliyon.2020.e03346

150. Invernizzi DC, Locatelli G, Velenturf A, Love PE, Purnell P, Brookes NJ (2020) Developing policies for the end-of-life of energy infrastructure: coming to terms with the challenges of decommissioning. Energy Policy 144:111677. https://doi.org/10.1016/j.enpol.2020.111677
151. Erickson LE (2017) Reducing greenhouse gas emissions and improving air quality: two global challenges. Environ Prog Sustain Energy 36(4):982–988. https://doi.org/10.1002/ep.12665
152. Jiang Z, Xiao T, Kuznetsov VL, Edwards PP (2010) Turning carbon dioxide into fuel. Philos Trans R Soc A Math Phys Eng Sci 368(1923):3343–3364. https://doi.org/10.1098/rsta.2010.0119
153. Gambhir A, Tavoni M (2019) Direct air carbon capture and sequestration: how it works and how it could contribute to climate-change mitigation. One Earth 1(4):405–409. https://doi.org/10.1016/j.oneear.2019.11.006
154. Navas-Anguita Z, García-Gusano D, Iribarren D (2020) Long-term production technology mix of alternative fuels for road transport: a focus on Spain. Energ Conver Manage 226:113498. https://doi.org/10.1016/j.enconman.2020.113498
155. Peters DR, Schnell JL, Kinney PL, Naik V, Horton DE (2020) Public health and climate benefits and trade-offs of U.S. Vehicle electrification. GeoHealth 4(10):e2020GH000275. https://doi.org/10.1029/2020GH000275
156. Prologis (2023) What's next: four forces shaping the logistics real estate cycle. https://www.prologis.com/news-research/global-insights/whats-next-four-forces-shaping-logistics-real-estate-cycle
157. Cromar KR, Duncan BN, Bartonova A, Benedict K, Brauer M, Habre R, Hagler GSW, Haynes JA, Khan S, Kilaru V, Liu Y, Pawson S, Peden DB, Quint JK, Rice MB, Sasser EN, Seto E, Stone SL, Thurston GD, Volckens J (2019) Air pollution monitoring for health research and patient care. An official American Thoracic Society workshop report. Ann Am Thorac Soc 16(10):1207–1214. https://doi.org/10.1513/AnnalsATS.201906-477ST
158. Khreis H, Johnson J, Jack K, Dadashova B, Park ES (2022) Evaluating the performance of low-cost air quality monitors in Dallas, Texas. Int J Environ Res Public Health 19(3):1647. https://doi.org/10.3390/ijerph19031647
159. NASA (2023) New instrument to track air pollution hourly, shed light on disparities – NASA. https://www.nasa.gov/missions/tempo/new-instrument-to-track-air-pollution-hourly-shed-light-on-disparities/
160. Vanian J (2015) Here's why IBM will acquire The Weather Company | Fortune. Fortune.Com. https://fortune.com/2015/10/28/ibm-weather-company-acquisition-data/
161. Mannan M, Al-Ghamdi SG (2021) Indoor air quality in buildings: a comprehensive review on the factors influencing air pollution in residential and commercial structure. Int J Environ Res Public Health 18(6):3276. https://doi.org/10.3390/ijerph18063276
162. Kumar A, Thakur AK, Gaurav GK et al (2023) A critical review on sustainable hazardous waste management strategies: a step towards a circular economy. Environ Sci Pollut Res 30:105030–105055. https://doi.org/10.1007/s11356-023-29511-8
163. Ferronato N, Torretta V (2019) Waste mismanagement in developing countries: a review of global issues. Int J Environ Res Public Health 16(6):1060. https://doi.org/10.3390/ijerph16061060
164. Project Drawdown (2020, February 6) Landfill methane capture. Project Drawdown. https://drawdown.org/solutions/landfill-methane-capture
165. Honest A, Manyele S, Saria J, Mbuna J (2020) Assessment of air pollutant emissions from healthcare waste incinerators of different design features. Afr J Environ Sci Technol 14(10):311–328
166. Circle Economy Foundation (2024) The circularity gap report 2024. https://www.circularity-gap.world/2024
167. Schmidt CW (2009) Carbon offsets: growing pains in a growing market. Environ Health Perspect 117(2):A62–A68. https://doi.org/10.1289/ehp.117-a62
168. Kaplan RS, Ramanna K, Roston M (2023, July 1) Accounting for carbon offsets. Harv Bus Rev. https://hbr.org/2023/07/accounting-for-carbon-offsets

169. George TE, Karatu K, Edward A (2020) An evaluation of the environmental impact assessment practice in Uganda: challenges and opportunities for achieving sustainable development. Heliyon 6(9):e04758. https://doi.org/10.1016/j.heliyon.2020.e04758
170. Winters M (2021, April 21) The rise of sustainability consulting. Conservice ESG. https://esg.conservice.com/the-rise-of-sustainability-consulting/
171. Aboagye EM, Zeng C, Owusu G, Mensah F, Afrane S, Ampah JD, Brenyah SA (2023) A review contribution to emission trading schemes and low carbon growth. Environ Sci Pollut Res Int 30(30):74575–74597. https://doi.org/10.1007/s11356-023-27673-z
172. Cushing L, Blaustein-Rejto D, Wander M, Pastor M, Sadd J, Zhu A, Morello-Frosch R (2018) Carbon trading, co-pollutants, and environmental equity: evidence from California's cap-and-trade program (2011-2015). PLoS Med 15(7):e1002604. https://doi.org/10.1371/journal.pmed.1002604
173. Almaqtari FA, Elsheikh T, Abdelkhair F, Mazrou YSA (2023) The impact of corporate environmental disclosure practices and board attributes on sustainability: empirical evidence from Asia and Europe. Heliyon 9(8):e18453. https://doi.org/10.1016/j.heliyon.2023.e18453
174. Akudugu MA, Ogwu MC (2024) Sustainable development policies and interventions: a bibliometric analysis of the contributions of the academic community. J Clean Prod. Article # 139919. https://doi.org/10.1016/j.jclepro.2023.139919
175. Ross K, Chmiel JF, Ferkol T (2012) The impact of the clean air act. J Pediatr 161(5):781–786. https://doi.org/10.1016/j.jpeds.2012.06.064
176. Nethery RC, Mealli F, Sacks JD, Dominici F (2020) Evaluation of the health impacts of the 1990 clean air act amendments using causal inference and machine learning. J Am Stat Assoc 1:1–12. https://doi.org/10.1080/01621459.2020.1803883
177. Bayer P, Aklin M (2020) The European Union emissions trading system reduced CO_2 emissions despite low prices. Proc Natl Acad Sci U S A 117(16):8804–8812. https://doi.org/10.1073/pnas.1918128117
178. McClellan RO (2012) Role of science and judgment in setting national ambient air quality standards: how low is low enough? Air Qual Atmos Health 5(2):243–258. https://doi.org/10.1007/s11869-011-0147-2
179. Faridi S, Krzyzanowski M, Cohen AJ, Malkawi M, Moh'd Safi HA, Yousefian F, Azimi F, Naddafi K, Momeniha F, Niazi S, Amini H, Künzli N, Shamsipour M, Mokammel A, Roostaei V, Hassanvand MS (2023) Ambient air quality standards and policies in eastern Mediterranean countries: a review. Int J Public Health 68:1605352. https://doi.org/10.3389/ijph.2023.1605352
180. Kim Y, Tanaka K, Matsuoka S (2020) Environmental and economic effectiveness of the Kyoto protocol. PLoS One 15(7):e0236299. https://doi.org/10.1371/journal.pone.0236299
181. Zhou J, Jin S (2023) Corporate environmental protection behavior and sustainable development: the moderating role of green investors and green executive cognition. Int j Environ Res Public Health 20(5):4179. https://doi.org/10.3390/ijerph20054179
182. Singh S, Kulshrestha MJ, Rani N, Kumar K, Sharma C, Aswal DK (2023) An overview of vehicular emission standards. Mapan 38(1):241–263. https://doi.org/10.1007/s12647-022-00555-4

Social Justice and Equity Issues of Air Pollution

Thi Phuoc Lai Nguyen

Contents

Abstract The impacts of air pollution on humans continue to receive significant attention from academia, practitioners, and policymakers. However, along with the visible effects of air pollution on human health, little is known about how they intersect with inequalities. This chapter attempts to review the social injustices and inequities related to air pollution and the connection between social deprivation and exposure to air pollution which results in environmental injustices. To support the review findings, a case study of Bangkok City, Thailand, one among the most air-polluted cities in the Asia Pacific region, was used to demonstrate disparities in air pollution exposure at the intersection of gender, age, occupation, and social background in the world of work where people are exposed to air pollution. The

T. P. L. Nguyen (✉)
Department of Development and Sustainability, School of Environment, Resources and Development, Asian Institute of Technology, Khlong Luang, Thailand
e-mail: phuoclai@ait.asia

Matthew Chidozie Ogwu and Sylvester Izah (eds.), *Sustainable Strategies for Air Pollution Mitigation: Development, Economics and Technologies*,
Hdb Env Chem (2024) 133: 173–200, DOI 10.1007/698_2024_1108,

Published online: 20 August 2024

findings of the case study unequivocally show that various socioeconomic classes suffered from varying health impacts from air pollution at work. When it came to health risks, workers in the informal economy were exposed more than those in the formal economy, and those who worked near streets, highways, and industrial areas had more health problems associated with bad air quality. Because they are frequently impoverished women and young people engaged in informal work that is hazardous, arduous, dirty, and low-paid, as well as lacking social assistance and insurance, workplace safety measures, pension contributions, and paid sick leave, informal laborers are particularly vulnerable to air pollution. In conclusion, the chapter illustrates how social injustices and inequities, such as racism and discrimination against particular racial or ethnic groups, unequal access to public services like healthcare, and limited opportunities for employment and education, have an impact on social and economic disparities in air pollution exposure.

Keywords Environmental equity, Environmental justice, Social deprivation, Social inequities, Social injustices, Socioeconomic status, Structural and geographic inequalities, Unequal distribution

Abbreviation

ALA	American Lung Association
CAA	Clean Air Act
COVID-19	Coronavirus disease of 2019
EEA	European Environmental Agency
MONRE	Ministry of Natural Resources and Environment
N95	Efficiency of at least 95% for non-oil-based particulates
OECD	Organisation for Economic Co-operation and Development
$PM_{2.5}$	Fine particulate matter with dimensions of 2.5 μm or less
SES	Spatial-economic-social
UNESCAP	United Nations Economic and Social Commission for Asia and the Pacific
UN-HABITAT	United Nations Human Settlements Programme
UCAR	University Corporation for Atmospheric Research
WHO	World Health Organization
$\mu g/m^3$	Microgram per cubic meter

1 Introduction

Air pollution is the outcome of both natural disasters like smoke and ash from wildfires, volcanic activity dust storms, and biological decomposition and human activity like urbanization, industry, electricity, agriculture, and transportation

[1]. Although air pollution trends have been decreasing in high-income countries, it is increasing in low- to middle-income ones. In cities with air quality monitoring, over 80% of the population is exposed to pollution levels above WHO guidelines. Air pollution affects all regions globally; people in low-income regions are the most impacted. 98% of cities in low- and middle-income countries with populations of more than 100,000 fail to meet the air quality standards established by the World Health Organization (WHO). However, this data drops to 56% in high-income nations [2]. While conventional household pollutants have decreased, there has been a sharp increase in pollution from cars and mass industry, especially in developing countries. The rate of air pollution is rising in these countries as they continue to industrialize rapidly. Around seven million premature deaths every year are caused by air pollution, with a disproportionate amount of these deaths taking place in poorer socioeconomic communities of low- to middle-income countries [3]. Air pollution poses serious health dangers to people at all stages of life. No matter how long they are exposed, breathing in contaminated air can cause several health problems, such as chronic obstructive pulmonary disease, lower respiratory tract infections, and bronchus and stroke [4, 5]. Numerous detrimental health impacts, such as respiratory [6] and cognitive problems [7], are associated with poor air quality. However, the health effects of air pollution exposure can be profound, especially for people from lower socioeconomic status. Additional risk factors that these people frequently experience include poverty, malnutrition, and restricted access to healthcare. Lower socioeconomic status and discrimination both increase psychological and social stress making these groups of people more vulnerable to diseases caused by air pollution.

Environmental threats and their impacts on human health have been the research and policy interests all over the world [8, 9]. Disparities in air pollution exposure have different health implications. According to the United Nations Economic and Social Commission for Asia and the Pacific (UNESCAP), socially and economically disadvantaged and deprived groups are more vulnerable to air pollution. The health risks associated with bad air quality are different among socioeconomic and demographic communities [10]. This is because these groups have no adequate resources and ability to protect themselves against air pollution which has a disproportionately negative impact on their health and productivity. Low-income and socially deprived people are often more vulnerable to air pollution due to higher levels of daily exposure as well as higher prevalence of preexisting chronic health conditions that are made worse by air pollution. Workers from informal economy, and people those living in underdeveloped areas, close to highways, landfills, industrial zones, and those without access to sanitation, clean water, and inadequate cooking facilities are all more exposed to the harmful impacts of air pollution [11]. In addition, these disadvantaged groups are often excluded from public healthcare systems and their low income is unable to afford private healthcare services. Health risks related to air pollution and poverty are interconnected. However, these health risks related to air pollution may differ between countries and communities. For example, according to the WHO [11], Southeast Asian, Western Pacific, and Sub-Saharan African regions are the regions most affected by health risks brought on by pollution linked to cooking energy.

This chapter aims to review the social injustices and inequities associated with air pollution and the connection between social deprivation and air pollution exposure, which leads to environmental injustices, in response to this background. The review's outcomes are supported by an intersectionality study carried out in Bangkok City, Thailand, one of Asia's most air-polluted nations. The case study looks at how persons are exposed to air pollution due to variances in the intersection of gender, age, occupation, social background, and place of employment.

The number of premature deaths in the Asia Pacific region that could be linked to ambient air pollution, particularly fine particulate matter ($PM_{2.5}$), increased significantly between 1990 and 2015 [12]. Thailand stands out as one of the most heavily polluted countries in the Asia Pacific. Thai people are seriously threatened by the high levels of air pollution. The annual exposure to fine particulate matter with dimensions of 2.5 μm or less ($PM_{2.5}$) exceeds by more than four times the worldwide threshold established by the WHO for 2021. $PM_{2.5}$ concentrations, especially in Bangkok, during several months of the year, can exceed the WHO's annual air quality guideline levels by up to eight times. For example, Bangkok's $PM_{2.5}$ values reached 47.4 microgram per cubic meter (μg/m^3) in January 2019 [13]. There are many different sources of air pollution in Thailand, such as automobile emissions in urban areas, biomass burning in rural areas, transboundary haze near borders, industrial emissions, and especially from heavy traffic [10, 14]. $PM_{2.5}$ is the main source of air pollution. It enters the respiratory system and causes long-term respiratory and other health problems [15]. Extended exposure to air pollution can lead to persistent mental health issues and socioeconomic disadvantages in addition to physical health difficulties [16, 17].

Furthermore, Thailand globally has one of the highest rates of socioeconomic inequality [18], which affects access to critical healthcare for the illnesses caused by air pollution exposure. In Bangkok, the informal sector employs about 28 percentage of the labor force and includes jobs such as street vendors, motorbike taxi drivers, and home workers. Remarkably, workers in the informal economy contribute 42% of occupations countrywide and 56% of urban jobs in Thailand [19]. These informal works, which are frequently filled by young people, poor women, and migrants, pay poorly and expose workers to a variety of risks. In addition, informal workers do not have access to social services or insurance plans that cover retirement, unemployment, healthcare, workplace safety, and disability [20]. They are often vulnerable socioeconomic groups [21].

Although academics, professionals, and policymakers are concerned about the effects of air pollution on human health, the general population knows very little about these effects and how they intersect with other disparities [9]. The case study examined the different effects of air pollution exposure on persons who work in the same workplace. It also investigated how various socioeconomic groups have been affected by social and labor protection laws as well as public health services against diseases and issues linked to air pollution. Thus, this chapter provides implications for policies and practices in addressing air pollution, considering the social injustices and inequities associated with it. It also highlights the importance of adopting a human well-being-centric approach in addressing air pollution. It suggests enhancing air pollution awareness among the public, stakeholders, and policymakers about the disparities and inequality in low air quality exposure and its effects on different socioeconomic groups. It aims to press for immediate policy changes that will improve the social policy and healthcare for these urban vulnerable socioeconomic groups. The ultimate goal of the chapter is to call for the integration of a spatial-economic-social (SES) inclusion approach into planning in order to ensure everyone's access to adequate housing, healthcare services, and decent occupation opportunities and clean living environment for mitigating social injustices and inequities associated with air pollution and guarantee environmental justice for all.

2 Theoretical Backgrounds

2.1 Intersectionality and Environmental Pollution and Hazards

The term "intersectionality" was coined in 1989 to address how color women are portrayed in feminist legal and political discourses. It was first used by renowned feminist activist Kimberlé Crenshaw [22]. A major turning point in the changing relationship between activism and academia is represented by Crenshaw's scholarly writings. However, earlier academic endeavors in 1983 like Maria Stewart's work in Boston and 1969 Frances Beal's piece "Double Jeopardy: To Be Black and Female" might be used to trace the roots of intersectional perspectives. Perhaps the most important thing to take away from an intersectionality perspective is that gender is never a stand-alone indication of inequity. The concept of intersectionality draws attention to the overlapping aspects of sexual orientation, caste, class, race, and other social characteristics that require consideration [23]. In the middle of the twentieth century, social movements promoted institutional change in sectors like housing, healthcare, education, and employment [22].

Some other intersectional research has brought attention to the significance of structural [24] and geographical [25] inequality in health outcomes, in addition to inequality at the individual level. These approaches highlight the impact of a number of factors, including socioeconomic and demographic characteristics, residential segregation, decent occupation opportunities and labor market that affect the outcomes of environmental health [24]. Therefore, intersectionality theory is being applied in an increasing number of studies to examine and elucidate health disparities arising from variations in exposure to environmental risks that cross social identities such as gender, race, and class [26, 27]. The world environmental quality varies significantly because of variations in the distribution of benefits and risks from interactions with the surrounding environments. It is anticipated that as the number and variety of synthetic chemicals utilized increases, chemical emissions would increase. When these persistent pollutants accumulate in the environment and food systems, future generations can experience challenges. People benefiting from a better local environment or being negatively impacted by pollution will depend on where they live and work.

By examining complex intersections, researchers can understand how different socioeconomic population groups experience different environmental impacts and exposure to environmental risks in the context of environmental pollution and hazards. For example, Osman-Elasha [28] pointed out that women are more vulnerable than men to the effects of climate change. This is mostly because women make up a disproportionate share of the impoverished and are largely dependent on natural resources, which are in danger of being depleted. Ganzleben and Kazmierczak [29] show that lower socioeconomic classes in several European countries are disproportionately exposed to air, noise pollution, and heat waves in urban areas. In Europe, the general population is not usually as susceptible to environmental health dangers as the elderly, children, and people with poor health are. For example, during heat waves, death rates among the elderly are higher than those of the general population. The quantity and quality of the findings, however, differ widely, with Western Europe having the most substantial evidence of environmental disparities. Furthermore, the impacts of noise on health and the interplay of many environmental dangers are poorly understood. The study by Nguyen and Pattanrsi [30] on urban marginalized people's vulnerability and severity of COVID-19 in Thailand shows that vulnerability to COVID-19 is significantly affected by several demographic factors, including age, gender, ethnicity, and residential occupancy types (including vacant spaces, beneath tollways, and near railroads). Furthermore, occupation and residential occupancy type appear to be risk factors for the severity of COVID-19, especially for daily wage workers. In addition, people who lived in impoverished areas and had access to the Internet demonstrated a greater understanding of self-defense strategies against COVID-19. However, there were some barriers that prevented women, migrant workers, and the elderly from receiving COVID-19 information online [31].

2.2 *Environmental Justice and Equity of Air Pollution*

Environmental injustice occurs when disadvantaged communities face unfair exposure to harmful pollution, often due to failures in political or market systems. It's not just about uneven exposure to pollutants; it also encompasses resulting health disparities [32, 33]. Environmental injustice highlights the need for specific measures to address the unjust treatment of these disadvantaged communities, whereas environmental inequity focuses on unequal allocation of environmental burdens. The sources and effects of environmental inequality linked to air pollution have a major role in contributing to socioeconomic inequities. The concept of "environmental justice"was first introduced during the mid-1980s American racial equality movement. It highlights how different racial and ethnic groups are exposed differently to environmental threats like pollution and toxic waste, and how they are underrepresented in the creation and implementation of environmental policy. These acts of injustice and exclusion have been referred to as instances of "environmental racism" [5]. Furthermore, through air pollution control policies the USA has seen reductions in air pollution emissions from different sources since the Clean Air Act (CAA) was put into effect in 1970 [5]. Similarly, in the last few decades, several European nations have seen a decrease in urban air pollution, primarily as a result of national laws that are in line with European directives (1999/30/EC, 2008/50/EC) [34]. The Directives impose stringent limits on air pollution concentrations for the most dangerous and difficult-to-manage chemicals, like fine particles, in its Member States. Law enforcement uses a monitoring system to notify the public because it is everyone's right to know about dangerous air quality. Since first developed in the USA and Canada, environmental justice now becomes a significant component of public health and environmental policy [35]. In Europe, Scotland and subsequently England were the first countries to successfully incorporate environmental justice concerns into social policy during the early 2000s [36].

Even though there has been a national decline in air pollution trends as well as environmental justice of air pollution has been integrated into laws and policies, research indicates that socioeconomic and racial/ethnic disparities in air pollution exposure still exist in Europe and the USA [37, 38]. When it comes to environmental justice, there are questions regarding whether some socioeconomic groups – particularly those that are marginalized in politics and the economy – bear a disproportionate amount of the burden. It challenges whether the distribution of social services and welfare, including green areas, is equitable and whether policies and actions about pollution sources are also fair [39].

Numerous research that has examined socioeconomic traits and pollution exposure in communities at various geographic levels have focused on environmental equity. Exposure to air pollution and toxins is disproportionately higher in people from lower socioeconomic backgrounds [40]. This is typically the result of their residences or places of employment being close to pollution sources, such as heavy traffic roads, industrial areas, waste disposal facilities, and landfills [10, 41]. Environmental inequality encompasses more than just exposure to air pollution; it also involves differences in the availability of public services, healthcare, and a healthy environment [29]. Exposure to low-quality air can have serious health repercussions because there is no public healthcare system in place. Having access to green spaces in cities can lessen health disparities and promote social cohesiveness. Additionally, environmental justice guarantees that all groups are equally represented in legal proceedings and decision-making processes pertaining to their environment. Equal access to environmental knowledge, the ability to seek justice, and participation in the decision-making process should be granted to all socioeconomic groups. Public education can address the potential negative health effects of environmental quality [42]. Making the public aware of bad air quality is essential to putting pressure on policymakers to deal with air pollution. When it comes to environmental quality, the public should be involved, taking legal action to defend their legal right in national courts.

2.3 Social Injustices and Inequities Associated with Air Pollution

Social injustice refers to racism, poverty, and social risk factors which are often experienced by disadvantaged and marginalized communities[43]. Air pollution-related social injustice draws attention to the disproportionate impact that air pollution has on these groups, including those with lower socioeconomic level which includes both a person's economic and social position [44]. Some populations—such as low-income, racial, and ethnic and marginalized communities—are disproportionately impacted by air pollution [37, 45]. Low socioeconomic status population is likely to have a higher burden of chronic sicknesses than higher socioeconomic status population [46]. Depending on the type of neighborhood a person lives in, the influence of socioeconomic status on health might be also different. Social inequality, which has mostly been researched in connection to racial, ethnic, and gender discrepancies, is referred to as "disparities." Different settings have focused on the distinctions between wealthy and poor countries, as well as within countries between social groupings categorized by occupation, income, education, or immigration status [47].

Because of a confluence of environmental, social, and economic factors, low-income people are more vulnerable to negative air pollution-related health effects. Exposure to air pollution is generally higher for the poor and advantaged racial and ethnic groups, and these groups may react negatively to air pollution. Poor air quality results in immediate and long-term human health problems, especially for pregnant women, children, and older persons [48, 49]. The different effects of air pollution on racial or ethnic groups, as well as those who have low socioeconomic status, little education, and low-paying jobs or live in close proximity to significant sources of pollution, have been noted. For instance, in the USA, a higher number of lower-class and racial/ethnic minorities are found to die from $PM_{2.5}$ exposure than other demographics and higher-income groups [37, 50]. Those who live in locations with poor air quality are also more vulnerable to the health problems related to air pollution. Living close to pollution sources such as high traffic roads, ports, and factories frequented by diesel trucks, low-income communities are frequently more exposed [10]. Living near industrial sites, ports, railway yards, and main thoroughfares can have a negative impact on one's health [41], increasing the risk of cardiovascular disease, asthma attacks, and early mortality [51].

Many of the underlying causes of why low socioeconomic groups often reside near significant sources of pollution and experience unequal distribution of environmental burdens, such as exclusionary policies and practices, are, for example, systemic racism which has traditionally been the driving force behind policies like residential segregation in various countries, which forces racial, ethnic, and low-income groups into areas with limited resources, inadequate urban amenities, elevated community stress, and heightened exposure to pollution. This compounds the unequal access to urban green infrastructure and exacerbates inequities in environmental health [52]. Despite evidence supporting the health advantages of vegetation and reduced air pollution impacts, disparities in urban greenery based on race and income raise concerns about restricted ecosystem services provided by green infrastructure in marginalized communities globally, for example in the regions like the UK [53], Canada [54], and China [55].

In fact, greater exposure to air pollution is associated with social deprivation [9]. Pollution levels, including $PM_{2.5}$, are typically correlated with lower economic status and higher social deprivation indices [44]. Social deprivation refers to living conditions that fall below the average standard [56]. Indicators of social deprivation include low income, low education, unemployment, informal occupations, non-house ownership, or living in informal or slum settlements. Low socioeconomic status results in poor housing conditions, low living standard, malnutrition, and lack of community facilities. Examples include increased exposure to indoor air pollution from cooking fuels or smoking and lack of home air-purifying devices. People living in low-income neighborhoods, elderly people, people with preexisting heart diseases, and children are more vulnerable to the negative health impacts of local air pollution [57]. These groups suffer from higher levels of air pollution exposure and more severe health effects because of factors such as lower general health conditions and lack of access to adequate healthcare. Moreover, discrimination against those from lower socioeconomic backgrounds is widespread and can result in social stress, psychological strain, and heightened susceptibility to diseases and illnesses. Stress increases the likelihood that exposure to air pollution may have negative health impacts [8].

Social injustice includes discrimination against lower and racial or ethnic groups, unequal access to public services like healthcare, and limited opportunity for education and quality jobs. These factors contribute to social and economic inequities in air pollution exposure and healthcare access. In sum, several forms of social injustice can be linked to socioeconomic differences and social inequities in air pollution exposure and health outcomes. Because of racism, discrimination based on class, socioeconomic status, and types of jobs, the housing policy, and the cost of land, disadvantaged groups frequently live close to sources of air pollution. This proximity increases their exposure to dangerous toxins. These communities might experience increased vulnerabilities because of poverty, limited access to public amenities, healthcare service, and good jobs, along with poor working conditions brought on by their lower social status. They may already be dealing with health problems that increase their risk. Many times, these diseases are the outcome of chronic illnesses brought on by poor living conditions, malnutrition, and limited access to healthcare and services.

3 Social Injustices and Inequity of Air Pollution: Case Study of Bangkok City

3.1 Bangkok City

Bangkok, one of the most polluted cities in Thailand, was chosen as the study area. Bangkok's pollution comes from several sources, such as burning agricultural waste, heavy traffic, and diesel emissions from industries and construction. Informal economy plays an important role in Bangkok City which accounts for about 28% of the workforce. Many informal laborers, such as street sellers, market vendors, motorbike taxi drivers, and domestic helpers, are women, and many of them put in more than 40 h a week. They frequently encounter black carbon released by diesel engines in cars, trucks, and boats. Unsafe $PM_{2.5}$ air quality levels are a common occurrence in Bangkok and are a significant environmental element that contributes to premature mortality.

Five districts were the subject of the study: Thon Buri, Khlong Toei, Din Daeng, Nong Chok, and Bang Khun Thian. These neighborhoods represent a range of sociocultural and economic groups and have a varied range of socioeconomic activities. Thon Buri is renowned for having a high concentration of tourist destinations, like historical and cultural landmarks. Khlong Toei is home to a variety of small-scale industries, businesses, and tourists. Dense zones of industry, services, and commerce define Din Daeng. These three districts are in Bangkok's downtown. In contrast, peri-urban areas include Nong Chok and Bang Khun Thian. Whereas Nong Chok works in agriculture, rice cultivation, and the automotive sector, Bang Khun Thian's main economic activities are aquaculture, fisheries, and processing industries. The purpose of selecting these contrasted districts is to investigate differences in the health effects of air pollution and how different socioeconomic classes perceive the quality of the air in varied geographical and pollution-prone areas.

3.2 Data Collection and Analysis

The study was carried out through a questionnaire survey with 400 people in the five selected districts. The sample of 384 people was calculated from the total working-age population of 451,418 people from 15 to 60 years by using Yamane's [58] formula with a 95% confidence level, but the total number of respondents was oversampled to 400. Other criteria of sample selection were also set such as a balanced number of both males and females, indoor and outdoor workers, formal and informal sector workers, all types of jobs, and geographical location of the workplace. Respondents were randomly approached at their workplace and requested their voluntary participation in the survey. The questionnaire survey was complemented with 30 semi-structured interviews of people of all categories against the sample criteria. An efficiency of at least 95% for non-oil-based particulate (N95) mask package was provided to each respondent as humble compensation for their time participating in the survey. The questions made during the surveys included (1) demographic and socioeconomic characteristics such as gender, age, marital status, family, nationality, job, education, and geographical location of working and living places; (2) any health symptoms they had over time; and (3) their perceptions of bad air exposures around their working places.

Both quantitative and qualitative data analyses were performed. Descriptive statistical analysis was used to analyze the data obtained from the questionnaire surveys, where frequency was calculated to capture the tendency of air pollution-related health symptoms among geographic locations, worker groups, workplaces, and between working environments and sectors. The mean was calculated to assess the levels of agreement on the frequency of exposure to different sources of air pollution. Chi-square analysis was also performed to examine differences among perceived exposures to air pollution among socioeconomic and demographic groups. Descriptive phenomenological analysis was applied to the interview information to capture the air pollution-related lived experiences of the interviewees.

3.3 Results and Discussion

3.3.1 Respondents' Profiles and Air Quality of the Study Area

Figure 1 illustrates the locations of the districts of Thon Buri, Khlong Toei, and Din Daeng in the downtown area of Bangkok and Nong Chok and Bang Khun Thian in the peri-urban portions of the city. Thon Buri and Din Daeng have the densest populations, whereas Khong Loei and Thon Buri have the greatest number of slum

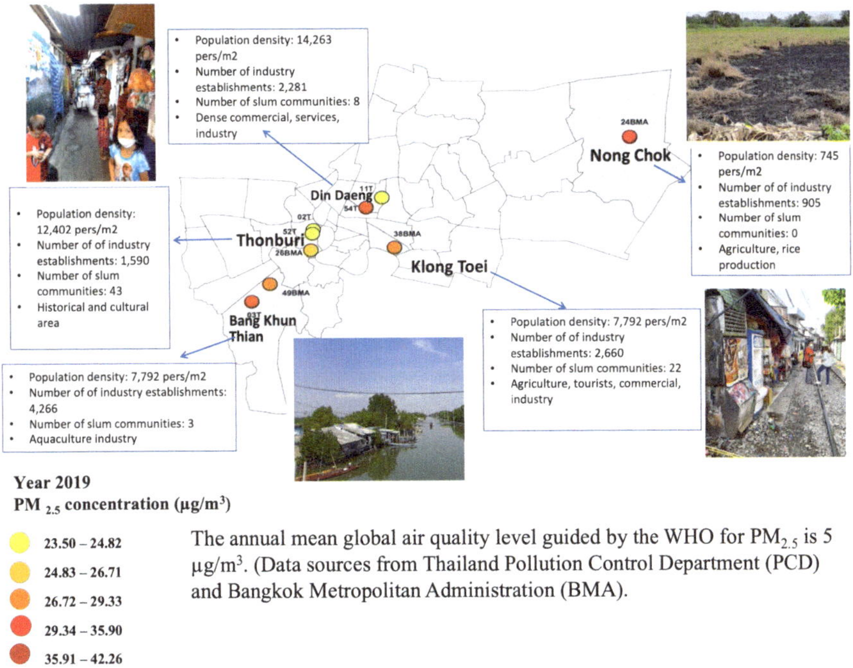

Fig. 1 Socioeconomic characteristics and air quality indexes of the five selected districts

settlements. There is a wide variety of socioeconomic activity in the neighborhoods of Thon Buri, Khlong Toei, and Din Daeng, which represent a range of sociocultural and economic groupings.

Thon Buri is well known for having a lot of tourist attractions, including historical and cultural sites. Numerous small-scale companies, visitors, and industries may be found in Khlong Toei. Din Daeng is defined as congested areas of business, services, and industry. On the other hand, Bang Khun Thian's primary economic endeavors include aquaculture, fisheries, and the processing industries, while Nong Chok works in agriculture, rice farming, and the automobile industry. According to observed air quality time series data from the Bangkok Metropolitan Administration and Thai Pollution Control Department, Din Daeng, Nong Chok, and Bang Khun Thian had high annual average $PM_{2.5}$ levels in 2019. These levels ranged from 35.90 to 42.26 $\mu g/m^3$. On the other hand, $PM_{2.5}$ levels in Khlong Toei and Thon Buri were lower, ranging from 23.60 to 24.92 $\mu g/m^3$, suggesting less pollution.

The respondents consist of an almost equal number of males and females, spanning a wide age range from 15 to over 45 years old. There is no illiteracy among them, and the majority have received education up to primary and secondary levels. They represent a diverse array of job categories, including official workers, street vendors, factory workers, farmers, and service providers in both indoor and outdoor working environments, across both formal and informal sectors. The majority of them work near open markets, along streets or highways, with a few employed in town or community areas, industrial zones, and slum areas.

3.3.2 Perceived Exposure to Low Air Quality Among Different Socioeconomic Groups and Workplaces

Table 1 shows the respondents' perceptions of low air quality exposure among the five districts. A majority of surveyed people in the districts located in the center of Bangkok perceived they had been exposed to low air quality, namely Thon Buri (97%), Din Daeng (93%), and Khlong Toei (88%). Most participants from these three districts in the semi-structured interviews expressed their concerns about low air quality in their districts, for example,

My surrounding air condition was mixed with fog and smoke. Outdoor air quality was not good for my customers and me.

The air pollution comes from vehicles and construction. I think air quality is worse than in the past.

I think the air quality is not good. There is much dust in the morning.

However, many surveyed people in the peri-urban districts of Bang Khun Thian and Nong Chok didn't think they had been exposed to the low air quality, especially a very high percentage of surveyed people (69%) in Nong Chok, stated by a motorbike driver in Nong Chok "*I live in the outskirts of Bangkok. The air quality is ok if compared with the inner city of Bangkok.*

Table 1 Perceived exposure to low air quality among different socioeconomic groups and workplaces

	N	No. of perceived exposure	%	*Df*	$\chi 2$	*p*-value
Districts				4.00	124.7956	**≤0.001**
Bang Khun Thian	90	56	62.22			
Thon Buri	70	68	97.14			
Khlong Toei	80	70	87.50			
Nong Chok	80	25	31.25			
Din Daeng	80	74	92.50			
Working location				4.00	68.203	**≤0.001**
Town or community	7	7	100.00			
Industrial area	20	16	80.00			
Slum area	27	23	85.19			
Close to markets	75	63	84.00			
Near the street or highway	227	154	67.84			
Job types				4.00	32.5791	**≤0.001**
Office workers	66	55	83.33			
Street vendors	112	77	68.75			
Factory workers	86	64	74.42			
Farmers	61	30	49.18			
Service providers	75	67	89.33			
Work sector				1.00	21.5925	**≤0.001**
Formal	164	138	84.15			
Informal	236	155	65.68			
Work environment						
Indoor	153	115	75.16	1.00	0.462	0.496
Outdoor	247	178	72.06			
Income				1.00	615.17	**≤0.001**
5,000–10,000 baht	72	48	66.67			
10,000–15,000 baht	251	53	21.12			
15,000–20,000 baht	77	6	7.79			
Age				3.00	3.2089	0.201
15–30 years	103	71	68.93			
31–45 years	171	133	77.78			
45 years above	126	89	70.63			
Gender				1.00	1.3066	0.253
Male	228	162	71.05			
Female	172	131	76.16			
Education						
Illiterate	4	3	75.00	4.00	12.67	**0.013**
Primary level	136	94	69.12			
Secondary level	185	129	69.73			
High level/vocational	45	39	86.67			
University	30	28	93.33			

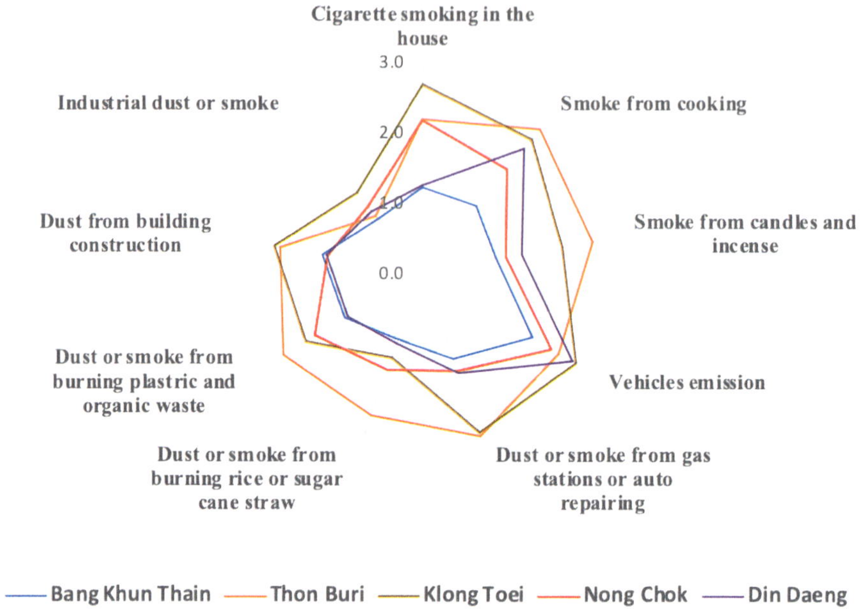

Fig. 2 Perceived sources of air pollution among the districts. Mean calculated from 4 scales of 1 = Never, 2 = Sometimes, 3 = Often, and 4 = Daily

More people who work within the town, in the center, or near markets perceive their exposure to low quality than those working near streets and highways. Farmers consider themselves less exposed to air pollution than others. Street vendors are perceived as experiencing less exposure to low air quality compared to office workers and service providers. Formal workers perceive greater exposure to air pollution than informal workers, and many respondents with the lowest income perceive themselves to be exposed to air pollution. There is no difference in perceptions of air pollution exposure between indoor and outdoor workers, as well as between male and female workers.

Figure 2 illustrates the perceived sources of air pollution exposure among the respondents. High levels of perception regarding air pollution from various sources were observed in Thon Buri and Khlong Toei. In Thon Buri, respondents reported being sometimes or often exposed to smoke and dust from construction works, plastic and organic waste, rice or sugarcane straws, candles and incense, cooking, and smoking cigarettes. Meanwhile, in Khlong Toei, perceived sources of exposure included dust and smoke from construction works, gas stations and automobile repair shops, vehicle emissions, and smoke from cooking and cigarettes. In Din Daeng, the main sources of perceived air pollution exposure were vehicle emissions and smoke from cooking, which respondents reported being exposed to sometimes or often. Conversely, respondents in Nong Chok claimed to never or only sometimes be exposed to smoke/dust from burning rice or sugarcane straw and plastic and organic waste, and sometimes exposed to smoke from cooking, cigarettes, and vehicle emissions. Meanwhile, residents of Bang Khun Thian believed they were never exposed to any sources of air pollution, except vehicle emissions.

The above findings showed the respondents' perceptions of air pollution exposure are asymmetric with the air monitoring. Most people living in Din Daeng, Khlong Toei, and Thon Buri believed themselves to be exposed to bad air quality rather than in Nong Chok and Bang Khun Thian. Higher awareness of air pollution sources among workers in urban areas compared to those in peri-urban areas was observed. Respondents in urban areas tended to have a better understanding of air pollution sources than those in peri-urban areas. The findings showed that the workers' perception of risk is distorted because their judgment about a threat is influenced by knowledge, personal preferences, attitudes, and individual assessments of their environment [59].

Furthermore, people in Thon Buri and Din Daeng believed they were exposed to air pollution from various sources, whereas those from Nong Chok and Bang Khun Thian declared not to be exposed to air pollution, only occasionally to smoke and dust from construction sites in Nong Chok and vehicle emissions in Bang Khun Thian. This difference may be attributed to the greater accessibility of knowledge and information among urban dwellers compared to peri-urban residents, influencing their perception of risk exposure. Urban residents generally have more access to communication channels such as TV, computers, and the Internet than rural residents.

According to Walters [60], air pollution is a silent killer that is hard for people to notice. However, communication on air quality is frequently lacking [61] and public awareness of the sources of air pollution is frequently inadequate [62]. Raising public awareness of air pollution and its effects on human health may encourage people to take steps to limit their exposure to health hazards, as knowledge influences behavior. Additionally, communication can support the development of skills that increase self-efficacy [30].

3.3.3 Air Pollution-Related Health Symptoms Among the Geographic Locations

Figure 3 shows Bang Khun Thian, Nong Chok, and Din Daeng exhibit the highest prevalence of individuals with one or more health symptoms relevant to bad air quality. The areas with the greatest incidence of burning or irritated eyes are Bang Khun Thian and Nong Chok where rates peak at 72.2% and 61%, respectively. Sneezing attacks are another noteworthy incidence reported from these two districts and are very prevalent in Di Daeng (86.25%). On the other hand, fewer people in Thon Buri and Khlong Toei report having symptoms, except a sizable proportion who suffer headaches (82.8% and 57.5%, respectively).

The results in Fig. 3 show that the common health problems that respondents linked to poor air quality and the data from air quality monitoring were correlated. The districts of Din Daeng, Nong Chok, and Bang Khun Thian had greater percentages of respondents reporting one or more health problems, and their $PM_{2.5}$

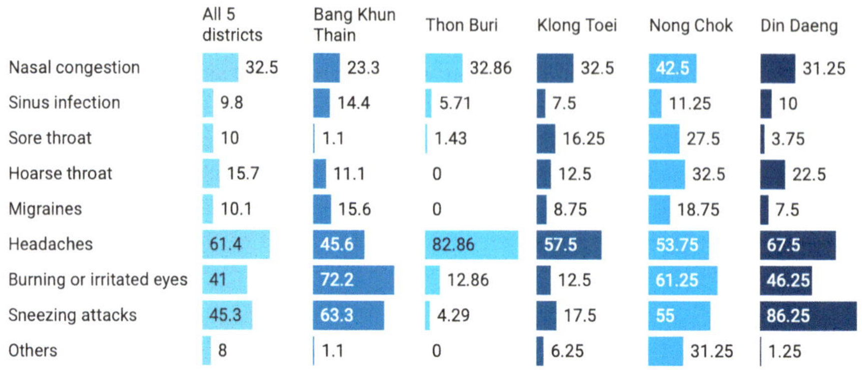

Fig. 3 Percentage of workers' air pollution-related health symptoms among the five districts

concentrations were dangerously high, ranging from 35.90 to 42.26 g/m^3. In contrast, the $PM_{2.5}$ concentrations in the Khlong Toei and Thon Buri districts ranged from 23.60 to 24.92 g/m^3, indicating lower pollution levels than other parts of the city and possibly fewer people experiencing health issues connected to poor air quality. These results are consistent with earlier research that showed the detrimental impacts of air pollution on human health and well-being, as shown by several studies carried out worldwide (e.g., [63]). Long- and short-term consequences of these effects include the development of chronic illnesses such as cancer, pulmonary insufficiency, cardiovascular disease, and asthma [64].

Furthermore, Fig. 4 shows people who work near streets and highways, as well as in industrial zones, were more likely to experience more health symptoms related to poor air quality than those in other workplaces. For example, a high number of people with migraines, burning eyes, sneezing attacks, and nasal congestion were found in industrial areas, while many individuals with burning eyes and sneezing attacks were observed near streets or highways (Fig. 4a).

There are not many differences in the prevalence of health symptoms related to low air quality between indoor and outdoor workers. However, outdoor workers tended to experience more issues related to burning or irritated eyes (Fig. 4b). On the other hand, informal workers tended to face more health problems associated with poor air quality than formal workers (Fig. 4c). Other results from the analysis also revealed that there are no differences in health symptoms between male and female workers, or among age groups. However, the lowest-income workers tended to have more health problems than other higher-income groups.

The above results illustrated the differences in the health effects of poor air quality workplace among socioeconomic groups. Because of the toxic industrial waste and pollutants that are harmful to human health, environmental pollution is a persistent problem in industrial zones [65]. On the other hand, those who work near streets and highways are at risk due to large traffic volumes linked to air pollution in urban areas [66]. Furthermore, there was no difference in health concerns between

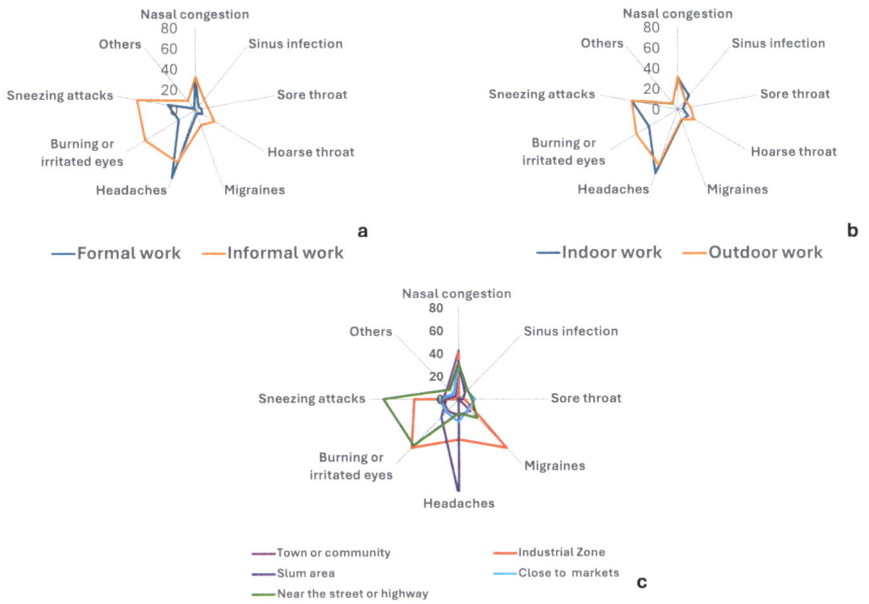

Fig. 4 Percentage of health symptoms among worker groups and workplaces and between working environments and sectors

indoor and outdoor work according to the data. This is true since indoor air quality is greatly impacted by outdoor air quality [67]. The outdoor worker group reported more eye irritation than the indoor worker group, which could have been caused by direct contact with dust. Compared to the formal employment worker group, more respondents from the informal employment group said they had experienced one or more symptoms such as headaches, hoarse throats, burning or irritated eyes, sneezing, and nasal congestion. Furthermore, compared to the other higher-income categories, the lower-income groups experienced more serious health issues. This proved that the population's vulnerability to diseases caused by air pollution is more significantly influenced by poverty and low socioeconomic status [68]. Sometimes laborers in the informal sector of the economy work in hazardous, unsafe environments for pitiful wages and no benefits. They are not provided with toolkits for labor safety protection or access to training on health protection measures. Their inability to obtain insurance and healthcare prevents their health problems from being promptly diagnosed and treated.

In conclusion, through an intersectionality study of the impacts of air pollution on various socioeconomic groups and workplaces in Bangkok City, Thailand, this case study is a comprehensive view of environmental inequalities in air pollution across both structural and geographical dimensions, affecting health outcomes and awareness of air pollution exposure. The intersectionality theory has been employed to illustrate disparities arising from differences in exposure to air pollution and its impacts across socioeconomic groups and geographical areas. The findings reveal that various segments of workers in different workplaces have limited awareness of air pollution around their work environments, yet the health symptoms experienced by individuals working in districts or environments with high air pollution indices are correlated. The disparities in pollution distribution among locations, sectors, and socioeconomic groups stem from social inequities in accessing public services such as healthcare, welfare, education, labor safety, and labor and social protection.

Thailand is exerting more effort than other countries in the same region to strengthen the legal framework on air quality control. This was evident as the Draft Management for Clean Air Act "Clean Air Bill," proposed by the Ministry of Natural Resources and the Environment (MONRE), was approved in principle by the Cabinet approved on November 28, 2023. However, Thailand still faces pressing issues concerning current and future social policies, at the intersection of demand for social and environmental justices. Various international organizations, such as the Canadian International Development Research Centre and the Stockholm Environmental Institute, along with the UNs and universities, are drawing policymakers' attention to the concepts of environmental justice and environmental inequalities. However, challenges to equality and equity in the environmental issues are numerous within Thailand and similar countries. Greater investment in environmental and air pollution communication is urgently needed, as every citizen has a right to know, seek justice, and participate in decisions for social policies that fairly safeguard all facets of society from environmental threats.

4 Developmental Implications of the Social Injustices and Inequities Associated with Air Pollution

The above literature review and the case study confirmed that different forms of social injustice and inequities are the root cause of the persistence of inequality in air pollution exposure. Due to issues such as social deprivation, socioeconomic status, informal occupation, racism, discrimination based on class, ethnicity, race, and migrant status, housing market dynamics, and property costs, disadvantaged communities frequently reside close to sources of pollution, which increases their exposure to toxins. Their vulnerability is increased by their limited access to decent jobs, healthcare, and public amenities, as well as by the poor working conditions brought on by their lower social position. Furthermore, their risks are further increased by preexisting health disorders, which are frequently brought on by low living standards, poverty, malnutrition, and restricted access to treatment. Given the differential exposure to air pollution across various socioeconomic groups worldwide, researchers and policymakers should understand the social injustice and inequities associated with the unequal distribution of air pollution. This understanding is crucial to establishing development policies and practices that address and rectify air pollution-induced environmental injustice. Policies aimed at reducing air pollution should prioritize resolving disparities. Expanding our knowledge of how environmental exposures impact health disparities across various racial, ethnic, occupational, and socioeconomic groups is crucial.

Inequality in air pollution exposure is greatly correlated with social deprivation. Air pollution exposure is highly driven by low socioeconomic status including low income, informal occupation, low education, and migrant status. Therefore, while addressing pollution-induced environmental injustice, it is necessary to address these associated social deprivation factors linked to pollution exposure. This implies that problems with low education, low-skill employment, limited access to decent jobs, and low income could not be solved independently. Reducing air pollution-induced environmental injustice would be more successful with an integrated policy approach that prioritizes the social equities for bettering the financial, educational, and vocational support given to people, especially youths from disadvantaged social backgrounds. Reducing air pollution-related diseases and death rates as well as bringing these disadvantaged groups more in line with higher-income communities can be achieved by improving access to quality, reasonably priced healthcare in low- and middle-income communities. Healthcare policies must give priority to combating air pollution since it directly contributes to poorer health among socially deprived groups. Putting in place economic initiatives and labor laws protecting informal and migrant workers offers social protection to underprivileged areas, and supplementing these actions with voluntary ones reduces the social disparities.

The literature and case study's findings also show that air pollution is among the main causes of death worldwide, which has a disproportionately negative impact on lower-income groups since they frequently reside in high-pollution areas, making them more vulnerable and exposed. Therefore, planning decisions that must account transparently for environmental and health externalities can help shift pollution sources, such as industrial zones, landfills, or power plants, out of low-income areas. Furthermore, lower-income groups can access incentives for the elimination of excessive fossil fuel consumption. The financial benefits of fossil fuel subsidies are disproportionately felt by wealthier households, yet low-income households are most affected by the resulting air pollution. Reversing these policy biases can benefit low-income communities in terms of finances and health. Furthermore, it is critical to improve environmental justice through compensation schemes to ensure that governments or corporations that harm the environment respect the fundamental rights of impacted communities. Socially disadvantaged populations should get compensation based on fairness to address environmental injustice. This compensation might be determined by taking into account the expenditures incurred by individuals of impacted communities in relation to environmental or health issues [9].

In many Asian developing countries, including Bangladesh, Thailand, China, India, and others, where there are a significant number of slums and urban informal settlements, high $PM_{2.5}$ pollution concentrations are also observed in high socioeconomic deprivation cities [69]. These slum communities frequently face poor living conditions, inadequate housing, overcrowding, bad sanitation, limited access to clean water, and a lack of community facilities and healthcare [70]. These communities are more sensitive to the detrimental effects of air pollution due to their high population density and inadequate housing standards and infrastructure. In combating air pollution, this emphasizes how crucial it is to consider housing policies for low-income communities as well as equitable access to urban green infrastructure like city parks.

Lastly, the underlying social injustices and inequalities are brought to light by the structural and geographical disparities in air pollution that led to environmental injustice. This implies that future planning needs to take justice- and location-focused policymaking into account. Air pollution control plans should prioritize human health over pollution alone to address areas with high concentrations of pollution and social deprivation indices. Future environmental policymaking can consider the entire effects of pollution on communities, the majority of which reside in socially disadvantaged neighborhoods, by utilizing social deprivation indices and scientific facts.

5 Strategies for Mitigating the Social Injustices and Inequities Associated with Air Pollution

The undeniable existence of social injustices and inequities intertwined with air pollution serves as proof of the urgent need for effective solutions to alleviate environmental injustice and lessen the disparities in exposure to air pollutants. In addition to measures aimed at reducing and controlling air pollution, a variety of strategies are required to address the related issues of unequal exposure to air pollution, such as:

Improve data exchange, research, and air monitoring related to health inequities caused by air pollution.

This requires identifying socioeconomic factors that affect people's exposure to air pollution and how it affects their health, especially in areas or sectors that are known to produce pollution. This strategy seeks to increase knowledge on the relationship between poor air quality and health disparities by identifying the most vulnerable communities. The knowledge collected can direct pollution reduction plans and initiatives to lessen the harmful impacts of air pollution on human health. However, as data are frequently fragmented, creating silos and impeding institutional cooperation, it is necessary to create interdisciplinary teams or organizations that are entrusted with establishing shared goals and assisting with data collection.

Enhance citizen science, democracy, and air pollution communication for environmental justice.

To enable the public to assess and monitor their personal air quality for self-protection, communication with data from air pollution monitoring and preventive measures should be encouraged. Empowering the public in using online tools kits and smartphone applications, citizen science initiatives are promoted to monitor air pollution. Furthermore, enhancing democracy guarantees the rights of disadvantaged groups to information access, justice, and involvement in the decision making of social and environmental policies that protect the environment and society.

Improve housing policies and incentives for low-income and socioeconomically marginalized populations.

Government and development organizations should place a high priority on housing policies and incentives to guarantee low-income communities have access to safe homes. Prioritizing communities living close to sources of air pollution is necessary. A notable example is Thailand's Baan Mankong slum upgrading program, which addresses concerns of insecure housing and substandard living conditions through a people-centered approach [71].

Promote community health services and provide free public health access of socioeconomically disadvantaged and low-income groups.

Critically increasing the capacity of primary healthcare organizations and community healthcare centers to serve the needs of air pollution-related patients should be a priority in health and social protection programs. Everybody should have access to primary care, with the capacity to address the population of low-income and socially deprived groups. A national medical program providing free access to public healthcare or at least biannual medical check-ups for low-income and socially deprived groups including informal workers should be established.

Create decent job opportunities and enhance labor and social protection for workers of the informal economy.

Establish vocational training programs for individuals, particularly youth from disadvantaged social backgrounds, to enhance their skills for accessing decent jobs. Prioritize opportunities for individuals from disadvantaged backgrounds in these decent job opportunities. Social policies should focus on providing incentives and labor safety toolkits (e.g., masks, gloves) to informal sector workers to promote self-protection against air pollution. Additionally, consider providing informal health insurance for workers who may be subcontracted and left out of existing health schemes. Provide training programs on health and labor safety protection at work in community centers for informal sector workers. These programs can be delivered by volunteers or community-based organizations and associations of informal sector workers with support from governmental agencies. Promote the training of trainers (TOT) within the informal sector worker population.

Introduce incentives to reduce fossil fuel consumption among low-income and disadvantaged households.

Targeted subsidies and incentives should be put in place to reduce the consumption of fossil fuels and mitigate the negative effects of air pollution on low-income and disadvantaged households. This improves their financial situation and has positive effects on their health. Providing these populations with access to renewable energy programs, such as solar panels, is also crucial.

Promote green urban infrastructure and amenities to promote air pollution justice.

Ensuring equitable access to parks and green spaces for socioeconomically disadvantaged and low-income communities can be achieved through increasing the covering of vegetation and investing urban green infrastructure. Incorporating residents of socioeconomically deprived areas into grant proposals that suggest intelligent ways to mitigate air pollution – such as planting trees or pedestrianizing – can help to foster community involvement. In addition, including communities with varying exposure levels and involving the public in the restoration process increases their sense of ownership.

Create compensation plans based on environmental justice.

Governments ought to take steps to identify the causes and sources of air pollution and the resulting health costs incurred by impacted communities. It is imperative that government action be taken to ensure environmental justice and promote socially responsible behavior in improving air quality while also offering incentives for such action to reduce the risks of air pollution exposure among disadvantaged populations. It is crucial to enhance environmental justice through compensation programs to ensure that those that harm the environment respect the fundamental rights of affected people. To lessen environmental injustice, socially disadvantaged groups should receive compensation based on justice. The costs incurred by impacted communities due to environmental or health concerns may be used to calculate this compensation.

6 Conclusion

This chapter uses a case study in Bangkok, one among the most air-polluted cities in the Asia Pacific region, and literature reviews to highlight the social inequalities and disparities caused by air pollution. It illustrates how social deprivation and unequal exposure to air pollution are related, and how the structurally and geographically uneven distribution of environmental consequences from air pollution can be explained by intersectional socioeconomic and demographic factors. Societal and economic disparities in air pollution exposure are a result of several social injustices and inequities, such as discrimination against certain racial or ethnic groups, unequal access to public services like healthcare, and limited opportunities for education and work. Disadvantaged populations frequently live close to sources of air pollution due to factors such as residential segregation, which results from racism and class-based discrimination, socioeconomic status, the housing market, and land costs. Due to poverty, lack of access to public goods and healthcare, limited access to decent jobs, and unfavorable working circumstances brought on by their lower social status and maybe preexisting medical issues, they are more susceptible to air pollution. Considering the social injustices and disparities connected to air pollution, the findings have implications for policy and practices in this area. They also stress how crucial it is to tackle air pollution from the perspective of human health.

The chapter suggests several strategies lessen the social injustices and disparities related to air pollution, since decreasing air pollution sources alone will not be sufficient to alleviate the unequal health implications of air pollution. These include advancing citizen science and democracy for environmental justice, as well as data sharing on health disparities related to air pollution, air monitoring, and communication about air pollution. It is essential to develop compensation plans based on environmental justice and to support green infrastructure and services in urban areas to further advance the cause of air pollution justice. Incentives to reduce fossil fuel consumption among low-income and disadvantaged households can help mitigate social injustices associated with air pollution, in addition to offering decent employment opportunities and enhancing labor and social protection for workers in the informal economy. In addition, an all-encompassing plan should prioritize improving housing policies for low-income and socioeconomically disadvantaged groups, promoting community health services, and providing free public health access to these populations.

In conclusion, the SES – the spatial-economic and social inclusion approach – should incorporate strategies to reduce air pollution as well as the social injustices and inequities linked to it [69]. This will enhance everyone's access to affordable land, housing, and services, as well as opportunities to share in urban prosperity and the conditions that allow the poor to participate in governance. To guarantee environmental justice for all, these strategies will uphold rights and participation by emphasizing community-driven development, participatory planning, and rights-based approaches.

Acknowledgments

This research was funded by the Stockholm Environmental Institute, from the funding of the International Development Research Centre, Canada (IDRC). SEI Project No/work order no. 10035600. The author acknowledged the voluntary participation of 400 workers in the case study and the support of the Thailand Pollution Control Department (PCD).

Ethical Approval The Research Ethics Review Committee (RERC) at the Asian Institute of Technology (AIT), Thailand, approved the interview and questionnaire survey protocol. Ref No. 2022/003 dated March 4, 2022.

References

1. UCAR (2024) Air pollution: how we're changing the air. Centre for Science Education. Retrieved 10/05 from https://scied.ucar.edu/learning-zone/air-quality/air-pollution
2. WHO (2016) Air pollution levels rising in many of the world's poorest cities. World Health Organization Retrieved 10/04 from https://www.who.int/news/item/12-05-2016-air-pollution-levels-rising-in-many-of-the-world-s-poorest-cities
3. Landrigan PJ, Fuller R, Acosta NJR, Adeyi O, Arnold R, Basu NN, Baldé AB, Bertollini R, Bose-O'Reilly S, Boufford JI, Breysse PN, Chiles T, Mahidol C, Coll-Seck AM, Cropper ML, Fobil J, Fuster V, Greenstone M, Haines A et al (2018) The lancet commission on pollution and health. Lancet 391(10119):462–512. https://doi.org/10.1016/s0140-6736(17)32345-0
4. EEA (2023) How air pollution affects our health. European Enviornmental Agency. Retrieved 25/03 from https://www.google.com/search?client=safari&rls=en&q=air+polltuion+and+health&ie=UTF-8&oe=UTF-8#ip=1
5. Laurent É (2011) Issues in environmental justice within the European Union. Ecol Econ 70(11): 1846–1853. https://doi.org/10.1016/j.ecolecon.2011.06.025

6. La Grutta S, Indinnimeo L, di Coste A, Ferrante G, Landi M, Pelosi U, Rusconi F (2013) Environmental risk factors and lung diseases in children: from guidelines to health effects. Early Hum Dev 89:S59–S62. https://doi.org/10.1016/j.earlhumdev.2013.07.025
7. Roberts S, Arseneault L, Barratt B, Beevers S, Danese A, Odgers CL, Moffitt TE, Reuben A, Kelly FJ, Fisher HL (2019) Exploration of NO2 and PM2.5 air pollution and mental health problems using high-resolution data in London-based children from a UK longitudinal cohort study. Psychiatry Res 272:8–17. https://doi.org/10.1016/j.psychres.2018.12.050
8. Hajat A, Hsia C, O'Neill MS (2015) Socioeconomic disparities and air pollution exposure: a global review. Curr Environ Health Rep 2(4):440–450. https://doi.org/10.1007/s40572-015-0069-5
9. Li VOK, Han Y, Lam JCK, Zhu Y, Bacon-Shone J (2018) Air pollution and environmental injustice: are the socially deprived exposed to more PM2.5 pollution in Hong Kong? Environ Sci Policy 80:53–61. https://doi.org/10.1016/j.envsci.2017.10.014
10. Nguyen TP, Virdis SGP, Winjikul E (2022) Inequality of low air quality-related health impacts among socioeconomic groups in the world of work. Int J Environ Res Public Health 19(19)
11. WHO (2024) Air quality, energy and health. World Health Organization. Retrieved 25/03 from https://www.who.int/teams/environment-climate-change-and-health/air-quality-energy-and-health/health-impacts/exposure-air-pollution
12. UNESCAP (2018) On environmental degradation and its impact on inequality in Asia and the Pacific. ESCAP. Retrieved 25/03 from https://www.unescap.org/blog/on-environmental-degradation-and-its-impact-on-inequality-in-asia-and-the-pacific
13. IQAir (2021) World air quality report: region & city PM2.5 ranking. https://www.iqair.com/world-most-polluted-cities/world-air-quality-report-2021-en.pdf
14. Nguyen TPL, Winjikul E, Virdis SGP (2023) Air pollution in Bangkok: addressing unequal exposure and enhancing public understanding of the risks. SEI Brief Stockholm Environment Institute. https://doi.org/10.51414/sei2023.12
15. Kim Y, Knowles S, Manley J, Radoias V (2017) Long-run health consequences of air pollution: evidence from Indonesia's forest fires of 1997. Econ Hum Biol 26:186–198. https://doi.org/10.1016/j.ehb.2017.03.006
16. Currie J, Zivin JG, Mullins J, Neidell M (2014) What do we know about short- and long-term effects of early-life exposure to pollution? Annu Rev Resour Econ 6:217–247. https://doi.org/10.1146/annurev-resource-100913-012610
17. Kim Y, Manley J, Radoias V (2020) Air pollution and long term mental health. Atmos 11(12)
18. Thongsawang S, Rehbein B, Chantavanich S (2020) Inequality, sociocultures and habitus in Thailand. Sojourn J Soc Issues Southeast Asia 35(3):493–524. https://www.jstor.org/stable/26937815
19. Poonsab W, Vanek J, Carré F (2019) Informal workers in urban Thailand: a statistical snapshot. WIEGO Stat Br 20:2–4
20. Komin W, Thepparp R, Subsing B, Engstrom D (2021) Covid-19 and its impact on informal sector workers: a case study of Thailand. Asia Pac J Soc Work Dev 31(1–2):80–88. https://doi.org/10.1080/02185385.2020.1832564
21. Mehrotra S (2009) The impact of the economic crisis on the informal sector and poverty in East Asia. Glob Soc Policy 9(1_suppl):101–118. https://doi.org/10.1177/1468018109106887
22. Collins PH (2019) Intersectionality as critical social theory. Duke University Press. https://doi.org/10.2307/j.ctv11hpkdj
23. Garcia A, Tschakert P (2022) Intersectional subjectivities and climate change adaptation: an attentive analytical approach for examining power, emancipatory processes, and transformation. Trans Inst Br Geogr 47(3):651–665. https://doi.org/10.1111/tran.12529
24. Homan P, Brown TH, King B (2021) Structural intersectionality as a new direction for health disparities research. J Health Soc Behav 62(3):350–370. https://doi.org/10.1177/00221465211032947
25. Bambra C (2022) Placing intersectional inequalities in health. Health Place 75:102761. https://doi.org/10.1016/j.healthplace.2022.102761

26. Alvarez CH, Calasanti A, Evans CR, Ard K (2022) Intersectional inequalities in industrial air toxics exposure in the United States. Health Place 77:102886. https://doi.org/10.1016/j.healthplace.2022.102886
27. Cannon CEB (2021) Intersectional and entangled risks: an empirical analysis of disasters and landfills [original research]. Front Clim 3. https://doi.org/10.3389/fclim.2021.709439
28. Osman-Elasha B (2009) Women...in the shadow of climate change. United Nations Retrieved 25/03 from https://www.un.org/en/chronicle/article/womenin-shadow-climate-change
29. Ganzleben C, Kazmierczak A (2020) Leaving no one behind – understanding environmental inequality in Europe. Environ Health 19(1):57. https://doi.org/10.1186/s12940-020-00600-2
30. Nguyen TPL, Pattanrsi S (2022) The influence of ethnic, social, and demographic factors on urban slum Dwellers' threat appraisal, awareness, and protective practices against COVID-19 in Thailand. Am J Trop Med Hyg 107(1):169–174. https://doi.org/10.4269/ajtmh.21-1096
31. Pattanasri S, Nguyen TP, Vu TB, Winijkul E, Ahmad MM (2022) Access to digital information and protective awareness and practices towards COVID-19 in urban marginalized communities. Healthcare 10(6)
32. Morello-Frosch R, Pastor M, Sadd J (2001) Environmental justice and Southern California's "Riskscape": the distribution of air toxics exposures and health risks among diverse communities. Urban Aff Rev 36(4):551–578. https://doi.org/10.1177/10780870122184993
33. Morello-Frosch R, Zuk M, Jerrett M, Shamasunder B, Kyle AD (2011) Understanding the cumulative impacts of inequalities in environmental health: implications for policy. HealthAffairs 30(5):879–887. https://doi.org/10.1377/hlthaff.2011.0153
34. Marco G, Bo X (2013) Air quality legislation and standards in the European Union: background, status and public participation. Adv Clim Chang Res 4(1):50–59. https://doi.org/10.3724/SP.J.1248.2013.050
35. Padilla CM, Kihal-Talantikite W, Vieira VM, Rossello P, Nir GL, Zmirou-Navier D, Deguen S (2014) Air quality and social deprivation in four French metropolitan areas – a localized spatio-temporal environmental inequality analysis. Environ Res 134:315–324. https://doi.org/10.1016/j.envres.2014.07.017
36. Stephens C, Willis R, Walker GP (2007) Addressing environmental inequalities: cumulative environmental impacts. Environment Agency
37. Jbaily A, Zhou X, Liu J, Lee T-H, Kamareddine L, Verguet S, Dominici F (2022) Air pollution exposure disparities across US population and income groups. Nature 601(7892):228–233. https://doi.org/10.1038/s41586-021-04190-y
38. Laurent É (2022) Air (ine)quality in the European Union. Curr Environ Health Rep 9(2):123–129. https://doi.org/10.1007/s40572-022-00348-6
39. Braubach M (2013) Benefits of environmental inequality assessments for action. J Epidemiol Community Health 67(8):625–628. https://doi.org/10.1136/jech-2012-201390
40. Namdeo A, Stringer C (2008) Investigating the relationship between air pollution, health and social deprivation in Leeds, UK. Environ Int 34(5):585–591. https://doi.org/10.1016/j.envint.2007.12.015
41. Laurian L, Funderburg R (2014) Environmental justice in France? A spatio-temporal analysis of incinerator location. J Environ Plan Manag 57(3):424–446. https://doi.org/10.1080/09640568.2012.749395
42. OECD (2017) OECD environmental performance reviews: Korea. https://doi.org/10.1787/9789264268265-en
43. Mathiarasan S, Hüls A (2021) Impact of environmental injustice on children's health-interaction between air pollution and socioeconomic status. Int J Environ Res Public Health 18(2). https://doi.org/10.3390/ijerph18020795
44. Fairburn J, Schüle SA, Dreger S, Karla Hilz L, Bolte G (2019) Social inequalities in exposure to ambient air pollution: a systematic review in the WHO European region. Int J Environ Res Public Health 16(17). https://doi.org/10.3390/ijerph16173127
45. AIA (2023) Disparities in the impact of air pollution. Amercian Lung Association. https://www.lung.org/clean-air/outdoors/who-is-at-risk/disparities

46. Niessen LW, Mohan D, Akuoku JK, Mirelman AJ, Ahmed S, Koehlmoos TP, Trujillo A, Khan J, Peters DH (2018) Tackling socioeconomic inequalities and non-communicable diseases in low-income and middle-income countries under the sustainable development agenda. Lancet 391(10134):2036–2046. https://doi.org/10.1016/s0140-6736(18)30482-3
47. Kawachi I, Subramanian SV, Almeida-Filho N (2002) A glossary for health inequalities. J Epidemiol Community Health 56(9):647–652. https://doi.org/10.1136/jech.56.9.647
48. Bell ML, Zanobetti A, Dominici F (2013) Evidence on vulnerability and susceptibility to health risks associated with short-term exposure to particulate matter: a systematic review and meta-analysis. Am J Epidemiol 178(6):865–876. https://doi.org/10.1093/aje/kwt090
49. Wang Y, Shi L, Lee M, Liu P, Di Q, Zanobetti A, Schwartz JD (2017) Long-term exposure to PM2.5 and mortality among older adults in the southeastern US. Epidemiology 28(2) https://journals.lww.com/epidem/fulltext/2017/03000/long_term_exposure_to_pm2_5_and_mortality_among.8.aspx
50. Di Q, Wang Y, Zanobetti A, Wang Y, Koutrakis P, Choirat C, Dominici F, Schwartz Joel D (2017) Air pollution and mortality in the medicare population. N Engl J Med 376(26): 2513–2522. https://doi.org/10.1056/NEJMoa1702747
51. Lee BJ, Kim B, Lee K (2014) Air pollution exposure and cardiovascular disease. Toxicol Res 30(2):71–75. https://doi.org/10.5487/tr.2014.30.2.071
52. Jennings V, Reid CE, Fuller CH (2021) Green infrastructure can limit but not solve air pollution injustice. Nat Commun 12(1):4681. https://doi.org/10.1038/s41467-021-24892-1
53. Ferguson M, Roberts HE, McEachan RRC, Dallimer M (2018) Contrasting distributions of urban green infrastructure across social and ethno-racial groups. Landsc Urban Plan 175:136–148. https://doi.org/10.1016/j.landurbplan.2018.03.020
54. Pham T-T-H, Apparicio P, Séguin A-M, Landry S, Gagnon M (2012) Spatial distribution of vegetation in Montreal: an uneven distribution or environmental inequity? Landsc Urban Plan 107(3):214–224. https://doi.org/10.1016/j.landurbplan.2012.06.002
55. Zhu Z, Ren J, Liu X (2019) Green infrastructure provision for environmental justice: application of the equity index in Guangzhou, China. Urban For Urban Greening 46:126443. https://doi.org/10.1016/j.ufug.2019.126443
56. Townsend P (1987) Deprivation. J Soc Policy 16(2):125–146
57. Tran HM, Tsai F-J, Lee Y-L, Chang J-H, Chang L-T, Chang T-Y, Chung KF, Kuo H-P, Lee K-Y, Chuang K-J, Chuang H-C (2023) The impact of air pollution on respiratory diseases in an era of climate change: a review of the current evidence. Sci Total Environ 898:166340. https://doi.org/10.1016/j.scitotenv.2023.166340
58. Yamane T (1973) Statistics: An Introductory Analysis. 3rd Edition, Harper & Row Ltd., New York.
59. Cori L, Donzelli G, Gorini F, Bianchi F, Curzio O (2020) Risk perception of air pollution: a systematic review focused on particulate matter exposure. Int J Environ Res Public Health 17(17). https://doi.org/10.3390/ijerph17176424
60. Walters R (2014) Air pollution and invisible violence. In: Davies P, Francis P, Wyatt T (eds) Invisible crimes and social harms. Palgrave Macmillan, pp 142–160. https://doi.org/10.1057/9781137347824_8
61. Maione M, Mocca E, Eisfeld K, Kazepov Y, Fuzzi S (2021) Public perception of air pollution sources across Europe. Ambio 50(6):1150–1158. https://doi.org/10.1007/s13280-020-01450-5
62. Ramírez AS, Ramondt S, Van Bogart K, Perez-Zuniga R (2019) Public awareness of air pollution and health threats: challenges and opportunities for communication strategies to improve environmental health literacy. J Health Commun 24(1):75–83. https://doi.org/10.1080/10810730.2019.1574320
63. Ghorani-Azam A, Riahi-Zanjani B, Balali-Mood M (2016) Effects of air pollution on human health and practical measures for prevention in Iran. J Res Med Sci 21:65. https://doi.org/10.4103/1735-1995.189646

64. Manisalidis I, Stavropoulou E, Stavropoulos A, Bezirtzoglou E (2020) Environmental and health impacts of air pollution: a review [review]. Front Public Health 8. https://doi.org/10.3389/fpubh.2020.00014
65. Aiman N, Gulnaz S, Alena M (2018) The characteristics of pollution in the big industrial cities of Kazakhstan by the example of Almaty. J Environ Health Sci Eng 16(1):81–88. https://doi.org/10.1007/s40201-018-0299-1
66. Kobza J, Geremek M (2017) Do the pollution related to high-traffic roads in urbanised areas pose a significant threat to the local population? Environ Monit Assess 189(1):33. https://doi.org/10.1007/s10661-016-5697-1
67. Leung DYC (2015) Outdoor-indoor air pollution in urban environment: challenges and opportunity [review]. Front Environ Sci 2. https://doi.org/10.3389/fenvs.2014.00069
68. Forastiere F, Stafoggia M, Tasco C, Picciotto S, Agabiti N, Cesaroni G, Perucci CA (2007) Socioeconomic status, particulate air pollution, and daily mortality: differential exposure or differential susceptibility. Am J Ind Med 50(3):208–216. https://doi.org/10.1002/ajim.20368
69. UN-Habitat (2022) World cities report 2022. Envisaging the future of cities
70. UN-Habitat (2003) The challenge of slums: global report on human settlements 2003
71. Archer D (2012) Baan Mankong participatory slum upgrading in Bangkok, Thailand: community perceptions of outcomes and security of tenure. Habitat Int 36(1):178–184. https://doi.org/10.1016/j.habitatint.2011.08.006

Air Pollution and the Sustainable Development Goals

Matthew Chidozie Ogwu and Sylvester Chibueze Izah

Contents

Abstract The interconnectedness of air pollution with climate change and sustainable development underscores the need for integrated and collaborative approaches to address complex environmental challenges. The air pollution and climate change feedback loops and amplification mechanisms exacerbate each other's impacts and contribute to environmental degradation. For example, air pollution can accelerate the melting of glaciers and ice caps through the deposition of dark particles (such as black carbon) on snow and ice surfaces, reducing their reflectivity and increasing the absorption of solar radiation. Therefore, recognizing and addressing these interlinkages will foster the development of holistic solutions that promote clean

M. C. Ogwu (✉)
Goodnight Family Department of Sustainable Development, Appalachian State University, Boone, NC, USA
e-mail: ogwumc@appstate.edu

S. C. Izah
Department of Microbiology, Faculty of Science, Bayelsa Medical University, Yenagoa, Bayelsa, Nigeria

Matthew Chidozie Ogwu and Sylvester Izah (eds.), *Sustainable Strategies for Air Pollution Mitigation: Development, Economics and Technologies*, Hdb Env Chem (2024) 133: 201–234, DOI 10.1007/698_2024_1118,

Published online: 17 September 2024

air, mitigate climate change, and advance sustainable development goals for the benefit of people and the planet. This chapter aims to discuss the connections between air pollution and sustainable development goals. Air pollution is a critical global environmental challenge that poses risks to human health, ecosystems, economies, and sustainable development. For instance, concerns about air pollution-related health risks and environmental degradation are driving community activism, advocacy, and demands for environmental justice. Addressing air pollution requires concerted efforts at the local, national, and international levels to reduce emissions, promote clean technologies, and safeguard public health and the environment. The Sustainable Development Goals (SDGs) provide a comprehensive framework for addressing air pollution and its associated challenges. By integrating efforts to reduce air pollution into broader sustainable development strategies, countries can promote public health, protect the environment, and advance progress toward achieving the SDGs and creating a more sustainable future for all. By prioritizing clean air as a fundamental right and adopting evidence-based policies and strategies, societies can work toward achieving cleaner, healthier, and more sustainable environments for the present and promote sustainable development for future generations.

Keywords Clean air, Climate change, Environmental justice, Environmental sustainability, Evidence-based policies, Public health

Abbreviations

PM	Particulate matter
PM10	Coarse particulate matter
PM2.5	Fine particulate matter
PM0.1	Ultrafine particulate matter
O_3	Ozone
NOx	Nitrogen oxide
CO	Carbon monoxide
VOCs	Volatile organic compounds
CO_2	Carbon dioxide
CH_4	Methane
NO_2	Nitrogen dioxide
SO_2	Sulfur dioxide
NH_3	Ammonia
SDGs	Sustainable development goals
PAHs	Polycyclic aromatic hydrocarbons
MDGs	Millennium development goals
EU	European Union
NCAP	National Clean Air Programme
NAAQS	National Ambient Air Quality Standards

GHGs	Greenhouse gases
CLRTAP	Convention on long-range transboundary air pollution
UNECE	United Nations Economic Commission for Europe
SOx	Sulfur oxides
IMO	International Maritime Organization
MARPOL	The international convention for the prevention of pollution from ships
GAHP	Global alliance on health and pollution
GEMS	Global environment monitoring system
UNEP	Air: The United Nations Environment Programme
UNFCCC	United Nations Framework Convention on Climate Change
ODA	Official development assistance
LEZs	Low-emission zones
°C	Degree celsius
EVs	Electric vehicles
%	Percent or percentage

1 Introduction

Air pollution is a pressing global environmental challenge with far-reaching implications for human health, ecosystems, economies, and sustainable development. It encompasses the presence of harmful substances in the Earth's atmosphere, originating from natural and anthropogenic sources, that degrade air quality and pose risks to living organisms. Air pollution refers to the presence of harmful or excessive quantities of substances in the Earth's atmosphere, which can be detrimental to human health, ecosystems, and the environment [1, 2]. These pollutants can be either naturally occurring or human-made and can originate from various sources, including industrial activities, transportation, agriculture, energy production, and natural events such as wildfires and volcanic eruptions. Particulate matter (PM) that is, tiny particles suspended in the air, ranging in size from coarse particles (PM10) to fine particles (PM2.5) and ultrafine particles (PM0.1). These particles can penetrate deep into the respiratory system and cause respiratory and cardiovascular diseases [3–5]. Gases produced from combustion processes, primarily from vehicles, power plants, and industrial facilities, also contribute to air pollution [6]. Nitrogen oxide (NOx) contributes to the formation of ground-level ozone (O_3) and fine particulate matter and can exacerbate respiratory conditions such as asthma. A gas emitted from burning fossil fuels containing sulfur, such as coal and oil, as well as industrial processes, is a well-known air pollutant. SO_2 can cause respiratory issues, particularly in people with preexisting respiratory conditions, and contribute to the formation of acid rain [7]. Another air pollutant, carbon monoxide (CO), interferes with the blood's ability to carry oxygen and can cause headaches, dizziness, and even death in high concentrations [8]. In the same way, volatile organic compounds (VOCs) can react with other pollutants in the atmosphere to form ground-level ozone and fine particulate matter, contributing to air pollution and smog formation [9, 10]. Ozone is considered a secondary pollutant formed from chemical reactions between NOx and VOCs in the presence of sunlight. Ground-level ozone can irritate the respiratory system, worsen asthma symptoms, and reduce lung function, especially in vulnerable populations.

Air pollution can have significant impacts on human health, including respiratory and cardiovascular diseases, cancer, and premature death. It also affects ecosystems, leading to biodiversity loss, soil and water contamination, and damage to crops and vegetation. Additionally, air pollution contributes to climate change by emitting greenhouse gases such as carbon dioxide (CO_2) and methane (CH_4), which trap heat in the Earth's atmosphere and contribute to global warming [11, 12]. Air pollution is a leading cause of morbidity and mortality worldwide, contributing to a range of respiratory and cardiovascular diseases, cancers, and other adverse health effects [13]. Exposure to pollutants such as PM, nitrogen dioxide (NO_2), sulfur dioxide (SO_2), O_3, and VOCs can lead to respiratory infections, asthma exacerbations, heart attacks, strokes, and premature mortality. Vulnerable populations, including children, the elderly, pregnant women, and individuals with preexisting health conditions, are particularly at risk. Air pollution contributes to environmental degradation and ecosystem disruption, affecting terrestrial and aquatic ecosystems, biodiversity, and ecological processes [14]. Pollutants such as NOx, SO_2, ammonia (NH_3), and CO can lead to acid rain, soil and water contamination, deforestation, and habitat destruction, harming wildlife, and vegetation [15]. Ozone depletion and climate change are also linked to air pollution, exacerbating environmental threats and global challenges. Air pollution imposes significant economic costs on societies, including healthcare expenditures, lost productivity, and damage to infrastructure and property [16]. Healthcare costs associated with air pollution-related illnesses strain healthcare systems and household budgets, leading to financial hardship and economic instability [17, 18]. Reduced productivity, absenteeism, and healthcare expenses for affected workers and businesses contribute to economic losses and decreased competitiveness in polluted areas.

Air pollution has social implications, including reduced quality of life, social inequalities, and community disruptions. Poor air quality limits outdoor recreational opportunities, undermines community well-being, and erodes social cohesion. Disadvantaged communities, low-income populations, and minority groups bear a disproportionate burden of pollution-related health risks and environmental injustices, exacerbating social inequalities and disparities in health outcomes [19–21]. Air pollution is a global concern that transcends national boundaries and affects populations worldwide. Transboundary air pollution, international trade, and global supply chains contribute to the global spread of pollutants and the interconnectedness of air pollution issues [22]. Addressing air pollution requires coordinated action, multilateral cooperation, and shared responsibility among countries to reduce emissions, protect public health, and mitigate environmental impacts.

The Sustainable Development Goals (SDGs) are a set of global objectives adopted by the United Nations in 2015 as part of the 2030 Agenda for Sustainable Development. These 17 interconnected goals aim to address the world's most pressing social, economic, and environmental challenges and provide a framework for achieving a more sustainable and equitable future for all [23, 24]. The SDGs cover a wide range of issues, including poverty eradication, health and well-being, education, gender equality, clean water and sanitation, affordable and clean energy, sustainable cities and communities, climate action, and life below water and on land. Each goal is accompanied by specific targets and indicators to measure progress toward achieving sustainable development outcomes. Among the SDGs, Goal 3 (Good Health and Well-being), Goal 7 (Affordable and Clean Energy), Goal 11 (Sustainable Cities and Communities), and Goal 13 (Climate Action) are particularly relevant to addressing air pollution. These goals recognize the importance of clean air for human health, environmental sustainability, and overall well-being. For instance, Goal 3 aims to ensure healthy lives and promote well-being for all ages. One of the targets of this goal is to substantially reduce the number of deaths and illnesses from hazardous chemicals and air, water, and soil pollution and contamination by 2030 [25, 26]. By addressing air pollution, countries can improve public health outcomes, reduce the burden of disease, and enhance the well-being of communities. Goal 7 seeks to ensure access to affordable, reliable, sustainable, and modern energy for all. Clean energy sources such as renewable energy technologies play a crucial role in reducing air pollution by replacing fossil fuels and minimizing emissions of harmful pollutants [6, 27]. By investing in clean energy solutions, countries can mitigate air pollution, promote energy efficiency, and advance sustainable development objectives. Goal 11 focuses on making cities and human settlements inclusive, safe, resilient, and sustainable. Air pollution is often more pronounced in urban areas due to high population density, industrial activities, and traffic congestion. Sustainable urban planning improved public transportation, green infrastructure, and pollution control measures are essential for reducing air pollution levels and creating healthier, more livable cities. Goal 13 calls for urgent action to combat climate change and its impacts. Air pollution is closely linked to climate change as many of the same pollutants contribute to both issues. Reducing emissions of greenhouse gases and air pollutants will contribute to mitigating climate change, improve air quality, and protect public health and the environment [28].

This chapter aims to discuss the connections between air pollution and sustainable development goals. Air pollution is a critical global environmental challenge that poses risks to human health, ecosystems, economies, and sustainable development. Addressing air pollution requires concerted efforts at the local, national, and international levels to reduce emissions, promote clean technologies, and safeguard public health and the environment. The Sustainable Development Goals provide a comprehensive framework for addressing air pollution and its associated challenges. By integrating efforts to reduce air pollution into broader sustainable development strategies, countries can promote public health, protect the environment, and advance progress toward achieving the SDGs and creating a more sustainable future for all. By prioritizing clean air as a fundamental right and adopting evidence-based policies and strategies, societies can work toward achieving cleaner, healthier, and more sustainable environments for the present and promote sustainable development for future generations.

2 Understanding Air Pollution

Air pollution is the presence of harmful or excessive quantities of substances in the Earth's atmosphere, which can be detrimental to human health, ecosystems, and the environment. It is a complex and multifaceted issue influenced by various factors, including natural processes, human activities, and atmospheric conditions. Air pollution can originate from both natural and human-made sources. Natural sources include volcanic eruptions, wildfires, dust storms, and biological emissions. Human-made sources encompass industrial processes, transportation, energy production, agriculture, waste disposal, and residential activities. Air pollutants can take various forms, including gases, particulate matter, and volatile organic compounds [29]. Common pollutants include

- **Particulate matter:** Tiny particles suspended in the air, categorized by size as PM10, PM2.5, and PM0.1.
- **Nitrogen oxides**: Gases produced from combustion processes, primarily from vehicles, power plants, and industrial facilities.
- **Sulfur dioxide**: A gas emitted from burning fossil fuels containing sulfur, such as coal and oil.
- **Carbon monoxide**: A colorless, odorless gas produced from incomplete combustion of fossil fuels.
- **Volatile organic compounds**: Organic chemicals emitted as gases from various sources.
- **Ozone**: A secondary pollutant formed from chemical reactions between NOx and VOCs in the presence of sunlight.

Once emitted into the atmosphere, these air pollutants can undergo dispersion and transport over long distances, affecting air quality regionally and globally [30, 31]. Factors such as wind patterns, atmospheric stability, and topography influence the dispersion and distribution of pollutants in the air [32]. Air pollution has significant impacts on human health, ecosystems, and the environment. Short-term effects include respiratory and cardiovascular diseases, asthma exacerbations, and eye and throat irritation while long-term exposure to air pollution is associated with chronic health conditions, reduced lung function, and premature mortality [15, 33]. Environmental impacts of air pollution include biodiversity loss, soil and water contamination, acid rain, and climate change. Pollutants can harm vegetation, aquatic ecosystems, and wildlife, leading to ecosystem disruption and habitat degradation. Air quality monitoring and measurement are essential for assessing the extent and severity of air pollution and informing decision-making and policy development [34]. Monitoring networks measure concentrations of key pollutants at various locations, providing data on air quality trends, sources of pollution, and compliance with air quality standards [35]. Strategies for mitigating and controlling air pollution include pollution prevention, emission reduction technologies, regulatory measures, and public awareness campaigns. Sustainable solutions such as transitioning to clean energy sources, promoting energy efficiency, improving transportation infrastructure, and implementing pollution control measures are critical for reducing air pollution and protecting public health and the environment. Certain air pollutants, such as benzene, formaldehyde, and polycyclic aromatic hydrocarbons (PAHs), are classified as carcinogens. These PAHs are associated with an elevated risk of cancer, including lung cancer and leukemia. Emerging research suggests that air pollution may have adverse effects on cognitive function and neurological development, particularly in children [36, 37]. Prenatal exposure to pollutants such as PM2.5 and PAHs has been linked to neurodevelopmental disorders and cognitive deficits [38]. Long-term exposure to air pollution is a significant risk factor for premature mortality. Studies have estimated that air pollution contributes to millions of premature deaths worldwide each year, making it one of the leading causes of mortality globally [39, 40].

Air pollution can harm terrestrial and aquatic ecosystems, leading to biodiversity loss and ecosystem disruption. Pollutants such as NOx and SO_2 can damage vegetation, soil quality, and aquatic habitats, affecting wildlife populations and ecological balance. Airborne pollutants can deposit onto the land and water bodies, contaminating soil, surface water, and groundwater [41]. Acid deposition from SO_2 and nitrogen oxides NOx contributes to soil acidification and water acidification, leading to nutrient imbalances, decreased soil fertility, and aquatic ecosystem degradation. Emissions of SO_2 and NOx can react with water vapor and atmospheric gases to form acid rain. Acid rain can damage vegetation, corrode buildings and infrastructure, and acidify soil and water bodies, impacting agricultural productivity, biodiversity, and ecosystem health. Air pollution contributes to climate change by emitting greenhouse gases such as CO_2, CH_4, and N_2O [42]. These gases trap heat in the Earth's atmosphere, leading to global warming, rising temperatures, altered weather patterns, and more frequent and intense extreme weather events.

Air pollution is a major public health concern, contributing to a range of respiratory and cardiovascular diseases, cancers, and other adverse health outcomes. The health burden of air pollution includes increased morbidity, mortality, and healthcare expenditures, straining healthcare systems and household budgets [43]. Vulnerable populations, including children, the elderly, pregnant women, and individuals with preexisting health conditions, are disproportionately affected by air pollution, exacerbating health disparities and inequalities. Air pollution diminishes the quality of life for affected individuals and communities by limiting outdoor recreational opportunities, reducing access to green spaces, and impairing the enjoyment of natural environments [44]. Poor air quality can lead to discomfort, stress, and anxiety, undermining overall well-being and social cohesion [45]. Residents of polluted areas may experience adverse effects on mental health, including depression, anxiety, and reduced life satisfaction, due to concerns about air pollution-related health risks and environmental degradation [46, 47]. Air pollution imposes significant economic costs on societies through lost productivity, healthcare expenditures, and damage to infrastructure and property. The economic burden of air pollution includes healthcare costs associated with treating air pollution-related illnesses, as well as indirect costs such as absenteeism, reduced labor productivity, and decreased consumer spending. Businesses may incur costs related to employee absenteeism, healthcare expenses, and disruptions to operations caused by air pollution-related issues [48, 49]. Additionally, air pollution can lead to decreased property values, increased insurance premiums, and higher healthcare and environmental cleanup costs for governments and taxpayers. Air pollution exacerbates social inequities and environmental injustices by disproportionately affecting disadvantaged communities, low-income populations, and minority groups [20, 50]. These communities often bear a disproportionate burden of pollution-related health risks and environmental hazards due to factors such as proximity to pollution sources, inadequate access to healthcare, and limited resources to mitigate exposure. Environmental racism, socioeconomic disparities, and unequal distribution of environmental benefits and burdens contribute to environmental injustices, perpetuating social inequalities and disparities in health outcomes, economic opportunities, and quality of life [51]. Air pollution can negatively impact community well-being by eroding social cohesion, disrupting community networks, and undermining trust in institutions and authorities responsible for environmental protection and public health. Concerns about air pollution-related health risks and environmental degradation may lead to community activism, advocacy, and demands for environmental justice. Community-based organizations, grassroots movements, and citizen science initiatives play a crucial role in raising awareness about air pollution issues, mobilizing communities, and advocating for policies and actions to address environmental concerns and protect public health.

3 Linkages Between Air Pollution and Sustainable Development Goals

The Sustainable Development Goals (SDGs) are a set of 17 interconnected global objectives adopted by the United Nations in 2015 as part of the 2030 Agenda for Sustainable Development [52]. These goals and their targets build upon the Millennium Development Goals (MDGs) and aim to address the world's most pressing social, economic, and environmental challenges [53, 24]. According to Chopra et al. [23], Crespo Cuaresma et al. [54], and Mattaini et al. [55], the SDGs provide a framework for collective action to achieve a more sustainable and equitable future (Box 1).

Box 1 The SDG Framework for a More Sustainable and Equitable Future

Goal 1 (No Poverty): End poverty in all its forms everywhere by eradicating extreme poverty and reducing inequalities.

Goal 2 (Zero Hunger): End hunger, achieve food security and improved nutrition, and promote sustainable agriculture.

Goal 3 (Good Health and Well-being): Ensure healthy lives and promote well-being for all ages by addressing major health issues, improving healthcare access, and reducing mortality rates.

Goal 4: (Quality Education): Ensure inclusive and equitable quality education and promote lifelong learning opportunities for all.

Goal 5 (Gender Equality): Achieve gender equality and empower all women and girls by eliminating discrimination, violence, and harmful practices.

Goal 6 (Clean Water and Sanitation): Ensure availability and sustainable management of water and sanitation for all, aiming to achieve universal access to safe drinking water and adequate sanitation.

Goal 7 (Affordable and Clean Energy): Ensure access to affordable, reliable, sustainable, and modern energy for all while promoting renewable energy sources and energy efficiency.

Goal 8 (Decent Work and Economic Growth): Promote sustained, inclusive, and sustainable economic growth, full and productive employment, and decent work for all.

Goal 9 (Industry, Innovation, and Infrastructure): Build resilient infrastructure, promote inclusive and sustainable industrialization, and foster innovation.

Goal 10 (Reduced Inequalities): Reduce inequality within and among countries by promoting social, economic, and political inclusion and ensuring equal opportunities for all.

Goal 11 (Sustainable Cities and Communities): Make cities and human settlements inclusive, safe, resilient, and sustainable by promoting sustainable urban planning and management.

Goal 12 (Responsible Consumption and Production): Ensure sustainable consumption and production patterns by promoting resource efficiency, sustainable consumption, and production practices.

Goal 13 (Climate Action): Take urgent action to combat climate change and its impacts by reducing greenhouse gas emissions, enhancing resilience to climate-related hazards, and promoting sustainable practices.

Goal 14 (Life Below Water): Conserve and sustainably use the oceans, seas, and marine resources for sustainable development.

(continued)

Box 1 (continued)

Goal 15 (Life on Land): Protect, restore, and promote sustainable use of terrestrial ecosystems, halt biodiversity loss, and combat desertification and land degradation.

Goal 16 (Peace, Justice, and Strong Institutions): Promote peaceful and inclusive societies for sustainable development, provide access to justice for all, and build effective, accountable, and inclusive institutions at all levels.

Goal 17 (Partnerships for the Goals): Strengthen the means of implementation and revitalize the global partnership for sustainable development by mobilizing resources, enhancing policy coherence, and fostering collaboration among stakeholders.

These goals are interconnected and indivisible, recognizing that progress in one area often depends on progress in others [56, 57]. By working toward achieving all 17 SDGs, countries can promote sustainable development, address global challenges, and build a better future for present and future generations. Air pollution is also intricately linked to several Sustainable Development Goals (SDGs), reflecting its broad impacts on human health, ecosystems, economies, and overall well-being. The key linkages between air pollution and selected SDGs are (Fig. 1).

Goal 3 (Good Health and Well-Being) For example, Target 3.9: By 2030, substantially reduce the number of deaths and illnesses from hazardous chemicals and air, water, and soil pollution and contamination [58]. Air pollution significantly affects public health, contributing to respiratory and cardiovascular diseases, cancers, and premature mortality. Achieving clean air aligns with SDG 3's targets to reduce the number of deaths and illnesses from hazardous chemicals and air, water, and soil pollution.

Goal 7 (Affordable and Clean Energy) Through Target 7.2: By 2030, increase substantially the share of renewable energy in the global energy mix [59].

Target 7.3: By 2030, double the global rate of improvement in energy efficiency [60, 61].

Transitioning to clean energy sources such as renewable energy technologies is essential for reducing air pollution from fossil fuel combustion. SDG 7 emphasizes access to affordable, reliable, sustainable, and modern energy, aligning with efforts to mitigate air pollution while promoting clean and sustainable energy systems.

Goal 11 (Sustainable Cities and Communities) Through Target 11.6: By 2030, reduce the adverse per capita environmental impact of cities, including by paying special attention to air quality and municipal and other waste management [62, 63].

Target 11.7: By 2030, provide universal access to safe, inclusive, and accessible, green, and public spaces, in particular for women and children, older persons, and persons with disabilities [64].

Air pollution is often more severe in urban areas due to population density, industrial activities, and traffic congestion. Sustainable urban planning, improved

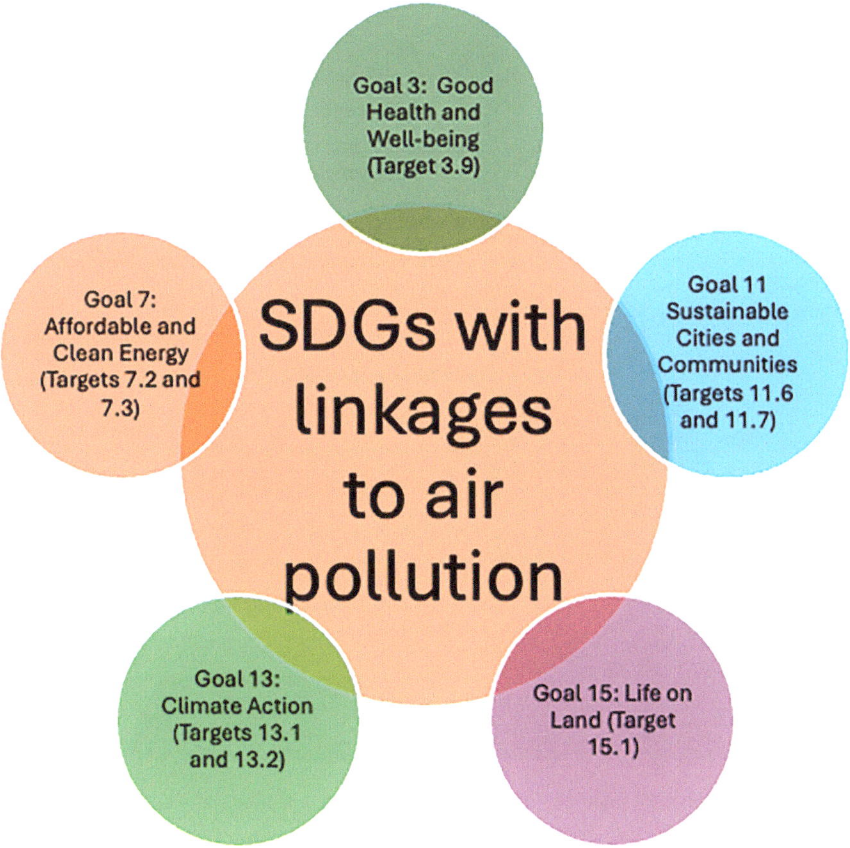

Fig. 1 United Nations Sustainable Development Goals with linkages to air pollution

public transportation, and green infrastructure can help reduce air pollution while promoting SDG 11's objectives of making cities inclusive, safe, resilient, and sustainable.

Goal 13 (Climate Action) Through Target 13.1: Strengthen resilience and adaptive capacity to climate-related hazards and natural disasters in all countries ([65]; Leah [66]).

Target 13.2: Integrate climate change measures into national policies, strategies, and planning [59, 60].

Many air pollutants, such as CO_2, CH_4, and black carbon, contribute to both air pollution and climate change. Addressing air pollution is essential for mitigating climate change and achieving SDG 13's targets to strengthen resilience and adaptive capacity to climate-related hazards.

Goal 15 (Life on Land) For example, Target 15.1: By 2020, ensure the conservation, restoration, and sustainable use of terrestrial and inland freshwater ecosystems and their services, in particular, forests, wetlands, mountains, and drylands, in line

Fig. 2 Some avenues for improving the linkages between air pollution and the SDGs

with obligations under international agreements [67]. Air pollution can have detrimental effects on terrestrial ecosystems, including biodiversity loss, soil degradation, and damage to vegetation. Protecting ecosystems and biodiversity aligns with SDG 15's objectives to conserve and sustainably use terrestrial and marine resources, highlighting the importance of reducing air pollution for ecosystem health.

These SDGs (SDG 3, 7, 11, 13, and 15) encompass targets related to reducing deaths and illnesses from air pollution, promoting clean and renewable energy sources, improving energy efficiency, enhancing urban air quality, strengthening resilience to climate-related hazards, and conserving terrestrial ecosystems. Through recognizing the interconnectedness of air pollution with various SDGs, policymakers, practitioners, and stakeholders can develop integrated strategies and initiatives to address air pollution while advancing progress toward sustainable development objectives. Collaborative efforts across sectors and stakeholders are essential for achieving cleaner air, healthier communities, and a more sustainable future for all.

Some avenues for improving the linkages between air pollution and SDGs include (Fig. 2).

Strengthening Policy Integration Enhance coordination and coherence between policies addressing air pollution and those related to health, energy, climate, and sustainable cities. Integrated policy approaches can maximize synergies and minimize trade-offs between air pollution mitigation and other sustainable development objectives [68].

Establishing Monitoring and Reporting Mechanisms The development of robust monitoring and reporting systems is needed to track progress toward SDG targets related to air pollution reduction [26, 69]. Regular data collection and analysis are essential for assessing the effectiveness of policies and interventions, identifying emerging trends, and informing evidence-based decision-making.

Promoting Cross-Sectoral Collaboration Foster collaboration and partnerships across sectors, disciplines, and stakeholders to address the complex and interconnected nature of air pollution and its impacts. Engage governments, civil society organizations, academia, industry, and communities in joint efforts to develop and implement integrated solutions.

Enhancing Public Awareness and Participation Raise awareness about the linkages between air pollution and sustainable development goals among policymakers, practitioners, and the public [70]. Promote education, communication, and outreach campaigns to mobilize support for action on air quality improvement and foster public participation in decision-making processes.

Investing in Research and Innovation Support research, innovation, and technological advancements to develop cleaner technologies, pollution control measures, and sustainable practices. Invest in interdisciplinary research initiatives to better understand the drivers, impacts, and solutions to air pollution and identify opportunities for transformative change.

Scaling Up Financing Mechanisms Mobilize financial resources and investment to support air pollution mitigation efforts and sustainable development goals. Explore innovative financing mechanisms, such as green bonds, carbon pricing, and public–private partnerships, to leverage funding and incentivize investments in clean air initiatives [71, 72].

Strengthening International Cooperation Foster multilateral cooperation and collaboration to address transboundary air pollution and global challenges. Support international agreements, initiatives, and partnerships aimed at reducing emissions, sharing best practices, and building capacity for air quality management at the regional and global levels [73].

Incorporating Equity and Social Justice Ensure that air pollution mitigation efforts prioritize the needs and interests of vulnerable and marginalized communities, who often bear a disproportionate burden of pollution-related health risks and environmental injustices [74]. Promote equity, social inclusion, and environmental justice in decision-making processes and policy implementation.

4 Policy and Governance Frameworks for Sustainable Clean Air

Air pollution poses significant threats to public health, ecosystems, and the environment, necessitating robust policy and governance frameworks to address this global challenge.

4.1 National Policy Frameworks

Examples of some National Policies and Regulations:

Clean Air Act (United States) The Clean Air Act (CAA) is a comprehensive federal law in the United States aimed at protecting and improving air quality. The CAA sets National Ambient Air Quality Standards (NAAQS) for six criteria pollutants, including particulate matter, ozone, nitrogen dioxide, sulfur dioxide, carbon monoxide, and lead [15]. The law also regulates emissions from major stationary sources, mobile sources, and hazardous air pollutants.

National Clean Air Programme (India) The National Clean Air Programme (NCAP) is a government initiative in India launched in 2019 to tackle air pollution in urban areas across the country. The program aims to reduce air pollution levels by 20–30% by 2024 through a combination of regulatory measures, technology interventions, and public awareness campaigns [75]. It focuses on city-specific action plans for air quality improvement and capacity building at the national, state, and local levels.

European Union Ambient Air Quality Directives (EU) The European Union (EU) has established Ambient Air Quality Directives to protect human health and the environment from harmful air pollution. These directives set air quality standards for pollutants such as particulate matter, nitrogen dioxide, sulfur dioxide, ozone, and lead, with member states required to monitor and report air quality data and develop air quality management plans to achieve compliance with the standards [76, 77].

National Clean Air Programme (China) China's National Clean Air Programme (NCAP) was launched in 2013 to address air pollution and improve air quality in urban areas across the country. The program includes measures to reduce emissions from industrial facilities, power plants, transportation, and residential sources, as well as initiatives to promote clean energy, enhance monitoring and enforcement, and engage stakeholders in air quality management efforts [78]. The usefulness of national policy frameworks for sustainable clean air include:

Establishing Air Quality Standards National governments play a crucial role in setting air quality standards and regulations to limit emissions of key pollutants such as particulate matter, nitrogen oxides, sulfur dioxide, and volatile organic compounds. These standards provide clear targets for pollution reduction and guide regulatory efforts to improve air quality.

Implementing Emission Reduction Measures Governments implement a range of policies and measures to reduce emissions from major sources of air pollution, including industrial facilities, power plants, transportation, and residential heating. These measures may include emission limits, technology standards, pollution control technologies, and incentives for cleaner production and energy efficiency.

Strengthening Monitoring and Enforcement Effective air quality management requires robust monitoring and enforcement mechanisms to track pollution levels, assess compliance with regulatory standards, and hold polluters accountable. Governments invest in air quality monitoring networks, deploy advanced monitoring technologies, and establish enforcement mechanisms to ensure compliance with air quality regulations.

Promoting Clean Energy and Sustainable Transportation Transitioning to clean energy sources and promoting sustainable transportation options are essential strategies for reducing air pollution. Governments implement policies to support renewable energy development, improve energy efficiency, and promote public transit, cycling, and walking as alternatives to private vehicle use.

4.2 International Governance Frameworks

Some useful international governance policy frameworks for sustainable clean air include.

Multilateral Agreements and Protocols International cooperation is vital for addressing transboundary air pollution and global environmental challenges. Multilateral agreements and protocols, such as the Paris Agreement, the Convention on Long-Range Transboundary Air Pollution, and regional agreements like the European Union Ambient Air Quality Directives, establish binding commitments and targets for reducing emissions and improving air quality [79].

Capacity Building and Technical Assistance International organizations provide capacity-building support and technical assistance to help countries strengthen their air quality management capabilities. Initiatives such as the World Health Organization's Urban Ambient Air Pollution Database and the United Nations Environment Programme's Air Quality Monitoring Training Program help countries enhance their monitoring, reporting, and regulatory enforcement capacities [80, 81].

Knowledge Sharing and Best Practices International forums and platforms facilitate knowledge sharing, exchange of best practices, and peer learning among countries facing similar air quality challenges [82]. Platforms like the Clean Air Asia Network and the International Network of Environmental Compliance and

Enforcement support collaboration and information exchange to enhance air quality management efforts worldwide.

4.3 Multilateral Cooperation and Partnerships for Sustainable Clean

Examples of Multilateral Cooperation and Partnerships for Sustainable Clean Air:

1. **Convention on Long-Range Transboundary Air Pollution**: The Convention on Long-Range Transboundary Air Pollution (CLRTAP), established under the United Nations Economic Commission for Europe (UNECE), is a multilateral treaty aimed at addressing air pollution across national and international boundaries in Europe and North America [79, 83]. The convention facilitates cooperation among member countries to reduce emissions of air pollutants that can cause transboundary pollution and harm human health and the environment. It includes protocols addressing specific pollutants such as sulfur dioxide, nitrogen oxides, and persistent organic pollutants, with member countries required to implement measures to achieve emission reduction targets and improve air quality.
2. **International Maritime Organization**: The International Maritime Organization (IMO) is a specialized agency of the United Nations responsible for regulating international shipping to ensure safety, security, and environmental protection [84]. The IMO has adopted regulations to reduce air pollution from ships by limiting emissions of sulfur oxides (SOx), NOx, and PMs [85]. Initiatives such as the International Convention for the Prevention of Pollution from Ships (MARPOL) Annex VI set standards for ship emissions and require member states to implement measures to reduce air pollution from maritime activities.
3. **Clean Air Asia**: Clean Air Asia is a regional network dedicated to promoting clean air and sustainable urban development across Asia. The organization facilitates multilateral cooperation and partnerships among governments, cities, academic institutions, civil society organizations, and private sector stakeholders to address air pollution and improve air quality in the region. Clean Air Asia provides technical assistance, capacity building, and knowledge-sharing initiatives to support member countries in developing and implementing effective air quality management strategies and policies.
4. **Global Alliance on Health and Pollution**: The Global Alliance on Health and Pollution (GAHP) is a collaborative partnership of governments, international organizations, civil society groups, and academic institutions working together to address the health impacts of pollution worldwide. GAHP aims to raise awareness about the links between pollution and public health, mobilize resources for pollution control and remediation efforts, and build capacity for pollution management in low- and middle-income countries [86]. The alliance supports

multilateral cooperation and knowledge sharing to tackle pollution-related health challenges and promote sustainable development.

5. **United Nations Environment Programme Global Environment Monitoring System**: The United Nations Environment Programme (UNEP) Global Environment Monitoring System (GEMS)/Air program is a global network of monitoring stations collecting air quality data to assess regional and global air pollution trends. GEMS/Air facilitates multilateral cooperation among participating countries to monitor air quality, share data and information, and assess the effectiveness of air pollution control measures. The program supports capacity building and technical assistance initiatives to strengthen air quality monitoring and management capabilities worldwide.

Policy and governance frameworks are essential for achieving sustainable clean air and protecting public health and the environment from the adverse impacts of air pollution. National governments play a central role in establishing air quality standards, implementing emission reduction measures, and strengthening monitoring and enforcement mechanisms. International cooperation through multilateral agreements, capacity-building initiatives, and knowledge-sharing platforms is critical for addressing transboundary air pollution and advancing global efforts to achieve cleaner and healthier air for all. By implementing comprehensive and effective policy and governance frameworks, countries can mitigate air pollution, promote environmental sustainability, and create a healthier and more prosperous future for current and future generations. Examples of some International Policies and Agreements:

1. **Paris Agreement** (United Nations Framework Convention on Climate Change): The Paris Agreement is an international treaty adopted in 2015 under the United Nations Framework Convention on Climate Change (UNFCCC) aimed at limiting global warming to well below 2°C above pre-industrial levels [87]. The agreement includes commitments by countries to reduce greenhouse gas emissions, including those from sources that contribute to air pollution, and to enhance resilience to climate change impacts.
2. **Gothenburg Protocol**: The Gothenburg Protocol is a multilateral agreement under UNECE aimed at reducing emissions of air pollutants in Europe and North America. The protocol sets emission reduction targets for sulfur dioxide, nitrogen oxides, volatile organic compounds, and ammonia, with member countries required to implement measures to achieve the targets and improve air quality [88].
3. **Convention on Long-Range Transboundary Air Pollution**: The Convention on Long-Range Transboundary Air Pollution is an international treaty under UNECE aimed at addressing air pollution across national and international boundaries. The convention focuses on reducing emissions of air pollutants that can cause transboundary pollution and harm human health and the environment, with protocols addressing specific pollutants such as sulfur dioxide, nitrogen oxides, and persistent organic pollutants.

Table 1 WHO AQGs for various pollutants in the years 2005 and 2021

Pollutant	Averaging time	2005 AQGs	2021 AQGs
$PM_{2.5}$ (µg/m³)	Annual	10	5
	24-h	25	15
PM_{10} (µg/m³)	Annual	20	15
	24-h	50	45
O_3 (µg/m³)	Peak season	Not specified	60
	8-h	100	100
NO_2 (µg/m³)	Annual	40	10
	24-h	Not specified	25
SO_2 (µg/m³)	24-h	20	40
CO (mg/m³)	24-h	Not specified	4

Note: Annual and peak season is long-term exposure, while 24-h and 8-h is short-term exposure
Source: Adapted from WHO [80]

The above examples illustrate the variety of national and international policies and regulations aimed at achieving sustainable clean air by reducing air pollution emissions, improving air quality, and protecting public health and the environment. Collaboration and cooperation among governments, stakeholders, and international organizations are essential for implementing effective policies and achieving collective goals for cleaner and healthier air.

Development of Air Quality Standards and Regulations Establish stringent air quality standards and regulations to limit emissions of key pollutants such as PM, NOx, SO_2, VOCs, and O_3. These standards should be based on scientific evidence and health-based criteria, with regular monitoring and enforcement mechanisms in place. Regional and national regulations can be based on the World Health Organization (WHO) Air Quality Guidelines (AQGs). The WHO AQGs are designed to protect public health by providing global benchmarks for key air pollutants, including $PM_{2.5}$, PM_{10}, O_3, NO_2, and SO_2 as outlined in Table 1 [80]. These guidelines are based on extensive scientific evidence and are periodically updated to reflect the latest research on the health effects of air pollution. The most recent update in 2021 introduced stricter guidelines, and in line the SDGs emphasizes the need for urgent action to reduce air pollution levels and mitigate associated health risks. WHO's AQGs serve as a critical tool for policymakers, helping to inform national and international air quality standards and regulations aimed at safeguarding public health.

Implementation of Emission Reduction Targets Set ambitious targets for reducing emissions from major sources of air pollution, including industrial facilities, power plants, transportation, and residential heating. Implement policies such as emission limits, technology standards, and pollution control measures to achieve these targets and improve air quality.

Promotion of Clean Energy and Renewable Sources Encourage the transition to clean energy and renewable sources such as solar, wind, hydroelectric, and geothermal power to reduce reliance on fossil fuels and lower emissions of greenhouse gases and air pollutants. Support policies and incentives for renewable energy development, energy efficiency, and carbon pricing to incentivize clean energy adoption.

Integration of Air Quality Considerations into Urban Planning Incorporate air quality considerations into urban planning and development processes to minimize exposure to air pollution and promote healthier, more sustainable cities. Implement measures such as land use zoning, green infrastructure, public transportation, and pedestrian-friendly design to reduce vehicle emissions and improve air quality in urban areas.

Adoption of Clean Transportation Policies This will require the promotion of sustainable transportation solutions such as public transit, cycling, walking, and electric vehicles to reduce emissions from the transportation sector, which is a major source of air pollution. Implement measures such as fuel efficiency standards, vehicle emissions testing, low-emission zones, and investment in alternative transportation infrastructure.

Implementation of Pollution Control Technologies Invest in pollution control technologies and the best available techniques to reduce emissions from industrial processes, power plants, and other stationary sources of pollution. Require industries to adopt cleaner production methods, install pollution abatement equipment, and implement emission monitoring and reporting systems to comply with regulatory requirements.

Strengthening Air Quality Monitoring and Reporting Enhance air quality monitoring and reporting systems to track pollution levels, identify hotspots, and assess compliance with air quality standards and regulations. Deploy monitoring stations equipped with advanced sensors and technology to collect real-time data on air pollutants and provide timely information to policymakers, stakeholders, and the public.

Capacity Building and Stakeholder Engagement Build institutional capacity for air quality management at the national, regional, and local levels through training programs, technical assistance, and knowledge-sharing initiatives. Foster collaboration and engagement with government agencies, industry partners, civil society organizations, academic institutions, and communities to develop and implement effective air quality policies and programs.

5 Gaps in Addressing Air Pollution with the Sustainable Development Goals

The SDGs represent a significant step forward in addressing global challenges, but they are not without imperfections. The SDGs consist of 17 goals, 169 targets, and numerous indicators, making the framework complex and challenging to implement and monitor effectively [89]. The broad scope and interlinkages between goals can lead to ambiguity and confusion regarding priorities, responsibilities, and accountability. The SDGs are nonbinding commitments, meaning that countries are not legally obligated to achieve the goals and targets outlined in the framework [90, 91]. This lack of enforceability may weaken accountability mechanisms and hinder progress toward achieving the SDGs. Financing the implementation of the SDGs remains a significant challenge, particularly for low- and middle-income countries [92]. The mobilization of financial resources, including official development assistance (ODA), domestic investment, and private sector financing, falls short of the estimated trillions of dollars needed to achieve the goals by 2030 [93]. The SDGs aim to leave no one behind; however, they often fail to address the root causes of inequality and marginalization adequately [24]. Vulnerable and marginalized populations, including women, children, persons with disabilities, indigenous peoples, and refugees, continue to face systemic barriers to accessing essential services and opportunities for development.

The SDGs are interconnected and interdependent, requiring integrated approaches and coherent policies to achieve sustainable development outcomes. However, in practice, there is often a lack of integration across sectors and levels of governance, leading to fragmented and siloed approaches that undermine progress toward the goals [94]. Data availability, quality, and disaggregation remain significant challenges for monitoring progress toward the SDGs. Many countries lack reliable data systems and capacity for collecting and analyzing data on key indicators, hindering efforts to track progress, identify disparities, and inform evidence-based decision-making [95]. Despite the emphasis on sustainability, the SDGs do not adequately address the environmental challenges facing the planet, such as climate change, biodiversity loss, and ecosystem degradation. The goals may prioritize short-term economic growth over long-term environmental sustainability, leading to trade-offs and compromises in achieving sustainable development objectives [96, 97]. The success of the SDGs depends on political will and leadership at the national and international levels. However, many governments lack the commitment and prioritization needed to implement and achieve the goals, particularly in the face of competing interests, geopolitical tensions, and domestic challenges [98].

Some SDG gaps connected to air pollution include (Fig. 3):

Lack of Explicit Targets While some SDGs indirectly address air pollution, such as Goal 3 (Good Health and Well-being) and Goal 7 (Affordable and Clean Energy), there is no specific SDG dedicated solely to addressing air pollution. This absence of a dedicated goal may lead to inadequate attention and prioritization of air quality issues within the broader sustainable development agenda.

Inadequate Monitoring and Reporting Despite the existence of global monitoring frameworks for tracking progress toward the SDGs, there is often limited data availability and consistency regarding air pollution indicators. This lack of comprehensive monitoring and reporting systems hinders efforts to assess the extent of air pollution problems, identify priority areas for action, and measure progress over time [99].

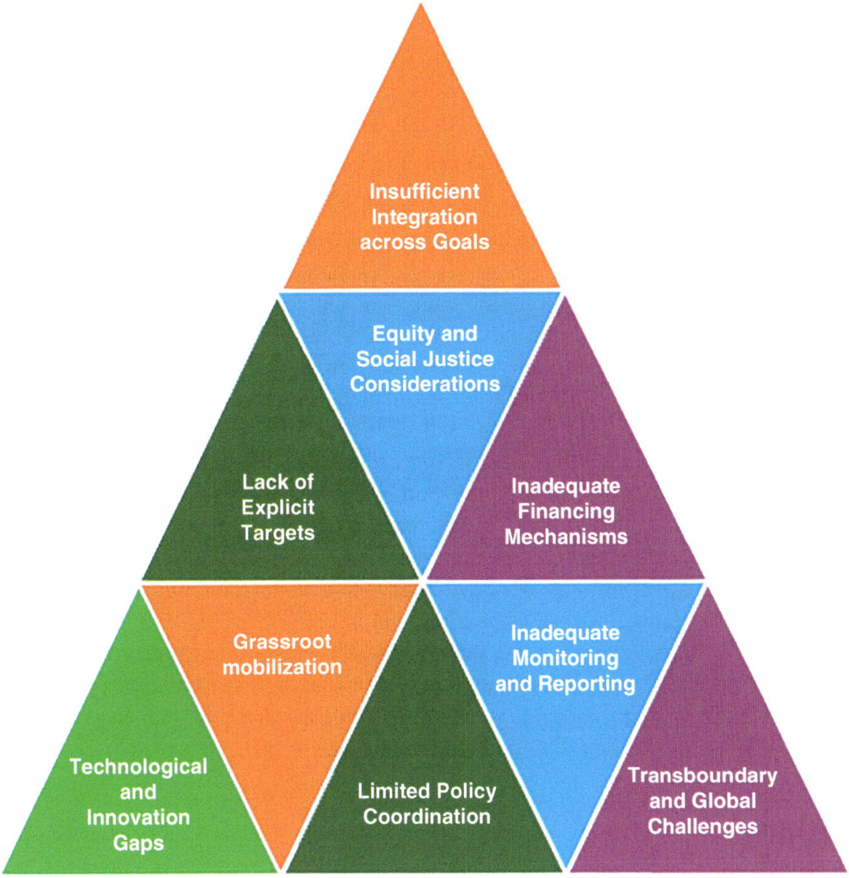

Fig. 3 SDG gaps with negative effects on air pollution

Insufficient Integration Across Goals There is a need for greater integration of air pollution considerations across multiple SDGs to ensure coherence and synergy in sustainable development efforts. While air pollution has direct implications for health, energy, cities, and climate, there is often a disconnect between sectors, leading to fragmented approaches and missed opportunities for addressing shared challenges.

Limited Policy Coordination Many countries lack coherent policy frameworks and coordination mechanisms for addressing air pollution in alignment with the SDGs [52]. Fragmented governance structures, overlapping mandates, and competing priorities among government agencies and stakeholders can impede effective policy formulation and implementation, resulting in suboptimal outcomes.

Inadequate Financing Mechanisms Mobilizing adequate financial resources to support air pollution mitigation efforts remains a challenge, particularly in low- and middle-income countries [15]. Limited funding for air quality monitoring, pollution control measures, and capacity-building initiatives hinders progress toward achieving SDG targets related to air pollution reduction and environmental sustainability.

Transboundary and Global Challenges Air pollution is a transboundary and global issue that requires coordinated action at the regional and international levels. However, existing mechanisms for addressing transboundary air pollution, such as the Convention on Long-Range Transboundary Air Pollution, may not adequately address emerging challenges, such as cross-border pollution from industrial activities and transportation [100].

Equity and Social Justice Considerations There is a need to ensure that efforts to address air pollution within the SDGs prioritize equity, social inclusion, and environmental justice. Vulnerable and marginalized communities, including low-income populations, ethnic minorities, and indigenous peoples, often bear a disproportionate burden of air pollution-related health risks and environmental degradation, exacerbating existing inequalities [21].

Technological and Innovation Gaps While technological advancements offer opportunities for improving air quality monitoring, pollution control, and emission reduction measures, there are gaps in technology transfer, capacity building, and innovation diffusion, particularly in developing countries. Bridging these technological gaps is essential for scaling up effective solutions and achieving SDG targets related to air pollution and environmental sustainability.

Addressing these gaps requires concerted efforts from governments, international organizations, civil society, and the private sector to mainstream air pollution considerations within the SDGs, strengthen policy coherence and coordination, mobilize resources, enhance monitoring and reporting mechanisms, promote equity and social justice, and foster technological innovation and collaboration across sectors and borders. By addressing these gaps, stakeholders can advance progress toward achieving cleaner air, healthier communities, and a more sustainable future for all.

6 Innovative Approaches to Air Quality Management

Innovative approaches to air quality management can lead to significant improvements in public health by reducing exposure to harmful air pollutants. Cleaner air can lower the incidence of respiratory diseases, cardiovascular ailments, and other health conditions associated with air pollution, resulting in better overall health outcomes for communities [101]. Implementing innovative air quality management strategies promotes environmental sustainability by reducing emissions of greenhouse gases and harmful pollutants. By adopting green infrastructure, promoting sustainable transportation, and implementing circular economy approaches, communities can minimize their environmental footprint and contribute to mitigating climate change and protecting ecosystems [102–104]. Investing in innovative air quality management initiatives can generate economic benefits by reducing healthcare costs, improving productivity, and stimulating economic growth [105]. Cleaner air enhances worker productivity, reduces absenteeism due to illness, and lowers healthcare expenditures associated with air pollution-related diseases, resulting in overall cost savings for governments, businesses, and individuals. Innovative approaches to air quality management can promote social equity and environmental justice by addressing disparities in exposure to air pollution and health outcomes among different socioeconomic and racial/ethnic groups. By targeting pollution hotspots and prioritizing interventions in vulnerable communities, policymakers can ensure equitable access to clean air and protect the health and well-being of all residents. Embracing innovative technologies and solutions for air quality management fosters technological advancements and innovation in the field. Developments in sensor technology, data analytics, remote sensing, and clean energy technologies drive progress toward more efficient and effective air quality monitoring, pollution control, and emission reduction measures, leading to continuous improvements in air quality management practices [106, 107].

Innovative approaches to air quality management often involve community engagement and empowerment, empowering citizens to participate in decision-making processes and take action to improve air quality in their neighborhoods. Citizen science initiatives, public education campaigns, and participatory planning processes build awareness, foster community ownership, and strengthen social cohesion around air quality issues. Many innovative approaches to air quality management yield co-benefits for climate change mitigation by reducing emissions of greenhouse gases and short-lived climate pollutants [108, 109]. Promoting clean energy, sustainable transportation, and green infrastructure not only improves air quality but also contributes to mitigating climate change and advancing progress toward global climate goals. Investing in innovative air quality management measures enhances resilience to environmental hazards and extreme weather events associated with air pollution. By improving air quality and reducing pollution levels, communities can mitigate the impacts of heatwaves, wildfires, and other climate-related disasters, safeguarding public health and safety during times of crisis [99, 110]. Some innovative approaches to air quality management include.

Smart Air Quality Monitoring Utilizing advanced sensor technologies, remote sensing, and data analytics to develop smart air quality monitoring systems. These systems provide real-time, high-resolution data on air pollution levels, enabling authorities to identify pollution hotspots, track trends, and make informed decisions for targeted interventions [111].

Citizen Science Initiatives Engaging citizens in air quality monitoring through citizen science initiatives and crowd-sourced data collection platforms [112]. Empowering communities to collect air quality data using low-cost sensors or smartphone apps not only increases public awareness but also enhances spatial coverage and complements official monitoring networks.

Green Infrastructure Implementing green infrastructure solutions such as green roofs, urban forests, and permeable surfaces to mitigate air pollution and improve air quality. Vegetation acts as a natural air filter, absorbing pollutants and particulate matter, while also providing additional environmental and social benefits such as carbon sequestration and urban heat island mitigation [113, 114].

Low-Emission Zones Establishing low-emission zones (LEZs) in urban areas to restrict the entry of high-emission vehicles and promote cleaner transportation. LEZs incentivize the use of low-emission vehicles, encourage modal shifts to public transit, cycling, and walking, and contribute to reducing traffic-related air pollution and improving air quality in city centers [115].

Vehicle Fleet Modernization Programs Implementing vehicle fleet modernization programs to incentivize the replacement of older, high-emission vehicles with cleaner and more fuel-efficient models. Providing financial incentives, subsidies, or tax breaks for the purchase of electric vehicles (EVs) or hybrid vehicles helps accelerate the transition to a cleaner transportation fleet and reduce emissions from the transportation sector [116].

Green Transport Infrastructure Investing in green transport infrastructure such as electric vehicle charging stations, bike lanes, and pedestrian-friendly infrastructure to promote sustainable modes of transportation and reduce reliance on fossil fuel-powered vehicles [117]. Developing integrated transportation networks and promoting non-motorized transportation options contribute to improving air quality and reducing traffic congestion.

Air Pollution Forecasting and Early Warning Systems Developing air pollution forecasting models and early warning systems to predict pollution episodes and inform the public and relevant authorities in advance [118]. These systems use meteorological data, air quality monitoring data, and predictive modeling techniques to anticipate high pollution events, allowing for the timely implementation of mitigation measures and public advisories [111].

Circular Economy Approaches Promoting circular economy approaches to reduce waste generation, minimize resource consumption, and mitigate air pollution from industrial processes [63]. Implementing circular economy principles such as recycling, reusing, and remanufacturing helps minimize the environmental footprint of production activities and reduce emissions of air pollutants and greenhouse gases.

7 Conclusion

The interconnectedness of air pollution with climate change and sustainable development is profound, as each of these issues is intertwined and has significant implications for the others. Understanding these connections is crucial for developing holistic strategies to address environmental challenges and promote sustainable development. Many of the activities that contribute to air pollution, such as burning fossil fuels for energy production, transportation, and industrial processes, also emit greenhouse gases (GHGs) that contribute to climate change. Common sources of air pollution, such as vehicle emissions, industrial smokestacks, and agricultural activities, release not only pollutants like particulate matter and nitrogen oxides but also GHGs like carbon dioxide and methane. Addressing these shared sources of emissions requires integrated approaches that simultaneously reduce air pollution and mitigate climate change, such as transitioning to cleaner energy sources, improving energy efficiency, and promoting sustainable transportation options. Both air pollution and climate change have adverse effects on human health, exacerbating respiratory diseases, cardiovascular problems, and other health conditions. Air pollution can worsen the health impacts of climate change by increasing the frequency and severity of heatwaves, wildfires, and extreme weather events, which can release additional pollutants into the air and degrade air quality further. Improving air quality and mitigating climate change can yield co-benefits for human health, reducing the burden of respiratory illnesses, heat-related illnesses, and premature mortality associated with both air pollution and climate change.

Air pollution and climate change can create feedback loops and amplification mechanisms that exacerbate each other's impacts and contribute to environmental degradation. For example, air pollution can accelerate the melting of glaciers and ice caps through the deposition of dark particles (such as black carbon) on snow and ice surfaces, reducing their reflectivity and increasing the absorption of solar radiation. Similarly, climate change can exacerbate air pollution by altering atmospheric conditions and weather patterns, leading to stagnation of pollutants, increased formation of ground-level ozone, and changes in the distribution and composition of airborne particles. Air pollution and climate change pose significant challenges to achieving sustainable development goals related to environmental protection, public health, poverty alleviation, and economic prosperity. Poor air quality can undermine efforts to improve living standards, promote economic growth, and reduce inequalities, particularly in low- and middle-income countries where pollution levels are often highest and resources for pollution control are limited. Addressing air pollution and climate change is essential for building resilient and sustainable communities, enhancing environmental sustainability, and promoting equitable and inclusive development that benefits present and future generations.

The pervasive nature of air pollution stems from natural and anthropogenic sources alike and has far-reaching consequences on human health, ecosystems, and economic prosperity. The multifaceted issue of air pollution and its implications for sustainable development goals has been discussed in this chapter as well as an analysis of the sources, impacts, policy frameworks, and innovative approaches to air quality management, several key findings and insights have emerged. Air pollution poses significant challenges to public health, environmental sustainability, and socioeconomic development worldwide. From its origins in natural and anthropogenic sources to its far-reaching health, environmental, and economic impacts, air pollution requires urgent attention and concerted action. Although progress has been made in recognizing and addressing air pollution within the framework of sustainable development goals, some gaps and imperfections need to be addressed. These include the lack of explicit targets, inadequate monitoring and reporting mechanisms, fragmented policy frameworks, and insufficient financing for air quality management initiatives. Despite these challenges, there is reason for optimism. Through collaboration, innovation, and commitment, stakeholders at all levels can work together to advance efforts to address air pollution and achieve sustainable development goals. By integrating air pollution considerations across sectors, strengthening policy coherence and coordination, mobilizing resources, promoting equity and social justice, and fostering technological innovation and collaboration, we can build a healthier, more sustainable future for all. The importance of collective action and shared responsibility cannot be overstated. The time for action is now, and the stakes could not be higher. Let us seize this opportunity to build a brighter, cleaner, and more sustainable future for generations to come.

References

1. Bouza E, Vargas F, Alcázar B, Álvarez T, Asensio A, Cruceta G, Gracia D, Guinea J, Gil MA, Linares C, Muñoz P, Pastor P, Pedro-Botet ML, Querol X, Tovar J, Urrutia I, Villar F, Palomo E (2022) Air pollution and health prevention: a document of reflection. Rev Esp Quimioter 35(4):307–332. https://doi.org/10.37201/req/171.2021
2. Chandrappa R, Chandra Kulshrestha U (2015) Major issues of air pollution. Sustain Air Pollut Manag Theory Pract:1–48. https://doi.org/10.1007/978-3-319-21596-9_1
3. Izzotti A, Spatera P, Khalid Z, Pulliero A (2022) Importance of punctual monitoring to evaluate the health effects of airborne particulate matter. Int J Environ Res Public Health 19(17):10587. https://doi.org/10.3390/ijerph191710587
4. Schraufnagel DE (2020) The health effects of ultrafine particles. Exp Mol Med 52(3):311–317. https://doi.org/10.1038/s12276-020-0403-3
5. You R, Ho YS, Chang RC (2022) The pathogenic effects of particulate matter on neurodegeneration: a review. J Biomed Sci 29(1):15. https://doi.org/10.1186/s12929-022-00799-x
6. Perera F (2017) Pollution from fossil-fuel combustion is the leading environmental threat to global pediatric health and equity: solutions exist. Int J Environ Res Public Health 15(1):16. https://doi.org/10.3390/ijerph15010016
7. Lee YG, Lee PH, Choi SM, An MH, Jang AS (2021) Effects of air pollutants on airway diseases. Int J Environ Res Public Health 18(18):9905. https://doi.org/10.3390/ijerph18189905
8. Hanley ME, Patel PH. Carbon monoxide toxicity. [Updated 2023 Jan 23]. In: StatPearls [Internet]. StatPearls Publishing, Treasure Island 2024 Jan. Available from: https://www.ncbi.nlm.nih.gov/books/NBK430740/#

9. Cao L, Men Q, Zhang Z, Yue H, Cui S, Huang X, Zhang Y, Wang J, Chen M, Li H (2024) Significance of volatile organic compounds to secondary pollution formation and health risks observed during a summer campaign in an industrial urban area. Toxics 12(1):34. https://doi.org/10.3390/toxics12010034
10. Zhang JJ, Wei Y, Fang Z (2019) Ozone pollution: a major health hazard worldwide. Front Immunol 10:2518. https://doi.org/10.3389/fimmu.2019.02518
11. Cassia R, Nocioni M, Correa-Aragunde N, Lamattina L (2018) Climate change and the impact of greenhouse gasses: CO_2 and NO, friends and foes of plant oxidative stress. Front Plant Sci 9:273. https://doi.org/10.3389/fpls.2018.00273
12. WHO (2024) Climate impacts of air pollution. Available online: https://www.who.int/teams/environment-climate-change-and-health/air-quality-energy-and-health/health-impacts/climate-impacts-of-air-pollution
13. Turner MC, Andersen ZJ, Baccarelli A, Diver WR, Gapstur SM, Pope 3rd CA, Prada D, Samet J, Thurston G, Cohen A (2020) Outdoor air pollution and cancer: An overview of the current evidence and public health recommendations. CA Cancer J Clin. https://doi.org/10.3322/caac.21632. Advance online publication
14. Stevens CJ, Bell JNB, Brimblecombe P, Clark CM, Dise NB, Fowler D, Lovett GM, Wolseley PA (2020) The impact of air pollution on terrestrial managed and natural vegetation. Philos Transact A Math Phys Eng Sci 378(2183):20190317. https://doi.org/10.1098/rsta.2019.0317
15. Manisalidis I, Stavropoulou E, Stavropoulos A, Bezirtzoglou E (2020) Environmental and health impacts of air pollution: a review. Front Public Health 8:14. https://doi.org/10.3389/fpubh.2020.00014
16. Pervin T, Gerdtham UG, Lyttkens CH (2008) Societal costs of air pollution-related health hazards: a review of methods and results. Cost Eff Resour Alloc 6:19. https://doi.org/10.1186/1478-7547-6-19
17. Liu H, Hu T, Wang M (2021) Impact of air pollution on residents' medical expenses: a study based on the survey data of 122 cities in China. Front Public Health 9:743087. https://doi.org/10.3389/fpubh.2021.743087
18. Sturmberg JP, Bircher J (2019) Better and fulfilling healthcare at lower costs: the need to manage health systems as complex adaptive systems. F1000Research 8:789. https://doi.org/10.12688/f1000research.19414.1
19. Beard S, Freeman K, Velasco ML et al (2024) Racism as a public health issue in environmental health disparities and environmental justice: working toward solutions. Environ Health 23(8). https://doi.org/10.1186/s12940-024-01052-8
20. Hajat A, Hsia C, O'Neill MS (2015) Socioeconomic disparities and air pollution exposure: a global review. Curr Environ Health Rep 2(4):440–450. https://doi.org/10.1007/s40572-015-0069-5
21. Rentschler J, Leonova N (2023) Global air pollution exposure and poverty. Nat Commun 14(1):4432. https://doi.org/10.1038/s41467-023-39797-4
22. Zhang Q, Jiang X, Tong D, Davis SJ, Zhao H, Geng G, Feng T, Zheng B, Lu Z, Streets DG, Ni R, Brauer M, van Donkelaar A, Martin RV, Huo H, Liu Z, Pan D, Kan H, Yan Y, Lin J et al (2017) Transboundary health impacts of transported global air pollution and international trade. Nature 543(7647):705–709. https://doi.org/10.1038/nature21712
23. Chopra M, Singh DSK, Gupta A, Aggarwal K, Gupta BB, Colace F (2022) Analysis & prognosis of sustainable development goals using big data-based approach during COVID-19 pandemic. Sustain Technol Entrepreneur 1(2):100012. https://doi.org/10.1016/j.stae.2022.100012
24. Akudugu MA, Ogwu MC (2024) Sustainable development policies and interventions: A bibliometric analysis of the contributions of the academic community. J Clean Prod 434:139919. https://doi.org/10.1016/j.jclepro.2023.139919
25. GBD 2015 SDG Collaborators (2016) Measuring the health-related sustainable development goals in 188 countries: a baseline analysis from the global burden of disease study 2015. Lancet 388(10053):1813–1850. https://doi.org/10.1016/S0140-6736(16)31467-2

26. Racioppi F, Martuzzi M, Matić S, Braubach M, Morris G, Krzyżanowski M, Jarosińska D, Schmoll O, Adamonytė D (2020) Reaching the sustainable development goals through healthy environments: are we on track? Eur J Public Health 30(Suppl_1):i14–i18. https://doi.org/10.1093/eurpub/ckaa028
27. Azam W, Khan I, Ali SA (2023) Alternative energy and natural resources in determining environmental sustainability: a look at the role of government final consumption expenditures in France. Environ Sci Pollut Res Int 30(1):1949–1965. https://doi.org/10.1007/s11356-022-22334-z
28. Erickson LE (2017) Reducing greenhouse gas emissions and improving air quality: two global challenges. Environ Prog Sustain Energy 36(4):982–988. https://doi.org/10.1002/ep.12665
29. Maung TZ, Bishop JE, Holt E, Turner AM, Pfrang C (2022) Indoor air pollution and the health of vulnerable groups: a systematic review focused on particulate matter (PM), volatile organic compounds (VOCs) and their effects on children and people with pre-existing lung disease. Int J Environ Res Public Health 19(14):8752. https://doi.org/10.3390/ijerph19148752
30. Chang JH, Lee YL, Chang LT, Chang TY, Hsiao TC, Chung KF, Ho KF, Kuo HP, Lee KY, Chuang KJ, Chuang HC (2023) Climate change, air quality, and respiratory health: a focus on particle deposition in the lungs. Ann Med 55(2):2264881. https://doi.org/10.1080/07853890.2023.2264881
31. Watson AY, Bates RR, Kennedy D (eds) (1988) Air pollution, the automobile, and public health. Atmospheric transport and dispersion of air pollutants associated with vehicular emissions. National Academies Press, Washington. Available from: https://www.ncbi.nlm.nih.gov/books/NBK218142/
32. Wu M, Zhang G, Wang L, Liu X, Wu Z (2022) Influencing factors on airflow and pollutant dispersion around buildings under the combined effect of wind and buoyancy – a review. Int J Environ Res Public Health 19(19):12895. https://doi.org/10.3390/ijerph191912895
33. Tran VV, Park D, Lee YC (2020) Indoor air pollution, related human diseases, and recent trends in the control and improvement of indoor air quality. Int J Environ Res Public Health 17(8):2927. https://doi.org/10.3390/ijerph17082927
34. Bhat H, Jiawen T, Farzaneh G (2021) Air pollution health risk assessment (AP-HRA), principles and applications. Int. J Environ Res Public Health 18:1935. https://doi.org/10.3390/ijerph18041935
35. Deshmukh P, Kimbrough S, Krabbe S, Logan R, Isakov V, Baldauf R (2020) Identifying air pollution source impacts in urban communities using mobile monitoring. Sci Total Environ 715:136979. https://doi.org/10.1016/j.scitotenv.2020.136979
36. Chandra M, Rai CB, Kumari N, Sandhu VK, Chandra K, Krishna M, Kota SH, Anand KS, Oudin A (2022) Air pollution and cognitive impairment across the life course in humans: a systematic review with specific focus on income level of study area. Int J Environ Res Public Health 19(3):1405. https://doi.org/10.3390/ijerph19031405
37. Costa LG, Cole TB, Dao K, Chang YC, Garrick JM (2019) Developmental impact of air pollution on brain function. Neurochem Int 131:104580. https://doi.org/10.1016/j.neuint.2019.104580
38. Morgan ZEM, Bailey MJ, Trifonova DI, Naik NC, Patterson WB, Lurmann FW, Chang HH, Peterson BS, Goran MI, Alderete TL (2023) Prenatal exposure to ambient air pollution is associated with neurodevelopmental outcomes at 2 years of age. Environ Health 22(1):11. https://doi.org/10.1186/s12940-022-00951-y
39. McDuffie E, Martin R, Yin H, Brauer M (2021) Global burden of disease from major air pollution sources (GBD MAPS): a global approach. Res Rep Health Eff Inst 2021(210):1–45
40. van der Wall EE (2015) Air pollution: 6.6 million premature deaths in 2050! Neth Heart J 23(12):557–558. https://doi.org/10.1007/s12471-015-0763-9
41. Münzel T, Hahad O, Daiber A, Landrigan PJ (2023) Soil and water pollution and human health: what should cardiologists worry about? Cardiovasc Res 119(2):440–449. https://doi.org/10.1093/cvr/cvac082

42. Lin X, Yang R, Zhang W, Zeng N, Zhao Y, Wang G, Li T, Cai Q (2023) An integrated view of correlated emissions of greenhouse gases and air pollutants in China. Carbon Balance Manag 18(1):9. https://doi.org/10.1186/s13021-023-00229-x
43. Albright JF, Goldstein RA (1996) Airborne pollutants and the immune system. Otolaryngol Head Neck Surg 114(2):232–238. https://doi.org/10.1016/S0194-59989670173-0
44. Bikis A (2023) Urban air pollution and greenness in relation to public health. J Environ Public Health 2023:8516622. https://doi.org/10.1155/2023/8516622
45. Bhui K, Newbury JB, Latham RM, Ucci M, Nasir ZA, Turner B, O'Leary C, Fisher HL, Marczylo E, Douglas P, Stansfeld S, Jackson SK, Tyrrel S, Rzhetsky A, Kinnersley R, Kumar P, Duchaine C, Coulon F (2023) Air quality and mental health: evidence, challenges and future directions. BJPsych Open 9(4):e120. https://doi.org/10.1192/bjo.2023.507
46. Abed Al Ahad M, Demšar U, Sullivan F, Kulu H (2022) Air pollution and individuals' mental well-being in the adult population in United Kingdom: a spatial-temporal longitudinal study and the moderating effect of ethnicity. PLoS One 17(3):e0264394. https://doi.org/10.1371/journal.pone.0264394
47. Cao Z, Zhou J, Li M et al (2023) Urbanites' mental health undermined by air pollution. Nat Sustain 6:470–478. https://doi.org/10.1038/s41893-022-01032-1
48. Czechowski PO, Piksa K, Da Browiecki P, Oniszczuk-Jastrząbek AI, Czermański E, Owczarek T, Badyda AJ, Cirella GT (2022) Financing costs and health effects of air pollution in the Tri-City agglomeration. Front Public Health 10:831312. https://doi.org/10.3389/fpubh.2022.831312
49. Martinez GS, Spadaro JV, Chapizanis D, Kendrovski V, Kochubovski M, Mudu P (2018) Health impacts and economic costs of air pollution in the metropolitan area of Skopje. Int J Environ Res Public Health 15(4):626. https://doi.org/10.3390/ijerph15040626
50. Mathiarasan S, Hüls A (2021) Impact of environmental injustice on children's health-interaction between air pollution and socioeconomic status. Int J Environ Res Public Health 18(2):795. https://doi.org/10.3390/ijerph18020795
51. Shrestha R, Flacke J, Martinez J, van Maarseveen M (2016) Environmental health related socio-spatial inequalities: identifying "Hotspots" of environmental burdens and social vulnerability. Int J Environ Res Public Health 13(7):691. https://doi.org/10.3390/ijerph13070691
52. Leal Filho W, Wolf F, Lange Salvia A, Beynaghi A, Shulla K, Kovaleva M, Vasconcelos CRP (2020) Heading towards an unsustainable world: some of the implications of not achieving the SDGs. Discov Sustain 1(1):2. https://doi.org/10.1007/s43621-020-00002-x
53. Kumar S, Kumar N, Vivekadhish S (2016) Millennium development goals (MDGs) to sustainable development goals (SDGs): addressing unfinished agenda and strengthening sustainable development and partnership. Indian J Commun Med 41(1):1–4. https://doi.org/10.4103/0970-0218.170955
54. Crespo Cuaresma J, Fengler W, Kharas H et al (2018) Will the sustainable development goals be fulfilled? Assessing present and future global poverty. Palgrave Commun 4:29. https://doi.org/10.1057/s41599-018-0083-y
55. Mattaini MA, Roose KM, Fawcett SB (2022) Behavioral interventions contributing to reducing poverty and inequities. Behav Soc Issues:1–24. Advance online publication. https://doi.org/10.1007/s42822-022-00114-1
56. Breuer A, Janetschek H, Malerba D (2019) Translating sustainable development goal (SDG) interdependencies into policy advice. Sustain For 11:2092. https://doi.org/10.3390/su11072092
57. Nilsson M, Chisholm E, Griggs D, Howden-Chapman P, McCollum D, Messerli P, Neumann B, Stevance AS, Visbeck M, Stafford-Smith M (2018) Mapping interactions between the sustainable development goals: lessons learned and ways forward. Sustain Sci 13(6):1489–1503. https://doi.org/10.1007/s11625-018-0604-z
58. Tóth G, Hermann T, da Silva MR, Montanarella L (2018) Monitoring soil for sustainable development and land degradation neutrality. Environ Monit Assess 190(2):57. https://doi.org/10.1007/s10661-017-6415-3

59. Peng K, Feng K, Chen B, Shan Y, Zhang N, Wang P, Fang K, Bai Y, Zou X, Wei W, Geng X, Zhang Y, Li J (2023) The global power sector's low-carbon transition may enhance sustainable development goal achievement. Nat Commun 14(1):3144. https://doi.org/10.1038/s41467-023-38987-4
60. Arora NK, Mishra I (2022) Progress of sustainable development goal 7: clean and green energy for all as the biggest challenge to combat climate crisis. Environ Sustain 5(4):395–399. https://doi.org/10.1007/s42398-022-00257-2
61. Walesiak M, Dehnel G (2024) Progress on SDG 7 achieved by EU countries in relation to the target year 2030: a multidimensional indicator analysis using dynamic relative taxonomy. PLoS One 19(2):e0297856. https://doi.org/10.1371/journal.pone.0297856
62. Fuldauer LI, Thacker S, Haggis RA, Fuso-Nerini F, Nicholls RJ, Hall JW (2022) Targeting climate adaptation to safeguard and advance the sustainable development goals. Nat Commun 13(1):3579. https://doi.org/10.1038/s41467-022-31202-w
63. Sharma HB, Vanapalli KR, Samal B, Cheela VRS, Dubey BK, Bhattacharya J (2021) Circular economy approach in solid waste management system to achieve UN-SDGs: solutions for post-COVID recovery. Sci Total Environ 800:149605. https://doi.org/10.1016/j.scitotenv.2021.149605
64. Baena-Morales S, Jerez-Mayorga D, Delgado-Floody P, Martínez-Martínez J (2021) Sustainable development goals and physical education. A proposal for practice-based models. Int J Environ Res Public Health 18(4):2129. https://doi.org/10.3390/ijerph18042129
65. Jia H, Chen F, Du E (2021) Adaptation to disaster risk – an overview. Int J Environ Res Public Health 18(21):11187. https://doi.org/10.3390/ijerph182111187
66. Filho WL, Wall T, Salvia AL, Dinis MAP, Mifsud M (2023) The central role of climate action in achieving the United Nations' sustainable development goals. Sci Rep 13(1):20582. https://doi.org/10.1038/s41598-023-47746-w
67. Szantoi Z, Geller GN, Tsendbazar NE, See L, Griffiths P, Fritz S, Gong P, Herold M, Mora B, Obregón A (2020) Addressing the need for improved land cover map products for policy support. Environ Sci Policy 112:28–35. https://doi.org/10.1016/j.envsci.2020.04.005
68. Seddon N, Chausson A, Berry P, Girardin CAJ, Smith A, Turner B (2020) Understanding the value and limits of nature-based solutions to climate change and other global challenges. Philos Trans R Soc Lond B Biol Sci 375(1794):20190120. https://doi.org/10.1098/rstb.2019.0120
69. Akhtar-Schuster M, Stringer LC, Barger N (2024) Fast-tracking action on the sustainable development goals by enhancing national institutional arrangements. PLoS One 19(3): e0298855. https://doi.org/10.1371/journal.pone.0298855
70. Jabakhanji SB, Arnold SR, Aunan K, Chersich MF, Jakobsson K, McGushin A, Kelly I, Roche N, Stauffer A, Stanistreet D (2022) Public health measures to address the impact of climate change on population health-proceedings from a stakeholder workshop. Int J Environ Res Public Health 19(20):13665. https://doi.org/10.3390/ijerph192013665
71. Liu P (2023) An assessment of financial mechanisms for green financial recovery and climate change mitigation: the case of China. Econ Chang Restruct 56(3):1567–1584. https://doi.org/10.1007/s10644-023-09483-y
72. Jian Y (2023) Green bonds and green environment: exploring innovative financing mechanisms for environmental project sustainability. Environ Sci Pollut Res Int 30(58): 122293–122303. https://doi.org/10.1007/s11356-023-30580-y
73. Salvador Costa MJ, Leitão A, Silva R, Monteiro V, Melo P (2022) Climate change prevention through community actions and empowerment: a scoping review. Int J Environ Res Public Health 19(22):14645. https://doi.org/10.3390/ijerph192214645
74. Hooper LG, Kaufman JD (2018) Ambient air pollution and clinical implications for susceptible populations. Ann Am Thorac Soc 15(Suppl 2):S64–S68. https://doi.org/10.1513/AnnalsATS.201707-574MG
75. Kansal A, Subuddhi SP, Pandey P, Gupta D, Rawat T, Gautam AS, Gautam S (2023) Investigating the impression of national clean air programme in enhancement of air quality

characteristics for non-attainment cities of Uttarakhand. Aerosol Sci Eng:1–11. Advance online publication. https://doi.org/10.1007/s41810-023-00181-w
76. Koolen CD, Rothenberg G (2019) Air pollution in Europe. ChemSusChem 12(1):164–172. https://doi.org/10.1002/cssc.201802292
77. Boogaard H, Andersen ZJ, Brunekreef B, Forastiere F, Forsberg B, Hoek G, Krzyzanowski M, Malmqvist E, Nieuwenhuijsen M, Hoffmann B (2023) Clean air in Europe for all: a call for more ambitious action. Environ Epidemiol 7(2):e245. https://doi.org/10.1097/EE9.0000000000000245
78. Chen ZY, Petetin H, Méndez Turrubiates RF, Achebak H, Garcia-Pando CPM, Ballester J (2024) Population exposure to multiple air pollutants and its compound episodes in Europe. Nat Commun 15:2094. https://doi.org/10.1038/s41467-024-46103-3
79. Yamineva Y, Romppanen S (2017) Is law failing to address air pollution? Reflections on international and EU developments. Rev Eur Comp Int Environ Law 26(3):189–200. https://doi.org/10.1111/reel.12223
80. World Health Organization (2021) Global air quality guidelines: particulate matter (PM2.5 and PM10), ozone, nitrogen dioxide, sulfur dioxide and carbon monoxide. Geneva: WHO. Available online https://www.who.int/news-room/feature-stories/detail/what-are-the-who-air-quality-guidelines
81. Shairsingh K, Ruggeri G, Krzyzanowski M, Mudu P, Malkawi M, Castillo J, Soares da Silva A, Saluja M, Martínez KC, Mothe J, Gumy S (2023) WHO air quality database: relevance, history and future developments. Bull World Health Organ 101(12):800–807. https://doi.org/10.2471/BLT.23.290188
82. Taghizadeh F, Mokhtarani B, Rahmanian N (2023) Air pollution in Iran: the current status and potential solutions. Environ Monit Assess 195(6):737. https://doi.org/10.1007/s10661-023-11296-5
83. Bull K, Johansson M, Krzyzanowski M (2008) Impacts of the convention on long-range transboundary air pollution on air quality in Europe. J Toxicol Environ Health A 71(1):51–55. https://doi.org/10.1080/15287390701557883
84. Dooley EE (2006) EHPnet: international maritime organization. Environ Health Perspect 114(4):A219
85. Silveira RS, Corrêa SM, de M Neto, N. (2022) Possible influence of shipping emissions on metals in size-segregated particulate matter in Guanabara Bay (Rio de Janeiro, Brazil). Environ Monit Assess 194(11):828. https://doi.org/10.1007/s10661-022-10517-7
86. Crane M, Lloyd S, Haines A, Ding D, Hutchinson E, Belesova K, Davies M, Osrin D, Zimmermann N, Capon A, Wilkinson P, Turcu C (2021) Transforming cities for sustainability: a health perspective. Environ Int 147:106366. https://doi.org/10.1016/j.envint.2020.106366
87. Maslin MA, Lang J, Harvey F (2023) A short history of the successes and failures of the international climate change negotiations. UCL Open Environ 5:e059. https://doi.org/10.14324/111.444/ucloe.000059
88. Liu Z, Rieder HE, Schmidt C, Mayer M, Guo Y, Winiwarter W, Zhang L (2023) Optimal reactive nitrogen control pathways identified for cost-effective $PM_{2.5}$ mitigation in Europe. Nat Commun 14(1):4246. https://doi.org/10.1038/s41467-023-39900-9
89. Nilashi M, Keng Boon O, Tan G, Lin B, Abumalloh R (2023) Critical data challenges in measuring the performance of sustainable development goals: solutions and the role of big-data analytics. Harv Data Sci Rev 5(3). https://doi.org/10.1162/99608f92.545db2cf
90. Delaney K, Maguire A (2022) Implementing the sustainable development goals post-COVID-19: a study of Australia and Sweden. Altern Law J 47(3):168–172. https://doi.org/10.1177/1037969X221095562
91. Friedman EA (2016) An independent review and accountability mechanism for the sustainable development goals: the possibilities of a framework convention on global health. Health Hum Rights 18(1):129–140

92. Arora RU, Sarker T (2023) Financing for sustainable development goals (SDGs) in the era of COVID-19 and beyond. Eur J Dev Res 35(1):1–19. https://doi.org/10.1057/s41287-022-00571-9
93. Zhan JX, Santos-Paulino AU (2021) Investing in the sustainable development goals: mobilization, channeling, and impact. J Int Bus Policy 4(1):166–183. https://doi.org/10.1057/s42214-020-00093-3
94. Stafford-Smith M, Griggs D, Gaffney O, Ullah F, Reyers B, Kanie N, Stigson B, Shrivastava P, Leach M, O'Connell D (2017) Integration: the key to implementing the sustainable development goals. Sustain Sci 12(6):911–919. https://doi.org/10.1007/s11625-016-0383-3
95. Etowa J, Hyman I, Dabone C, Mbagwu I, Ghose B, Sano Y, Osman M, Mohamoud H (2021) Strengthening the collection and use of disaggregated data to understand and monitor the risk and burden of COVID-19 among racialized populations. Can Stud Popul 48(2–3):201–216. https://doi.org/10.1007/s42650-021-00050-2
96. Basheer M, Nechifor V, Calzadilla A, Ringler C, Hulme D, Harou JJ (2022) Balancing national economic policy outcomes for sustainable development. Nat Commun 13(1):5041. https://doi.org/10.1038/s41467-022-32415-9
97. Vavoura C, Vavouras I (2022) Sustainable economic development in the European Union and COVID-19. Evol Inst Econ Rev 19(1):449–467. https://doi.org/10.1007/s40844-021-00217-1
98. Gyimah J, Saalidong BM, Nibonmua LKM (2023) The battle to achieve sustainable development goal two: the role of environmental sustainability and government institutions. PLoS One 18(9):e0291310. https://doi.org/10.1371/journal.pone.0291310
99. Kelly FJ, Fussell JC (2015) Air pollution and public health: emerging hazards and improved understanding of risk. Environ Geochem Health 37(4):631–649. https://doi.org/10.1007/s10653-015-9720-1
100. Grennfelt P, Engleryd A, Forsius M, Hov Ø, Rodhe H, Cowling E (2020) Acid rain and air pollution: 50 years of progress in environmental science and policy. Ambio 49(4):849–864. https://doi.org/10.1007/s13280-019-01244-4
101. Raju S, Siddharthan T, McCormack MC (2020) Indoor air pollution and respiratory health. Clin Chest Med 41(4):825–843. https://doi.org/10.1016/j.ccm.2020.08.014
102. Kumar S, Darshna A, Ranjan D (2023) A review of literature on the integration of green energy and circular economy. Heliyon 9(11):e21091. https://doi.org/10.1016/j.heliyon.2023.e21091
103. Štrbac S, Kašanin-Grubin M, Pezo L, Stojić N, Lončar B, Ćurčić L, Pucarević M (2023) Green infrastructure designed through nature-based solutions for sustainable urban development. Int J Environ Res Public Health 20(2):1102. https://doi.org/10.3390/ijerph20021102
104. Yang M, Chen L, Wang J et al (2023) Circular economy strategies for combating climate change and other environmental issues. Environ Chem Lett 21:55–80. https://doi.org/10.1007/s10311-022-01499-6
105. Imran M, Khan S, Nassani AA, Haffar M, Khan HUR, Zaman K (2023) Access to sustainable healthcare infrastructure: a review of industrial emissions, coal fires, and particulate matter. Environ Sci Pollut Res Int 30(26):69080–69095. https://doi.org/10.1007/s11356-023-27218-4
106. Morawska L, Thai PK, Liu X, Asumadu-Sakyi A, Ayoko G, Bartonova A, Bedini A, Chai F, Christensen B, Dunbabin M, Gao J, Hagler GSW, Jayaratne R, Kumar P, Lau AKH, Louie PKK, Mazaheri M, Ning Z, Motta N, Mullins B et al (2018) Applications of low-cost sensing technologies for air quality monitoring and exposure assessment: how far have they gone? Environ Int 116:286–299. https://doi.org/10.1016/j.envint.2018.04.018
107. Uddin W (2006) Air quality management using modern remote sensing and spatial technologies and associated societal costs. Int J Environ Res Public Health 3(3):235–243. https://doi.org/10.3390/ijerph2006030028
108. Limaye VS, Magal A, Joshi J, Maji S, Dutta P, Rajput P, Pingle S, Madan P, Mukerjee P, Bano S, Beig G, Mavalankar D, Jaiswal A, Knowlton K (2023) Air quality and health co-benefits of climate change mitigation and adaptation actions by 2030: an interdisciplinary

modeling study in Ahmedabad, India. Environ Res Health 1(2):021003. https://doi.org/10.1088/2752-5309/aca7d8

109. West JJ, Smith SJ, Silva RA, Naik V, Zhang Y, Adelman Z, Fry MM, Anenberg S, Horowitz LW, Lamarque JF (2013) Co-benefits of global greenhouse gas mitigation for future air quality and human health. Nat Clim Chang 3(10):885–889. https://doi.org/10.1038/NCLIMATE2009
110. Madani Hosseini M, Zargoush M, Ghazalbash S (2024) Climate crisis risks to elderly health: strategies for effective promotion and response. Health Promot Int 39(2):daae031. https://doi.org/10.1093/heapro/daae031
111. Bobulski J, Szymoniak S, Pasternak K (2024) An IoT system for air pollution monitoring with safe data transmission. Sensors 24(2):445. https://doi.org/10.3390/s24020445
112. Sîrbu A, Becker M, Caminiti S, De Baets B, Elen B, Francis L, Gravino P, Hotho A, Ingarra S, Loreto V, Molino A, Mueller J, Peters J, Ricchiuti F, Saracino F, Servedio VD, Stumme G, Theunis J, Tria F, Van den Bossche J (2015) Participatory patterns in an international air quality monitoring initiative. PLoS One 10(8):e0136763. https://doi.org/10.1371/journal.pone.0136763
113. Ogwu MC (2019) Towards sustainable development in Africa: the challenge of urbanization and climate change adaptation. In: Cobbinah PB, Addaney M (eds) The geography of climate change adaptation in urban Africa. Palgrave Macmillan, Cham. https://doi.org/10.1007/978-3-030-04873-0_2
114. Ogwu MC (2024) Plants as monitors and managers of pollution. Role of green chemistry in ecosystem restoration to achieve environmental sustainability, 51–60. https://doi.org/10.1016/B978-0-443-15291-7.00022-5
115. Ogilvie D, Egan M, Hamilton V, Petticrew M (2004) Promoting walking and cycling as an alternative to using cars: systematic review. BMJ 329(7469):763. https://doi.org/10.1136/bmj.38216.714560.55
116. Ou Y, Kittner N, Babaee S, Smith SJ, Nolte CG, Loughlin DH (2021) Evaluating long-term emission impacts of large-scale electric vehicle deployment in the US using a human-earth systems model. Appl Energy 300:1–117364. https://doi.org/10.1016/j.apenergy.2021.117364
117. Din AU, Ming J, Rahman IU, Han H, Yoo S, Alhrahsheh RR (2023) Green road transportation management and environmental sustainability: the impact of population density. Heliyon 9(9):e19771. https://doi.org/10.1016/j.heliyon.2023.e19771
118. Mendes L, Monjardino J, Ferreira F (2022) Air quality forecast by statistical methods: application to Portugal and Macao. Front Big Data 5:826517. https://doi.org/10.3389/fdata.2022.826517

Part III
Technological and Other Sustainable Strategies for Air Pollution Mitigation

Eco-Friendly Methods for Combating Air Pollution

Matthew Chidozie Ogwu, Bokjin Ro, and Bhuwan Thapa

Contents

Abstract Air pollution, the presence of harmful or excessive concentrations of pollutants in the Earth's atmosphere, results in severe health impacts, including respiratory and cardiovascular diseases, lung cancer, asthma, and premature death especially among vulnerable populations such as children, the elderly, and

M. C. Ogwu (✉)
Goodnight Family Department of Sustainable Development, Appalachian State University, Boone, NC, USA
e-mail: ogwumc@appstate.edu

B. Ro
Interdisciplinary Program for Disaster, Crisis, and Risk Management, Sungkyunkwan University, Seoul, South Korea

B. Thapa
Department of Geography and Planning, College of Arts and Sciences, Appalachian State University, Boone, NC, USA
e-mail: thapab@appstate.edu

Matthew Chidozie Ogwu and Sylvester Izah (eds.), *Sustainable Strategies for Air Pollution Mitigation: Development, Economics and Technologies*,
Hdb Env Chem (2024) 133: 237–278, DOI 10.1007/698_2024_1119,

Published online: 12 September 2024

individuals with preexisting health conditions. It also affects ecosystems, leading to biodiversity loss, damage to vegetation, soil degradation, and contamination of water bodies. Given the significant health and environmental impacts of air pollution, the implementation of effective mitigation strategies is crucial. Despite growing awareness of the detrimental effects of air pollution, several challenges hinder efforts to achieve sustainable clean air. The chapter aims to explore sustainable strategies for mitigating air pollution and promoting cleaner, healthier environments. It examines various aspects of air pollution, including its sources, impacts, and mitigation measures, with a focus on sustainability and long-term solutions. The chapter highlights the roles of policy and governance frameworks, technological innovations, behavioral changes, sustainable urban planning, industrial and agricultural practices, climate change mitigation, and economic and social considerations to inform and inspire action toward achieving cleaner air and healthier communities. Efforts to address air pollution require both local and regional interventions and global cooperation. Promoting the use of renewable energy sources, such as solar, wind, and hydroelectric power, will reduce the use and reliance on fossil fuels and decrease emissions from power generation. In the same way, the implementation of pollution prevention and control measures in industries will help minimize emissions of pollutants such as sulfur dioxide, nitrogen oxides, particulate matter, and volatile organic compounds. More investment in public transportation infrastructure, the use of electric and hybrid vehicles, and clear emission standards will reduce emissions from transportation sources. National governments also need to establish monitoring networks, enforcement mechanisms, and reporting requirements to track air quality levels, assess compliance with regulatory standards, and take corrective actions to address noncompliance. Education and outreach play a pivotal role in raising awareness about air pollution issues and empowering individuals to act.

Keywords Best Available Technology, Best environmental practices, Clean air, Policy framework, Pollutants, Public health

Abbreviations

NH_3	Ammonia
NO_2	Nitrogen dioxide
O_3	Ozone
VOCs	Volatile organic compounds
CO	Carbon monoxide
SO_2	Sulfur dioxide
PM	Particulate matter
NH_3	Ammonia
NOx	Nitrogen oxides
CH_4	Methane
COPD	Chronic obstructive pulmonary disease

CCS	Carbon capture and storage
EVs	Electric vehicles
CAA	Clean Air Act
EPA	Environmental Protection Agency
HC	Hydrocarbons
CFCs	Chlorofluorocarbons
HCFCs	hydrochlorofluorocarbons
NAAQS	National Ambient Air Quality Standards
N_2O	Nitrous oxide
IPM	Integrated pest management
BAT	Best available techniques
BEP	Best environmental practices
BRT	Bus rapid transit
UGB	Urban growth boundary
TOD	Transit-oriented development
GIS	Geographic information systems
ITS	Intelligent transportation systems
EIAs	Environmental impact assessments
HVAC	Heating, ventilation, and air conditioning
CBPR	Community-based participatory research
CLRTAP	Convention on long-range transboundary air pollution
$	United States dollar

1 Introduction

From densely populated urban areas to remote rural communities, the impacts of air pollution are far-reaching and require urgent attention. Air pollution refers to the presence of harmful or excessive concentrations of pollutants in the Earth's atmosphere. These pollutants can originate from various sources, including industrial activities, transportation, agriculture, and natural phenomena. Common air pollutants include particulate matter (PM), nitrogen dioxide (NO_2), sulfur dioxide (SO_2), ozone (O_3), carbon monoxide (CO), and volatile organic compounds (VOCs). Depending on their composition and concentration levels, these pollutants can have adverse effects on human health, such as respiratory diseases, cardiovascular disorders, and even premature death [1, 2]. Additionally, air pollution can harm ecosystems, damage crops, and contribute to climate change.

Given the significant health and environmental impacts of air pollution, the implementation of effective mitigation strategies is crucial. Mitigation strategies aim to reduce or prevent the emission of pollutants into the atmosphere, thereby improving air quality and protecting public health [3]. These strategies may involve technological innovations, policy interventions, behavioral changes, and community engagement efforts. By addressing the root causes of air pollution and promoting cleaner, more sustainable practices, mitigation strategies can help mitigate the adverse effects of air pollution and create healthier environments for current and future generations [4, 5].

Clean air is an essential component of a healthy environment, vital for sustaining life on Earth. However, in recent years, air pollution has emerged as a significant global challenge, posing serious threats to human health, ecosystems, and the planet's overall well-being. Clean air is fundamental to human health as polluted air can lead to a myriad of respiratory and cardiovascular diseases, including asthma, lung cancer, and heart attacks [6, 7]. Additionally, air pollution adversely affects ecosystems, leading to biodiversity loss, acid rain, and damage to crops and forests. Furthermore, polluted air exacerbates climate change, contributing to global warming and extreme weather events [8]. Therefore, ensuring sustainable clean air is crucial for safeguarding public health, preserving ecosystems, and mitigating climate change impacts. Despite growing awareness of the detrimental effects of air pollution, several challenges hinder efforts to achieve sustainable clean air. These challenges include industrial emissions, vehicular pollution, agricultural practices, inadequate waste management, and urbanization [9]. Additionally, socioeconomic factors such as poverty, lack of access to clean energy, and inadequate regulatory frameworks exacerbate air pollution issues, particularly in developing countries [10]. Moreover, the interconnected nature of air pollution requires coordinated efforts at local, national, and global levels, posing governance and implementation challenges.

Achieving sustainable clean air requires a multifaceted approach that addresses the root causes of air pollution while promoting sustainable development [11]. For instance, investing in renewable energy sources such as solar, wind, and hydroelectric power can reduce reliance on fossil fuels and mitigate greenhouse gas emissions. In urban centers, promoting public transportation, cycling, and walking, along with adopting electric and hybrid vehicles, can reduce vehicular emissions and alleviate traffic congestion [12, 13]. Implementing cleaner production techniques, adopting pollution control measures, and enforcing stringent emission standards can minimize industrial pollution. Also, designing cities with green spaces, efficient public transportation systems, and pedestrian-friendly infrastructure can reduce urban air pollution and enhance livability [14]. Educating the public about the health impacts of air pollution and promoting sustainable lifestyle choices, such as energy conservation and waste reduction, can also foster behavioral changes [15, 16]. By embracing sustainable practices, promoting clean technologies, and fostering collaboration, we can strive toward a future where everyone breathes clean air and thrives in harmony with nature.

The primary objective of this chapter is to explore sustainable strategies for mitigating air pollution and promoting cleaner, healthier environments. It will examine various aspects of air pollution, including its sources, impacts, and mitigation measures, with a focus on sustainability and long-term solutions. The chapter will cover policy and governance frameworks, technological innovations, behavioral changes, sustainable urban planning, industrial and agricultural practices, climate change mitigation, and economic, and social considerations to inform and inspire action toward achieving cleaner air and healthier communities.

2 Mitigating Air Pollution: Strategies for Cleaner Air and Healthier Communities

Air pollution poses significant threats to human health, ecosystems, and the environment, highlighting the urgent need for effective mitigation strategies. Mitigating air pollution requires a multifaceted approach that integrates technological innovations, policy interventions, behavioral changes, and community engagement efforts [17, 18]. These combined strategies aimed at reducing emissions from various sources, controlling pollution levels, and promoting sustainable practices [19, 20]. By adopting holistic strategies and collaborative approaches, societies can achieve cleaner air, improve public health, and create healthier and more sustainable communities for future generations.

Promote the use of renewable energy sources, such as solar, wind, and hydroelectric power, to reduce reliance on fossil fuels and decrease emissions from power generation. The implementation of pollution prevention and control measures in industries to minimize emissions of pollutants such as sulfur dioxide, nitrogen oxides, particulate matter, and volatile organic compounds is necessary [21]. Investment in public transportation infrastructure, the use of electric and hybrid vehicles, and the implementation of vehicle emission standards are necessary to reduce emissions from transportation sources, especially in urban areas. In the same way, encouraging the adoption of cleaner technologies and practices in sectors such as agriculture, construction, and manufacturing to reduce emissions of pollutants and minimize environmental impact [22]. Enforcing air quality standards, emission limits, and pollution control regulations, and establishing monitoring and enforcement mechanisms can help to ensure compliance with environmental regulations. Implement energy efficiency measures in buildings, appliances, and industrial processes to reduce energy consumption and decrease emissions associated with energy production [23–25]. Designing cities with compact, mixed-use developments, green spaces, and efficient transportation systems can contribute to reducing vehicle emissions and minimizing air pollution hotspots [26]. There is a need to educate the public about the health impacts of air pollution, promote sustainable lifestyle choices, and encourage behavioral changes such as reducing energy consumption, using public transportation, and practicing eco-friendly habits. Invest in research and development of new technologies, pollution control methods, and monitoring techniques to improve air quality management and advance understanding of air pollution sources and impacts [27, 28]. Collaboration between neighboring countries, international organizations, and global initiatives is needed to address transboundary air pollution issues, share best practices, and promote global efforts to mitigate air pollution and combat climate change [29]. By implementing these strategies in a coordinated and comprehensive manner, governments, businesses, communities, and individuals can work together to mitigate air pollution and create cleaner and healthier environments for current and future generations.

Mitigating air pollution offers a wide range of benefits that positively impact human health, environmental sustainability, and socioeconomic well-being. Some of these benefits and their examples are presented in Table 1. Reducing air pollution

Table 1 Strategies for mitigating air pollution and their benefits

Strategy	Benefits	Examples
Transition to clean energy	Reduces greenhouse gas emissions, improves air quality, and promotes sustainable development	Installation of solar panels on rooftops, wind farm development, and hydroelectric power generation
Improvement of industrial practices	Reduces emissions of pollutants, protects public health, and enhances environmental sustainability	Installation of pollution control devices, implementation of emission standards, and adoption of cleaner production technologies
Enhance transportation systems	Decreases vehicle emissions, improves urban air quality, and reduces traffic congestion	Expansion of public transportation networks, promotion of electric and hybrid vehicles, and implementation of vehicle emission standards
Adaptation of cleaner technologies	Minimizes emissions from various sources, enhances energy efficiency, and supports sustainable practices	Deployment of advanced air pollution control technologies, installation of low-emission industrial equipment, and adoption of clean manufacturing processes
Strengthening of regulatory frameworks	Ensures compliance with air quality. Standards and emission limits, protect public health, and promote environmental sustainability	Implementation of emission standards and regulations, enforcement of pollution control measures, establishment of monitoring and enforcement mechanisms, and the integration of air quality considerations into land-use planning
Promote energy efficiency	Reduces energy consumption, lowers emissions associated with energy, and saves costs through efficiency measures	Retrofitting buildings with energy-efficient systems, implementing energy efficiency standards, and the adoption of energy-efficient appliances
Encourage sustainable urban planning	Reduces vehicle emissions in urban areas, improves air quality in densely populated areas, and enhances the quality of life for residents	Designing cities with efficient transportation systems, zoning regulations to minimize pollution hotspots, and the creation of green spaces and pedestrian-friendly areas
Raise public awareness and behavioral changes	Increases awareness of air pollution, promotes sustainable behaviors, and empowers individuals to take action	Educational campaigns on air quality and health, encouraging the use of public transportation, and advocating for clean air policies and regulations

levels leads to significant improvements in public health by decreasing the incidence of respiratory and cardiovascular diseases, such as asthma, lung cancer, heart attacks, and strokes [30]. Lowering exposure to harmful pollutants, such as particulate matter, nitrogen oxides, and sulfur dioxide, reduces the risk of respiratory infections, chronic obstructive pulmonary disease (COPD), and other respiratory ailments [31, 32]. Cleaner air contributes to a better quality of life for individuals and communities by reducing the prevalence of health problems associated with air pollution. Improved air quality creates safer outdoor environments for physical activity, recreation, and social interaction, fostering overall well-being and community cohesion. Mitigating air pollution can extend life expectancy by reducing the risk of premature death and chronic health conditions associated with long-term exposure to polluted air [33]. Studies have shown that improvements in air quality led to longer life spans and better health outcomes for populations exposed to cleaner air [16, 34]. Lowering air pollution levels benefits ecosystems and wildlife by minimizing damage to vegetation, soil, water bodies, and wildlife habitats. Reduced emissions of pollutants such as nitrogen oxides and sulfur dioxide help mitigate acid rain, forest decline, soil acidification, and eutrophication, preserving biodiversity and ecosystem services essential for ecosystem health and resilience [35–37]. Many air pollutants, such as carbon dioxide, methane, and black carbon, are also greenhouse gases that contribute to global warming and climate change. By reducing emissions of these pollutants through air pollution mitigation efforts, societies can help mitigate climate change impacts, such as rising temperatures, extreme weather events, and sea-level rise, and promote environmental sustainability. Mitigating air pollution generates economic benefits through healthcare cost savings, increased labor productivity, and reduced damage to infrastructure and natural resources. Cleaner air leads to lower healthcare expenditures associated with treating air pollution-related illnesses, fewer missed workdays due to sickness, and higher productivity levels among workers. Additionally, investments in technological innovations like air quality monitoring systems, emission control technologies, clean energy solutions, green infrastructures, and others listed in Table 2 as pollution control measures can create job opportunities, stimulate innovation, and drive economic growth in sectors such as renewable energy, clean transportation, and environmental remediation. Mitigating air pollution promotes social equity and environmental justice by reducing disparities in exposure to harmful pollutants and protecting vulnerable populations, such as children, the elderly, low-income communities, and minority groups, from disproportionate health risks associated with air pollution [56, 57]. Addressing environmental inequalities and promoting equitable access to clean air fosters social cohesion, inclusivity, and resilience in communities.

Table 2 Strategies for mitigating air pollution and their benefits

Technological innovations	Description	References
Air quality monitoring systems	Advanced sensor networks and monitoring devices for real-time tracking of air pollution levels	Cromar et al. [38], Saini et al. [39, 40]
Emission control technologies	Innovative techniques and devices for reducing emissions from industrial processes, vehicles, and power plants	Liu et al. [41, 42]
Clean energy solutions	Renewable energy sources such as solar, wind, and hydropower as alternatives to fossil fuels	Gallagher and Holloway [43], Xie et al. [44]
Green infrastructure and low-impact development	Nature-based solutions such as green roofs, permeable pavements, and urban forests improve air quality and reduce urban heat islands	Hewitt et al. [45], Jennings et al. [46]
Electric vehicles (EVs) and alternative transportation	Development and adoption of electric vehicles, public transit systems, and cycling infrastructure to reduce vehicular emissions and promote sustainable mobility	Carey [47]
Waste management technologies	Advanced waste-to-energy systems, recycling technologies, and landfill gas capture for minimizing emissions from waste disposal	Anuardo et al. [48], Tanveer et al. [49]
Indoor air quality solutions	Ventilation systems, air purifiers, and indoor plants to improve indoor air quality and reduce exposure to indoor pollutants	Nassikas et al. [50]
Smart city solutions	Integration of digital technologies, data analytics, and smart infrastructure for optimizing urban systems and reducing environmental impacts	Toma et al. [51], Wang et al. [52], Neo et al. [53]
Carbon capture and storage (CCS)	Technologies for capturing carbon dioxide emissions from industrial processes and storing them underground or repurposing them for industrial use	Chen et al. [54]
Green construction practices	Sustainable building materials, energy-efficient designs, and passive cooling/heating systems for reducing emissions from the construction sector	Venter et al. [55]

3 Policy and Governance Frameworks for Air Pollution Mitigation

Air pollution mitigation policies refer to the set of rules, regulations, strategies, and measures implemented by governments, organizations, and institutions to reduce air pollution levels and minimize its adverse effects on human health, the environment, and society. International agreements and protocols provide a framework for cooperation among countries to address transboundary air pollution and global environmental challenges. Examples include the United Nations Framework Convention on Climate Change (UNFCCC), the Paris Agreement, and the Convention on Long-Range Transboundary Air Pollution (CLRTAP). These agreements set targets, guidelines, and principles for reducing greenhouse gas emissions, mitigating air pollution, and promoting sustainable development on a global scale [58, 59]. They facilitate information exchange, technology transfer, capacity building, and financial support to help countries implement effective air pollution mitigation measures and achieve common environmental objectives [60, 61]. In the same way, national legislation and regulations establish legal frameworks for air quality management and pollution control at the country level. Governments enact laws, statutes, and regulations to protect public health, safeguard the environment, and regulate emissions from various sources. These policies set air quality standards, emission limits, and compliance requirements for industries, vehicles, power plants, and other sources of pollution [62]. They mandate the use of pollution control technologies, cleaner fuels, and best practices to minimize emissions and mitigate air pollution. National governments can also establish monitoring networks, enforcement mechanisms, and reporting requirements to track air quality levels, assess compliance with regulatory standards, and take corrective actions to address noncompliance. Local governments, communities, and stakeholders play a vital role in addressing air pollution through grassroots initiatives, local regulations, and community engagement efforts [63]. Local initiatives may include the development of air quality management plans, emission reduction programs, and sustainable transportation strategies tailored to the specific needs and challenges of local communities [64–66]. Community engagement activities raise awareness about air pollution issues, empower residents to take action to improve air quality, and foster collaboration between government agencies, businesses, academia, and civil society organizations. Participatory approaches, citizen science projects, and public consultations enable communities to voice their concerns, contribute to decision-making processes, and hold authorities accountable for implementing effective air pollution mitigation measures [67–69]. These policies aim to control emissions of pollutants, promote clean technologies and practices, and encourage sustainable development. Some examples of air pollution mitigation policies are (Fig. 1).

Emission Standards Governments set limits on the amount of pollutants that industries, vehicles, and other sources are allowed to emit into the atmosphere. Emission standards specify the maximum concentration of pollutants permissible from specific sources, such as power plants, factories, vehicles, and construction sites [70]. Compliance with these standards is enforced through monitoring, inspections, and penalties for noncompliance.

Clean Air Act The Clean Air Act (CAA), enacted by many countries, including the United States, sets national air quality standards and mandates regulations to control emissions of criteria pollutants such as SO_2, NOx, PM, CO, lead, and ozone. An

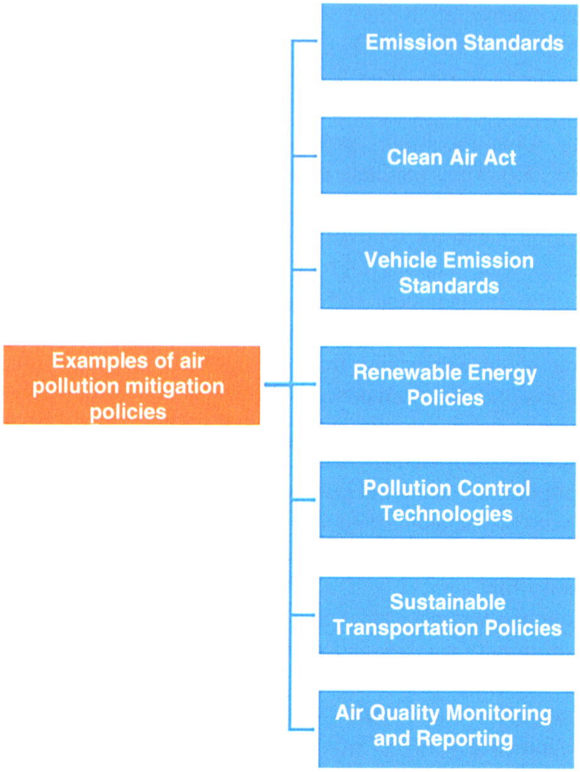

Fig. 1 Examples of air pollution mitigation policies and programs

analysis of the relevance of the United States CAA is presented in Box 1. The CAA requires the Environmental Protection Agency (EPA) to develop and implement regulations to achieve and maintain air quality standards, protect public health, and prevent air pollution [69].

Box 1 Analysis of Some of the Benefits and Challenges of the United States Clean Air Act

United States Clean Air Act of 1970

The Clean Air Act (CAA) of the United States was enacted in 1970 to protect and enhance air quality for promoting public health and welfare. It establishes the national standards, also called National Ambient Air Quality Standards (NAAQS), for six major air pollutants, including particulate matter, ozone, carbon monoxide, sulfur dioxide, nitrogen dioxide, and lead. The standards are based on the latest scientific information.

The states are required to develop plans to meet the NAAQS. These plans include air emissions inventory, air monitoring, emission control strategies,

(continued)

Box 1 (continued)
and air modeling to demonstrate how the state plans can satisfy federal standards. States issue permits and make rules to implement the state plans. The CAA has been revised multiple times to tighten transportation emission standards, establish the acid rain control program, and implement the Montreal Protocol to phase out ozone-depleting chemicals [71].

There have been multiple legal court battles to regulate carbon dioxide under CAA. However the Inflation Reduction Act of 2022 "officially established CO2 emissions from burning of fossil fuels as an "air pollutant" within the text of CAA" [72]. The CAA has significantly reduced major air pollutants like sulfur dioxide, volatile organic compounds, PM-10 and lead across many states [73, 74]. These reductions have provided billions of economic benefits by reducing health costs and improving the visibility of national parks [75].

Vehicle Emission Standards Governments may implement regulations to limit vehicle emissions by setting standards for vehicle engines, fuel quality, and emission control systems [76]. These standards require manufacturers to produce vehicles that meet certain emission limits for pollutants such as CO_2, NOx, hydrocarbons (HC), and PM. Measures include the promotion of electric vehicles, hybrid vehicles, fuel-efficient technologies, and alternative fuels.

Renewable Energy Policies Governments may choose to incentivize the adoption of renewable energy sources, such as solar, wind, hydroelectric, and biomass, to reduce reliance on fossil fuels and lower greenhouse gas emissions [77]. Renewable energy policies may include financial incentives, tax credits, feed-in tariffs, renewable portfolio standards, and subsidies for renewable energy projects and investments.

Pollution Control Technologies Governments promote the use of pollution control technologies and practices to reduce emissions from industrial processes, power generation, and other sources [41, 42]. Examples include the installation of scrubbers, catalytic converters, electrostatic precipitators, and flue gas desulfurization systems to capture and remove pollutants before they are released into the atmosphere.

Sustainable Transportation Policies Governments may implement measures to promote sustainable transportation options, such as public transit, cycling, walking, and carpooling, to reduce vehicle emissions and congestion [78, 79]. Policies may include the expansion of public transportation networks, the construction of bike lanes and pedestrian pathways, congestion pricing, and incentives for the use of clean vehicles and alternative transportation modes.

Market-Based Mechanisms Market-based mechanisms to regulate air pollutants are increasingly popular among policymakers because they are more flexible and economically efficient than the command-and-control approach. Under the market-based mechanism, the polluter can purchase air permits from entities that emit fewer air pollutants. Regional Greenhouse Gas Initiative and California Cap-and-Trade Program are some of the market-based pollutant measurements (Box 2).

Box 2 Regional Greenhouse Gas Initiative

The Regional Greenhouse Gas Initiative (RGGI) is a mandatory cap-and-trade program that limits CO_2 emissions from the power sector in selected US states. As of 2024, 11 Eastern states are participating in the program: Connecticut, Delaware, Maine, Maryland, Massachusetts, New Hampshire, New Jersey, New York, Vermont, Virginia, and Pennsylvania [80].

The program requires fossil fuel power plants producing more than 25 MW of electricity to obtain an allowance for each ton of annual CO_2 emissions. The allowance is purchased from auctions, other power generators, or offset projects. The offset projects must be located within one of the RGGI states and be preapproved. The number of allowance permits is lowered each year. Since its inception in 2005, the program has reduced power sector emissions by 50% [80].

The RGGI provides both economic and environmental benefits. Part of the revenue from CO_2 trading is channeled back to communities on energy efficiency, renewable energy, and other energy-related projects. The program has saved $618 million in energy bills and contributed to $5.7 billion in health benefits by reducing air pollutants like soot and smog alongside CO_2 [81].

Air Quality Monitoring and Reporting Governments establish air quality monitoring networks to track pollution levels, identify sources of emissions, and assess compliance with air quality standards [82, 83]. Real-time air quality data is made available to the public through air quality indexes, websites, and mobile applications, enabling individuals and communities to take action to protect their health and reduce exposure to air pollution.

These policies and governance frameworks play a crucial role in mitigating air pollution by establishing regulations, standards, and enforcement mechanisms to control emissions, protect public health, and promote environmental sustainability (Table 3). Some key components and considerations for effective policy and governance frameworks for air pollution mitigation include (Fig. 2).

Regulatory Framework It involves the establishment of national air quality standards and emission limits for key pollutants based on scientific evidence and health guidelines as well as the development of comprehensive air quality management plans outlining strategies, targets, and timelines for reducing pollution levels [93]. The implementation of regulatory mechanisms, such as permits, licenses, and compliance requirements, will ensure industries and other sources of pollution adhere to emission standards [94]. Integration of air quality considerations into

Table 3 Relevant policies and governance frameworks for managing air pollution

Policy and governance framework	Examples	References
International agreements and protocols	**The Paris Agreement:** A legally binding international treaty on climate change, aiming to limit global warming to well below 2°C above pre-industrial levels. It includes commitments from participating countries to reduce greenhouse gas emissions, which contribute to air pollution and climate change.	Markandya et al. [84], Hamilton et al. [85]
	The Montreal Protocol: An international treaty designed to protect the ozone layer by phasing out the production and use of ozone-depleting substances, such as chlorofluorocarbons (CFCs) and hydrochlorofluorocarbons (HCFCs). While not directly addressing air pollution, it indirectly contributes to improving air quality by reducing emissions of harmful pollutants.	Andersen et al. [86], Barnes et al. [87]
National legislation and regulations	**Clean Air Act (United States):** A comprehensive federal law aimed at protecting and improving air quality by regulating emissions of air pollutants from stationary and mobile sources. It sets National Ambient Air Quality Standards (NAAQS) for criteria pollutants and establishes regulatory programs to control emissions from industries, vehicles, and other sources.	Ross et al. [69], Manisalidis et al. [88]
	Air Quality Directive (European Union): A key piece of legislation governing ambient air quality and emissions of air pollutants within the European Union. It sets limits and objectives for pollutants such as particulate matter, nitrogen dioxide, sulfur dioxide, and ozone, and requires member states to develop national air quality management plans to achieve these targets.	Laurent [89], Boogaard et al. [90]
Local initiatives and community engagement	**Community Air Quality Monitoring Programs**: Many local communities and organizations implement air quality monitoring initiatives to assess pollution levels in their vicinity and raise awareness about the health impacts of poor air quality. These programs often involve partnerships between community groups, government agencies, and academic institutions, with volunteers collecting data and advocating for cleaner air policies. An example is the Youth Engaged Participatory Air Monitoring Program in California.	Johnston et al. [91], Ward et al. [92]
	Urban Planning and Zoning Regulations: Local governments can influence air quality through land-use planning and zoning	Venter et al. [55]

(continued)

Table 3 (continued)

Policy and governance framework	Examples	References
	regulations that promote sustainable development, reduce vehicle emissions, and minimize exposure to pollution. Measures such as promoting public transportation, enhancing green spaces, and incentivizing energy-efficient buildings can contribute to improving air quality in urban areas.	

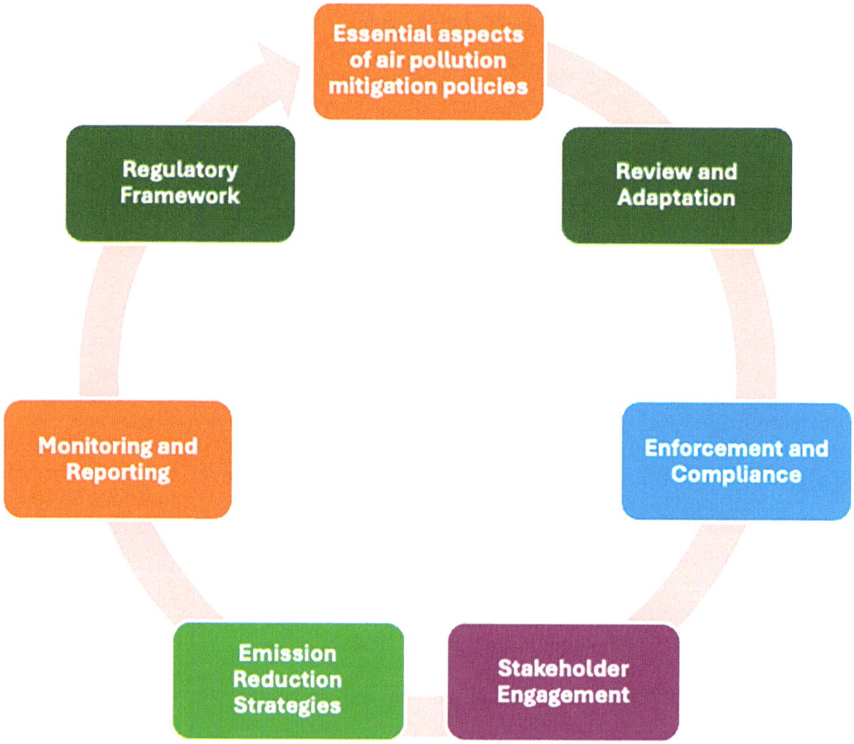

Fig. 2 Key characteristics of air pollution mitigation policies and regulatory frameworks

land-use planning, transportation policies, and industrial zoning regulations will minimize pollution hotspots and promote sustainable development.

Realtime Monitoring and Reporting This requires the establishment of a robust air quality monitoring network to track pollution levels, identify sources of emissions, and assess compliance with air quality standards [95]. Regular reporting of air quality data to the public, policymakers, and relevant stakeholders to raise awareness, guide decision-making, and evaluate the effectiveness of mitigation efforts. Use of advanced monitoring technologies, such as remote sensing, satellite imagery, and low-cost sensors, to enhance spatial and temporal coverage of air quality monitoring and improve data accuracy [96].

Emission Reduction Strategies Adoption of pollution prevention and control measures, such as cleaner production technologies, fuel switching, and pollution abatement equipment, to reduce emissions from industrial processes, power generation, transportation, and other sources [97, 98]. Promotion of energy efficiency measures, renewable energy sources, and low-carbon technologies to minimize greenhouse gas emissions and mitigate climate change impacts. Implementation of vehicle emission standards, vehicle inspection and maintenance programs, and incentives for adopting cleaner transportation options, such as electric vehicles and public transit.

Stakeholder Engagement Engagement of government agencies, industry stakeholders, civil society organizations, academic institutions, and the public in the development, implementation, and evaluation of air quality policies and programs. Collaboration with international organizations, neighboring countries, and regional initiatives to address transboundary air pollution issues and promote regional cooperation in air quality management. Provision of technical assistance, capacity-building initiatives, and funding support to facilitate participation and empower stakeholders to contribute to air pollution mitigation efforts.

Enforcement and Compliance Strengthening of enforcement mechanisms, including inspections, audits, and penalties for noncompliance with air quality regulations, to ensure accountability and deter pollution violations [99, 100]. Enhancement of regulatory enforcement capacity through training programs, provision of resources, and collaboration with law enforcement agencies and regulatory authorities. Implementation of incentive-based approaches, such as pollution taxes, emissions trading schemes, and green procurement policies, to incentivize compliance and encourage pollution reduction efforts [101].

Review and Adaptation Periodic review and evaluation of air quality policies, regulations, and programs to assess their effectiveness, identify gaps and opportunities for improvement, and incorporate new scientific knowledge and technological advancements. Flexibility and adaptability in policy design and implementation to respond to emerging air pollution challenges, changing socioeconomic conditions, and evolving environmental priorities [102, 103]. Continuous engagement with stakeholders and regular communication of policy updates, progress reports, and best practices to foster transparency, accountability, and public trust in air quality governance.

4 Behavioral Changes and Public Awareness for Air Pollution Mitigation

Behavioral changes and public awareness are vital components of environmental conservation efforts. As societies grapple with pressing environmental challenges such as air and water pollution, deforestation, and climate change, fostering a culture of sustainability becomes increasingly imperative [104]. Behavioral changes refer to alterations in individual actions and lifestyle choices that contribute to more sustainable practices [105]. These changes can encompass a wide range of actions, from reducing energy consumption and waste generation to adopting eco-friendly transportation methods and supporting sustainable products and businesses. Behavioral changes are crucial because they directly influence resource consumption, pollution levels, and overall environmental impact. By encouraging individuals to adopt more sustainable behaviors, human societies can collectively reduce our ecological footprint and mitigate the negative effects of human activities on the environment. Public awareness plays a key role in driving behavioral changes and fostering a culture of environmental conservation. Awareness refers to the understanding and knowledge that individuals have about environmental issues, their causes, consequences, and potential solutions. When people are informed about environmental challenges such as climate change, habitat destruction, and pollution, they are more likely to take action to address these issues. Public awareness campaigns, educational programs, and media outreach efforts are essential for disseminating information, raising consciousness, and mobilizing support for environmental conservation initiatives.

Behavioral science provides valuable insights into understanding human behavior and decision-making processes, which can inform strategies for promoting sustainable behaviors related to air pollution mitigation [106]. Psychological factors such as risk perception, cognitive biases, and social norms play a significant role in influencing individual actions toward air quality improvement. By understanding these factors, policymakers and environmental advocates can design interventions and communication strategies that resonate with people's values, beliefs, and motivations. Behavior change techniques such as goal setting, feedback, incentives, and social norms can be effective in encouraging sustainable behaviors [107]. For example, setting achievable goals for reducing personal emissions, providing feedback on air quality levels in real time, offering incentives for adopting clean transportation options, and leveraging social norms to make sustainable behaviors more socially desirable can all contribute to fostering a culture of clean air stewardship.

Education and outreach play a pivotal role in raising awareness about air pollution issues and empowering individuals to act. Environmental education programs in schools, universities, and communities can provide people with knowledge and understanding of the causes and effects of air pollution. An example of such a citizen science initiative is the PurpleAir initiative in Pennsylvania in the United States for monitoring and improving air quality, which is analyzed in Box 3. These programs can teach individuals about the sources of air pollution, the health impacts associated with exposure, and the importance of sustainable practices [18]. Furthermore, public awareness campaigns and outreach initiatives can amplify the message of air pollution mitigation to a broader audience. Through various communication channels such as social media, television, and community events, these campaigns can disseminate information about air quality, pollution hotspots, and simple actions individuals can take to reduce their emissions. By engaging with the public through interactive workshops, educational materials, and citizen science projects, education and outreach efforts can inspire behavioral changes and foster a sense of collective responsibility for air quality improvement.

Box 3 Low-cost Purple Air Sensor to Monitor Air Quality in Pennsylvania, USA

Like many areas in the United States, Pennsylvania faces significant air quality challenges due to industrial activity, vehicle emissions, and geography that traps pollution [112]. And existing monitoring systems have limitations in coverage and cost-effectiveness [113].

PurpleAir utilizes a network of low-cost air quality sensors that offer real-time data collection for monitoring and improving air quality, especially in low-income neighborhoods. Compared to traditional systems, PurpleAir's approach is more cost-effective and provides timely information. PurpleAir has a significant presence in Pennsylvania, with numerous sensors deployed across the state [114]. Its data has helped in understanding local air quality issues, identifying pollution sources, and guiding public health initiatives.

A local university is collaborating with local government, environmental organizations, and community groups to install Purple Air Sensors to monitor air quality. Joint projects focus on improving air quality and raising awareness about environmental health issues [115]. Stakeholder testimonials underscore PurpleAir's value in improving air quality in Pennsylvania. Challenges associated with PurpleAir's approach include sensor accuracy and data interpretation. Future directions may involve addressing these challenges and expanding initiatives to further enhance air quality monitoring and improvement efforts in other parts of the United States.

Empowering communities to take ownership of air pollution issues is essential for achieving meaningful and lasting change. Communities are often at the forefront of air pollution impacts and have firsthand knowledge of local pollution sources and concerns. By involving communities in decision-making processes, policymakers can ensure that policies and interventions are responsive to local needs and priorities. Participatory approaches such as community-based participatory research (CBPR), co-design workshops, and participatory budgeting can facilitate meaningful engagement and collaboration between stakeholders [108, 109]. These approaches enable community members to contribute their expertise, perspectives, and resources to air pollution mitigation efforts. By building partnerships with local organizations, grassroots movements, and citizen scientists, communities can mobilize collective action, advocate for policy reforms, and implement local interventions to improve air quality in their neighborhoods.

Promoting behavioral changes and public awareness is essential for mitigating air pollution and safeguarding public health and the environment. Education and outreach efforts can empower individuals with the knowledge and motivation to adopt sustainable practices. Behavioral science insights can inform the design of effective interventions that resonate with people's values and motivations. Community empowerment and participation can mobilize collective action and drive meaningful change at the local level. Government policies and regulations also play a crucial role in shaping behaviors and incentivizing sustainable practices. Policies such as carbon pricing, renewable energy incentives, and waste management regulations can create economic incentives for businesses and individuals to reduce their environmental impact [110, 111]. By aligning policies with environmental goals and priorities, governments can encourage and facilitate the adoption of sustainable behaviors.

5 Sustainable Urban Planning for Air Pollution Mitigation

With the rapid pace of urbanization, cities around the world are grappling with the challenges of air pollution, which poses significant threats to public health, environmental sustainability, and overall quality of life. Sustainable urban planning offers a holistic approach to addressing these challenges by integrating environmental considerations into the design and management of urban spaces [116]. Urban areas face significant challenges related to air pollution, which can have detrimental effects on public health and the environment. Urban air pollution is a complex and multifaceted issue, arising from a variety of sources such as vehicular emissions, industrial activities, construction dust, and residential energy use [117]. These pollutants include PM, NOx, SO_2, VOCs, and CO, among others. The health impacts of urban air pollution are significant, contributing to respiratory diseases, cardiovascular problems, and premature mortality. Moreover, air pollution also has environmental consequences, including ecosystem degradation, acid rain formation, and climate change exacerbation. Some pertinent principles of sustainable urban planning include.

Compact and Mixed-Use Development Compact and mixed-use development aims to reduce urban sprawl and promote higher-density, walkable neighborhoods. By concentrating development in compact urban centers, this approach minimizes the need for long-distance commuting and encourages alternative transportation modes such as walking, cycling, and public transit [118]. Compact and mixed-use development also fosters vibrant, diverse communities with access to amenities, services, and green spaces within walking distance.

Public Transportation and Active Mobility Public transportation systems play a critical role in reducing air pollution by providing an efficient and sustainable alternative to private vehicle use. Investments in mass transit, including buses, trains, trams, and subways, can help alleviate traffic congestion and reduce emissions from cars and trucks. Active mobility options such as walking and cycling are also prioritized in sustainable urban planning, with the development of pedestrian-friendly infrastructure, dedicated bike lanes, and safe routes for nonmotorized transportation.

Green Infrastructure and Urban Green Spaces Green infrastructure refers to the strategic use of natural and semi-natural elements to manage environmental challenges and enhance urban resilience [119]. Examples of green infrastructure include parks, green roofs, rain gardens, urban forests, and permeable pavements. These features help mitigate air pollution by capturing and filtering pollutants, reducing heat island effects, and improving overall air quality. Urban green spaces also provide numerous co-benefits, including biodiversity conservation, recreational opportunities, and mental health benefits.

Sustainable Building Design and Energy Efficiency Sustainable building design emphasizes energy efficiency, resource conservation, and indoor environmental quality. Green building practices such as passive solar design, energy-efficient HVAC systems, insulation, and renewable energy integration help reduce energy consumption and emissions from buildings. Green roofs and living walls can further enhance air quality by absorbing pollutants and CO_2, while also providing insulation and habitat for wildlife. Additionally, building codes and standards play a crucial role in promoting sustainable construction practices and ensuring the long-term environmental performance of buildings [120].

Strategies for the implementation of sustainable urban planning may include.

Land-Use Zoning and Regulations Land-use zoning and regulations guide the spatial distribution of development and influence the built environment's form and function. Zoning ordinances can promote mixed-use development, transit-oriented development (TOD), and higher-density, pedestrian-friendly neighborhoods [121]. Incentives and bonuses can encourage developers to incorporate green building features, open space, and affordable housing into their projects. Regulatory tools such as environmental impact assessments (EIAs), development agreements, and design guidelines help ensure that new developments are compatible with air quality objectives and community needs.

Transportation Planning and Infrastructure Investment Transportation planning focuses on improving mobility, accessibility, and safety while minimizing environmental impacts and promoting sustainable modes of transportation. Integrated transportation plans prioritize investments in public transit, active transportation, and smart mobility solutions. Transit-oriented development (TOD) seeks to locate high-density, mixed-use development near transit stations, reducing dependence on cars and encouraging transit ridership [122]. Complete streets policies aim to design streets for all users, including pedestrians, cyclists, public transit users, and motorists, with features such as sidewalks, bike lanes, transit stops, and traffic calming measures.

Community Engagement and Participation Community engagement is essential for ensuring that urban planning processes are inclusive, transparent, and responsive to community needs and priorities. Meaningful engagement fosters trust, builds social capital, and empowers residents to shape their neighborhoods and advocate for positive change. Participatory planning approaches such as charrettes, workshops, focus groups, and citizen advisory committees enable stakeholders to contribute local knowledge, expertise, and perspectives to planning decisions [123]. Digital tools and online platforms can also facilitate broader public participation and reach diverse audiences.

5.1 *Green Infrastructure and Urban Design*

Green Spaces and Urban Forests Incorporating green spaces, parks, and urban forests into city design helps absorb pollutants, reduce heat island effects, and improve air quality. Trees and vegetation act as natural filters, capturing particulate matter and absorbing pollutants such as carbon dioxide and nitrogen dioxide [124]. Urban greenery also provides numerous co-benefits, including habitat creation, noise reduction, and aesthetic enhancement.

Sustainable Stormwater Management Sustainable stormwater management practices such as green roofs, rain gardens, and permeable pavements help reduce runoff and prevent pollution of waterways [125]. By capturing and filtering rainwater, these green infrastructure solutions improve water quality and reduce the transport of pollutants into rivers, lakes, and oceans. Additionally, green stormwater infrastructure contributes to flood mitigation and climate resilience.

Compact and Mixed-Use Development Compact and mixed-use development patterns promote walkability, reduce vehicle miles traveled, and minimize air pollution from transportation [126]. By locating housing, businesses, and amenities in proximity, cities can encourage active transportation modes such as walking, cycling, and public transit. Mixed-use neighborhoods also support vibrant, diverse communities and reduce dependence on cars for daily activities.

5.2 *Smart City Solutions*

Air Quality Monitoring and Management Smart city technologies enable real-time monitoring of air quality parameters such as particulate matter, nitrogen oxides, and ozone levels [127]. By collecting and analyzing data from sensors deployed throughout the city, authorities can identify pollution hotspots, assess trends, and develop targeted mitigation strategies. Public access to air quality information empowers citizens to make informed decisions about their health and activities.

Intelligent Transportation Systems Intelligent transportation systems (ITS) leverage technology to optimize traffic flow, reduce congestion, and minimize emissions from vehicles. Solutions such as adaptive traffic signal control, dynamic route guidance, and congestion pricing help improve traffic efficiency and reduce air pollution [128]. Integrated multimodal transportation networks provide travelers with convenient, sustainable alternatives to private car use, such as public transit, bike sharing, and carpooling.

Digital Planning and Decision Support Tools Digital planning tools and decision support systems assist urban planners and policymakers in optimizing land use, infrastructure investment, and environmental management [129]. Geographic information systems (GIS), simulation models, and scenario planning software enable data-driven decision-making and stakeholder engagement. By simulating different scenarios and assessing their potential impacts, cities can design more sustainable and resilient urban environments.

5.3 Transportation Planning and Management

Public Transit and Active Mobility Investing in public transit infrastructure and promoting active transportation modes are key strategies for reducing air pollution from transportation. High-capacity transit systems such as buses, trams, and metro rail networks provide efficient, affordable alternatives to private car travel [130]. Supporting infrastructure such as bike lanes, pedestrian paths, and bike-sharing schemes encourages walking and cycling for short trips, further reducing vehicle emissions.

Transit-Oriented Development Transit-oriented development (TOD) principles promote compact, mixed-use development around transit nodes, encouraging higher-density, walkable neighborhoods. By locating housing, employment centers, and amenities within walking distance of public transit stations, cities can reduce automobile dependency and promote sustainable travel patterns. TOD also supports economic development, social equity, and community well-being.

Sustainable Freight and Logistics Sustainable freight and logistics practices help minimize emissions from the movement of goods within urban areas. Strategies such as consolidation centers, last-mile delivery solutions, and electric and hybrid vehicles reduce the environmental impact of freight transport. Encouraging off-peak deliveries, optimizing routing, and implementing low-emission zones further contribute to air quality improvement and congestion reduction.

Sustainable urban planning plays a critical role in mitigating air pollution and creating healthier, more livable cities. By integrating green infrastructure and urban design, smart city solutions, and transportation planning and management strategies, cities can reduce emissions, improve air quality, and enhance the quality of life for residents. Policymakers, urban planners, and stakeholders must collaborate to implement sustainable practices and achieve cleaner, more sustainable urban environments for present and future generations. Some cases of best practices include Curitiba, Brazil, which is renowned for its innovative transit system, known as the Bus Rapid Transit (BRT) system, which prioritizes public transportation over private cars [131, 132]. The city's integrated land-use and transportation planning approach, combined with pedestrian-oriented design and green spaces, has made it a model for sustainable urban development. Another example is Copenhagen, Denmark, which is a global leader in sustainable transportation, with a comprehensive network of cycling infrastructure, pedestrian-friendly streets, and efficient public transit. The city's commitment to active mobility, combined with green initiatives such as green roofs, district heating, and renewable energy, has helped reduce air pollution and improve overall quality of life. In the same way, Portland, OR, USA, is known for its urban growth boundary (UGB) policy, which limits urban sprawl and encourages compact, infill development [133]. The city's emphasis on mixed-use neighborhoods, public transit, and bicycle infrastructure has resulted in high levels of transit ridership, cycling, and walking, reducing traffic congestion and greenhouse gas emissions.

6 Sustainable Industrial and Agricultural Practices for Air Pollution Mitigation

Industrial and agricultural activities are significant contributors to air pollution, emitting various pollutants that pose risks to human health and the environment. As the global population continues to grow, the demand for industrial products and agricultural goods is increasing, putting pressure on natural resources and exacerbating air pollution problems [134]. Sustainable practices in these sectors are essential for mitigating air pollution, reducing greenhouse gas emissions, and promoting environmental sustainability.

Industrial and agricultural activities release a wide range of pollutants into the atmosphere, including PM, NOx, SO_2, VOCs, NH_3, and CH_4. These pollutants can originate from various sources, such as manufacturing processes, combustion of fossil fuels, agricultural operations, and livestock farming. The health and environmental impacts of air pollution from industrial and agricultural sources are significant, contributing to respiratory diseases, cardiovascular problems, ecosystem degradation, and climate change. Some sustainable industrial practices for air pollution mitigation include.

Pollution Prevention and Control Technologies The adoption of pollution prevention measures and advanced control technologies is crucial for reducing emissions from industrial sources [135]. These technologies include scrubbers, filters, electrostatic precipitators, and catalytic converters, which capture pollutants before they are released into the atmosphere. Implementing the best available techniques (BAT) and best environmental practices (BEP) ensures that industrial facilities operate in compliance with air quality regulations and minimize their environmental footprint. Both BAT and BEP are relevant for setting air pollution emission standards.

Energy Efficiency and Renewable Energy Improving energy efficiency is essential for reducing greenhouse gas emissions and air pollutants associated with energy production. Industrial facilities can implement energy-saving measures such as upgrading equipment, optimizing processes, and implementing energy management systems. Transitioning to renewable energy sources such as solar, wind, and biomass can further reduce emissions and dependence on fossil fuels, leading to cleaner air and a more sustainable energy future.

Sustainable Supply Chain Management Sustainable supply chain management focuses on minimizing environmental impacts throughout the entire product lifecycle, from raw material extraction to manufacturing, distribution, and disposal [136]. By promoting resource efficiency, waste reduction, and recycling, companies can reduce emissions associated with production processes and minimize their environmental footprint. Collaborating with suppliers, customers, and other stakeholders to implement sustainable practices can create a ripple effect throughout the supply chain, leading to broader environmental benefits.

Some sustainable agricultural practices for air pollution mitigation include.

Precision Farming and Sustainable Land Management Precision farming techniques such as precision irrigation, variable rate fertilization, and controlled-release pesticides help optimize resource use and minimize environmental impacts [137]. By applying inputs more efficiently, farmers can reduce emissions of greenhouse gases and air pollutants such as ammonia and nitrous oxide. Sustainable land management practices such as conservation tillage, crop rotation, and agroforestry enhance soil health, reduce erosion, and sequester carbon, further mitigating air pollution and climate change [138].

Nutrient Management and Emissions Reduction Improving nutrient management practices is essential for reducing emissions of ammonia and nitrous oxide from agricultural activities. Strategies such as optimizing fertilizer application rates, timing, and placement, incorporating organic amendments, and implementing cover cropping can help minimize nutrient losses and greenhouse gas emissions [139]. Livestock management practices such as manure management, dietary modifications, and methane capture technologies also play a critical role in reducing emissions from animal agriculture.

Agroecological Approaches and Biodiversity Conservation Agroecological approaches promote ecological balance, biodiversity conservation, and sustainable farming practices. By enhancing ecosystem services such as natural pest control, soil fertility, and pollination, agroecological farming reduces the need for synthetic inputs and minimizes environmental impacts. Diverse cropping systems, mixed cropping, and integrated pest management (IPM) reduce reliance on chemical inputs and promote ecological resilience, contributing to cleaner air and healthier ecosystems.

Benefits of Sustainable Industrial and Agricultural Practices:

- Reduced emissions: Sustainable practices in industrial and agricultural sectors lead to reduced emissions of air pollutants and greenhouse gases, resulting in improved air quality and public health.
- Resource efficiency: By optimizing resource use and minimizing waste, sustainable practices contribute to resource efficiency and environmental conservation.
- Climate change mitigation: Sustainable practices help mitigate climate change by reducing emissions of greenhouse gases such as CO_2, CH_4, and nitrous oxide (N_2O).
- Economic benefits: Adopting sustainable practices can lead to cost savings, increased productivity, and enhanced competitiveness for businesses and farmers.
- Social benefits: Sustainable practices promote social equity, community resilience, and environmental justice by safeguarding public health and preserving natural resources for future generations.

7 Climate Change and Air Pollution Nexus

The interplay between climate change and air pollution is a complex and multifaceted phenomenon with far-reaching implications for human health, ecosystems, and the global environment [140]. Many of the same activities that contribute to climate change, such as burning fossil fuels for energy production, transportation, and industrial processes, also release air pollutants such as particulate matter, nitrogen oxides, sulfur dioxide, and volatile organic compounds. These pollutants not only degrade air quality but also contribute to the formation of greenhouse gases such as carbon dioxide, methane, and tropospheric ozone, exacerbating climate change [88, 141]. Climate change can influence atmospheric conditions, weather patterns, and natural processes that affect air quality. For example, rising temperatures can lead to increased ground-level ozone formation, prolonged wildfire seasons, and changes in precipitation patterns that impact the distribution of air pollutants. Climate-related events such as heatwaves, droughts, and extreme weather events can also exacerbate air pollution episodes and pose additional health risks to vulnerable populations. Feedback loops between climate change and air pollution can create amplifying effects that worsen both phenomena. For example, higher temperatures and increased humidity can accelerate the chemical reactions that produce ground-level ozone and particulate matter, leading to more severe air pollution episodes. Similarly, air pollutants such as black carbon and methane can deposit on snow and ice surfaces, reducing their albedo and accelerating melting, which in turn contributes to further warming and climate change.

Transitioning from fossil fuels to renewable energy sources such as solar, wind, and hydropower not only reduces greenhouse gas emissions but also improves air quality by eliminating pollution from combustion processes [142]. By investing in clean energy technologies and infrastructure, countries can simultaneously address climate change and air pollution, leading to significant public health benefits and environmental improvements. Promoting sustainable transportation modes such as public transit, walking, cycling, and electric vehicles reduces emissions of greenhouse gases and air pollutants from the transportation sector [143, 144]. By investing in efficient public transit systems, expanding cycling infrastructure, and incentivizing the adoption of zero-emission vehicles, cities can reduce traffic congestion, improve air quality, and enhance urban livability while mitigating climate change. Afforestation efforts and the expansion of green infrastructure, such as urban parks, green roofs, and tree planting initiatives, help sequester carbon dioxide from the atmosphere and mitigate climate change [145]. Additionally, green spaces provide ecosystem services such as air purification, heat island mitigation, and stormwater management, which contribute to improved air quality and human well-being. Building climate-resilient infrastructure that can withstand the impacts of climate change, such as extreme weather events, sea-level rise, and heat waves, is essential for protecting public health and minimizing air pollution risks. Incorporating green infrastructure features such as permeable surfaces, green roofs, and flood-resistant designs can enhance resilience and reduce vulnerability to climate-related hazards. Implementing health adaptation strategies that address the impacts of climate change on air quality and public health is critical for protecting vulnerable populations. These strategies may include heatwave preparedness plans, early warning systems for air pollution episodes, and targeted interventions to reduce exposure to indoor and outdoor pollutants. Enhancing healthcare capacity, community outreach, and public education can also improve resilience and adaptive capacity. Ecosystem-based adaptation approaches harness the protective and regulatory functions of natural ecosystems to enhance resilience and mitigate air pollution risks. Protecting and restoring ecosystems such as forests, wetlands, and coastal habitats can provide multiple benefits, including air purification, carbon sequestration, and natural flood control [146]. Integrating ecosystem-based solutions into climate adaptation strategies can enhance adaptive capacity and promote sustainable development.

8 Economic and Social Considerations of Mitigating Air Pollution

Mitigating air pollution is not only an environmental imperative but also a matter of economic and social importance. The detrimental impacts of air pollution on public health, ecosystems, and the economy underscore the urgent need for mitigation measures. However, addressing air pollution entails costs and benefits, raises issues of social equity and environmental justice, and presents economic opportunities in green industries [88].

Air pollution imposes significant economic costs on society, including healthcare expenditures, productivity losses, and environmental damages [147]. The World Bank estimates that the global cost of air pollution-related health issues exceeds $8.1 trillion annually, representing a substantial burden on healthcare systems and economies worldwide [148]. Implementing mitigation measures to reduce air pollution entails upfront costs, including investments in cleaner technologies, infrastructure upgrades, and regulatory compliance [98]. For industries and businesses, these costs may include retrofitting existing facilities, transitioning to cleaner fuels, and implementing pollution control measures, which can require significant capital investments. Despite the costs associated with mitigation measures, there are substantial health benefits to be gained from improving air quality. Studies have shown that reducing air pollution levels can lead to fewer cases of respiratory and cardiovascular diseases, fewer hospital admissions, and lower mortality rates, resulting in savings in healthcare expenditures and improved overall productivity [149, 150].

Cleaner air leads to economic benefits beyond healthcare savings, including increased labor productivity, reduced absenteeism, and enhanced quality of life. Improved air quality can also attract investment, spur economic growth, and create jobs in sectors such as renewable energy, clean transportation, and green infrastructure development [151, 152]. Air pollution disproportionately affects vulnerable communities, including low-income neighborhoods, minority populations, and marginalized groups. These communities often bear a higher burden of pollution due to proximity to industrial facilities, highways, and other pollution sources, leading to health disparities and social inequities. Addressing air pollution raises important environmental justice considerations, as disadvantaged communities may lack access to clean air, green spaces, and adequate healthcare services. Ensuring equitable distribution of the benefits and burdens of pollution control measures is essential for promoting social justice and protecting the rights of vulnerable populations [153]. Meaningful community engagement and participation are essential for addressing environmental justice concerns and ensuring that mitigation measures are responsive to local needs and priorities. Empowering communities to participate in decision-making processes, advocate for their rights, and access information and resources fosters social inclusion and promotes environmental justice.

The transition to clean energy sources such as solar, wind, and hydroelectric power presents significant economic opportunities for job creation, innovation, and sustainable development [77, 154]. Renewable energy industries have experienced rapid growth in recent years, driven by declining costs, supportive policies, and increasing demand for clean energy solutions. Investing in green infrastructure and technologies, such as energy-efficient buildings, public transportation systems, and sustainable urban development projects, can stimulate economic growth and promote environmental sustainability. Green industries offer employment opportunities across various sectors, including construction, manufacturing, and engineering, while reducing greenhouse gas emissions and improving air quality. Adopting sustainable practices in agriculture and forestry can contribute to air pollution mitigation efforts while generating economic benefits for rural communities. Practices such as organic farming, agroforestry, and land restoration enhance soil health, sequester carbon, and reduce emissions from agricultural activities, supporting livelihoods and enhancing resilience to climate change.

9 Asian Perspectives of Sustainable Air Pollution Mitigation Strategies

Urbanization has surged globally since World War II, concentrating populations in cities, with over half of the world's population now residing in urban areas. East Asia particularly stands out, hosting 16 megacities, predominantly in South Korea, China, and Japan [155]. This urbanization process, rooted in developmentalism, emphasizes industrialization and economic growth, leading to rapid and intensive growth of megacities. However, this compressed growth has resulted in environmental concerns, particularly air pollution, which is a significant challenge for megacities in East Asia [156, 9, 157].

According to the 2023 World Air Quality Report by IQAir, 19 of the top 100 cities with severe air quality issues regarding PM2.5, the particles with a diameter of 2.5 micrometers or less (e.g., black carbon and dust), are in East Asia, including 12 megacities [158]. These megacities heavily rely on fossil fuel-based energy, emitting substantial air pollutants. The emissions impact not only their urban environments but also neighboring regions and have global ramifications. Various air pollutants, such as NO_x, SO_x, O_3, and CH_4, contribute to climate change and threaten human health and the environment [159]. Efforts to reduce air pollution in East Asia focus on international collaboration, strategies for mitigating climate change related to air pollution, and case studies highlighting air pollution reduction strategies. These endeavors aim to address the interconnected challenges of air pollution and climate change, emphasizing the importance of comprehensive strategies and citizen participation in achieving sustainable urban environments [160–163].

In Northeast Asia, the prevailing winds and seasonal monsoons cause air pollutants to shift eastward. For example, yellow dust from northwest China can travel westward on the wind, affecting air quality in Korea and Japan, as both countries are located to the east of China. Collaborative efforts have been undertaken to address transboundary air pollution, exemplified by initiatives like the Long-range Transboundary Air Pollution Research Project and the Tripartite Environment Ministers Meeting. Additionally, the North-East Asia Clean Air Partnership and the East Asia Clean Air Cities Network aim to collate data, conduct research, and formulate policies to combat air pollution collectively [164, 165]. On top of the air pollution issues in the region, and the complex relationship between air pollution and climate change necessitates comprehensive strategies. Rising temperatures due to climate change sometimes disrupt atmospheric flow, causing the atmosphere to linger over certain areas for a longer time than usual. This can lead to an increase in PM2.5 levels in those areas. Therefore, policies aimed to address both air pollution and climate change simultaneously are essential [166, 167].

In East Asian countries, efforts to create low-carbon cities have gained momentum. China adopts a bottom-up approach, incentivizing cities to reduce coal consumption and transition to renewable energy sources. Japan aims for nationwide carbon neutrality by 2050, with coastal cities prioritizing renewable energy expansion. In Tokyo, comprehensive policies across multiple sectors target carbon reduction [168, 169]. In South Korea, Seoul exemplifies efforts to combat air pollution and mitigate climate change. Emergency measures and seasonal control systems target PM2.5 reduction, while the Clear Seoul 2010 program promotes eco-friendly transportation. Policies include restrictions on vehicles emitting pollutants, incentives for eco-friendly transportation, and initiatives to distribute eco-friendly boilers. Citizen engagement is fostered through online events and awareness campaigns, highlighting the importance of individual action in achieving sustainable urban environments [168].

10 Challenges and Opportunities in Mitigating Air Pollution

Mitigating air pollution is a pressing global challenge with profound implications for public health, environmental sustainability, and economic development. Although significant progress has been made in recognizing the urgency of addressing air pollution, various barriers hinder the effective implementation of mitigation measures [170]. However, amidst these challenges lie opportunities for innovation, collaboration, and transformative change. Inadequate or outdated regulatory frameworks, lax enforcement mechanisms, and competing political priorities pose significant barriers to implementing effective air pollution control measures [171, 172]. Policymakers may face resistance from vested interests, industry lobbyists, and short-term economic considerations, hindering the adoption of stringent regulations and emission standards [173]. The adoption of cleaner technologies and pollution control measures often requires significant investments in research, development, and infrastructure upgrades. However, many developing countries and resource-constrained communities lack the technical expertise, financial resources, and access to clean technologies necessary to address air pollution effectively [174]. High upfront costs and limited access to financing further impede progress in mitigation efforts. Overcoming societal inertia and fostering public awareness and behavioral change are key challenges in addressing air pollution. Misinformation, apathy, and complacency among the public may undermine support for pollution control measures and hinder efforts to promote sustainable lifestyles and consumption patterns. In many cases, individuals and communities may lack awareness of the health risks associated with air pollution and the actions they can take to mitigate its impacts.

The transition to renewable energy sources such as solar, wind, and hydropower presents a significant opportunity for reducing air pollution and mitigating climate change [175]. Advances in clean energy technologies, coupled with declining costs and supportive policies, have accelerated the shift away from fossil fuels and toward low-carbon alternatives. Investments in renewable energy infrastructure and grid modernization are driving innovation and unlocking new opportunities for decarbonizing the economy. Smart city technologies, data analytics, and digital solutions offer innovative approaches to monitoring, managing, and mitigating air pollution. Real-time air quality monitoring networks, sensor technologies, and predictive modeling tools enable authorities to identify pollution hotspots, assess trends, and develop targeted interventions [39, 40, 176]. Smart transportation systems, urban planning strategies, and mobility solutions help reduce emissions from vehicles and improve air quality in urban areas.

Nature-based solutions such as afforestation, reforestation, and green infrastructure development provide cost-effective means of mitigating air pollution while enhancing ecosystem resilience and biodiversity [177, 178]. Green spaces, urban forests, and vegetative barriers act as natural filters, capturing pollutants, reducing heat island effects, and improving air quality. Integrating nature-based solutions into urban planning and infrastructure development can yield multiple co-benefits, including climate adaptation, water management, and community well-being.

Enhancing regulatory frameworks, strengthening enforcement mechanisms, and mainstreaming air quality considerations into broader policy agendas are essential for overcoming barriers to implementation. Policymakers must prioritize air pollution control, set ambitious emission reduction targets, and implement comprehensive strategies that address the root causes of pollution while promoting sustainable development and social equity [179]. Mobilizing financial resources, fostering public–private partnerships, and leveraging international cooperation are critical for overcoming technological and financial constraints. Governments, multilateral institutions, and development agencies can support capacity-building efforts, provide technical assistance, and facilitate technology transfer to help countries transition to cleaner, more sustainable energy systems and industries. Empowering communities, engaging stakeholders, and raising public awareness are essential for driving behavioral change and fostering grassroots support for air pollution mitigation efforts. Education campaigns, public outreach initiatives, and community-based monitoring programs can empower individuals and communities to act, advocate for clean air policies, and adopt sustainable lifestyles and consumption patterns [180].

11 Conclusion

Air pollution poses significant threats to human health, environmental sustainability, and economic development, underscoring the urgent need for action. Addressing air pollution entails costs and challenges with substantial benefits to be gained from investing in clean technologies, promoting behavioral changes, and fostering public awareness. Importantly, a multifaceted approach that integrates policy interventions, technological innovations, community engagement, and international cooperation is essential for achieving sustainable air quality improvement.

There is a need for action from all stakeholders, including governments, industries, communities, and individuals, to prioritize sustainable air quality improvement. This requires ambitious policy commitments, robust regulatory frameworks, and targeted investments in clean energy, transportation, and urban planning. Additionally, fostering public awareness, promoting behavioral changes, and empowering communities to participate in decision-making processes are critical for driving collective action and fostering a culture of environmental stewardship.

Looking ahead, it is imperative to continue advancing efforts to mitigate air pollution and promote sustainable development. This includes scaling up successful initiatives, leveraging emerging technologies, and strengthening international cooperation to address transboundary pollution challenges. Moreover, integrating air quality considerations into broader sustainability agendas, such as the Sustainable Development Goals, can help mainstream air pollution mitigation efforts and maximize co-benefits across multiple sectors. Achieving sustainable air quality improvement requires concerted efforts, collaboration, and innovation at all levels. By adopting holistic strategies, embracing the principles of environmental justice and social equity, and prioritizing the well-being of present and future generations, we can create a healthier, more sustainable world for all.

References

1. Hamanaka RB, Mutlu GM (2018) Particulate matter air pollution: effects on the cardiovascular system. Front Endocrinol 9:680. https://doi.org/10.3389/fendo.2018.00680
2. Lee BJ, Kim B, Lee K (2014) Air pollution exposure and cardiovascular disease. Toxicol Res 30(2):71–75. https://doi.org/10.5487/TR.2014.30.2.071
3. Limaye VS, Magal A, Joshi J, Maji S, Dutta P, Rajput P, Pingle S, Madan P, Mukerjee P, Bano S, Beig G, Mavalankar D, Jaiswal A, Knowlton K (2023) Air quality and health co-benefits of climate change mitigation and adaptation actions by 2030: an interdisciplinary modeling study in Ahmedabad, India. Environ Res Health 1(2):021003. https://doi.org/10.1088/2752-5309/aca7d8
4. Awewomom J, Dzeble F, Takyi YD et al (2024) Addressing global environmental pollution using environmental control techniques: a focus on environmental policy and preventive environmental management. Discov Environ 2:8. https://doi.org/10.1007/s44274-024-00033-5

5. Vilcins D, Christofferson RC, Yoon JH, Nazli SN, Sly PD, Cormier SA, Shen G (2024) Updates in air pollution: current research and future challenges. Ann Glob Health 90(1):9. https://doi.org/10.5334/aogh.4363
6. Dondi A, Carbone C, Manieri E, Zama D, Del Bono C, Betti L, Biagi C, Lanari M (2023) Outdoor air pollution and childhood respiratory disease: the role of oxidative stress. Int J Mol Sci 24(5):4345. https://doi.org/10.3390/ijms24054345
7. Schraufnagel DE, Balmes JR, Cowl CT, De Matteis S, Jung SH, Mortimer K, Perez-Padilla R, Rice MB, Riojas-Rodriguez H, Sood A, Thurston GD, To, T, Vanker A, Wuebbles DJ (2019) Air pollution and noncommunicable diseases: a review by the forum of International Respiratory Societies' Environmental Committee, part 2: air pollution and organ systems. Chest 155(2):417–426. https://doi.org/10.1016/j.chest.2018.10.041
8. D'Amato G, Bergmann KC, Cecchi L, Annesi-Maesano I, Sanduzzi A, Liccardi G, Vitale C, Stanziola A, D'Amato M (2014) Climate change and air pollution: effects on pollen allergy and other allergic respiratory diseases. Allergo J Int 23(1):17–23. https://doi.org/10.1007/s40629-014-0003-7
9. Ogwu MC (2019) Towards sustainable development in Africa: the challenge of urbanization and climate change adaptation. In: Cobbinah PB, Addaney M (eds) The geography of climate change adaptation in urban Africa. Palgrave Macmillan, Cham. https://doi.org/10.1007/978-3-030-04873-0_2
10. Jessel S, Sawyer S, Hernández D (2019) Energy, poverty, and health in climate change: a comprehensive review of an emerging literature. Front Public Health 7:357. https://doi.org/10.3389/fpubh.2019.00357
11. Akudugu MA, Ogwu MC (2024) Sustainable development policies and interventions: a bibliometric analysis of the contributions of the academic community. J Clean Prod 434:139919. https://doi.org/10.1016/j.jclepro.2023.139919
12. Filigrana P, Levy JI, Gauthier J, Batterman S, Adar SD (2022) Health benefits from cleaner vehicles and increased active transportation in Seattle, Washington. J Expo Sci Environ Epidemiol 32(4):538–544. https://doi.org/10.1038/s41370-022-00423-y
13. Xia T, Zhang Y, Crabb S, Shah P (2013) Cobenefits of replacing car trips with alternative transportation: a review of evidence and methodological issues. J Environ Public Health 2013:797312. https://doi.org/10.1155/2013/797312
14. Lafrenz AJ (2022) Designing multifunctional urban green spaces: an inclusive public health framework. Int J Environ Res Public Health 19(17):10867. https://doi.org/10.3390/ijerph191710867
15. Carlsten C, Salvi S, Wong GWK, Chung KF (2020) Personal strategies to minimise effects of air pollution on respiratory health: advice for providers, patients and the public. Eur Respir J 55(6):1902056. https://doi.org/10.1183/13993003.02056-2019
16. Kelly FJ, Fussell JC (2015) Air pollution and public health: emerging hazards and improved understanding of risk. Environ Geochem Health 37(4):631–649. https://doi.org/10.1007/s10653-015-9720-1
17. McCarron A, Semple S, Braban CF, Swanson V, Gillespie C, Price HD (2023) Public engagement with air quality data: using health behaviour change theory to support exposure-minimising behaviours. J Expo Sci Environ Epidemiol 33(3):321–331. https://doi.org/10.1038/s41370-022-00449-2
18. Ramírez AS, Ramondt S, Van Bogart K, Perez-Zuniga R (2019) Public awareness of air pollution and health threats: challenges and opportunities for communication strategies to improve environmental health literacy. J Health Commun 24(1):75–83. https://doi.org/10.1080/10810730.2019.1574320
19. Mailloux NA, Henegan CP, Lsoto D, Patterson KP, West PC, Foley JA, Patz JA (2021) Climate solutions double as health interventions. Int J Environ Res Public Health 18(24):13339. https://doi.org/10.3390/ijerph182413339

20. Panepinto D, Riggio VA, Zanetti M (2021) Analysis of the emergent climate change mitigation technologies. Int J Environ Res Public Health 18(13):6767. https://doi.org/10.3390/ijerph18136767
21. Expert Consensus Task Force, Shi X, Duan G (2022) Recommendations of controlling and preventing acute health risks of fine particulate matter pollution – China, 2021. China CDC Weekly 4(16):329–341. https://doi.org/10.46234/ccdcw2022.078
22. Xia Y, Guo H, Xu S, Pan C (2024) Environmental regulations and agricultural carbon emissions efficiency: evidence from rural China. Heliyon 10(4):e25677. https://doi.org/10.1016/j.heliyon.2024.e25677
23. Berretta M, Furgeson J, Wu YN, Zamawe C, Hamilton I, Eyers J (2021) Residential energy efficiency interventions: a meta-analysis of effectiveness studies. Campbell Syst Rev 17(4): e1206. https://doi.org/10.1002/cl2.1206
24. Di Foggia G (2018) Energy efficiency measures in buildings for achieving sustainable development goals. Heliyon 4(11):e00953. https://doi.org/10.1016/j.heliyon.2018.e00953
25. Gillingham KT, Huang P, Buehler C, Peccia J, Gentner DR (2021) The climate and health benefits from intensive building energy efficiency improvements. Sci Adv 7(34):eabg0947. https://doi.org/10.1126/sciadv.abg0947
26. Liang L, Gong P (2020) Urban and air pollution: a multi-city study of long-term effects of urban landscape patterns on air quality trends. Sci Rep 10(1):18618. https://doi.org/10.1038/s41598-020-74524-9
27. Costa D (2011) Air quality in a changing climate. Environ Health Perspect 119(4):A154–A155. https://doi.org/10.1289/ehp.1103649
28. Kjellstrom T, Lodh M, McMichael T et al (2006) Air and water pollution: burden and strategies for control. In: Jamison DT, Breman JG, Measham AR et al (eds) Disease control priorities in developing countries2nd edn. The International Bank for Reconstruction and Development/The World Bank, Washington. Chapter 43. Available from: https://www.ncbi.nlm.nih.gov/books/NBK11769/Co-published
29. Yamineva Y, Liu Z (2019) Cleaning the air, protecting the climate: policy, legal and institutional nexus to reduce black carbon emissions in China. Environ Sci Policy 95:1–10. https://doi.org/10.1016/j.envsci.2019.01.016
30. Brauer M, Davaakhuu N, Escamilla Nuñez MC, Hadley M, Kass D, Miller M, Prabhakaran D, Sliwa K, Su TC, Vaartjes ICH, Vedanthan R, Mwangi J, Armstrong-Walenczak K (2021) Clean air, smart cities, healthy hearts: action on air pollution for cardiovascular health. Glob Heart 16(1):61. https://doi.org/10.5334/gh.1073
31. Bălă GP, Râjnoveanu RM, Tudorache E, Motișan R, Oancea C (2021) Air pollution exposure-the (in)visible risk factor for respiratory diseases. Environ Sci Pollut Res Int 28(16): 19615–19628. https://doi.org/10.1007/s11356-021-13208-x
32. Duan RR, Hao K, Yang T (2020) Air pollution and chronic obstructive pulmonary disease. Chronic Dis Transl Med 6(4):260–269. https://doi.org/10.1016/j.cdtm.2020.05.004
33. Luo H, Zhang Q, Yu K, Meng X, Kan H, Chen R (2022) Long-term exposure to ambient air pollution is a risk factor for trajectory of cardiometabolic multimorbidity: a prospective study in the UK Biobank. EBioMedicine 84:104282. https://doi.org/10.1016/j.ebiom.2022.104282
34. Li X, Lyu Y, Dong W, Xu A (2023) Exploring the relationship between air quality and health shocks to the elderly: a retrospective cross-sectional study in China. Front Public Health 11: 1087626. https://doi.org/10.3389/fpubh.2023.1087626
35. De Marco A, Sicard P, Feng Z, Agathokleous E, Alonso R, Araminiene V, Augustatis A, Badea O, Beasley JC, Branquinho C, Bruckman VJ, Collalti A, David-Schwartz R, Domingos M, Du E, Garcia Gomez H, Hashimoto S, Hoshika Y, Jakovljevic T, McNulty S et al (2022) Strategic roadmap to assess forest vulnerability under air pollution and climate change. Glob Chang Biol 28(17):5062–5085. https://doi.org/10.1111/gcb.16278
36. Grennfelt P, Engleryd A, Forsius M, Hov Ø, Rodhe H, Cowling E (2020) Acid rain and air pollution: 50 years of progress in environmental science and policy. Ambio 49(4):849–864. https://doi.org/10.1007/s13280-019-01244-4

37. Shin YJ, Midgley GF, Archer ERM, Arneth A, Barnes DKA, Chan L, Hashimoto S, Hoegh-Guldberg O, Insarov G, Leadley P, Levin LA, Ngo HT, Pandit R, Pires APF, Pörtner HO, Rogers AD, Scholes RJ, Settele J, Smith P (2022) Actions to halt biodiversity loss generally benefit the climate. Glob Chang Biol 28(9):2846–2874. https://doi.org/10.1111/gcb.16109
38. Cromar KR, Duncan BN, Bartonova A, Benedict K, Brauer M, Habre R, Hagler GSW, Haynes JA, Khan S, Kilaru V, Liu Y, Pawson S, Peden DB, Quint JK, Rice MB, Sasser EN, Seto E, Stone SL, Thurston GD, Volckens J (2019) Air pollution monitoring for health research and patient care. An official American Thoracic Society workshop report. Ann Am Thorac Soc 16(10):1207–1214. https://doi.org/10.1513/AnnalsATS.201906-477ST
39. Saini J, Dutta M, Marques G (2020) A comprehensive review on indoor air quality monitoring systems for enhanced public health. Sustain Environ Res 30:6. https://doi.org/10.1186/s42834-020-0047-y
40. Saini J, Dutta M, Marques G (2020) Indoor air quality monitoring systems based on internet of things: a systematic review. Int J Environ Res Public Health 17(14):4942. https://doi.org/10.3390/ijerph17144942
41. Liu C, Xin L, Li J (2022) Environmental regulation and manufacturing carbon emissions in China: a new perspective on local government competition. Environ Sci Pollut Res Int 29(24):36351–36375. https://doi.org/10.1007/s11356-021-18041-w
42. Liu Y, Tang L, Liu G (2022) Carbon dioxide emissions reduction through technological innovation: empirical evidence from Chinese provinces. Int J Environ Res Public Health 19(15):9543. https://doi.org/10.3390/ijerph19159543
43. Gallagher CL, Holloway T (2020) Integrating air quality and public health benefits in U.-S. Decarbonization strategies. Front Public Health 8:563358. https://doi.org/10.3389/fpubh.2020.563358
44. Xie Y, Xu M, Pu J, Pan Y, Liu X, Zhang Y, Xu S (2023) Large-scale renewable energy brings regionally disproportional air quality and health co-benefits in China. iScience 26(8):107459. https://doi.org/10.1016/j.isci.2023.107459
45. Hewitt CN, Ashworth K, MacKenzie AR (2020) Using green infrastructure to improve urban air quality (GI4AQ). Ambio 49(1):62–73. https://doi.org/10.1007/s13280-019-01164-3
46. Jennings V, Reid CE, Fuller CH (2021) Green infrastructure can limit but not solve air pollution injustice. Nat Commun 12(1):4681. https://doi.org/10.1038/s41467-021-24892-1
47. Carey J (2023) The other benefit of electric vehicles. Proc Natl Acad Sci U S A 120(3):e2220923120. https://doi.org/10.1073/pnas.2220923120
48. Anuardo RG, Espuny M, Costa ACF, Oliveira OJ (2022) Toward a cleaner and more sustainable world: a framework to develop and improve waste management through organizations, governments, and academia. Heliyon 8(4):e09225. https://doi.org/10.1016/j.heliyon.2022.e09225
49. Tanveer M, Khan SAR, Umar M, Yu Z, Sajid MJ, Haq IU (2022) Waste management and green technology: future trends in circular economy leading towards environmental sustainability. Environ Sci Pollut Res Int 29(53):80161–80178. https://doi.org/10.1007/s11356-022-23238-8
50. Nassikas NJ, McCormack MC, Ewart G, Balmes JR, Bond TC, Brigham E, Cromar K, Goldstein AH, Hicks A, Hopke PK, Meyer B, Nazaroff WW, Paulin LM, Rice MB, Thurston GD, Turpin BJ, Vance ME, Weschler CJ, Zhang J, Kipen HM (2024) Indoor air sources of outdoor air pollution: health consequences, policy, and recommendations: an official American Thoracic Society workshop report. Ann Am Thorac Soc 21(3):365–376. https://doi.org/10.1513/AnnalsATS.202312-1067ST
51. Toma C, Alexandru A, Popa M, Zamfiroiu A (2019) IoT solution for smart cities' pollution monitoring and the security challenges. Sensors 19(15):3401. https://doi.org/10.3390/s19153401
52. Wang L, Xie Q, Xue F, Li Z (2022) Does smart city construction reduce haze pollution? Int J Environ Res Public Health 19(24):16421. https://doi.org/10.3390/ijerph192416421

53. Neo EX, Hasikin K, Lai KW, Mokhtar MI, Azizan MM, Hizaddin HF, Razak SA, Yanto (2023) Artificial intelligence-assisted air quality monitoring for smart city management. PeerJ Comput Sci 9:e1306. https://doi.org/10.7717/peerj-cs.1306
54. Chen WY, Wu ZX, Wang WZ (2007) Huan jing ke xue. Huanjing kexue 28(6):1178–1182
55. Venter ZS, Hassani A, Stange E, Schneider P, Castell N (2024) Reassessing the role of urban green space in air pollution control. Proc Natl Acad Sci U S A 121(6):e2306200121. https://doi.org/10.1073/pnas.2306200121
56. Casey JA, Daouda M, Babadi RS, Do V, Flores NM, Berzansky I, González DJX, Van Horne YO, James-Todd T (2023) Methods in public health environmental justice research: a scoping review from 2018 to 2021. Curr Environ Health Rep 10(3):312–336. https://doi.org/10.1007/s40572-023-00406-7
57. Rigolon A, Browning MHEM, McAnirlin O, Yoon HV (2021) Green space and health equity: a systematic review on the potential of Green space to reduce health disparities. Int J Environ Res Public Health 18(5):2563. https://doi.org/10.3390/ijerph18052563
58. Cenci S, Burato M, Rei M, Zollo M (2023) The alignment of companies' sustainability behavior and emissions with global climate targets. Nat Commun 14(1):7831. https://doi.org/10.1038/s41467-023-43116-2
59. Erickson LE (2017) Reducing greenhouse gas emissions and improving air quality: two global challenges. Environ Prog Sustain Energy 36(4):982–988. https://doi.org/10.1002/ep.12665
60. Ma N, Liu P, Xiao Y, Tang H, Zhang J (2022) Can Green technological innovation reduce hazardous air pollutants?-An empirical test based on 283 cities in China. Int J Environ Res Public Health 19(3):1611. https://doi.org/10.3390/ijerph19031611
61. Ma Y, Wei X, Yan G, He X (2023) The impact of Fintech development on air pollution. Int J Environ Res Public Health 20(4):3387. https://doi.org/10.3390/ijerph20043387
62. Ou Y, West JJ, Smith SJ, Nolte CG, Loughlin DH (2020) Air pollution control strategies directly limiting national health damages in the US. Nat Commun 11(1):957. https://doi.org/10.1038/s41467-020-14783-2
63. Zikargae MH, Woldearegay AG, Skjerdal T (2022) Assessing the roles of stakeholders in community projects on environmental security and livelihood of impoverished rural society: a nongovernmental organization implementation strategy in focus. Heliyon 8(10):e10987. https://doi.org/10.1016/j.heliyon.2022.e10987
64. Mahadevia D, Mukhopadhyay C, Lathia S, Gounder K (2023) The role of urban transport in delivering sustainable development goal 11: learning from two Indian cities. Heliyon 9(9):e19453. https://doi.org/10.1016/j.heliyon.2023.e19453
65. Pisoni E, Christidis P, Thunis P, Trombetti M (2019) Evaluating the impact of "Sustainable Urban Mobility Plans" on urban background air quality. J Environ Manage 231:249–255. https://doi.org/10.1016/j.jenvman.2018.10.039
66. Slovic AD, de Oliveira MA, Biehl J, Ribeiro H (2016) How can urban policies improve air quality and help mitigate global climate change: a systematic mapping review. J Urban Health 93(1):73–95. https://doi.org/10.1007/s11524-015-0007-8
67. Davis LF, Ramírez-Andreotta MD (2021) Participatory research for environmental justice: a critical interpretive synthesis. Environ Health Perspect 129(2):26001. https://doi.org/10.1289/EHP6274
68. Mahajan S, Chung MK, Martinez J et al (2022) Translating citizen-generated air quality data into evidence for shaping policy. Humanit Soc Sci Commun 9:122. https://doi.org/10.1057/s41599-022-01135-2
69. Ross K, Chmiel JF, Ferkol T (2012) The impact of the clean air act. J Pediatr 161(5):781–786. https://doi.org/10.1016/j.jpeds.2012.06.064
70. Singh S, Kulshrestha MJ, Rani N, Kumar K, Sharma C, Aswal DK (2023) An overview of vehicular emission standards. Mapan 38(1):241–263. https://doi.org/10.1007/s12647-022-00555-4

71. Congressional Research Service (2022) Clean air act: a summary of the act and its major requirements (RL30853). Congressional Research Service, p. 24. https://crsreports.congress.gov/product/pdf/RL/RL30853
72. Hu S (2022, October 21) The Clean Air Act 101 [Non-profit organization]. The Clean Air Act 101. https://www.nrdc.org/stories/clean-air-act-101
73. Butler TJ, Likens GE, Stunder BJB (2001) Regional-scale impacts of phase I of the clean air act amendments in the USA: the relation between emissions and concentrations, both wet and dry. Atmos Environ 35(6):1015–1028. https://doi.org/10.1016/S1352-2310(00)00386-1
74. Hand JL, Prenni AJ, Copeland S, Schichtel BA, Malm WC (2020) Thirty years of the clean air act amendments: impacts on haze in remote regions of the United States (1990–2018). Atmos Environ 243:117865. https://doi.org/10.1016/j.atmosenv.2020.117865
75. Chestnut LG, Dennis RL (1997) Economic benefits of improvements in visibility: acid rain provisions of the 1990 clean air act amendments. J Air Waste Manage Assoc 47(3):395–402. https://doi.org/10.1080/10473289.1997.10464437
76. Ribeiro CB, Rodella FHC, Hoinaski L (2022) Regulating light-duty vehicle emissions: an overview of US, EU, China and Brazil programs and its effect on air quality. Clean Techn Environ Policy 24(3):851–862. https://doi.org/10.1007/s10098-021-02238-1
77. Akash FA, Shovon SM, Rahman W, Rahman MA, Chakraborty P, Monir MU (2024) Greening the grid: a comprehensive review of renewable energy in Bangladesh. Heliyon 10(5):e27477. https://doi.org/10.1016/j.heliyon.2024.e27477
78. Mundorf N, Redding CA, Paiva AL (2018) Sustainable transportation attitudes and health behavior change: evaluation of a brief stage-targeted video intervention. Int J Environ Res Public Health 15(1):150. https://doi.org/10.3390/ijerph15010150
79. Winkler L, Pearce D, Nelson J, Babacan O (2023) The effect of sustainable mobility transition policies on cumulative urban transport emissions and energy demand. Nat Commun 14(1): 2357. https://doi.org/10.1038/s41467-023-37728-x
80. RGGI (2024) The regional greenhouse gas initiative: an initiative of eastern states of the US. Regional Greenhouse Gas Initiative. https://www.rggi.org/sites/default/files/Uploads/Fact%20Sheets/RGGI_101_Factsheet.pdf
81. Morris RH, Counsell SJ, McGonnell IM, Thornton C (2021) Early life exposure to air pollution impacts neuronal and glial cell function leading to impaired neurodevelopment. BioEssays 43(9):2000288. https://doi.org/10.1002/bies.202000288
82. Pinder RW, Klopp JM, Kleiman G, Hagler GSW, Awe Y, Terry S (2019) Opportunities and challenges for filling the air quality data gap in low- and middle-income countries. Atmos Environ 215:116794. https://doi.org/10.1016/j.atmosenv.2019.06.032
83. Shairsingh K, Ruggeri G, Krzyzanowski M, Mudu P, Malkawi M, Castillo J, Soares da Silva A, Saluja M, Martínez KC, Mothe J, Gumy S (2023) WHO air quality database: relevance, history and future developments. Bull World Health Organ 101(12):800–807. https://doi.org/10.2471/BLT.23.290188
84. Markandya A, Sampedro J, Smith SJ, Van Dingenen R, Pizarro-Irizar C, Arto I, González-Eguino M (2018) Health co-benefits from air pollution and mitigation costs of the Paris agreement: a modelling study. Lancet Planet Health 2(3):e126–e133. https://doi.org/10.1016/S2542-5196(18)30029-9
85. Hamilton I, Kennard H, McGushin A, Höglund-Isaksson L, Kiesewetter G, Lott M, Milner J, Purohit P, Rafaj P, Sharma R, Springmann M, Woodcock J, Watts N (2021) The public health implications of the Paris agreement: a modelling study. Lancet Planetary Health 5(2):e74–e83. https://doi.org/10.1016/S2542-5196(20)30249-7
86. Andersen SO, Gao S, Carvalho S, Ferris T, Gonzalez M, Sherman NJ, Wei Y, Zaelke D (2021) Narrowing feedstock exemptions under the Montreal protocol has multiple environmental benefits. Proc Natl Acad Sci U S A 118(49):e2022668118. https://doi.org/10.1073/pnas.2022668118
87. Barnes PW, Bornman JF, Pandey KK, Bernhard GH, Bais AF, Neale RE, Robson TM, Neale PJ, Williamson CE, Zepp RG, Madronich S, Wilson SR, Andrady AL, Heikkilä AM,

Robinson SA (2021) The success of the montreal protocol in mitigating interactive effects of stratospheric ozone depletion and climate change on the environment. Glob Chang Biol 27(22):5681–5683. https://doi.org/10.1111/gcb.15841
88. Manisalidis I, Stavropoulou E, Stavropoulos A, Bezirtzoglou E (2020) Environmental and health impacts of air pollution: a review. Front Public Health 8:505570. https://doi.org/10.3389/fpubh.2020.00014
89. Laurent É (2022) Air (ine)quality in the European Union. Curr Environ Health Rep 9(2): 123–129. https://doi.org/10.1007/s40572-022-00348-6
90. Boogaard H, Andersen ZJ, Brunekreef B, Forastiere F, Forsberg B, Hoek G, Krzyzanowski M, Malmqvist E, Nieuwenhuijsen M, Hoffmann B (2023) Clean air in Europe for all: a call for more ambitious action. Environ Epidemiol 7(2):e245. https://doi.org/10.1097/EE9.0000000000000245
91. Johnston JE, Juarez Z, Navarro S, Hernandez A, Gutschow W (2019) Youth engaged participatory air monitoring: a 'Day in the Life' in urban environmental justice communities. Int J Environ Res Public Health 17(1):93. https://doi.org/10.3390/ijerph17010093
92. Ward F, Lowther-Payne HJ, Halliday EC, Dooley K, Joseph N, Livesey R, Moran P, Kirby S, Cloke J (2022) Engaging communities in addressing air quality: a scoping review. Environ Health 21(1):89. https://doi.org/10.1186/s12940-022-00896-2
93. World Health Organization (2021) WHO global air quality guidelines: Particulate matter (PM2.5 and PM10), ozone, nitrogen dioxide, sulfur dioxide and carbon monoxide [Internet]. Implementation of the guidelines. World Health Organization, Geneva, p 6. Available from: https://www.ncbi.nlm.nih.gov/books/NBK574587/
94. Wang M, Liao G, Li Y (2021) The relationship between environmental regulation, pollution and corporate environmental responsibility. Int J Environ Res Public Health 18(15):8018. https://doi.org/10.3390/ijerph18158018
95. Stevens KA, Bryer TA, Yu H (2021) Air quality enhancement districts: democratizing data to improve respiratory health. J Environ Stud Sci 11(4):702–707. https://doi.org/10.1007/s13412-021-00670-9
96. Morawska L, Thai PK, Liu X, Asumadu-Sakyi A, Ayoko G, Bartonova A, Bedini A, Chai F, Christensen B, Dunbabin M, Gao J, Hagler GSW, Jayaratne R, Kumar P, Lau AKH, Louie PKK, Mazaheri M, Ning Z, Motta N, Mullins B et al (2018) Applications of low-cost sensing technologies for air quality monitoring and exposure assessment: how far have they gone? Environ Int 116:286–299. https://doi.org/10.1016/j.envint.2018.04.018
97. Bistline JET, Blanford G, Grant J, Knipping E, McCollum DL, Nopmongcol U, Scarth H, Shah T, Yarwood G (2022) Economy-wide evaluation of CO_2 and air quality impacts of electrification in the United States. Nat Commun 13(1):6693. https://doi.org/10.1038/s41467-022-33902-9
98. Woollacott J, Alsufyani W, Beach RH, Morrison LTR, Bean de Hernández A, Rakic S, AlOmran M, Alsukait RF, Herbst CH, AlBalawi S (2022) Effective options for addressing air quality-related environmental public health burdens in Saudi Arabia. Heliyon 8(9):e10335. https://doi.org/10.1016/j.heliyon.2022.e10335
99. Zhang T, Qin H, Xu W (2022) Environmental regulation, greenwashing behaviour, and green governance of high-pollution enterprises in China. Int J Environ Res Public Health 19(19): 12539. https://doi.org/10.3390/ijerph191912539
100. He B, Qi M, Wang N, Zhang Z (2022) Avoiding real harm but false good: the influence mechanism of political relations on the effectiveness of environmental regulation policies. Int J Environ Res Public Health 19(23):15953. https://doi.org/10.3390/ijerph192315953
101. Shen Y, Zhang X (2022) Study on the impact of environmental tax on industrial Green transformation. Int J Environ Res Public Health 19(24):16749. https://doi.org/10.3390/ijerph192416749
102. Craig RK, Garmestani AS, Allen CR, Arnold CAT, Birgé H, DeCaro DA, Fremier AK, Gosnell H, Schlager E (2017) Balancing stability and flexibility in adaptive governance: an

analysis of tools available in U.S. environmental law. Ecol Soc 22(2):1–3. https://doi.org/10.5751/ES-08983-220203
103. Amolegbe SM, Lopez AR, Velasco ML, Carlin DJ, Heacock ML, Henry HF, Trottier BA, Suk WA (2022) Adapting to climate change: leveraging systems-focused multidisciplinary research to promote resilience. Int J Environ Res Public Health 19(22):14674. https://doi.org/10.3390/ijerph192214674
104. Wigand ME, Timmermann C, Scherp A, Becker T, Steger F (2022) Climate change, pollution, deforestation, and mental health: research trends, gaps, and ethical considerations. GeoHealth 6(11):e2022GH000632. https://doi.org/10.1029/2022GH000632
105. Espinosa-Salas S, Gonzalez-Arias M. Behavior Modification for Lifestyle Improvement. [Updated 2023 Apr 23]. In: StatPearls [Internet]. StatPearls Publishing, Treasure Island, 2024 Jan. Available from: https://www.ncbi.nlm.nih.gov/books/NBK592418/
106. Bergquist M, Thiel M, Goldberg MH, van der Linden S (2023) Field interventions for climate change mitigation behaviors: a second-order meta-analysis. Proc Natl Acad Sci U S A 120(13):e2214851120. https://doi.org/10.1073/pnas.2214851120
107. Nguyen-Van P, Stenger A, Tiet T (2021) Social incentive factors in interventions promoting sustainable behaviors: a meta-analysis. PLoS One 16(12):e0260932. https://doi.org/10.1371/journal.pone.0260932
108. Berge JM, Mendenhall TJ, Doherty WJ (2009) Using community-based participatory research (CBPR) to target health disparities in families. Fam Relat 58(4):475–488. https://doi.org/10.1111/j.1741-3729.2009.00567.x
109. Collins SE, Clifasefi SL, Stanton J, Board TLA, Straits KJE, Gil-Kashiwabara E, Rodriguez Espinosa P, Nicasio AV, Andrasik MP, Hawes SM, Miller KA, Nelson LA, Orfaly VE, Duran BM, Wallerstein N (2018) Community-based participatory research (CBPR): towards equitable involvement of community in psychology research. Am Psychol 73(7):884–898. https://doi.org/10.1037/amp0000167
110. Qin Z, Tu C, Han W, Jiang Q (2024) The impact of market-incentive environmental regulation policies on corporate environmental costs: evidence from China's carbon trading policy. PLoS One 19(2):e0297003. https://doi.org/10.1371/journal.pone.0297003
111. Saqib N, Radulescu M, Usman M, Balsalobre-Lorente D, Cilan T (2023) Environmental technology, economic complexity, renewable electricity, environmental taxes and CO2 emissions: implications for low-carbon future in G-10 bloc. Heliyon 9(6):e16457. https://doi.org/10.1016/j.heliyon.2023.e16457
112. Rose A, Neff R, Yarnal B, Greenberg H (2005) A greenhouse gas emissions inventory for Pennsylvania. J Air Waste Manag Assoc 55(8):1122–1133. https://doi.org/10.1080/10473289.2005.10464714
113. Sun Y, Mousavi A, Masri S, Wu J (2022) Socioeconomic disparities of low-cost air quality sensors in California, 2017-2020. Am J Public Health 112(3):434–442. https://doi.org/10.2105/AJPH.2021.306603
114. Tanzer R, Malings C, Hauryliuk A, Subramanian R, Presto AA (2019) Demonstration of a low-cost multi-pollutant network to quantify intra-urban spatial variations in air pollutant source impacts and to evaluate environmental justice. Int J Environ Res Public Health 16(14):2523. https://doi.org/10.3390/ijerph16142523
115. Hawkins A (2023) Case study: how PurpleAir is helping with Pennsylvania air quality. Available online: https://www2.purpleair.com/blogs/blog-home/case-study-how-purpleair-is-helping-with-pennsylvania-air-quality
116. Tefera Y, Soebarto V, Bishop C, Kandulu J, Williams C (2023) A scoping review of urban planning decision support tools and processes that account for the health, environment, and economic benefits of trees and greenspace. Int J Environ Res Public Health 21(1):48. https://doi.org/10.3390/ijerph21010048
117. Bikis A (2023) Urban air pollution and greenness in relation to public health. J Environ Public Health 2023:8516622. https://doi.org/10.1155/2023/8516622

118. Adlakha D, Hipp JA, Sallis JF, Brownson RC (2018) Exploring neighborhood environments and active commuting in Chennai, India. Int J Environ Res Public Health 15(9):1840. https://doi.org/10.3390/ijerph15091840
119. Pamukcu-Albers P, Ugolini F, La Rosa D, Grădinaru SR, Azevedo JC, Wu J (2021) Building green infrastructure to enhance urban resilience to climate change and pandemics. Landsc Ecol 36(3):665–673. https://doi.org/10.1007/s10980-021-01212-y
120. Callway R, Pineo H, Moore G (2020) Understanding the role of standards in the negotiation of a healthy built environment. Sustain For 12(23):9884. https://doi.org/10.3390/su12239884
121. Thrun E, Leider J, Chriqui JF (2016) Exploring the cross-sectional association between transit-oriented development zoning and active travel and transit usage in the United States, 2010-2014. Front Public Health 4:113. https://doi.org/10.3389/fpubh.2016.00113
122. Tehrani SO, Wu SJ, Roberts JD (2019) The color of health: residential segregation, light rail transit developments, and gentrification in the United States. Int J Environ Res Public Health 16(19):3683. https://doi.org/10.3390/ijerph16193683
123. Duea SR, Zimmerman EB, Vaughn LM, Dias S, Harris J (2022) A guide to selecting participatory research methods based on project and partnership goals. J Particip Res Methods 3(1). https://doi.org/10.35844/001c.32605
124. Isaifan RJ, Baldauf RW (2020) Estimating economic and environmental benefits of urban trees in desert regions. Urban For Urban Greening. https://doi.org/10.3389/fevo.2020.00016
125. Berland A, Shiflett SA, Shuster WD, Garmestani AS, Goddard HC, Herrmann DL, Hopton ME (2017) The role of trees in urban stormwater management. Landsc Urban Plan 162:167–177. https://doi.org/10.1016/j.landurbplan.2017.02.017
126. Mansfield TJ, Rodriguez DA, Huegy J, Gibson JM (2015) The effects of urban form on ambient air pollution and public health risk: a case study in Raleigh, North Carolina. Risk Anal 35(5):901–918. https://doi.org/10.1111/risa.12317
127. Bobulski J, Szymoniak S, Pasternak K (2024) An IoT system for air pollution monitoring with safe data transmission. Sensors 24(2):445. https://doi.org/10.3390/s24020445
128. Oladimeji D, Gupta K, Kose NA, Gundogan K, Ge L, Liang F (2023) Smart transportation: an overview of technologies and applications. Sensors 23(8):3880. https://doi.org/10.3390/s23083880
129. Huang A, Tian L, Li Q, Li Y, Yu J, Gao Y, Xia J (2023) Land-use planning serves as a critical tool for improving resources and environmental carrying capacity: a review of evaluation methods and application. Int J Environ Res Public Health 20(3):2370. https://doi.org/10.3390/ijerph20032370
130. El Ouadi J, Malhene N, Benhadou S, Medromi H (2021) Shared public transport within a physical internet framework: reviews, conceptualization and expected challenges under COVID-19 pandemic. Iatss Res 45(4):417–439. https://doi.org/10.1016/j.iatssr.2021.03.001
131. Larbi M, Kellett J, Palazzo E (2022) Urban sustainability transitions in the global south: a case study of Curitiba and Accra. Urban Forum 33(2):223–244. https://doi.org/10.1007/s12132-021-09438-4
132. Stewart OT, Moudon AV, Saelens BE (2017) The causal effect of bus rapid transit on changes in transit ridership. J Public Transp 20(1):91–103. https://doi.org/10.5038/2375-0901.20.1.5
133. Kruize H, van der Vliet N, Staatsen B, Bell R, Chiabai A, Muiños G, Higgins S, Quiroga S, Martinez-Juarez P, Aberg Yngwe M, Tsichlas F, Karnaki P, Lima ML, García de Jalón S, Khan M, Morris G, Stegeman I (2019) Urban Green space: creating a triple win for environmental sustainability, health, and health equity through behavior change. Int J Environ Res Public Health 16(22):4403. https://doi.org/10.3390/ijerph16224403
134. Tuomisto HL, Scheelbeek PFD, Chalabi Z, Green R, Smith RD, Haines A, Dangour AD (2017) Effects of environmental change on population nutrition and health: a comprehensive framework with a focus on fruits and vegetables. Wellcome Open Res 2:21. https://doi.org/10.12688/wellcomeopenres.11190.2
135. Lee S, Bi X (2019) Can adoption of pollution prevention techniques reduce pollution substitution? PLoS One 14(11):e0224868. https://doi.org/10.1371/journal.pone.0224868

136. Ahmad A, Ikram A, Rehan MF, Ahmad A (2022) Going green: impact of green supply chain management practices on sustainability performance. Front Psychol 13:973676. https://doi.org/10.3389/fpsyg.2022.973676
137. Monteiro A, Santos S, Gonçalves P (2021) Precision agriculture for crop and livestock farming-brief review. Animals 11(8):2345. https://doi.org/10.3390/ani11082345
138. Guidoboni MV, Duparque A, Boissy J, Mouny JC, Auberger J, van der Werf HM (2023) Conservation agriculture reduces climate change impact of a popcorn and wheat crop rotation. PLoS One 18(5):e0285586. https://doi.org/10.1371/journal.pone.0285586
139. Hassan MU, Aamer M, Mahmood A, Awan MI, Barbanti L, Seleiman MF, Bakhsh G, Alkharabsheh HM, Babur E, Shao J, Rasheed A, Huang G (2022) Management strategies to mitigate N_2O emissions in agriculture. Life 12(3):439. https://doi.org/10.3390/life12030439
140. Orru H, Ebi KL, Forsberg B (2017) The interplay of climate change and air pollution on health. Curr Environ Health Rep 4(4):504–513. https://doi.org/10.1007/s40572-017-0168-6
141. von Schneidemesser E, Driscoll C, Rieder HE, Schiferl LD (2020) How will air quality effects on human health, crops and ecosystems change in the future? Philos Transact A Math Phys Eng Sci 378(2183):20190330. https://doi.org/10.1098/rsta.2019.0330
142. Perera F (2017) Pollution from fossil-fuel combustion is the leading environmental threat to global pediatric health and equity: solutions exist. Int J Environ Res Public Health 15(1):16. https://doi.org/10.3390/ijerph15010016
143. Din AU, Ming J, Rahman IU, Han H, Yoo S, Alhrahsheh RR (2023) Green road transportation management and environmental sustainability: the impact of population density. Heliyon 9(9): e19771. https://doi.org/10.1016/j.heliyon.2023.e19771
144. Zhang L, Tao L, Yang F et al (2024) Promoting green transportation through changing behaviors with low-carbon-travel function of digital maps. Humanit Soc Sci Commun 11: 298. https://doi.org/10.1057/s41599-024-02737-8
145. Štrbac S, Kašanin-Grubin M, Pezo L, Stojić N, Lončar B, Ćurčić L, Pucarević M (2023) Green infrastructure designed through nature-based solutions for sustainable urban development. Int J Environ Res Public Health 20(2):1102. https://doi.org/10.3390/ijerph20021102
146. Zhu Z, Middleton B, Pindilli E, Johnson D, Johnson K, Covington S (2022) Conservation of carbon resources and values on public lands: a case study from the National Wildlife Refuge System. PLoS One 17(1):e0262218. https://doi.org/10.1371/journal.pone.0262218
147. Pervin T, Gerdtham UG, Lyttkens CH (2008) Societal costs of air pollution-related health hazards: a review of methods and results. Cost Effect Res Alloc 6:19. https://doi.org/10.1186/1478-7547-6-19
148. The World Bank (2023) Pollution. Available online: https://www.worldbank.org/en/topic/pollution
149. Leão MLP, Penteado JO, Ulguim SM, Gabriel RR, Dos Santos M, Brum AN, Zhang L, da Silva Júnior FMR (2021) Health impact assessment of air pollutants during the COVID-19 pandemic in a Brazilian metropolis. Environ Sci Pollut Res Int 28(31):41843–41850. https://doi.org/10.1007/s11356-021-13650-x
150. Martinez GS, Spadaro JV, Chapizanis D, Kendrovski V, Kochubovski M, Mudu P (2018) Health impacts and economic costs of air pollution in the metropolitan area of Skopje. Int J Environ Res Public Health 15(4):626. https://doi.org/10.3390/ijerph15040626
151. Akça EE (2024) Do renewable energy sources improve air quality? Demand- and supply-side comparative evidence from industrialized and emerging industrial economies. Environ Sci Pollut Res Int 31(1):293–311. https://doi.org/10.1007/s11356-023-30946-2
152. Qamruzzaman M, Karim S (2023) Clarifying the relationship between green investment, technological innovation, financial openness, and renewable energy consumption in MINT. Heliyon 9(11):e21083. https://doi.org/10.1016/j.heliyon.2023.e21083
153. Resnik DB (2022) Environmental justice and climate change policies. Bioethics 36(7): 735–741. https://doi.org/10.1111/bioe.13042

154. Babayomi OO, Dahoro DA, Zhang Z (2022) Affordable clean energy transition in developing countries: pathways and technologies. iScience 25(5):104178. https://doi.org/10.1016/j.isci.2022.104178
155. Demographia (2023) Demographia World Urban Areas: built up urban areas or world agglomerations. Retrieved 29 Apr 2024, from http://www.demographia.com/db-worldua.pdf
156. Dirlik A (2014) Developmentalism: a critique. Interventions 16(1):30–48
157. Ogwu MC (2024) Plants as monitors and managers of pollution. Role of Green Chemistry in Ecosystem Restoration to Achieve Environmental Sustainability 51–60. https://doi.org/10.1016/B978-0-443-15291-7.00022-5
158. IQAir (2023) 2023 world air quality report: region & city PM2.5 ranking. Retrieved 3 May 2024, from https://www.iqair.com/dl/2023_World_Air_Quality_Report.pdf
159. Fleck A (January 19, 2023) The world's next megacities [Digital image]. Retrieved 29 Apr 2024, from https://www.statista.com/chart/29152/the-worlds-next-megacities/
160. Jang AS (2011) Climate change and air pollution. J Korean Med Assoc 54(2):175–180
161. Kurokawa J, Ohara T (2020) Long-term historical trends in air pollutant emissions in Asia: regional emission inventory in ASia (REAS) version 3. Atmos Chem Phys 20:12761–12793
162. Sim CS, Jang IS, Kitano N (2015) International collaborative research project to improve air quality in East Asia megacities. Korea Environment Institute
163. Shim SB, Seo JB, Kwon SH, Lee JH, Sung HM, Bu KO, Byun YH, Yim YJ, Kim YH (2020) Impact of future air quality in East Asia under SSP scenarios. Korean Meteorol Soc 30(4): 439–454
164. NEIR (National Institute of Environmental Resaerch) (2019) Summary report of the 4th stage (2013–2017) LTP project. Joint research project for Long–range Transboundary Air Pollutans in Northeast Asia, National Institute of Environmental Research
165. PAGE (Partnership for Action on Green Economy) (2019) Green transformation of industrial parks in China's Jiangsu Province: a synthesis report. Retrieved 6 May 2024, from https://archive.un-page.org/files/public/page_jiangsu_gip_assessment_final_laid_out_17.04.2019_.pdf
166. Kim Y-A, Choi J, Son S-W (2023) Impacts of carbon neutrality and air quality control on near-term climate change in East Asia. Korea Meteorol Soc 33(5):505–517
167. NIMS (National Institute of Meteorological Sciences) (2022) Analysis of future changes in atmospheric congestion and the concentrations of fine particulate matter in winter and spring. National Institute of Meteorological Sciences
168. MoE (Ministry of the Environment) (March 29, 2024) 2050 zero carbon cities in Japan. https://wwwenvgojp/en/earth/cc/2050_zero_carbon_cities_in_japanhtml Last visited May 14, 2024
169. Wang KI, Seo TS, Jung JK, Jo JC, Sim W-B, Kwon TJ, Lee B-H, Cha JW, Noh KS (2010) A study on the applicability of low carbon green cities in East Asia. Korea research Institute for Human Settlement. 2010-45-212, Korea International Cooperation Agency
170. Bhui K, Newbury JB, Latham RM, Ucci M, Nasir ZA, Turner B, O'Leary C, Fisher HL, Marczylo E, Douglas P, Stansfeld S, Jackson SK, Tyrrel S, Rzhetsky A, Kinnersley R, Kumar P, Duchaine C, Coulon F (2023) Air quality and mental health: evidence, challenges and future directions. BJPsych Open 9(4):e120. https://doi.org/10.1192/bjo.2023.507
171. Jonidi Jafari A, Charkhloo E, Pasalari H (2021) Urban air pollution control policies and strategies: a systematic review. J Environ Health Sci Eng 19(2):1911–1940. https://doi.org/10.1007/s40201-021-00744-4
172. Wu Y, Tham J (2023) The impact of environmental regulation, environment, social and government performance, and technological innovation on enterprise resilience under a green recovery. Heliyon 9(10):e20278. https://doi.org/10.1016/j.heliyon.2023.e20278
173. Carattini S, Carvalho M, Fankhauser S (2018) Overcoming public resistance to carbon taxes. Wiley interdisciplinary reviews. Clim Change 9(5):e531. https://doi.org/10.1002/wcc.531
174. Filippelli GM, Taylor MP (2018) Addressing pollution-related global environmental health burdens. GeoHealth 2(1):2–5. https://doi.org/10.1002/2017GH000119

175. Dong F, Qin C, Zhang X, Zhao X, Pan Y, Gao Y, Zhu J, Li Y (2021) Towards carbon neutrality: the impact of renewable energy development on carbon emission efficiency. Int J Environ Res Public Health 18(24):13284. https://doi.org/10.3390/ijerph182413284
176. Nizeyimana E, Hanyurwimfura D, Hwang J, Nsenga J, Regassa D (2023) Prototype of monitoring transportation pollution spikes through the internet of things edge networks. Sensors 23(21):8941. https://doi.org/10.3390/s23218941
177. Anderson V, Gough WA (2022) Enabling nature-based solutions to build back better-an environmental regulatory impact analysis of Green infrastructure in Ontario, Canada. Buildings 12(1):61. https://doi.org/10.3390/buildings12010061
178. Seddon N, Chausson A, Berry P, Girardin CAJ, Smith A, Turner B (2020) Understanding the value and limits of nature-based solutions to climate change and other global challenges. Philos Trans R Soc Lond B Biol Sci 375(1794):20190120. https://doi.org/10.1098/rstb.2019.0120
179. Salvador Costa MJ, Leitão A, Silva R, Monteiro V, Melo P (2022) Climate change prevention through community actions and empowerment: a scoping review. Int J Environ Res Public Health 19(22):14645. https://doi.org/10.3390/ijerph192214645
180. Adabanya U, Awosika A, Moon JH, Reddy YU, Ugwuja F (2023) Changing a community: a holistic view of the fundamental human needs and their public health impacts. Cureus 15(8): e44023. https://doi.org/10.7759/cureus.44023

Technological Advancements for Sustainable Clean Air

Sylvester Chibueze Izah, Timinipre Amabie, and Ayobami O. Aigberua

Contents

S. C. Izah (✉)
Department of Microbiology, Faculty of Science, Bayelsa Medical University, Yenagoa, Bayelsa, Nigeria

Department of Community Medicine, Faculty of Clinical Sciences, Bayelsa Medical University, Yenagoa, Bayelsa, Nigeria
e-mail: sylvester.izah@bmu.edu.ng

T. Amabie
Department of Computer Science, Faculty of Science, Bayelsa Medical University, Yenagoa, Bayelsa, Nigeria

A. O. Aigberua
Quality Assurance Department, Infra Pipe Solutions Ltd, Saskatoon, SK, Canada

Matthew Chidozie Ogwu and Sylvester Izah (eds.), *Sustainable Strategies for Air Pollution Mitigation: Development, Economics and Technologies*,
Hdb Env Chem (2024) 133: 279–310, DOI 10.1007/698_2024_1120,

Published online: 12 September 2024

Abstract Technology advancements continue to have made a significant contribution to the search for sustainable clean air by offering a ground-breaking path toward environmental health and welfare. This chapter discusses key parts of the ongoing technological evolution for sustainable clean air with an emphasis on smart city initiatives, emission reduction approaches, predictive modeling for air quality forecasts, real-time air quality monitoring, and data analytics for pollution source identification. Real-time air quality monitoring enables the early identification of pollution sources because of sophisticated sensor technology. This information, when paired with data analytics, becomes a powerful tool for identifying sources and directing targeted mitigation efforts for air pollution. Predictive modeling improves air quality forecasts by providing communities and decision-makers with proactive information. Proposals to build "smart cities" make use of technology to incorporate environmental issues and urban planning into day-to-day life. The adoption of pollution reduction strategies backed by electric cars and renewable energy sources is necessary for a greener future. Public engagement and awareness campaigns reinforce the benefits of technology interventions and foster a sense of shared responsibility for environmental stewardship. International collaboration and data exchange provide a foundation for cross-national group action in the fight for clean air. The energy environment becomes more sustainable and the carbon footprint shrinks as renewable energy sources gain popularity. In conclusion, the cooperative use of technology is bringing us closer to a future where everyone will have access to clean air. Dedication to innovation, collaboration, and sustainable methods by researchers and government agencies may pave the way for a world that thrives in ecological balance.

Keywords Data analytics, Emission reduction, Predictive modeling, Real-time monitoring, Renewable energy, Smart City initiatives, Sustainable clean air, Technology advancements

Abbreviations

CCS	Carbon capture and storage
CO_2	Carbon dioxide
HAWT	Horizontal-axis wind turbine
VAWT	Vertical-axis wind turbine
EGS	Enhanced geothermal systems
MFCs	Microbial fuel cells
VOCs	Volatile organic compounds
EVs	Electric vehicles
HVAC	Heating, Ventilation, and Air Conditioning

1 Introduction

Air pollution involves pollutants of chemical, physical, or biological origin, and affects both indoor and outdoor environments [1]. Particulate matter, carbon monoxide, ozone, nitrogen dioxide, sulfur dioxide, etc., are major air pollutants that portend significant health concerns, contributing to various diseases and global mortality [2–5].

Globally, 99% of the population breathes air exceeding WHO-recommended pollution limits, with developing nations experiencing the most severe impacts [6]. The interconnection between air quality, climate, and ecosystems highlights the importance of regulations targeting air pollution, offering dual benefits for the environment and human health.

Symptoms of air pollution are observable in urban haze and indoor smoking, posing threats to both human health and ecosystems. Ambient air pollution, particularly fine particulate matter, is a root cause of health issues like lung cancer, cardiovascular diseases, and respiratory illnesses in urban and rural settings [6, 7]. Approximately 2.4 billion people face harmful indoor air pollution due to the use of polluting cooking stoves, resulting in around seven million premature deaths annually [6].

Clean air is a fundamental component of a healthy and sustainable environment [8]. Sustainably achieving clean air is an intricate endeavor that presents a complex array of challenges, varying across global, continental, and national scales. The challenges become even more formidable when considering the diverse environmental landscapes of the different parts of the world, especially in Africa, and Nigeria in particular.

On a global scale, clean air demands collaborative efforts from different countries to combat it [9]. Different countries largely have diversity in policies, economic growth rates, and priorities, which makes it challenging to apply standardized procedures across all countries. The issue of transboundary pollution further complicates matters as pollutants easily traverse national borders, requiring international cooperation to address this shared environmental problem. The surge in global industrial activities, driven by globalization, contributes significantly [10] to air pollution, posing a conundrum between economic growth and environmental preservation. Furthermore, the transmission of technological know-how to developing countries remains a hurdle, hindering their progress in adopting advanced clean technologies.

The challenges faced by Africa in achieving clean air are unique and multifaceted. Rapid urbanization and industrial expansion contribute to elevated emissions, particularly in areas with inadequate infrastructure and urban planning. The use of traditional biomass for heating and cooking in certain regions adds to indoor air pollution [5]. Agricultural practices, including vegetative fires and unsustainable methods, also contribute to air pollution. Ohimain et al. [7] and Ohimain and Izah [11] reported pollutant gas emissions during the boiling of palm fruit and digestion processes in smallholder palm-producing plants in Nigeria. Limited monitoring infrastructure in many African countries pose challenges in assessing and implementing efficient policies, emphasizing the need for comprehensive data.

Some countries, particularly in Africa such as Nigeria, grapple with distinct challenges in its pursuit of clean air. The growing industrial sector generates emissions that are difficult to monitor and control, especially in older buildings lacking modern pollution control technologies. Increased vehicular traffic, coupled with insufficient public transit infrastructure, contributes significantly to urban air pollution [12]. Poor waste management practices, including open-air burning [13], further degrade air quality and contribute to the spread of diseases [14]. Izah et al. [14] reported a diversity of fungal aerosols in the air environment around waste dumpsites. Sawyer et al. [1] reported that diverse microbial diversity capable of causing diseases in humans is found in diverse outdoor and indoor air environments. Furthermore, despite having environmental policies in place, some countries, particularly Nigeria, struggle with effective implementation and enforcement. This is primarily due to inconsistent regulatory requirements. Socioeconomic factors, including poverty, pose a barrier to the adoption of environmentally friendly activities and technologies in many countries, mostly within the African continent.

Technological innovation, robust regulatory frameworks, and considerations of social and economic factors are all crucial components of a comprehensive solution to air pollution. Therefore, recognizing the urgent need to address air pollution using technological advancements could play an essential role in developing innovative solutions for achieving sustainable clean air [15, 16].

This chapter focuses on technological advancements for achieving clean air. The chapter discusses the role of monitoring and data analytics, renewable energy sources, electric vehicles in sustainable transportation and environmental stewardship, green technologies, carbon capture and storage, and air purification technologies in driving sustainable clean air practices. The chapter further presents policy and regular framework work for effective implementation.

2 Advancements in Monitoring and Data Analytics for Clean Air

Addressing concerns related to air quality is a top priority. One of the key elements toward achieving sustainable clean air is the implementation of accurate real-time monitoring [17]. Recent technological advancements have enabled the creation of sophisticated monitoring systems to assess the quality of the air we breathe. These systems incorporate sensors, satellites, and data analytics, collectively providing a comprehensive approach to monitoring air quality. The instruments utilized can measure a wide range of pollutants, such as particulate matter, nitrogen dioxide, sulfur dioxide, and ozone, among others. These technologies (real-time air quality monitoring, data analytics for pollution source identification, predictive modeling for air quality forecasting, smart city initiatives, emission reduction strategies, public

Table 1 Integrating real-time monitoring, data analytics, and global collaboration for sustainable clean air solutions

Characteristics	Advancing air quality management
Real-time air quality monitoring	Sensor networks deployment can advance real-time data monitoring for major pollutants such as PM, NO_2, SO_2, CO, and O_3 [18, 19]
	Integration of artificial intelligence can enable continuous and remote monitoring, crucial for prompt response [20, 21]
Data analytics for pollution source identification	The use of machine learning algorithms can help analyze datasets and identify pollution sources [22]
	The application of source apportionment in air quality studies can help determine the percentage contribution of different sources [14, 23, 24]
Predictive modeling for air quality forecasting	The integration of meteorological and atmospheric models with historical data can enable accurate prediction [25]
	Putting the general public on alert based on predictive modeling and analytics can help residents prepare for episodes of poor air quality [26, 27]
Smart city initiatives	Building integrated platforms for possible use in air quality data can be helpful for sustainable urban planning and development [28]
	Policy decision support enables policymakers to formulate and adjust strategies for emission reduction and environmental sustainability [29, 30]
Emission reduction strategies	The identification of high-impact areas using analytics tools is helpful in order to assess the effectiveness of measures [31, 32]
Public awareness and engagement	The application of data visualization tools makes air quality information accessible, thereby fostering awareness and community engagement [33]
Global collaboration and data sharing	International cooperation can be facilitated by monitoring and analytics This is good for sharing air quality data
	It is also good for addressing transboundary air pollution and global environmental challenges through a collective approach [34, 35]

awareness and engagement, global collaboration, and data sharing) are summarized in Table 1 and Fig. 1.

Real-time air monitoring is essential for achieving and maintaining clean air [36]. It uses state-of-the-art equipment to continuously measure air quality. Early pollutant detection makes it easier to quickly identify the causes of pollution. With their array of sensors, modern monitoring systems can measure a large number of contaminants, providing comprehensive information on the composition of the air. Constant real-time data gathering facilitates data-driven decision-making, helping authorities design efficient plans for pollution management and regulatory oversight [37]. Public engagement and awareness are encouraged by the designation of pollution hotspots and the availability of real-time data on air quality. This is because actual-time monitoring makes it easier to regularly assess the success of clean air

Fig. 1 Overview of integrating real-time monitoring and data analytics for sustainable clean air solutions

efforts and improve tactics based on data from the actual world. To promote continuous technical innovation and advance sensor technology, satellite imagery, and data analytics is crucial.

Controlling air quality requires the use of data analytics, especially when applying machine learning methods [38]. These sophisticated algorithms can correctly identify the causes of pollution by analyzing big datasets involving a variety of factors, such as traffic, weather, and air quality [39, 40]. Through the evaluation of these interconnected elements, machine learning algorithms can precisely identify the sources of pollution, thereby establishing how much each source of pollution contributes proportionately.

For proactive pollution control, predictive modeling for air quality estimates is essential. Model creators can create more accurate forecast models by combining meteorological and atmospheric models with historical data on air quality. By projecting future air quality conditions, these models can help authorities take preventive action before pollution levels become dangerous. The timely release of alerts to the public can be made possible by the integration of data analytics and predictive modeling. This method could lessen health risks and make it easier for businesses and communities to plan for times when the air quality is poor by simplifying the application of necessary safety steps.

Programs for smart cities use data analytics and monitoring as essential elements. These programs first use integrated platforms to combine data on air quality with different urban factors. Cities can create plans for sustainable urban planning, infrastructure development, and transportation, thanks to this integration [41]. The comprehensive integration can help to promote a deep understanding of urban dynamics. In terms of policy decision support, decision-makers can then use analytics-derived insights to create and modify policies. The goals of this data-driven decision-making process are to improve environmental sustainability, reduce emissions, and improve public health. In smart city efforts, the combination of real-time data and analytics may help make strategic and educated decisions that will ultimately promote the well-being of humans as well as urban ecosystems [42, 43].

To effectively control air pollution, emission reduction methods need to be implemented. However, the efficiency of these initiatives depend on advanced analytics. Policymakers can determine high-impact locations with higher pollution levels using cutting-edge methodology. This accuracy enables focused evaluation of emission reduction initiatives, maximizing resource distribution to the most negatively impacted locations. Additionally, the combination of monitoring and analytics can provide a continuous feedback loop, making it easier to assess activities that have been put into effect [44, 45].

3 Renewable Energy Technologies

Since renewable energy technologies provide a greener option to conventional fossil fuel-based energy sources [46], they are essential to preserving clean air and promoting sustainable development. To combat climate change, lower greenhouse gas emissions, and enhance air quality, solar and wind power are major contributors. A cleaner and healthier environment is a result of the removal of hazardous pollutants. Furthermore, the decentralization of energy production, improvements in energy efficiency, the creation of jobs, and continuous technological innovation all highlight the diverse ways in which renewable energy contributes to the development of a cleaner and more sustainable energy landscape.

Technological developments in geothermal, hydro, wind, and solar energy have increased the economic viability of these energy sources while also improving their efficiency [47, 48]. Widespread use of renewable energy contributes to a reduction in the dependency on combustion-based power generation, which lowers greenhouse gas and air pollution emissions significantly.

3.1 Solar Power

Clean air and sustainable development can depend on solar electricity. One can use solar thermal systems or photovoltaic (PV) cells to harness solar radiation, creating electricity [49, 50]. Photovoltaic cells directly turn sunlight into electricity, while solar thermal systems use the heat of the sun to create steam that powers turbines coupled to energy generators [51–53]. Figure 2 illustrates the key elements involved in a solar power plant and a hybrid system configuration. It involves various components essential for harnessing solar energy, such as photovoltaic panels, inverters, and batteries for energy storage [54]. Additionally, the hybrid system integrates renewable energy sources with traditional power systems to enhance efficiency and reliability [55]. By combining these technologies, the system can optimize energy production and supply, making it a crucial part of modern sustainable energy solutions.

One of the key benefits of solar electricity is that it is environmentally friendly. Solar energy generation contributes significantly to mitigating climate change. By reducing dependency on fossil fuels, solar power plays a key role in addressing air quality issues and achieving clean air, especially in urban areas where air pollution poses major health risks. Beyond simply cutting emissions, solar energy is sustainable in other ways as well. Solar energy is a reliable and durable energy source because of its abundance and renewable nature [56, 57]. This characteristic makes the transition to sustainable energy sources essential for maintaining economic growth and development.

Solar energy is also suitable for both off-grid and grid-connected applications due to its adaptability and decentralization [58–60]. Distributed solar installations like rooftop solar panels on houses and businesses lessen reliance on centralized power networks by empowering individuals and communities to generate their electricity. In remote or off-grid areas, solar power provides access to electricity, fostering economic development and improving the quality of life.

The solar industry is currently driving innovation and employment development in the economy, particularly in Nigeria. As costs decrease due to economies of scale and technological improvements, solar energy is starting to look more and more like a viable alternative to existing energy sources. Therefore, investing in solar projects can stimulate economic growth by creating jobs in production, installation,

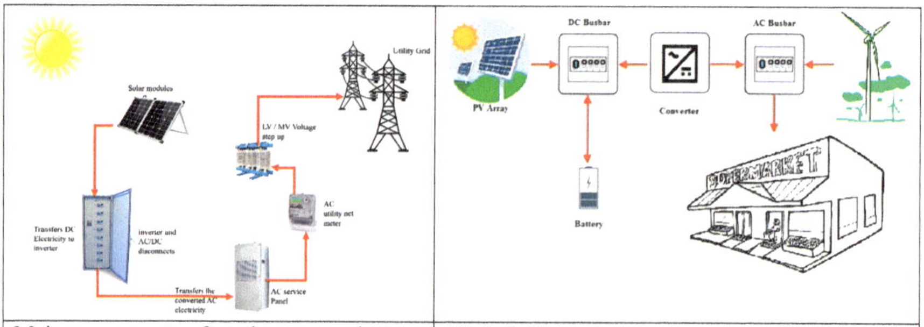

Main components of a solar power plant. (Obaideen et al. 54)	The hybrid system's arrangement. Ghadirinejad et al. 55

Fig. 2 Major component of solar power plant and hybrid system arrangement. Source: [54, 55]. https://creativecommons.org/licenses/by/4.0/

maintenance, and even research and development; furthermore, solar energy has the potential to overtake other energy sources in areas with an abundance of sunshine, making it less susceptible to supply interruptions and price swings that occur with more conventional energy sources.

3.2 Wind Power

Clean air and sustainable development are fundamentally dependent on wind power, particularly when it comes to the integration of smart grid technology and turbine technology [61–63]. Wind energy turbines are now a popular renewable energy source with several advantages that promote both socioeconomic development and environmental preservation.

Figure 3 shows the basic types of wind turbines including horizontal-axis wind turbine (HAWT) and vertical-axis wind turbines (VAWT) [64]. HAWT has blades that rotate around a horizontal axis. It is the most common and widely used design for large-scale wind energy production. HAWTs are typically characterized by their tall towers and three-bladed rotors that face into the wind. In contrast to HAWTs, VAWTs have blades that rotate around a vertical axis. This design allows them to capture wind from any direction without needing to be oriented toward it. VAWTs are often used in smaller-scale applications and in environments where wind direction is highly variable.

The main components of offshore wind turbines (OWTs) include the towers, support structures, generators, transmissions, blades, and other required machinery (Fig. 4). The entire offshore wind farm electricity generation, operation, maintenance, transmission, and grid connection process is depicted in Fig. 4. When

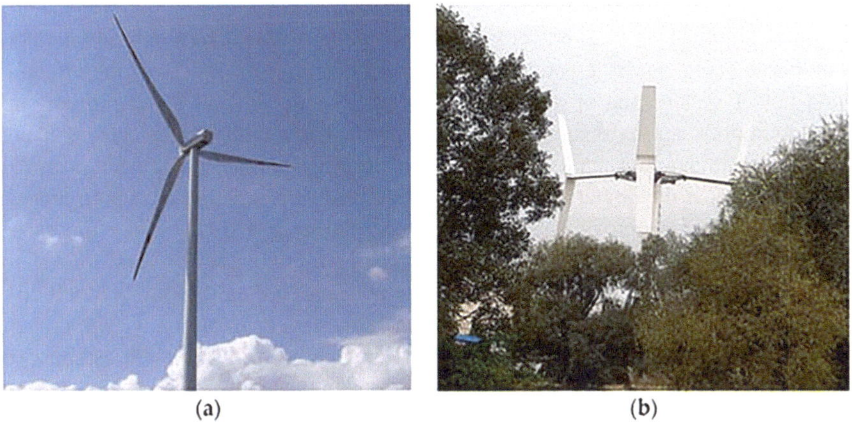

Fig. 3 Basic types of wind turbines: (**a**) the horizontal-axial wind turbine (HAWT) turbine; (**b**) the vertical-axial wind turbine (VAWT) turbine. Source: [64]; https://creativecommons.org/licenses/by/4.0/

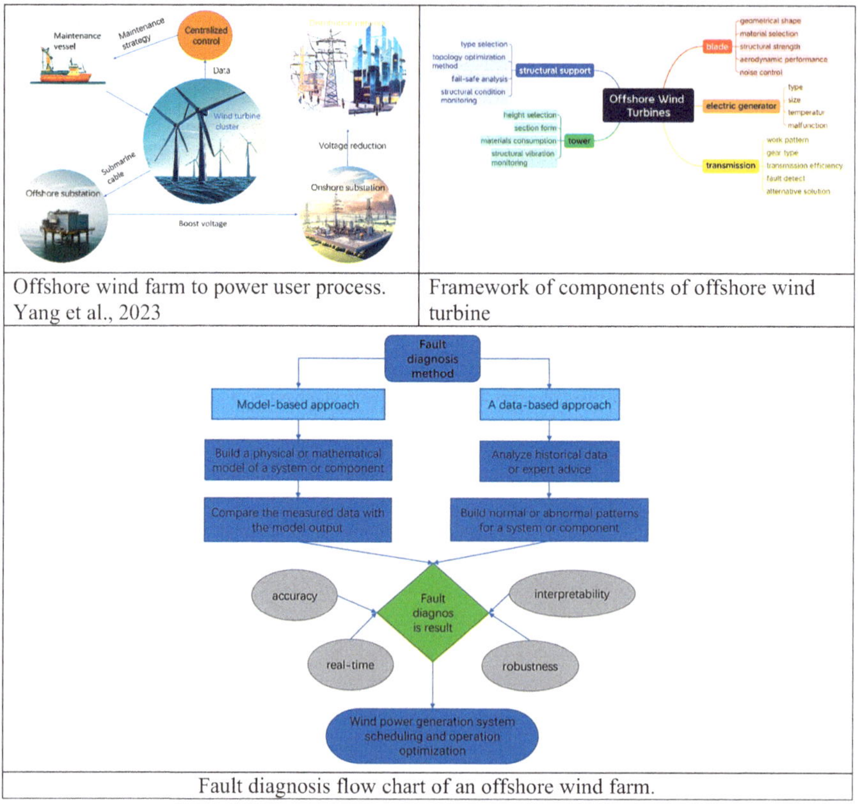

Offshore wind farm to power user process. Yang et al., 2023

Framework of components of offshore wind turbine

Fault diagnosis flow chart of an offshore wind farm.

Fig. 4 Comprehensive framework of offshore wind farms: from power generation to user process and fault diagnosis. Source: [65]; https://creativecommons.org/licenses/by/4.0/

diagnosing faults in a wind power system or its components, a model-based method is employed (Fig. 4). By creating a mathematical or physical model of the offshore wind power system, the operating characteristics of the system under normal and fault conditions are explained [65]. The output of the model is calculated using measured data as input, and any differences are detected by comparing it with the observed data. If the deviation exceeds a predefined threshold, an alert is triggered, indicating the presence of an abnormality or malfunction [65]. When assessing diagnostic results, factors like accuracy, robustness, interpretability, and real-time performance are taken into consideration. In addition, a data-driven approach collects current or historical data on several variables, such as rotational speed, power, vibration, temperature, and pressure [65]. By comparing this data with established guidelines, historical data, and expert opinions, anomalies, and failures' causes and effects are determined. This analysis yields maintenance or control recommendations.

The environmental sustainability of wind power is one of its main advantages. In contrast to traditional fossil fuel-based energy sources, wind power generates electricity without releasing any damaging pollutants or greenhouse gases into the atmosphere [66, 67]. Concerns about water scarcity can be addressed by the fact that wind power uses less water than other conventional power production techniques, which highlights its environmental impact.

One significant development in smart grid technology is the incorporation of wind power [62]. By monitoring and controlling the flow of electricity using digital communication technology, smart grids improve the efficiency and dependability of energy distribution. The smooth integration of fluctuating renewable energy sources, such as wind power, into the larger energy infrastructure is made possible by this consolidation.

The financial advantages of wind energy are significant. Jobs in manufacture, installation, maintenance, and research and development are created by the wind sector. Furthermore, the use of wind power contributes to energy portfolio diversification by lowering reliance on limited and politically sensitive fossil fuel resources. Through the construction of wind farms, the benefits of wind power are extended to nearby towns.

3.3 Hydro Power

Hydropower, particularly tidal and wave energy, harnesses the force of water to generate clean, renewable electricity [68], contributing to clean air and sustainable development. These aquatic energy resources provide economic, environmental, and societal benefits that extend beyond merely supplying power. Tidal energy, generated by the gravitational pull of the moon, and seas on Earth's waters [69, 70], is a reliable and consistent energy source. Tidal power stations utilize turbines to convert the kinetic energy of moving water into electricity. The predictability of tidal cycles ensures effective power generation, reducing the demand for fossil fuels and enhancing grid stability. Similarly, wave energy, derived from the motion of ocean waves, represents a vast, largely untapped renewable resource [71, 72]. Wave energy converters transform the kinetic energy of waves into electrical power. Unlike some intermittent renewable sources, wave energy offers a consistent and predictable power supply, further supporting reliable renewable energy generation.

One of the key benefits of wave and tidal energy is its environmental sustainability. Compared to traditional fossil fuels, marine energy (wave and tidal) sources produce fewer greenhouse emissions, making them an important weapon in the battle against climate change and for improving air quality. Furthermore, because tidal and wave energy extraction has less of an impact on the ecosystem and the visual landscape than other terrestrial renewable energy sources, it is more socially and environmentally acceptable.

The development and implementation of tidal and wave energy plants can foster economic growth and job creation. Furthermore, investments in technology, infrastructure, and research related to marine energy can promote the development of a sustainable sector. In coastal communities, the manufacturing, installation, and upkeep of maritime energy systems can boost local economies and provide jobs.

Moreover, the decentralization of energy production through wave and tidal energy might enhance energy security by broadening the energy mix [73, 74]. Decreasing over-dependency on power plants and transmission lines located in distant communities, and improving localized power generation can benefit coastal towns. This decentralization can promote resilience and reliability in the case of natural disasters or disturbances to the centralized energy system.

3.4 Geothermal Energy

Geothermal energy is an essential part of clean air and sustainable development since enhanced geothermal systems (EGS) offer a renewable and environmentally beneficial alternative to conventional energy sources [75]. Through innovative use of Earth's inherent heat reserves, EGS provides a stable and reliable power source.

Enhanced geothermal systems use technology to increase the productivity and permeability of hot rock reservoirs located deep beneath the Earth's crust, where conventional geothermal resources are less abundant [76, 77]. By injecting water into these reservoirs and creating artificial channels for fluid flow, EGS enhances geothermal potential, enabling power generation even in areas [78] where natural geothermal resources are scarce.

Geothermal energy has a lot of benefits, especially when it is produced with EGS. One of its primary advantages is that it is environmentally sustainable. Compared to fossil fuels, geothermal energy emits fewer greenhouse gases, hence it is a clean, sustainable resource. The extraction technique uses the natural heat of the earth to bypass combustion and the associated air pollutants that deteriorate air quality and fuel climate change.

Furthermore, because it never stops, geothermal energy provides a steady and reliable source of electricity. Unlike wind and solar energy, which are subject to weather fluctuations, geothermal power plants can operate continuously, making them a dependable source of baseload power. Geothermal energy can make a substantial contribution to grid stability and energy security due to its dependability [73, 79].

In addition to being dependable and good for the environment, geothermal energy can promote sustainable development. In rural areas with abundant geothermal resources, EGS projects can contribute to the development of rural communities, provide jobs, and strengthen local economies. Geothermal power plants' infrastructure aligns with the principles of sustainable development as they utilize preexisting geological features, reduce land use, and have minimum impact on local ecosystems.

Therefore, utilizing geothermal energy also helps to reduce dependency on non-renewable resources and advance energy independence. States with significant geothermal potential can diversify their energy sources and reduce their reliance on fluctuations in the price of fossil fuels, thereby enhancing energy security. This diversification also lessens the geopolitical dangers associated with an excessive reliance on foreign energy resources.

3.5 Energy Storage

Energy storage, which includes advanced battery technology and grid-scale storage devices [80, 81], is crucial to achieving clean air and sustainable growth in the rapidly evolving energy landscape. This transformative capability reduces the environmental impact of conventional energy sources while enabling a stable and reliable power supply by addressing the inherently unpredictable nature of renewable energy sources like wind and solar power [82].

Advancements in battery technology have resulted in a surge in the demand for electric cars and portable gadgets [83, 84]. Beyond personal use, these developments are critical to decentralized energy systems. Effective energy storage has been linked to lithium-ion batteries in particular because of their high energy density, long cycle life, and low environmental impact [85–87], together with other cutting-edge inventions like solid-state batteries [88], are essential to the implementation of renewable energy solutions.

The seamless integration of renewable energy sources into the current electrical grid requires larger-scale grid-scale storage devices. These systems can serve as reservoirs, storing extra energy during times of high generation and releasing it during times of high demand or in the absence of renewable energy sources. By bridging the gap between energy supply and consumption, grid-scale storage encourages system stability, reduces reliance on fossil fuels, and facilitates the transition to a low-carbon energy mix.

As a cornerstone of clean air and sustainable development, energy storage offers many benefits. First of all, it allows for the optimization of renewable energy sources, unleashing the full potential of solar and wind power. Extra energy can be stored for later use. This flexibility in the energy supply leads to a more resilient and flexible energy infrastructure.

Secondly, energy storage plays a major role in load balancing and peak shaving [89]. Grid-scale storage stores excess energy during off-peak hours and releases it during peak demand, reducing the need for additional power units to manage brief spikes in demand [90, 91]. By doing this, emissions associated with the construction and operation of fossil fuel-based power plants are reduced and energy efficiency is increased.

Additionally, energy storage increases the system's dependability and resilience [92]. As a safeguard against power outages, it could ensure a steady flow of power throughout disruptions. This is particularly crucial in places that are susceptible to severe weather or natural catastrophes since emergency services and community resilience rely on a consistent supply of electricity.

Furthermore, the widespread application of energy storage technology can promote the decentralization of the electrical grid. Furthermore, localized storage systems reduce transmission losses caused by centralized power generation and encourage energy independence [93, 94] by empowering communities to generate, store, and manage their energy.

3.6 Microbial Fuel Cells Technology

Microbial fuel cells (MFCs), presenting themselves as a pillar of clean air and sustainable development, are a revolutionary technology that can revolutionize energy production and address environmental issues at the same time [95]. These gadgets generate power by microbial breakdown of organic materials.

The ability of MFCs to convert a range of organic substrates, including wastewater and organic waste, into electricity in an environmentally friendly manner is one of its key advantages [95]. MFCs provide an appealing alternative for waste treatment plants because of their innate capacity to both generate power and remediate waste [95]. MFCs use the microbial populations present in these systems to generate cleaner water and sustainable electricity.

MFCs are also crucial in lowering air pollution since they offer an alternative to conventional energy sources, which have a significant effect on greenhouse gas emissions. By using microbial processes rather than burning fossil fuels to produce electricity, MFCs reduce emissions of pollutants such as carbon dioxide, nitrogen oxides, and sulfur dioxide. Thus, MFCs are positioned as a sustainable energy solution that can lessen the environmental impact of conventional power generation because of this property.

MFCs' scalability and versatility further increase their potential impact on sustainable development ([96]). These gadgets can be applied in a wide range of situations, from small-scale uses in isolated locations to large-scale installations in metropolitan areas. Their adaptability can allow them to be used for decentralized energy generation, providing communities and enterprises with additional clean and sustainable energy sources.

In addition to generating electricity, MFCs can aid in the cleanup of contaminated air by lowering volatile organic compounds (VOCs) [97]. The microorganisms included in MFCs may effectively oxidize and break down VOCs [97], hence enhancing the air quality. MFCs are positioned as flexible tools for achieving sustainability goals related to energy and the environment because of their capacity to produce power and purify air.

MFCs also align with the principles of the circular economy because they use energy from organic waste streams that would otherwise contribute to environmental degradation [95]. By solving the issue of waste management and creating a closed-loop system where resources are used efficiently, this technique encourages the development of a more resilient and sustainable ecosystem.

4 Role of Electric Vehicles in Fostering Sustainable Clean Air

The main source of air pollution in the transportation industry is the burning of fossil fuels. The development of electric vehicles (EVs) is one innovative approach to reducing vehicle emissions [98]. The growing public appeal and accessibility of electric vehicles can be attributed to developments in battery technology, charging infrastructure, and vehicle design. A shift toward more environmentally friendly and sustainable forms of transportation is being aided by an increase in the amount of money that corporations and governments are investing in electrification.

EVs and sustainable mobility are essential for improving air quality, mitigating climate change, and achieving long-term environmental sustainability [99]. The multifaceted impact of electric vehicles on sustainability is shown in Fig. 5.

EVs are crucial to the global effort to reduce greenhouse gas emissions from the transportation sector [100]. Unlike traditional internal combustion engine cars [100], which are fueled by fossil fuels, EVs run on electricity. By significantly lowering carbon emissions, this provides a more environmentally responsible choice, particularly when powered by renewable energy, making the transition to EVs a step toward reducing the environmental effect of transportation and combating climate change [101, 102].

When EVs replace conventional automobiles, the amount of exhaust emissions they emit from burning fossil fuels is reduced, which significantly improves air quality. Particularly in urban areas where traffic congestion is a significant problem, the reduced emissions from EVs are essential for fostering healthier living conditions.

The environmental benefits of EVs are maximized when combined with renewable energy sources. Emissions from the EV lifespan are minimized by infrastructure for EV charging that is powered by solar, wind, or other renewable energy sources [103, 104]. This strategy is a comprehensive approach to sustainable transportation because it tackles both the direct emissions from vehicle operation and the indirect emissions connected to energy production.

Governments of some countries are enacting legislation and offering incentives to speed up the adoption of electric vehicles. Policies like tax credits, subsidies, and regulations that push automakers to produce more electric vehicles are crucial to overcoming early pricing barriers and promoting widespread adoption. These regulatory measures can foster an environment that is conducive to the expansion of the electric vehicle market.

For EVs to succeed in the future, battery technology must continue to advance. Therefore, expanding the infrastructure for charging EVs is necessary for their widespread adoption [105, 106]. Governments, businesses, and communities are investing in a broad network of charging stations, including fast chargers. This reduces adoption obstacles and improves the convenience of EV ownership by addressing range restrictions and promoting the smooth integration of electric vehicles into daily life.

Fig. 5 The multifaceted impact of electric vehicles toward sustainability

Sustainable urban planning is crucial to the effective adoption of electric vehicles. Globally, cities are putting into practice regulations that are conducive to EVs, like parking incentives, designated lanes for EVs, and urban planning that encourages the expansion of EV charging infrastructure. These actions encourage the widespread adoption of electric vehicles and help to create a cleaner, more effective urban transportation system.

The use of electric vehicles is driving innovation in transportation technologies [107]. This includes advancements in connectedness, autonomous driving, and energy management systems. These advancements could enhance both the user experience and the sustainability and efficiency of transportation. Because of the innovative technology they incorporate, EVs can act as catalysts for good changes in the transportation sector as a whole.

Completing a lifecycle analysis is essential to comprehending how EVs affect the environment [108]. Therefore, every step of the process, from manufacture to use to end-of-life disposal, needs to be optimized to reduce the environmental impact. By ensuring that the whole life cycle of EVs is in line with sustainability objectives, this all-involving approach offers a more precise evaluation of their environmental advantages.

Encouraging consumer education and awareness is essential to EV adoption on a large scale [109]. Increased adoption is a result of dispelling myths, reducing range anxiety, and emphasizing the long-term financial benefits of electric vehicles. Thus, by cultivating an awareness and admiration for the financial and ecological advantages of electric vehicles, customers take an active role in creating a more sustainable transportation landscape.

5 Roles of Green Technologies in Ensuring Clean Air and Sustainability

Advances in technology in architecture and construction have led to the development of green building technologies, which prioritize energy efficiency and indoor air quality [110]. Energy-efficient heating, ventilation, and air conditioning (HVAC) systems [110], intelligent building designs, and sustainable building materials can all help reduce energy use and emissions. Additionally, green walls and roofs enhance the quality of the air in cities by absorbing pollutants and supporting biodiversity.

Enhancing sustainable practices and clean air quality requires the use of green building technologies. Enhancing buildings' energy efficiency, resource conservation, and environmental performance are the primary objectives of these technologies [111].

Green building techniques have evolved into ground-breaking innovations that prioritize energy efficiency while also significantly improving air quality. These techniques involve a broad spectrum of innovative strategies intended to lessen the adverse impacts of buildings on the environment and provide sustainable constructions. Figure 6 illustrates sustainable building practices that enhance air quality and promote environmental well-being. These practices include using eco-friendly materials and smart building technologies, optimizing energy efficiency, integrating renewable energy, incorporating green roofs and walls, adhering to certifications and standards, conserving water, selecting sustainable materials, improving indoor air quality with good ventilation systems, and considering public health impacts.

Green building technologies that prioritize energy efficiency include high-performance insulation, energy-efficient windows, and smart building systems [111]. Moreover, using renewable energy sources, including solar and wind turbines on-site is a crucial part of green building practices. Buildings' overall carbon footprint is reduced and their need for traditional power sources is downsized by enabling them to generate renewable energy [112–114]. This approach aligns with sustainable energy practices and advances the broader goal of environmental preservation.

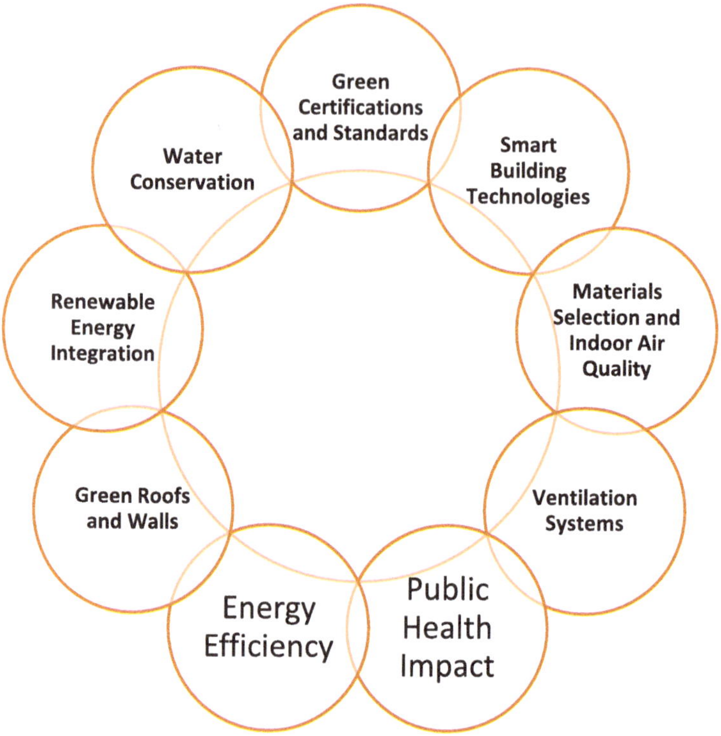

Fig. 6 Sustainable building practices for improved air quality and environmental well-being

Green architecture techniques need the use of lush walls and roofs. They serve as organic air filters, capturing pollutants and particles, thus actively mitigating the urban heat island effect in addition to providing insulation. This double goal enhances the local air quality, encouraging a more salubrious urban environment and shedding light on the connections between sustainable building practices and broader environmental goals.

Green building techniques reduce the environmental impact of construction by utilizing sustainable and low-emission materials. This material selection process also improves indoor air quality, leading to healthier and more ecologically friendly living and working environments by lowering the release of VOCs and other contaminants. The integration of smart technologies into green buildings enables the optimization of energy consumption [115]. In addition to using energy-efficient construction methods and recycled materials, green building technologies also give waste reduction priority. In addition to resolving environmental challenges, reducing construction waste minimizes the environmental impact of landfills and associated air pollution.

Leadership in Energy and Environmental Design certificates and other green building certifications significantly facilitate the uptake of sustainable solutions [116]. The recommendations from the report could ensure that buildings are designed and constructed with an emphasis on minimizing their environmental impact, thereby encouraging the building industry to adopt a sustainable mindset.

The influence of green building technologies extends beyond the structures. The general public's health benefits instantly from an environment with good indoor air quality and a healthy atmosphere. This is because green building solutions address environmental factors that might lead to respiratory problems and other health issues associated with poor air quality [117, 118], creating spaces that support inhabitants' general well-being. This comprehensive approach highlights the significance of sustainable construction methods in cultivating a more conducive and environmentally conscious future.

6 Vital Role of Air Purification Technologies in Sustainable Clean Air Practices

Sophisticated air filtration equipment has been developed at an exponential rate as urban air quality degrades. Modern filtration methods such as high-efficiency particulate air filters and electrostatic precipitators are used to remove pollutants from both indoor and outdoor environments [119]. Furthermore, the integration of artificial intelligence (AI) into air purifiers could enable real-time monitoring and change based on conditions related to air quality.

Air filtration technologies are crucial to achieving and maintaining sustainable clean air, given the challenges presented by air pollution [119], increasing urbanization, and industry. These technologies involve a wide range of devices and setups meant to remove poisons, pollutants, and allergens from the air [120]. Figure 7 illustrates the primary ways in which air filtration technologies progress the development of clean, sustainable air.

Air purifiers are a vital instrument for eliminating pollutants from the environment and offering significant health benefits because they target a wide range of airborne pollutants. Particulate matter, sulfur dioxide, nitrogen dioxide, and ozone can all be removed by air purifiers. These devices are especially crucial in urban areas where elevated pollutant levels can negatively affect respiratory and cardiovascular health.

The increasing use of air purifiers in both residential and business settings contributes to the improvement of indoor air quality [2, 3]. Through the effective reduction of allergens, pollutants, and airborne pollutants, these devices enhance the standard of living and working spaces [1]. The majority of people spend a significant amount of time indoors, therefore enhancing indoor air quality is essential for overall health.

Technological innovations in air purifiers include advanced filtration systems like electrostatic precipitators, high-efficiency particulate air filters, and activated carbon filters [119, 121, 122]. These devices can successfully capture and neutralize a wide variety of airborne particles and pollutants.

Fig. 7 Key aspects of the role of air purification technologies in advancing sustainable clean air

Integrating HVAC systems with whole-house air filtration can also ensure comprehensive coverage throughout a building [123]. This strategy can offer a more useful approach to controlling indoor air quality. Furthermore, wearable and portable air purifiers can satisfy individual needs by offering filtered air zones on their own. These handheld devices provide a targeted and user-friendly solution for individuals who are always searching for clean air, and they are particularly beneficial in crowded or unclean outdoor environments.

Technologies for air filtration received more attention during global health crises, such as the COVID-19 pandemic [124]. Certain systems are designed to filter and neutralize airborne pathogens to assist limit the spread of infectious diseases [125].

Growing public awareness of the detrimental effects of air pollution on health has increased the need for air filtration systems. The increasing acknowledgment of the benefits air purifiers offer to environmental and public health is seen in the policies and initiatives put in place by organizations to promote the use of this technology [126, 127].

7 The Role of Carbon Capture and Storage as Technological Advancements for Sustainable Clean Air

Reducing carbon dioxide (CO_2) emissions from factories and power plants requires the use of carbon capture and storage technologies [128]. By capturing CO_2 emissions at their source and burying them underground, these devices prevent greenhouse gases from entering the environment. Continuous advancements in carbon capture and storage (CCS) supports more environmentally friendly industrial processes and aid in meeting climate targets [129, 130]. Carbon capture and storage is one of the key technologies in the global fight to slow down climate change and produce clean, sustainable air. CO_2 emissions from burning fossil fuels for energy production and industrial activities will be caught and prevented from entering the atmosphere with the aid of this innovative technique. Incorporating CCS into the broader context of sustainable energy strategies has several significant benefits. Figure 8 highlights the key role of carbon capture and storage as technological advancements for sustainable clean air CCS is a crucial element of emission reduction and climate mitigation [129, 130]. Its primary goal is to reduce greenhouse gas emissions, mainly CO_2. Therefore, to meet the global commitment to minimize global warming and lessen its negative repercussions, it is imperative to absorb and store CO_2 before releasing it into the atmosphere, as it is a primary source of global warming.

Regardless of its impact on the climate, the implementation of CCS immediately preserves air quality. Through the cessation of emissions of pollutants associated with fossil fuel combustion, including particulate matter, sulfur dioxide, and nitrogen oxides, CCS becomes an effective means of improving local air quality and mitigating the detrimental health effects of air pollution.

From a sustainability perspective, CCS acts as a bridge toward a more sustainable energy future by enabling the continued use of fossil fuels in a form that is more environmentally beneficial [131]. This acknowledges the transitional role that CCS can play and is particularly relevant in sectors where it is challenging to phase out fossil fuels, such as heavy industry and some modes of transportation.

Beyond just producing electricity, CCS is also used in industrial operations like the production of steel and cement, which result in large CO_2 emissions [132]. By capturing and storing CO_2, CCS reduces emissions from industries that are usually challenging to decarbonize through alternative means.

The fact that CCS can eliminate and offset carbon is evidence of its adaptability, particularly about Direct Air Capture technologies [133]. This specific subtype of CCS enables direct extraction of CO_2 from the atmosphere, making it a valuable instrument for balancing emissions from sources such as aviation and some industrial activities that are challenging to fully eliminate. Also, improvements in storage technology have a big impact on the CCS scene because they increase the security and effectiveness of underground CO_2 storage [134]. Governments and international organizations provide financial incentives and regulatory support for carbon capture

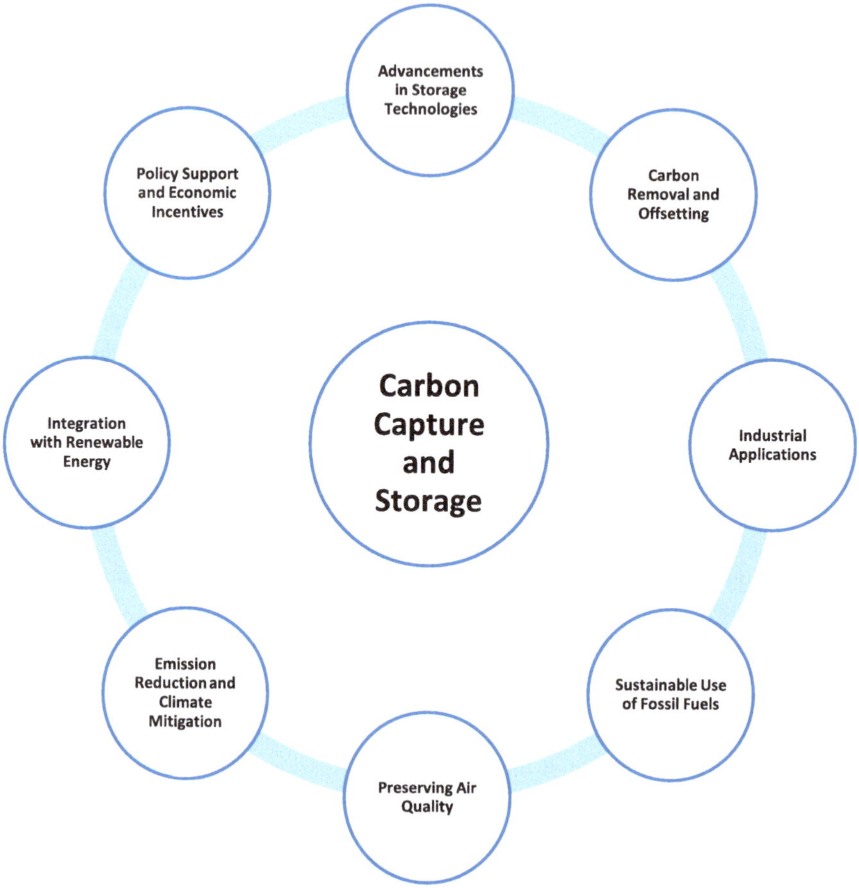

Fig. 8 Key role of carbon capture and storage as technological advancements for sustainable clean air

and storage (CCS) in recognition of its importance in achieving climate objectives [135]. To encourage the uptake and development of CCS technologies, these might include tax breaks and carbon pricing.

The combination of CCS and renewable energy sources, known as Bioenergy with Carbon Capture and Storage, opens up the possibility of zero emissions [136]. This innovative approach involves absorbing CO_2 during the growth of bioenergy crops, burning the biomass to produce energy, and subsequently capturing and storing the released emissions. This leads to a net decrease of CO_2 in the atmosphere, showing how CCS can cooperate with renewable energy sources to build a future that is more climate resilient and sustainable.

8 Policies and Regulations for Driving Sustainable Technological Advances for Clean Air Worldwide

Technical advancements are vital, but implementing sustainable practices necessitates the backing of strict rules and regulations. Diverse governments are pushing for eco-friendly technologies, imposing stricter emission limits, and offering incentives for sustainable practices. Together, legislators, business executives, and environmental organizations also need to develop a regulatory framework that promotes the development and use of cleaner technologies.

Sustainable clean air can be promoted by technological breakthroughs, but doing so calls for a complex web of local, national, and international laws and regulations. The focus of this is to control and reduce air pollution while encouraging the development and adoption of greener technologies. Figure 9 illustrates the pertinent legislation and policies that are encouraging sustainable technological breakthroughs for clean air.

International agreements such as the Paris Agreement, which operates within the framework of the United Nations Framework Convention on Climate Change, set global targets for reducing greenhouse gas emissions. Technological advancements that result in cleaner air are consistent with the goals of this agreement. Furthermore, air quality is indirectly enhanced by the Montreal Protocol, which mainly targets ozone-depleting compounds, by encouraging the phase-out of particular chemicals that cause air pollution.

National air quality rules are defined by governments and include vehicle types and respective emission restrictions. These rules, which set the maximum allowable emissions of pollutants, promote the development and use of more environmentally friendly technologies. Because of technological advancements in reporting and monitoring that enhance air quality, the public is better informed about air quality in many countries through the use of an Air Quality Index.

Enacting renewable energy policies is crucial to addressing air quality problems. Renewable Portfolio limits set limits for the percentage of energy from renewable sources in order to reduce air pollution from traditional energy sources [137, 138]. A few examples of incentives that hasten the adoption of clean energy technologies and enhance air quality are financial incentives, tax credits, and subsidies.

Transportation regulations that enhance air quality, like vehicle emission standards and public transportation promotion, are driving the development of cleaner engines, alternative fuels, and lower emissions from individual vehicles [46]. Some governments and organizations in developed countries offer incentives to encourage the concept of car pooling, where colleagues with similar commutes share a vehicle for their trips, thereby reducing overall carbon footprints.

Waste management policies, such as waste-to-energy programs and recycling incentives, assist reduce air pollution brought on by reckless garbage disposal and open burning [139, 140]. Also, the encouragement of recycling lessens the need for resource-intensive industrial processes, which in turn reduces associated air pollution.

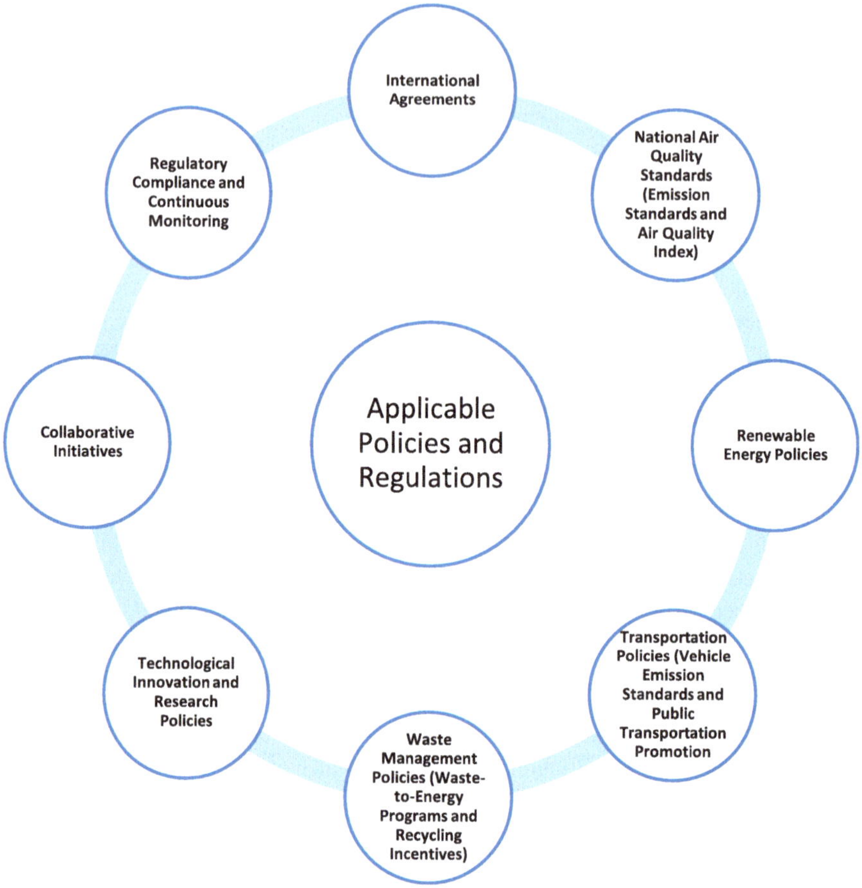

Fig. 9 Key aspects of the applicable policies and regulations driving sustainable technological advances for clean air

Collaborations in the form of public–private partnerships and information exchange can facilitate the development and application of sustainable technology. Additionally, collaborating with corporations, charitable organizations, and governments fosters innovation and accelerates the adoption of clean air technologies.

Enforcement and monitoring are essential components of effective air quality policies. Strict enforcement of rules ensures that emission standards are fulfilled, and continuous monitoring of air quality is maintained. This will enable technological advancements in monitoring systems and boost the overall effectiveness of programs. These many strategies and laws support international efforts to reduce air pollution and provide cleaner air.

9 Conclusion

Air pollution poses serious dangers to the environment, human health, and the economy. It causes respiratory issues, cardiac ailments, and other health problems that lead to millions of premature deaths worldwide. Air pollution reduces biodiversity and causes financial losses in addition to damaging forests, agricultural, and aquatic habitats. It also contributes to reduced vision, acid rain, and climate change, all of which lower our quality of life. The annual expenses of air pollution are estimated to be in the trillions of dollars, indicating its tremendous financial impact. Low-income neighborhoods, children, and the elderly are among the vulnerable groups that are disproportionately affected by air pollution. All things considered, air pollution is a significant global issue that requires prompt attention as well as a concerted response.

The advancement of technology is making it possible to maintain clean air, which is crucial for the environment's well-being. Air pollution is causing many problems for the world, yet one area where progress and hope are consistently seen is in inventive technology. The intricate web of strategies, involving global accords as well as regional enforcement initiatives, demonstrates the global dedication to combating air pollution and its detrimental consequences. Some of these technologies that can reduce emissions into the atmosphere when properly utilized include solar and hydro energy, energy storage systems, microbial fuel cells, and utilization of EVs. Furthermore, international accords like the Paris Agreement support global efforts to cut greenhouse gas emissions and restrict pollutants that degrade air quality. These agreements highlight the interdependence of environmental issues and the need for global cooperation in finding solutions.

References

1. Sawyer WE, Izah SC, Richard G, Ogwu MC (2023) Microbial air contaminants: diversity and health effects. Ann Public Health Res 10(1):1120
2. Richard G, Izah SC, Morufu OR, Austin-Asomeji I (2023) Public and environmental health implications of artisanal petroleum refining and risk reduction strategies in The Niger Delta region of Nigeria. Bio-Research 21(1):1896–1910
3. Richard G, Izah SC, Ibrahim M (2023) Air pollution in The Niger Delta region of Nigeria: sources, health effects, and strategies for mitigation. J Environ Sci 29(1):1–15
4. Richard G, Izah SC, Sawyer EW, Ogwu MC (2023) Seasonal variations in the concentration of some atmospheric pollutants along a major highway in southern Nigeria. J Environ Stud 30(1): 10–25
5. Richard G, Izah SC, Ogwu MC (2022) Implications of artisanal crude oil refining on sustainable food production in The Niger Delta region of Nigeria. J Environ Bioremediat Toxicol 5(2):69–77
6. WHO (2023) Air pollution. https://www.who.int/health-topics/air-pollution#tab=tab_1. Accessed 15 Nov 2023
7. Ohimain EI, Izah SC, Abah SO (2013) Air quality impacts of smallholder oil palm processing in Nigeria. J Environ Prot 4:83–98

8. World Health Organization (2019) Healthy environments for healthier populations: why do they matter and what can we do?. (No. WHO/CED/PHE/DO/19.01). World Health Organization
9. Barman A, Tiwari AK, Dhar P (2024) Breath of the future: advancing air pollution solutions through innovative technologies and global collaboration. In: A comprehensive exploration of soil, water, and air pollution in agriculture. BFC Publications, p 366
10. Mol AP (2003) Globalization and environmental reform: the ecological modernization of the global economy. MIT Press
11. Ohimain EI, Izah SC (2013) Gaseous emissions from a semi-mechanized oil palm processing mill in Bayelsa state, Nigeria. Continent J Water Air Soil Pollut 4(1):15–25
12. Beaudoin J, Farzin YH, Lawell CYCL (2015) Public transit investment and sustainable transportation: a review of studies of transit's impact on traffic congestion and air quality. Res Transp Econ 52:15–22
13. Seiyaboh EI, Izah SC (2019) Impacts of Soil Pollution on Air Quality under Nigerian Setting. Journal of Soil and Water Science, 3(1):45–53
14. Izah SC, Uzoekwe S, Aigberua A (2021) Source, geochemical spreading and risks of trace metals in particulate matter 2.5 within a gas flaring area in Bayelsa State, Nigeria. Adv Environ Technol 7(2):101–118
15. Rafaj P, Kiesewetter G, Gül T, Schöpp W, Cofala J, Klimont Z, Purohit P, Heyes C, Amann M, Borken-Kleefeld J, Cozzi L (2018) Outlook for clean air in the context of sustainable development goals. Glob Environ Chang 53:1–11
16. Awewomom J, Dzeble F, Takyi YD, Ashie WB, Ettey ENYO, Afua PE, Sackey LN, Opoku F, Akoto O (2024) Addressing global environmental pollution using environmental control techniques: a focus on environmental policy and preventive environmental management. Discover Environ 2(1):8
17. Kumar P, Skouloudis AN, Bell M, Viana M, Carotta MC, Biskos G, Morawska L (2016) Real-time sensors for indoor air monitoring and challenges ahead in deploying them to urban buildings. Sci Total Environ 560:150–159
18. Idrees Z, Zheng L (2020) Low cost air pollution monitoring systems: a review of protocols and enabling technologies. J Ind Inf Integr 17:100123
19. Aziz ZAA, Ameen SYA (2021) Air pollution monitoring using wireless sensor networks. J Inf Technol Inform 1(1):20–25
20. Arroyo P, Herrero JL, Suárez JI, Lozano J (2019) Wireless sensor network combined with cloud computing for air quality monitoring. Sensors 19(3):691
21. Sun AY, Scanlon BR (2019) How can big data and machine learning benefit environment and water management: a survey of methods, applications, and future directions. Environ Res Lett 14(7):073001
22. Masih A (2019) Machine learning algorithms in air quality modeling. Glob J Environ Sci Manag 5(4):515–534
23. Belis CA, Larsen BR, Amato F, El Haddad I, Favez O, Harrison RM, Hopke PK, Nava S, Paatero P, Prevot A, Viana M (2014) European guide on air pollution source apportionment with receptor models. JRC reference reports EUR26080 EN
24. Uzoekwe SA, Izah SC, Aigberua AO (2021) Environmental and human health risk of heavy metals in atmospheric particulate matter (PM 10) around gas flaring vicinity in Bayelsa State, Nigeria. Toxicol Environ Heal Sci 13(4):323–335
25. Zhang Y, Bocquet M, Mallet V, Seigneur C, Baklanov A (2012) Real-time air quality forecasting, part I: history, techniques, and current status. Atmos Environ 60:632–655
26. Baawain MS, Al-Serihi AS (2014) Systematic approach for the prediction of ground-level air pollution (around an industrial port) using an artificial neural network. Aerosol Air Qual Res 14(1):124–134
27. Lee M, Lin L, Chen CY, Tsao Y, Yao TH, Fei MH, Fang SH (2020) Forecasting air quality in Taiwan by using machine learning. Sci Rep 10(1):4153

28. Rathore MM, Ahmad A, Paul A, Rho S (2016) Urban planning and building smart cities based on the internet of things using big data analytics. Comput Netw 101:63–80
29. Wilkinson SJ, James K, Reed R (2009) Using building adaptation to deliver sustainability in Australia. Struct Surv 27(1):46–61
30. Phdungsilp A (2010) Integrated energy and carbon modeling with a decision support system: policy scenarios for low-carbon city development in Bangkok. Energy Policy 38(9): 4808–4817
31. Ma J, Ding Y, Cheng JC, Jiang F, Tan Y, Gan VJ, Wan Z (2020) Identification of high impact factors of air quality on a national scale using big data and machine learning techniques. J Clean Prod 244:118955
32. Bera T, Mohanty G, Dwivedi R, Pradhan J (2024) Performance analysis and science mapping on the effect of air pollution on child health: a bibliometric study from 2003–2023. J Public Health 1:1–17
33. Degbelo A, Granell C, Trilles S, Bhattacharya D, Casteleyn S, Kray C (2016) Opening up smart cities: citizen-centric challenges and opportunities from GIScience. ISPRS Int J Geo Inf 5(2):16
34. Yarime M, Li A (2018) Facilitating international cooperation on air pollution in East Asia: fragmentation of the epistemic communities. Global Pol 9:35–41
35. Castells N, Ravetz J (2001) Science and policy in international environmental agreements: lessons from the European experience on transboundary air pollution. Int Environ Agreements 1:405–425
36. Kadri A, Yaacoub E, Mushtaha M, Abu-Dayya A (2013) Wireless sensor network for real-time air pollution monitoring. In: 2013 1st international conference on communications, signal processing, and their applications (ICCSPA). IEEE, pp 1–5
37. Kamyab H, Khademi T, Chelliapan S, SaberiKamarposhti M, Rezania S, Yusuf M, Farajnezhad M, Abbas M, Jeon BH, Ahn Y (2023) The latest innovative avenues for the utilization of artificial intelligence and big data analytics in water resource management. Results Eng 20:101566
38. Ameer S, Shah MA, Khan A, Song H, Maple C, Islam SU, Asghar MN (2019) Comparative analysis of machine learning techniques for predicting air quality in smart cities. IEEE Access 7:128325–128338
39. Huang W, Li T, Liu J, Xie P, Du S, Teng F (2021) An overview of air quality analysis by big data techniques: monitoring, forecasting, and traceability. Inf Fusion 75:28–40
40. Popa CL, Dobrescu TG, Silvestru CI, Firulescu AC, Popescu CA, Cotet CE (2021) Pollution and weather reports: using machine learning for combating pollution in big cities. Sensors 21(21):7329
41. Suzuki H, Cervero R, Iuchi K (2013) Transforming cities with transit: transit and land-use integration for sustainable urban development. World Bank Publications
42. Bibri SE (2023) Data-driven smart eco-cities of the future: an empirically informed integrated model for strategic sustainable urban development. World Futures 79(7–8):703–746
43. Appio FP, Lima M, Paroutis S (2019) Understanding smart cities: innovation ecosystems, technological advancements, and societal challenges. Technol Forecast Soc Chang 142:1–14
44. Patel S, Park H, Bonato P, Chan L, Rodgers M (2012) A review of wearable sensors and systems with application in rehabilitation. J Neuroeng Rehabil 9:1–17
45. Porter ME, Heppelmann JE (2015) How smart, connected products are transforming companies. Harv Bus Rev 93(10):96–114
46. Izah SC, Ohimain EI (2013) The challenge of biodiesel production from oil palm feedstock in Nigeria. Greener J Biol Sci 3(1):1–12
47. Evans A, Strezov V, Evans TJ (2009) Assessment of sustainability indicators for renewable energy technologies. Renew Sust Energ Rev 13(5):1082–1088
48. Østergaard PA, Duic N, Noorollahi Y, Mikulcic H, Kalogirou S (2020) Sustainable development using renewable energy technology. Renew Energy 146:2430–2437

49. Trinh AK, González I, Fournier L, Pelletier R, Lesage FJ (2014) Solar thermal energy conversion to electrical power. Appl Therm Eng 70(1):675–686
50. Tabassum M, Kashem SBA, Siddique MBM (2017) Feasibility of using photovoltaic (PV) technology to generate solar energy in Sarawak. In: 2017 international conference on computer and drone applications (IConDA). IEEE, pp 11–16
51. El Bassam N (2021) Solar energy: technologies and options. In: Distributed renewable energies for off-grid communities. Elsevier, pp 123–147
52. Powell KM, Rashid K, Ellingwood K, Tuttle J, Iverson BD (2017) Hybrid concentrated solar thermal power systems: a review. Renew Sust Energ Rev 80:215–237
53. Philibert C (2005) The present and future use of solar thermal energy as a primary source of energy. International Energy Agency, pp 1–16
54. Obaideen K, Olabi AG, Al Swailmeen Y, Shehata N, Abdelkareem MA, Alami AH, Rodriguez C, Sayed ET (2023) Solar energy: applications, trends analysis, bibliometric analysis and research contribution to sustainable development goals (SDGs). Sustainability 15(2):1418
55. Ghadirinejad N, Ottermo F, Nowzari R, Alsaadi N, Ghadiri Nejad M (2023) Optimizing a green and sustainable off-grid energy-system design: a real case. Sustainability 15(17):12800
56. Hayat MB, Ali D, Monyake KC, Alagha L, Ahmed N (2019) Solar energy—a look into power generation, challenges, and a solar-powered future. Int J Energy Res 43(3):1049–1067
57. Kannan N, Vakeesan D (2016) Solar energy for future world: a review. Renew Sust Energ Rev 62:1092–1105
58. Kumar RR, Kumar A, Ranjan R, Shrivastava S (2017) Off-grid and on-grid connected power generation: a review. Int J Comput Appl 164(9):12–16
59. Bhattacharyya S (2013) Rural electrification through decentralised off-grid systems in developing countries. Springer, London
60. Babayomi OO, Olubayo B, Denwigwe IH, Somefun TE, Adedoja OS, Somefun CT, Olukayode K, Attah A (2023) A review of renewable off-grid mini-grids in Sub-Saharan Africa. Front Energy Res 10:1089025
61. Colak I, Fulli G, Bayhan S, Chondrogiannis S, Demirbas S (2015) Critical aspects of wind energy systems in smart grid applications. Renew Sust Energ Rev 52:155–171
62. Tuballa ML, Abundo ML (2016) A review of the development of smart grid technologies. Renew Sust Energ Rev 59:710–725
63. Buchholz BM, Styczynski Z (2014) Smart grids-fundamentals and technologies in electricity networks, vol 396. Springer, Heidelberg
64. Augustyn M, Lisowski F (2023) Comparison of single and dual coherent blades for a vertical Axis carousel wind rotor using CFD and wind tunnel testing. Appl Sci 13(23):12624
65. Yang C, Jia J, He K, Xue L, Jiang C, Liu S, Zhao B, Wu M, Cui H (2023) Comprehensive analysis and evaluation of the operation and maintenance of offshore wind power systems: a survey. Energies 16(14):5562
66. Jaber S (2013) Environmental impacts of wind energy. J Clean Energy Technol 1(3):251–254
67. Nazir MS, Mahdi AJ, Bilal M, Sohail HM, Ali N, Iqbal HM (2019) Environmental impact and pollution-related challenges of renewable wind energy paradigm–a review. Sci Total Environ 683:436–444
68. Nguyen PQP (2019) Ocean energy-a clean energy source. Eur J Eng Technol Res 4(1):5–11
69. Gorlov AM (2001) Tidal energy. In: Dhanak MR, Xiros NI (eds) Elements of physical oceanography. Springer, Cham, pp 103–108
70. Lim YS, Koh SL (2009) Marine tidal current electric power generation: state of art and current status. Renew Energy:211–226
71. Gonzalez AT, Dunning P, Howard I, McKee K, Wiercigroch M (2021) Is wave energy untapped potential? Int J Mech Sci 205:106544
72. Farrok O, Ahmed K, Tahlil AD, Farah MM, Kiran MR, Islam MR (2020) Electrical power generation from the oceanic wave for sustainable advancement in renewable energy technologies. Sustainability 12(6):2178

73. Cherp A, Adenikinju A, Goldthau A, Hernandez F, Hughes L, Jewell J, Olshanskaya M, Jansen J, Soares R, Vakulenko S (2012) Energy and security. In: Global energy assessment: toward a sustainable future. Cambridge University Press, pp 325–383
74. Almoghayer MA, Woolf DK, Kerr S, Davies G (2022) Integration of tidal energy into an Island energy system–a case study of Orkney islands. Energy 242:122547
75. Kumari WGP, Ranjith PG (2019) Sustainable development of enhanced geothermal systems based on geotechnical research–a review. Earth Sci Rev 199:102955
76. Loewer M, Keim M (2022) Tapping hot rocks: a review of petrothermal energy and enhanced geothermal systems (EGSs). In: Jeguirim M (ed) Renewable energy production and distribution. Academic Press, pp 273–297
77. Olasolo P, Juárez MC, Morales MP, Liarte IA (2016) Enhanced geothermal systems (EGS): a review. Renew Sust Energ Rev 56:133–144
78. Lu SM (2018) A global review of enhanced geothermal system (EGS). Renew Sust Energ Rev 81:2902–2921
79. Shortall R, Davidsdottir B, Axelsson G (2015) Geothermal energy for sustainable development: a review of sustainability impacts and assessment frameworks. Renew Sust Energ Rev 44:391–406
80. Gür TM (2018) Review of electrical energy storage technologies, materials and systems: challenges and prospects for large-scale grid storage. Energy Environ Sci 11(10):2696–2767
81. Fan X, Liu B, Liu J, Ding J, Han X, Deng Y, Lv X, Xie Y, Chen B, Hu W, Zhong C (2020) Battery technologies for grid-level large-scale electrical energy storage. Trans Tianjin Univ 26: 92–103
82. Wang Y, Wang Y, Zhao Z, Zhou Z, Hou Z (2023) Multi-timescale optimal operation strategy for renewable energy power systems based on inertia evaluation. Energies 16(8):3577
83. Martins LS, Guimarães LF, Junior ABB, Tenório JAS, Espinosa DCR (2021) Electric car battery: an overview on global demand, recycling and future approaches towards sustainability. J Environ Manag 295:113091
84. Liu W, Placke T, Chau KT (2022) Overview of batteries and battery management for electric vehicles. Energy Rep 8:4058–4084
85. Wen J, Zhao D, Zhang C (2020) An overview of electricity powered vehicles: Lithium-ion battery energy storage density and energy conversion efficiency. Renew Energy 162:1629–1648
86. Di Lecce D, Verrelli R, Hassoun J (2017) Lithium-ion batteries for sustainable energy storage: recent advances towards new cell configurations. Green Chem 19(15):3442–3467
87. Thackeray MM, Wolverton C, Isaacs ED (2012) Electrical energy storage for transportation—approaching the limits of, and going beyond, lithium-ion batteries. Energy Environ Sci 5(7): 7854–7863
88. Anand D, Kumar M (2024) Study of types of battery technologies in electric vehicles. Darpan Int Res Anal 12(1):16–21
89. Uddin M, Romlie MF, Abdullah MF, Abd Halim S, Kwang TC (2018) A review on peak load shaving strategies. Renew Sust Energ Rev 82:3323–3332
90. Byrne RH, Nguyen TA, Copp DA, Chalamala BR, Gyuk I (2017) Energy management and optimization methods for grid energy storage systems. IEEE Access 6:13231–13260
91. Denholm P, Ela E, Kirby B, Milligan M (2010) Role of energy storage with renewable electricity generation (No. NREL/TP-6A2-47187). National Renewable Energy Lab (NREL), Golden, CO
92. Ameli MT, Jalilpoor K, Amiri MM, Azad S (2021) Reliability analysis and role of energy storage in resiliency of energy systems. In: Energy Storage in Energy Markets. Academic Press, pp 399–416
93. Bhatti HJ, Danilovic M (2018) Making the world more sustainable: enabling localized energy generation and distribution on decentralized smart grid systems. World J Eng Technol 6(2): 350–382

94. Koirala BP, van Oost E, van der Windt H (2018) Community energy storage: a responsible innovation towards a sustainable energy system? Appl Energy 231:570–585
95. Izah SC, Aigberua AO, Srivastav AL (2022) Microbial fuel cells: potentially sustainable technology for bioelectricity production using palm oil mill effluents. In: Dubey AK, Narang SK, Srivastav AL, Kumar A, Garcia-Diaz V (eds) Artificial intelligence for renewable energy systems. Woodhead Publishing Series in Energy. Elsevier, pp 105–129. https://doi.org/10.1016/B978-0-323-90396-7.00014-6
96. Kurniawan TA, Othman MHD, Liang X, Ayub M, Goh HH, Kusworo TD, Mohyuddin A, Chew KW (2022) Microbial fuel cells (MFC): a potential game-changer in renewable energy development. Sustainability 14(24):16847
97. Zhang S, You J, Kennes C, Cheng Z, Ye J, Chen D, Chen J, Wang L (2018) Current advances of VOCs degradation by bioelectrochemical systems: a review. Chem Eng J 334:2625–2637
98. Wu Y, Zhang L (2017) Can the development of electric vehicles reduce the emission of air pollutants and greenhouse gases in developing countries? Transp Res Part D: Transp Environ 51:129–145
99. The Tech Vortex (2024) Energy-efficient transportation systems: the no. 1 positive evolution in transportation for the world. https://the-tech-vortex.com/2023/05/26/efficient-transportation-systems/
100. Adzo K (2024) Earning carbon credits through EV charging: C+ charge's vision. https://startup.info/earning-carbon-credits-through-ev-charging-c-charges-vision/
101. Berkeley N, Bailey D, Jones A, Jarvis D (2017) Assessing the transition towards battery electric vehicles: a multi-level perspective on drivers of, and barriers to, take up. Transp Res A Policy Pract 106:320–332
102. Juan AA, Mendez CA, Faulin J, De Armas J, Grasman SE (2016) Electric vehicles in logistics and transportation: a survey on emerging environmental, strategic, and operational challenges. Energies 9(2):86
103. Delucchi MA, Yang C, Burke AF, Ogden JM, Kurani K, Kessler J, Sperling D (2014) An assessment of electric vehicles: technology, infrastructure requirements, greenhouse-gas emissions, petroleum use, material use, lifetime cost, consumer acceptance and policy initiatives. Philos Trans R Soc A Math Phys Eng Sci 372(2006):20120325
104. Arvesen A, Völler S, Hung CR, Krey V, Korpås M, Strømman AH (2021) Emissions of electric vehicle charging in future scenarios: the effects of time of charging. J Ind Ecol 25(5):1250–1263
105. Farinloye T, Oluwatobi O, Ugboma O, Dickson OF, Uzondu C, Mogaji E (2024) Driving the electric vehicle agenda in Nigeria: the challenges, prospects and opportunities. Transp Res Part D Transp Environ 130:104182
106. Saha D, Carlock G, Shrestha R, Feldmann J, Leslie-Bole H (2021) Building blocks for a low-carbon economy: catalytic policy and infrastructure for decarbonizing the United States by 2050. World Resources Institute
107. Patil P (2021) Innovations in electric vehicle technology: a review of emerging trends and their potential impacts on transportation and society. Rev Contemp Bus Anal 4(1):1–13
108. Egede P, Dettmer T, Herrmann C, Kara S (2015) Life cycle assessment of electric vehicles–a framework to consider influencing factors. Procedia Cirp 29:233–238
109. Jin L, Slowik P (2017) Literature review of electric vehicle consumer awareness and outreach activities. International Council on Clean Transportation. Available frominternet: https://www.theicct.org/sites/default/files/publications/Consumer-EV-Awareness_ICCT_Working-Paper_23032017_vF.pdf
110. Cho DJ (2005) Scaling back the mineral problem. https://eponline.com/articles/2005/11/01/scaling-back-the-mineral-problem.aspx
111. Dodd N, Donatello S, Garbarino E, Gama-Caldas M (2015) Identifying macro-objectives for the life cycle environmental performance and resource efficiency of EU buildings. European Commission

112. Kabeyi MJB, Olanrewaju OA (2022) Sustainable energy transition for renewable and low carbon grid electricity generation and supply. Front Energy Res 9
113. Dutton AG, Halliday JA, Blanch MJ (2005) The feasibility of building-mounted/integrated wind turbines (BUWTs): achieving their potential for carbon emission reductions. Energy Research Unit, CCLRC, pp 77–83
114. Omer AM (2008) Energy, environment and sustainable development. Renew Sust Energ Rev 12(9):2265–2300
115. Metallidou CK, Psannis KE, Egyptiadou EA (2020) Energy efficiency in smart buildings: IoT approaches. IEEE Access 8:63679–63699
116. US Green Building Council (2000) Leadership in energy and environmental design. US Green Building Council, Washington, DC. http://www.usgbc.org. Nov. 2009
117. Loftness V, Hakkinen B, Adan O, Nevalainen A (2007) Elements that contribute to healthy building design. Environ Health Perspect 115(6):965–970
118. Hanie O, Aryan A, MohammadReza L, Elham L (2010) Understanding the importance of sustainable buildings in occupants environmental health and comfort. J Sustain Dev 3(2):194
119. Liu G, Xiao M, Zhang X, Gal C, Chen X, Liu L, Pan S, Wu J, Tang L, Clements-Croome D (2017) A review of air filtration technologies for sustainable and healthy building ventilation. Sustain Cities Soc 32:375–396
120. Tran VV, Park D, Lee YC (2020) Indoor air pollution, related human diseases, and recent trends in the control and improvement of indoor air quality. Int J Environ Res Public Health 17(8):2927
121. Aditya ROY (2018) A review of general and modern methods of air purification. J Thermal Eng 5(2):22–28
122. Krigmont HV (2022) Development of the disinfecting air filter. In: International conference on electrostatic precipitation. Springer International Publishing, Cham, pp 202–226
123. Li T (2021) Air quality and energy implications of filtration system and operation strategies in residential buildings., Doctoral dissertation. University of Toronto
124. Elsaid AM, Ahmed MS (2021) Indoor air quality strategies for air-conditioning and ventilation systems with the spread of the global coronavirus (COVID-19) epidemic: improvements and recommendations. Environ Res 199:111314
125. Bolashikov ZD, Melikov AK (2009) Methods for air cleaning and protection of building occupants from airborne pathogens. Build Environ 44(7):1378–1385
126. Kelly FJ, Fussell JC (2019) Improving indoor air quality, health and performance within environments where people live, travel, learn and work. Atmos Environ 200:90–109
127. Shrivastava P (2018) Environmental technologies and competitive advantage. In: Business ethics and strategy, volumes I and II. Routledge, pp 317–334
128. Al-khuzaei M, Abushaikha A (2022) Numerical simulation of CO2 sequestration into depleted natural gas reservoirs of Qatar. In: Fourth EAGE WIPIC workshop. (Vol. 2022, No. 1). European Association of Geoscientists & Engineers, pp 1–6
129. Martin-Roberts E, Scott V, Flude S, Johnson G, Haszeldine RS, Gilfillan S (2021) Carbon capture and storage at the end of a lost decade. One Earth 4(11):1569–1584
130. Bui M, Adjiman CS, Bardow A, Anthony EJ, Boston A, Brown S, Fennell PS, Fuss S, Galindo A, Hackett LA, Mac Dowell N (2018) Carbon capture and storage (CCS): the way forward. Energy Environ Sci 11(5):1062–1176
131. Lai NYG, Yap EH, Lee CW (2011) Viability of CCS: a broad-based assessment for Malaysia. Renew Sust Energ Rev 15(8):3608–3616
132. Paltsev S, Morris J, Kheshgi H, Herzog H (2021) Hard-to-abate sectors: the role of industrial carbon capture and storage (CCS) in emission mitigation. Appl Energy 300:117322
133. Bates A, Lai M, Thao W (2023) Grab and gone: expert perspectives on innovation to diffusion of direct air carbon capture and storage technology. Carbon Manag 14(1):2235577
134. Fagorite VI, Onyekuru SO, Opara AI, Oguzie EE (2023) The major techniques, advantages, and pitfalls of various methods used in geological carbon sequestration. Int J Environ Sci Technol 20(4):4585–4614

135. Amir Rashidi MR, Dabbi EP, Abu Bakar ZA, Misnan MS, Pedersen C, Wong KY, Sallehud-Din MTM, Shamsudin MA, Wo S (2020) A field case study of modelling the environmental fate of leaked CO2 gas in the marine environment for carbon capture and storage CCS. In: SPE Asia Pacific oil and gas conference and exhibition. SPE, p D013S103R005
136. Vergragt PJ, Markusson N, Karlsson H (2011) Carbon capture and storage, bio-energy with carbon capture and storage, and the escape from the fossil-fuel lock-in. Glob Environ Chang 21(2):282–292
137. Berry T, Jaccard M (2001) The renewable portfolio standard:: design considerations and an implementation survey. Energy Policy 29(4):263–277
138. Zhou S, Solomon BD (2020) Do renewable portfolio standards in the United States stunt renewable electricity development beyond mandatory targets? Energy Policy 140:111377
139. Kumar A, Agrawal A (2020) Recent trends in solid waste management status, challenges, and potential for the future Indian cities–a review. Curr Res Environ Sustain 2:100011
140. Naworio NB (2022) Waste-to-energy project: an integrated sustainable waste management in Yenagoa Metropolis., Master Thesis. Savonia University of Applied Sciences

Health and Environmental Justice Strategies for Mitigating Air Pollution

Sylvester Chibueze Izah, Chinwe A. Onwudiegwu, Ligeiaziba Sylva, and Nsikak G. Etim

Contents

S. C. Izah (✉)
Department of Microbiology, Faculty of Science, Bayelsa Medical University, Yenagoa, Bayelsa State, Nigeria

Department of Community Medicine, Faculty of Clinical Sciences, Bayelsa Medical University, Yenagoa, Bayelsa State, Nigeria
e-mail: sylvester.izah@bmu.edu.ng

C. A. Onwudiegwu
African Institute for Science Policy and Innovation, Obafemi Awolowo University Ile-Ife, Osun State, Nigeria

L. Sylva
Department of Mathematics, Faculty of Science, Bayelsa Medical University, Yenagoa, Bayelsa State, Nigeria

N. G. Etim
Department of Medical Laboratory Science, Faculty of Basic Medical Science, Niger Delta University, Wilberforce Island, Bayelsa State, Nigeria

Matthew Chidozie Ogwu and Sylvester Izah (eds.), *Sustainable Strategies for Air Pollution Mitigation: Development, Economics and Technologies*, Hdb Env Chem (2024) 133: 311–342, DOI 10.1007/698_2024_1121,

Published online: 21 September 2024

Abstract Air pollution poses significant health risks, exacerbating asthma and adversely affecting respiratory and cardiovascular systems, while increasing the likelihood of heart disease and stroke. Marginalized populations, often in underprivileged communities, frequently experience disproportionate exposure to toxins due to socioeconomic inequalities. The concepts in health and environmental justice that concerns air quality involve equity vs. equality, the cumulative impact of environmental and social stressors, participatory approaches in environmental decision-making, and the preventive and precautionary principle. Strategies rooted in environmental justice and health seek to rectify these disparities by advocating for a fair distribution of environmental benefits and adverse burdens. This chapter focuses on Health and Environmental Justice Strategies for Mitigating Air Pollution. The strategies, which have their origins in the civil rights movements, strive to ensure universal access to clean air and a healthy environment, regardless of financial status. These efforts often employ various tactics, including health impact assessments that integrate health considerations into decision-making processes for policies and programs. Cumulative risk assessments often evaluate compounded risks from multiple sources, guiding the prioritization of interventions in high-risk areas. Social determinants of health theory often inform strategies by addressing factors influencing susceptibility to pollution exposure. Participatory action research often empowers communities to advocate for tailored solutions, enhancing environmental conditions collaboratively. Climate justice frameworks can link climate change mitigation with improved air quality, advocating for equitable distribution of benefits. The precautionary principle guides proactive measures in uncertain situations to protect public health amidst evolving scientific understanding. Therefore, effective strategies involve implementing energy-saving practices, promoting sustainable transportation, and enhancing urban green spaces to mitigate air pollution. Also, policies should ensure equitable access to clean air, enforce stricter emissions regulations, and prioritize renewable energy sources. Integrated health and environmental justice measures can foster resilient and equitable communities by addressing the main causes of air pollution and minimizing immediate health impacts.

Keywords Air pollutants, Air pollution, Environmental impacts, Environmental justice, Health impacts, Sustainable systems

Abbreviations

HIA	Health impact assessment
CRA	Cumulative risk assessments
SDH	Social determinants of health
PAR	Participatory action research
HEJ	Health and environmental justice
HIA	Health impact assessment
SO_2	Sulfur dioxide
PM	Particulate matter
VOCs	Volatile organic compounds
NOx	Nitrogen oxides
CO	Carbon monoxide
EVs	Electric vehicles

1 Introduction

Air pollution is a major environmental issue that poses serious threat to human health and the ecology [1–3]. Several factors contribute to the complex mix of pollutants in the air we breathe, including household activities, car exhaust, and industrial emissions [4, 5]. It has been well established that air pollution negatively affects health, contributing to cardiovascular and respiratory disorders; [6–8] often leading to unfavorable pregnancy outcomes and early mortality. Furthermore, air pollution harms ecosystems, endangers species, and disrupts climate stability [9]. As urbanization and industrialization increase, the challenge of reducing air pollution becomes more complex. Thus, there is an urgent need to consider public health and environmental justice in efforts to mitigate the impacts of air pollution.

Air pollution has clear detrimental effects on health, disproportionately harming vulnerable populations such as children, the elderly, and those with underlying medical conditions [4, 10, 11]. This is because, in some regions it is primarily the poor who work as scavengers and are thus exposed to pollutants, especially from waste dumpsites. Additionally, low-income workers who are employed in factories and industrial areas often live nearby to cut down on transportation costs. The intersection of social and environmental inequalities emphasizes the need for targeted measures that prioritize health and environmental justice.

Understanding environmental justice concepts and perspectives is crucial for effectively addressing the unequal distribution of environmental risks and benefits [12, 13]. Regardless of race, color, country of origin, or income level, it is important to treat all individuals equally and to involve all stakeholders in the meaningful development, implementation, and observance of environmental laws, rules, and policies. This suggests that every community has access to clean air and that no area is subjected to pollution levels that exceed acceptable limits. Thus, it is vital to pursue environmental justice which entails reducing emissions and granting local communities the power to influence policies that affect their health and environment [14, 15].

Seeing that air pollution is one of the most urgent global environmental health challenges, posing significant risks to human health and the environment [5, 9, 16], there is immediate need for effective management of this issue. This involves a thorough understanding of the complex associations between socioeconomic, public health, and pollution sources. Thus, frameworks for environmental justice and health are important pathways through which government can formulate policies that guarantee inclusive and equitable solutions [17] while also efficiently reducing air pollution.

Achieving a sustainable environment, requires inclusive plans that incorporate new technologies and regulatory policies to effectively minimize air pollution. Additionally, enforcing compliance and implementing stringent emission regulations are essential strategies for reducing pollution at its source. Moreover, pollution can be significantly reduced through the use of technological solutions such as advanced filtration systems, clean energy, and electric vehicles [18–20]. However, these strategies must also include initiatives that promote social equity such as enhancing green spaces, improving public transportation, and supporting neighborhood-led environmental advocacy and monitoring.

Automobile emissions are one of the primary sources of air pollution, especially in urban areas. This pollution can be significantly reduced by increasing the use of public transportation. By reducing the number of private automobiles on the road through the development and enhancement of public transportation options, cities can lower pollution and traffic congestion. Additionally, infrastructure development for active transportation modes like walking and bicycling can be reduce pollution while also promoting healthier lifestyles [21]. These modes of transportation not only improve air quality but also promote social justice by offering affordable mobility options to all residents.

Investing in environmentally friendly infrastructure is another crucial strategy for reducing air pollution. Urban green spaces such as parks and urban woodlands can absorb pollutants, mitigate urban heat islands, and provide recreational opportunities that enhance both mental and physical well-being [22–24]. Green infrastructure not only can helps to manage stormwater, reduce runoff, and improve water quality but also addresses environmental challenges and enhances community resilience and well-being through urban development that incorporates green spaces [25, 26].

Community involvement and education are essential components of effective air pollution mitigation strategies. When locals are informed about the sources of air pollution and its effects on health, they can advocate for cleaner air and participate in local decision making [27]. An example of a community-led initiative that can provide valuable data to inform policy and hold polluters accountable is a neighborhood air monitoring program [28, 29]. Additionally, fostering partnerships between governmental entities, nonprofit organizations, and community associations can improve the effectiveness and sustainability of air pollution reduction programs.

This chapter explores various strategies for reducing air pollution and addressing related environmental injustices and health disparities. It offers an overview of the technological and regulatory approaches to pollution control, as well as investments in green infrastructure and community-based initiatives. The chapter also highlights the importance of integrating health and environmental justice considerations in air pollution mitigation programs to ensure that the benefits of cleaner air are equitably distributed. A key benefit of this study is that it provides information on strategies for reducing air pollution with a focus on both public health and environmental justice.

2 Concepts in Health and Environmental Justice

The Health and Environmental Justice (HEJ) framework is illustrates the link between social justice and environmental health. It seeks to ensure equitable access to a clean and healthy environment for all communities, particularly focusing on disadvantaged and marginalized populations. Figure 1 summarizes HEJ ideas related

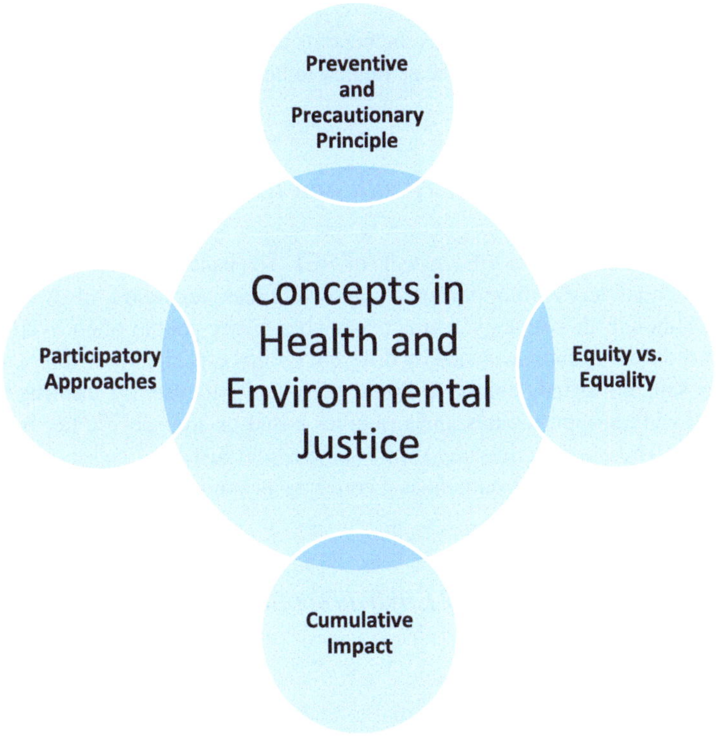

Fig. 1 Major concepts in health and environmental justice

Table 1 The benefits and implications of major concepts in health and environmental justice as related to air quality

Concept	Air quality benefits	Implications
Equity vs. equality in health and environmental justice	Targeted interventions enhance air quality in the most affected areas, thereby reducing health disparities	It can ensure that resources are allocated based on need, promoting fair outcomes and reducing pollution-related health risks
Cumulative impact of environmental and social stressors	Comprehensive strategies that address multiple pollution sources and stressors, lead to significant air quality improvements	It acknowledges the interconnectedness of various factors, leading to more holistic and effective pollution mitigation efforts
Participatory approaches in environmental decision-making	Community input leads to tailored, context-specific solutions that are more likely to be both effective and sustainable	It enhances policy relevance and effectiveness by incorporating the experiences and needs of those most impacted by air pollution
Preventive and precautionary principle in health and environmental justice	Proactive measures prevent air pollution, safeguarding public health before harm occurs	It emphasizes the importance of taking action despite uncertainty to protect health and the environment

to air pollution. Table 1 summarizes the benefits and implications of major concepts in health and environmental justice as related to air quality.

2.1 Equity vs. Equality in Health and Environmental Justice

Equity and equality are two vital aspects of HEJ. To ensure equality, all individuals must have equal access to opportunities and resources, regardless of their circumstances. Although this strategy seems reasonable, many people often overlook the diverse needs and challenges faced by different groups especially in countries where regulations are not strictly enforced. Conversely, equity focuses on ensuring fair and just by allocating opportunities and resources based on the specific needs of each community. Therefore, addressing the systematic and historical causes of the inequities in health and the environment as it concerns air pollution is necessary [30, 31].

2.2 Cumulative Impact of Environmental and Social Stressors

The cumulative effect theory is essential to understanding the complex interactions between social and environmental factors that impact marginalized communities. These communities often encounter multiple environmental hazards, including emissions from automobiles, industrialization, urbanization, population growth, and inadequate housing. These risks are often compounded by social factors such as poverty, institutional discrimination, and limited access to healthcare. The simultaneous exposure to these factors can lead to interaction in ways that exacerbate preexisting health disorders and potentially lead to the development of newhealth concerns. For example, in a low-income neighborhood with high levels of air pollution, chronic stress, poor nutrition, and limited access to healthcare can contribute to a higher prevalence of respiratory ailments.

2.3 Participatory Approaches in Environmental Decision-Making

Participatory techniques, which emphasize the importance of involving affected individuals in the decision-making processes that impact their environment and health, are a crucial component of health and environmental justice [32–34]. This approach ensures that the perspectives of those impacted most by environmental hazards are taken into consideration when developing policies, regulations, and actions. Participatory techniques can be used in a variety of ways, including public hearings, collaborative research initiatives, and community advisory boards. Through proactive resident involvement, these methodologies can help build trust and ensure that interventions are tailored to the community's specific needs that are culturally appropriate. Therefore, by involving the community, for instance, in the identification of pollution sources, consistent air quality monitoring can lead to more accurate and widely acceptable solutions. Participatory models can also empower communities with the capacity to demand accountability from legislators and polluters, thus enabling them to defend their rights [35–37].

2.4 Preventive and Precautionary Principle in Health and Environmental Justice

The precautionary and preventative principle is a fundamental component of health and environmental justice [38–40]. This principle advocates for proactive measures to safeguard the environment and public health from potential harm. This principle also supports the idea that preventing harm is preferable to responding to it after it has already occurred. It suggests implementing policies and legislation to reduce or eliminate exposure to environmental hazards even before conclusive evidence of their detrimental health impact is established.

3 Advancing Equity and Health: Perspectives of Air Pollution Mitigation Frameworks

In order to sustainably integrate frameworks to mitigate air pollution it is necessary to assess the complex associations that exist between the precautionary principle, environmental justice, participatory action research (PAR), cumulative risk assessment (CRA), social determinants of health (SDH), and health impact assessment (HIA). Reducing disparities in air pollution exposure, which disproportionately affects impoverished groups is crucial for achieving sustainable clean air. By incorporating these frameworks into policy and practice, the program aims to strengthen public health safeguards, empower communities in decision-making, and promote equitable distribbution of environmental benefits and responsibilities. This section explores some of these theories and frameworks.

3.1 Theories of Environmental Justice

The concept of environmental justice highlights the importance of equitable distribution of environmental benefits and expenses [41, 42]. These theories emphasize that disadvantaged populations often face disproportionate environmental concerns such as air pollution. These communities are often located near industrial regions, highways, and other sources of pollution, especially in developing countries.

3.2 Health Impact Assessment (HIA) Framework

The HIA framework is a critical tool for assessing the potential health impacts of initiatives, programs, and policies on human populations [43, 44]. Screening, scoping, evaluation, recommendations, reporting, and monitoring are integral to this systematic framework. By incorporating health concerns related to air pollution exposure into decision-making processes, HIAs ensure that health impacts particularly those affecting vulnerable groups are explicitly considered. This approach aligns with environmental justice theories and promotes legislation that protects public health and reduces health disparities.

3.3 Cumulative Risk Assessment (CRA) Framework

The cumulative risk assessment (CRA) paradigm is another essential tool for managing the combined risks resulting from numerous stressors and sources of pollution [45, 46]. Traditional risk assessments usually focus on individual pollutants, whereas CRA considers the combined impact of exposure to multiple toxins on health. This method provides a more comprehensive understanding of risk and is particularly relevant for communities facing various environmental stresses.

3.4 Social Determinants of Health Theory (SDH)

The social determinants of health (SDH) theory states that social, economic, and environmental factors influence health outcomes [47, 48]. According to this theory, a person's income, employment, education, and neighborhood conditions all have a significant influence on their overall health and well-being. From an air pollution perspective, SDH theory highlights how socioeconomic disparities result in varying degrees of exposure and sensitivity to pollution. Targeting these fundamental characteristics can lead to initiatives that enhance community resilence to air pollution and reduce health inequities.

3.5 Participatory Action Research (PAR) Methodology

The participatory action research (PAR) approach prioritizes the views and experiences of those most affected by air pollution when developing mitigation strategies [49]. PAR can promote context-specific, culturally appropriate solutions by involving communities in identifying problems, designing interventions, and evaluating outcomes. This inclusive approach aligns environmental justice values by empowering marginalized communities to advocate for improved environmental conditions.

3.6 Climate Justice Framework

The climate justice paradigm clarifies the connections between social justice, air pollution, and climate change [50]. This idea recognizes that existing inequalities can be exacerbated by climate change, with the impoverished communities suffering the most. Environmental and human health benefits can coexist when air pollution is reduced through climate-friendly measures lowering greenhouse gas emissions and transitioning to renewable energy sources [51].

3.7 The Principle of Precaution

The precautionary principle is a risk management strategy that advocates for preventive action in the face of uncertainty [52]. This approach is particularly relevant to air pollution, where the health effects of many toxins are still under study. Governments can reduce emissions and exposure, even when scientific data is inconclusive by adopting preventive measures. This proactive approach to preventing harm and protecting public health aligns with the goals of environmental justice and public health [39, 53].

Combining these frameworks and theories can provide a solid foundation for developing comprehensive strategies to lower air pollution. To create and enhance inclusive, effective, and equitable policies and programs, a combination of environmental justice, HIA, CRA, SDH, PAR, climate justice, and the precautionary principle can be utilized. These strategies address both the immediate health consequences of air pollution and the underlying social and environmental determinants of health.

4 Air Pollution and Environmental Justice

Air pollution is a key concern within the health and environmental justice (HEJ) paradigm because of its extensive impacts and disproportionate effects on vulnerable communities. Vulnerable communities and cities are often located near high concentrations of pollution sources, such as busy highways, industrial facilities, and agricultural operations [54, 55]. Many pollutants, including sulfur dioxide (SO_2), particulate matter (PM), volatile organic compounds (VOCs), nitrogen oxides (NOx), and carbon monoxide (CO) [6] are released into the atmosphere by industrial facilities such as factories and refineries. Air pollution is primarily caused by emissions from transportation sources, such as trucks, buses, and cars; agricultural activities including, the use of fertilizer and insecticides that release ammonia and other harmful substances; and the burning of fossil fuels for home heating and wood stoves which can increase indoor air pollution.

4.1 Health Impacts of Air Pollution

Air pollution is harmful to bad for your health, increases the risk of cancer, heart disease, lung problems, developmental abnormalities, and other diseases [56–58] (Fig. 2). This section of the chapter discusses the health impacts of air pollutants.

4.1.1 Respiratory Diseases

Air pollutants, like SO_2, NOx, and PM, are major contributors to respiratory illnesses. Particulate pollution, particularly fine particles, can penetrate the lungs, causing inflammation and worsening respiratory conditions including asthma, bronchitis (increased bronchial reactivity), increased susceptibility to infection and chronic obstructive pulmonary disease (COPD) [56, 58]. Automobile exhaust and industrial activities are the primary sources of NOx emissions. These irritants have the potential to aggravate respiratory conditions over time. Burning of fossil fuels releases SO_2, a gas that can cause both short-term and long-term respiratory problems. Vulnerable groups, including young children, the elderly, and those with pre-existing respiratory conditions, are particularly at risk to these comtaminants, which can lead to deteriorating lung function and increased hospital admissions.

4.1.2 Cardiovascular Diseases

Long-term exposure to air pollution is directly associated with various cardiovascular illnesses. Fine particulate matter ($PM_{2.5}$) can penetrate the lungs and enter the bloodstream causing systemic inflammation and oxidative stress [56]. This can play a role in the development of atherosclerosis, a disorder in which fatty deposits obstruct blood vessels. The resulting decrease in blood flow can lead to heart attacks and strokes. Pollutants like CO and NOx can also elevate blood pressure, impair heart function and increase the risk of cardiovascular events [59–61]. Studies have shown that residing in areas with high levels of air pollution is associated with an increased risk of heart disease and the related mortality. Specifically, Izzotti et al. [56] reported particulate matter can lead to myocardial infarction, heart failure, change in ventricular repolarization, and hypertension.

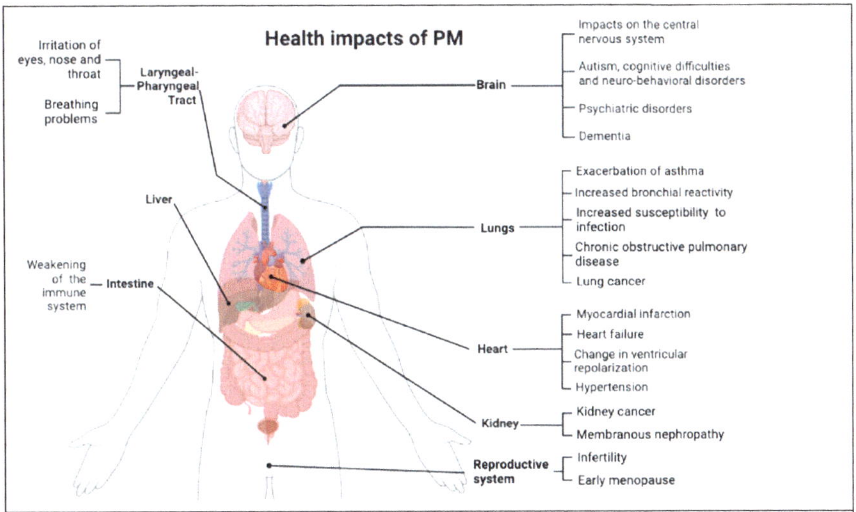

Impacts of air pollution on human health. Izzotti et al. 56.
https://creativecommons.org/licenses/by/4.0/

Fig. 2 Impacts of air pollution on human health. (Source: [56–58]; https://creativecommons.org/licenses/by/4.0/)

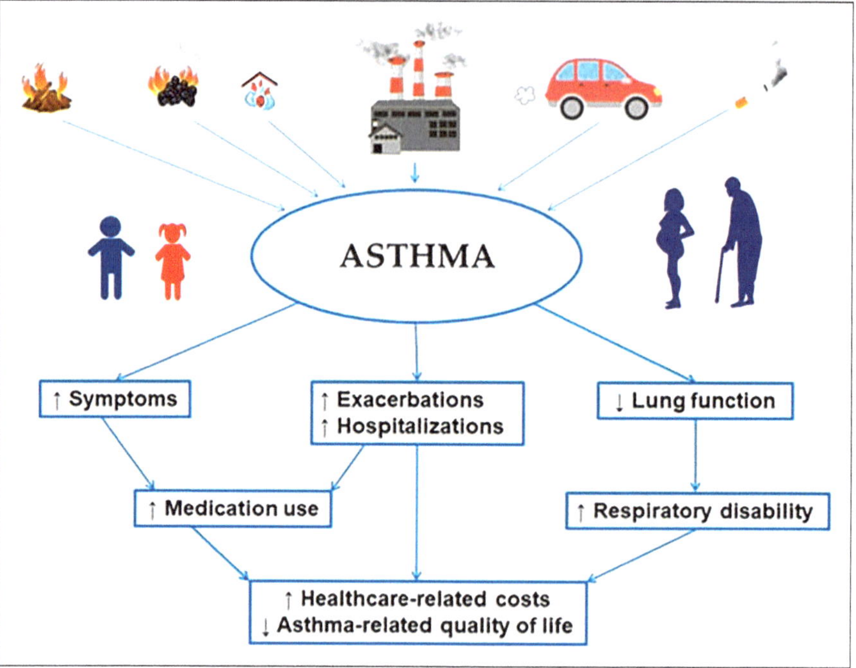

Impact of air pollution on asthma. Tiotiu et al. 58.

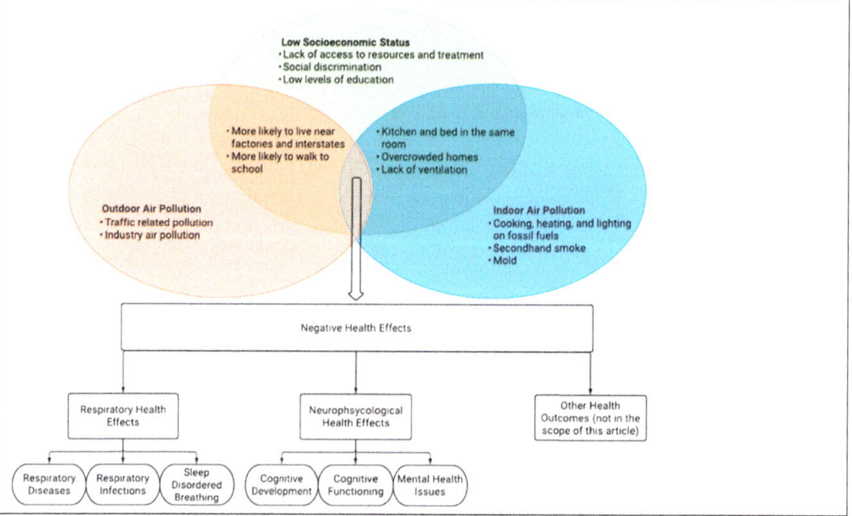

Environmental injustice. The detrimental intersection of outdoor and indoor air pollution with socioeconomic status and its consequences for children's health. Mathiarasan and Hüls (57)

Fig. 2 (continued)

4.1.3 Cancer

Certain air pollutants are known carcinogens and significantly increase the risk of developing cancer. For example, benzoene, found in car emissions and industrial processes, has been linked to leukemia and other blood cancers [62]. Formaldehyde, used in building and household products, has been associated with nasopharyngeal cancer and leukemia [63]. Polycyclic aromatic hydrocarbons, produced from the incomplete combustion of organic materials, have been linked to lung, bladder, and skin cancers [64–66]. Extended exposure to these carcinogens poses a significant health risk to individuals residing in affected area, particularly for those with limited access to healthcare and preventive measures.

4.1.4 Developmental and Neurological Effects

Air pollution significantly impacts fetal growth and nervous system health. Exposure to lead, mercury, and $PM_{2.5}$ during pregnancy can disrupt normal fetal development, leading to low birth weight, premature delivery, and developmental issues [67, 68]. Children exposed to high levels of air pollution are more likely to exhibit behavioral issues, lower IQs, and delayed cognitive development [69–72]. Lead, a neurotoxic pollutant, can cause long-term damage to the developing brain, leading to cognitive deficiencies and learning difficulties. Air pollution has also been linked to neurological diseases such as Parkinson's and Alzheimer's. Prolonged exposure to pollutants can accelerate cognitive decline and increase the risk of developing these conditions [73, 74]. Specifically, Izzotti et al. [56] reported that particulate matter could contribute to dementia, psychiatric disorders, autism, cognitive difficulties, and neuro-behavioral disorders as well as affect the central nervous system.

4.1.5 Premature Death

Children and the elderly are the most vulnerable groups when it comes to preventable deaths caused by air pollution, which is a serious global problem. The World Health Organization estimates that air pollution causes millions of preventable deaths annually by worsening conditions such as cancer, brain, heart, kidney and lung diseases, particularly in areas with air quality [56].

4.1.6 Developmental and Reproductive Issues

Pollutants like lead and mercury are especially harmful to fetal development and reproductive health [75]. Exposure to these dangerous substances during pregnancy can result in adverse birth outcomes, such as low birth weight, congenital defects, and preterm delivery [75]. Lead can cause developmental delays and learning problems in children even at low concentrations [76, 77]. Mercury exposure has been linked to brain damage that impairs movement and cognitive function [78]. Adults exposed to air pollution may experience hormonal abnormalities and weakened reproductive systems [79, 80], potentially leading to infertility and complications during pregnancy. For instance, Izzotti et al. [56] reported that exposure to particulate matter can result in infertility and early menopause.

4.2 Disproportionate Exposure

Due to institutional factors perpetuating environmental injustice, marginalized communities often live in areas with higher air pollution levels. These communities often live near industrial areas, highways, and landfills (Fig. 3), which are major sources of air pollution. Low-income and minority communities are often situated in high-exposure zones. Furthermore, their limited resources and political influence can result to inadequate regulatory oversight and poor enforcement of environmental protections in these areas. Financial constraints often limit these communities'

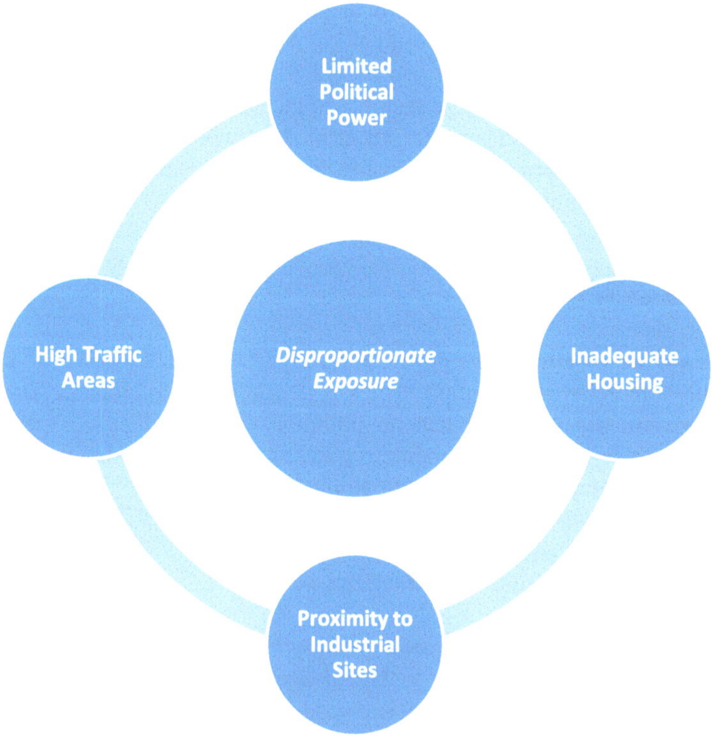

Fig. 3 Disproportionate exposure to air pollution, and health and environmental justice perspective

ability to relocate to areas with better air quality making residents more susceptible to respiratory conditions, heart issues, and other health issues associated with poor air quality.

In many countries in the Global South, communities are often located near multiple industrial facilities, such as power plants, refineries, and factories largely due to economic inequality. These industrial facilities emit a range of pollutants, including hazardous air pollutants (HAPs) such as formaldehyde and benzene as well as PM, SO_2, nitrogen NOx, and VOCs. Residents living close to these industrial plants are routinely exposed to toxic fumes, increasing their risk of cancer, heart problems, respiratory ailments, and other conditions. Futhermore, frequent releases of toxins during routine operations and industrial emissions can have immediate health effects and long-term negative impacts on the environment's quality of life and quality of life. The disproportionate placement of industrial facilities in these areas examplifies environmental injustice, as residents often lack the financial resources and political influence to combat or mitigate the effects of industrial pollution.

Poorer communities located near major roads, highways, and transit hubs, often face elevated pollution levels from vehicular emissions. Vehicle emissions include pollutants such as NO_2, O_3, CO, and fine particulate matter (PM2.5), which can deeply penetrate the lungs and bloodstream, leading to a range of health problems [81–83]. Living near busy highways is associated with an increased risk of asthma, especially in children, as well as other respiratory disorders and cardiovascular issues. Additionally, the constant noise and stress from heavy traffic can worsen preexisting health conditions, such as sleep difficulties.

Unhealthy living conditions can significantly worsen indoor air pollution, leading to serious health issues for residents. Inadequate housing often has poor ventilation systems, which can trap indoor pollutants such as fungi spores, dust particles, and VOCs from building materials and household goods. Poor-quality construction and maintenance materials can contaminate indoor air by releasing hazardous chemicals. Lead paint and asbestos, commonly found in older homes that pose serious health risks when disturbed [84, 85]. Insufficient housing may force residents to rely on poorly maintained heating systems or biomass fuels, which can worsen indoor air pollution [86, 87]. This can lead to allergies, respiratory illnesses, and other health problems particularly in vulnerable populations such as youngsters and the elderly.

Communities facing high levels of air pollution often lack the political power needed to influence environmental regulations and promote cleaner air [88]. This lack of power is driven by several factors, such as systemic racism, historical disenfranchisement, and socioeconomic challenges [89–91].

5 Strategies for Mitigating Air Pollution

To effectively mitigate air pollution and address health and environmental justice issues, robust strategies are required (Fig. 4). Key strategies for reducing car emissions include strengthening industry emissions regulations, promoting the use

Fig. 4 Strategies for mitigating air pollution

of electric and low-emission vehicles [92, 93], and improving public transit infrastructure. Additionally, increasing green space in cities can help absorb pollutants [94]. Policies that prioritize renewable energy sources over fossil fuels are also crucial. Many countries, especially in the Global North, are actively seeking renewable energy solutions. Community involvement and education are essential for ensuring that vulnerable groups are informed about and can participate in pollution control efforts. Equitable enforcement of environmental regulations contributes to cleaner air for all communities.

5.1 Regulatory Measures

Effective regulatory actions are necessary to reduce air pollution, especially when considering environmental justice and public health [95, 96]. Improved air quality and health outcomes will benefit all individuals in the area, particularly the most vulnerable and disadvantaged communities.

5.1.1 Stricter Emission Standards

Enforcing stringent emission regulations for automobiles, industries, and other sources of pollution is one of the most effective and direct ways to reduce air pollution [97, 98]. Regulatory agencies establish limits on pollutants, such SO_2, NO_2, PM, and VOCs, that can be released into the atmosphere [6, 7, 99, 100]. Implementing stringent emission laws can encourage the use of cleaner fuels and advanced emission-reduction technologies in vehicles. By targeting emissions at the source, these regulations can significantly improve air quality, especially in cities with heavy traffic and industrial activity [101]. Strict enforcement of these regulations ensures compliance, protecting public health and reducing the prevalence of illnesses associated with air pollution.

5.1.2 Air Quality Monitoring

Enhanced monitoring of air pollution can help identify pollution sources and hotspots, providing real-time data on pollutant levels. This information is crucial for developing targeted emission control strategies. Additionally, making this data publicly accessible allows communities to hold polluters accountable and advocate for cleaner air [102, 103]. By identifying problem areas and evaluating the effectiveness of regulatory activities, comprehensive monitoring supports the creation of policies that can lead to improved air quality.

5.1.3 Cap-and-Trade Programs

Market-based strategies such as "cap-and-trade" regimes can incentivize industries to reduce emissions by allowing companies to buy and sell pollution allowances [104, 105]. A regulatory authority sets a cap on the total amount of specific pollutants that can be emitted. Companies are then allocated emission allowances which they can trade with other businesses, creating financial incentives for reducing emissions [106]. Businesses with lower-cost options for reducing their emissions can sell their excess allowances to companies facing higher reductions costs. Cap-and-trade programs have proven effective in reducing emissions of pollutants, like sulfur dioxide and nitrogen oxides, which are major contributors to acid rain [107].

5.2 *Technological Innovations for Mitigating Air Pollution*

5.2.1 Clean Energy Transition

Transitioning to renewable energy sourcesis crucial for mitigating the harmful effects of air pollution on human health. Utilising renewable energy sources such as wind, solar, and hydro can significantly reduce our reliance on fossil fuels, which are a major source of air pollution [108–110]. Additionally, transitioning to sustainable energy is essential for advancing environmental justice. It can address the disproportionate exposure to air pollutants faced by vulnerable populations. Utilizing renewable energy can help decrease health disparities by reducing air pollution and improving air quality in vulnerable areas.

5.2.2 Electric Cars

Encouraging the use of electric vehicles (EVs) is another key strategy for reducing emissions from the transportation sector, a major source of urban air pollution. EVs improve urban air quality immediately because they produce no exhaust emissions. Widespread adoption of EVs, can significantly reduce pollutants ants like NOx and PM, which are associated with various health issues. From an environmental justice perspective, promoting EVs can benefit communities disproportionately affected by traffic-related air pollution [111]. Policy initiativessuch as subsidies and the development of a charging infrastructure can make EVs more accessible across different socioeconomic backgrounds, ensuring equitable benefits for all.

5.2.3 Technologies for Reducing Emissions

Modern emission control methods are essential for preventing pollutants from getting into the environment. Technologies such as scrubbers, filters, and catalytic converters can significantly reduce emissions of SO_2, NOx, VOCs, and PM [112–115]. These technologies not only help industries comply with environmental regulations but also protect public health by reducing exposure to hazardous contaminants. For te sake of environmental justice, it is crucial that these technologies are uniformly implemented across all industries, particularly in vulnerable areas, especially in the Global South.

5.3 *Community Engagement and Education*

Community engagement and education are vital for reducing air pollution and promoting environmental justice and health. Active participation and education can drive action, influence policy positively, and improve living conditions.

5.3.1 Public Awareness Campaigns

Public awareness campaigns are essential for educating people about the sources, consequences and mitigation of air pollution. These campaigns can utilize radio, television, social media, live events, and other media platforms to spread information. By highlighting the connection between air pollution and a various health issues, especially in vulnerable populations, these campaigns can enhance public understanding and drive action [116, 117]. Additionally, posters and billboards can offer practical advice on reducing personal exposure to pollutants, such as using air purifiers, staying indoors on high pollution days, and supporting clean energy initiatives [118]. Awareness programs can also drive policy reforms and encourage community involvement fostering a culture that prioritizes environmental and health-conscious.

5.3.2 Engagement of the Community

Involving communities in planning and decision-making processes is crucial to ensuring that their needs and concerns about clean air are addressed. This participatory approach fosters a sense of trust and ownership among community members [102]. At town hall meetings, citizen advisory committees, and public consultations, locals can voice their concerns, share their experiences, and propose solutions. By incorporating local expertise and perspectives, policymakers can create more equitable and effective air quality management strategies [119]. Additionally, participating in community-based air quality monitoring initiatives can offer valuable insights into pollution sources and distribution, facilitating more targeted mitigation measures.

5.3.3 Committees and Associations

Supporting community-led initiatives and organizations that advocate for environmental justice and cleaner air is necessary for inclusive and sustainable progress [120]. Grassroots efforts are often rooted in a deep understanding of local issues and are driven by a commitment to community welfare. These initiatives can take various forms such as campaigns for stricter pollution controls on nearby industrial plants or local air quality monitoring initiatives. They also have the potential to mobilize the public and draw lawmakers' attention to environmental justice issues.

5.4 Policy and Planning for Health, Environmental Justice, and Air Pollution

5.4.1 Integrated Environmental Policies

Effective air quality control requires the development of an integrated environmental policy. Air quality management plans should be aligned with other environmental and health regulations to create a cohesive strategy that addresses multiple issues simultaneously. For example, policies could integrate efforts to improve air quality with strtegies to reduce greenhouse gas emissions, addressing both climate change and air pollution simultaneously [121, 122]. This approach ensures that air quality improvement programs do not unintentionally harm soil fertility, water quality or food production.

5.4.2 Planning for Land Use

Effective land-use planning is essential for reducing air pollution and enhancing environmental justice [123]. Implementing legislation that restricts the placement of polluting industries near residential areas is one way planners can protect communities from harmful pollution. This is particularly important in low-income and vulnerable communities. Addtionally, expanding green spaces in urban areas can significantly enhance air quality [124].

5.4.3 Evaluations of Health Impact

Conduct health impact assessments (HIAs) on proposed projects is essential for safeguarding public health, particularly in vulnerable areas. HIAs assess potential health risks and benefits, enabling decision-makers to prioritize public health and make informed choices [125]. This is especially critical in areas with existing environmental injustices, as new programs could potentially exacerbate existing health disparities.

5.5 Economic and Social Interventions for Mitigating Air Pollution

Social and economic considerations are integral components of a comprehensive strategy for effective air pollution management. These actions ensures that everyone benefits equally from cleaner air while also reducing pollution.

5.5.1 Funding and Incentives

Financial incentives for businesses and individuals adopting greener technologies and practices are a key aspect of economic intervention [126]. Governments can provide tax exemptions, subsidies, and low-interest loans to promote the use of electric vehicles, renewable energy installations, and energy-efficient appliances. Financial incentives encourage businesses and families to invest in cleaner technologies, thereby reducing overall emissions. For exampe, tax breaks or holidays for electric cars can decrease the number of gas-powered vehicles on the road [127], while subsidies for solar and wind power plants can accelerate the shift to renewable energy sources. These financing mechanisms support both economic growth and environmental objectives by fostering innovation and creating new markets for eco-friendly technologies.

5.5.2 Employment Training Programs

Developing employment training programs for workers in polluting industries is essential for ensuring a just transition to a greener economy [128, 129]. As societies transition to more environmentally friendly practises, workers in industries like manufacturing, oil and gas extraction, and coal mining may face job losses. Job training programs can assist these individuals in acquiring the skills needed for employment in greener sectors such as renewable energy [130].

5.5.3 Equity in Policy Implementation

Protecting the most vulnerable communities requires the equitable implementation of policies and measures aimed at reducing air pollution [131, 132]. Air pollution has historically disproportionately affected communities, particularly in certain countries in the Global South. These areas often have fewer green spaces, higher levels of traffic-related pollution, and greater industrial emissions. Therefore, ensuring that policy implementation is equity-focused is crucial for prioritizing vulnerable groups in pollution reduction measures.

5.6 Green Spaces and Urban Forests

Investing in green spaces and urban forests is a critical component of green infrastructure and sustainable urban development [133, 134]. Parks, playgrounds, ans similar spaces not only offer leisure possibilities but also enhance mental and physical health by fostering social interaction, encouraging physical activity, and offering spaces for relaxation. Urban forests are essential for improving air quality due to their dense tree canopies, which absorb harmful pollutants such as carbon dioxide, sulfur dioxide, and nitrogen oxides [135, 136]. Trees also mitigate the urban heat island effect by providing shade and releasing water vapor through transpiration, which cools the surrounding air [137, 138].

5.6.1 Stormwater Management

Stormwater management is essential to mitigate the impact of urbanization on the natural water cycles, stormwater management is essential with rain gardens, permeable pavements, and green roofs being key components of this strategy [139, 140]. Rain gardens are designed to collect and filter runoff, reducing the burden on stormwater systems and preventing pollutants from entering water bodies [141, 142]. Permeable pavements reduce surface runoff and the likelihood of flooding by allowing water to seep through surfaces. Green roofs absorb rainfall and insulate buildings, and help mitigate the effects of urban heat islands.

5.6.2 Conservation and Management of Water Resources

Water resources must be carefully managed and protected to support sustainable urban development. Rainwater harvesting systems reduce the demand on municipal water supplies by collecting and storing rainwater for non-potable use, such as irrigation and toilet flushing [143, 144]. Constructed wetlands are an innovative approach to naturally filter wastewater and stormwater. They enhance water quality and biodiversity while offering communities with access to leisure and educational opportunities.

5.6.3 Energy Efficiency and Renewable Energy

Incorporating energy efficiency and renewable energy sources into urban infrastructure in essential to reduce reliance on fossil fuels and mitigate climate change [145]. Solar panels and wind turbines can be installed on buildings and in public spaces to generate renewable energy and reduce greenhouse gas emissions. Energy-efficient buildings, designed with materials and systems that minimize energy usage, help reduce energy demands and carbon footprints [146–148]. These buildings often feature high-performance insulation, energy-efficient windows, and advanced heating, ventilation, and air conditioning systems.

5.6.4 Environmentally Friendly Vehicles

Reducing air pollution and enhancing urban mobility require the promotion of ecologically sustainable transportation options. Public transportation systems can lower PM, NOx, and CO emissions by reducing the number of private vehicles on the road. Promoting non-motorized transportation through the construction of bike lanes, walkways, and other bicycle and pedestrian-friendly infrastructure helps reduce air pollution and traffic congestion [149–151]. Additionally, encouraging the usage of electric and low-emission vehicles can significantly lower emissions from the transportation sector.

5.6.5 Urban Agriculture

Urban agriculture initiatives, such as rooftop and community gardens, provide significant benefits to cities. Community gardens provide access to fresh vegetables and enhance local food security and foster community involvement. These gardens also serve as green spaces that support urban biodiversity and offer residents valuable learning opportunities. Rooftop gardens utilize underused to reduce stormwater runoff, grow food, and insulate buildings by making use of underused rooftop space [152]. These urban green spaces enhance air quality, and support sustainable food systems.

5.6.6 Biodiversity and Habitat Conservation

Preservation of natural habitats and conserving biodiversity are essential components of green infrastructure. Wildlife corridors enables animals to move safely across between ecosystems, supporting healthy populations [153–156]. By providing food and shelter for native species while requiring less water and maintenance, Native plants used in landscaping enhance the health of neighboring ecosystems by providing food and shelter for native species while requiring less water and maintenance. Such programs offer ecological services, protect biodiversity, and contribute to creating resilient urban environments.

5.6.7 Adaptability to Climate Change

To prepare cities for future climate challenges, urban development must incorporate climate resilience. Incorporating of climate projections into infrastructure development enhances the resilience of buildings and public places against extreme weather events and changing environmental conditions [157, 158]. This approach could involve elevating structures, enhancing drainage systems, and using materials resistant to temperature changes. By implementing climate-resilient policies, cities can protect their citizens, infrastructure, and ecosystems from the impacts of climate change.

6 Conclusion

Air pollution has numerous adverse effects on health, including worsening chronic conditions, increasing mortality rates, and impairing the respiratory and cardiovascular systems. Exposure to air pollution worsens these health effects for vulnerable population, highlighting the need for targeted interventions. Therefore, addressing air pollution from the perspectives of environmental justice and health requires a comprehensive understanding of several fundamental concepts. The debate between equality and equity in environmental justice and health underscores the need for targeted strategies that address the specific needs of marginalized populations. Equality involves providing everyone with the same resources, while equity requires allocating resources based on individual needs to achieve fair outcomes. This means that addressing air pollution, solutions should prioritize groups that are disproportionately affected by poor air quality. Furthermore, the cumulative impact of environmental and social stressors illustrates how racial disparities, social position, and environmental exposures intersect to exacerate health inequities. For example, low-income communities often have limited access to healthcare and are exposed to higher pollution levels, increasing health risks for residents. Implementing participatory approaches in environmental decision-making ensures that affected communities are involved in shaping public policy. This approach ensures that the experiences of people most affected by air pollution inform policy, leading to more equitable and effective solutions. Additionally, the precautionary principle underscores the importance of taking proactive measures to reduce air pollution and safeguard public health, especially in when complete scientific certainty is not available.

It is also crucial to use diverse theoretical frameworks and methodologies to address air pollution and promote equality and health. Environmental justice theories provide a foundational understanding of how environmental problems disproportionately affect marginalized communities and highlight the need for equitable solutions. To ensure that health considerations are integrated throughout the decision-making process, policies, projects, or programs are assessed for potential health impacts using the HIA framework. Additionally, the CRA framework offers a comprehensive perspective on environmental health issues by evaluating the cumulative dangers of several contaminants and stressors.

The SHD theory explores the connections between socioeconomic variables and health outcomes, including community environment, education, and economic stability. Applying this theory to air pollution reduction initiatives helps address the underlying causes of health disparities. The PAR approach allows communities to actively engage in research and policy formulation, more inclusive and practical solutions. The climate justice framework highlights the disproportionate impact of climate change on vulnerable populations, linking environmental justice and climate action. The precautionary principle, by prioritizing environmental preservation and public health, advocates for preventive measures in the face of uncertainty.

To reduce air pollution effectively, a comprehensive plan that considers both health and environmental justice is essential. Successful air pollution reduction programs integrate a range of actions including government initiatives, community involvement, and technological solitions. Stricter emissions regulations and the enforcement of air quality laws are crucial for minimizing sources of pollution. Sustainable emissions reduction can be achieved through technology advancements including improved pollution management and the use of greener energy sources. Community involvement and education are essential for raising awareness, supporting community-led activities, and ensuring that interventions are appropriate for the location context and culture.

To address the social and environmental determinants of health, cross-sectoral methods must be incorporated into policy and planning for environmental justice, health, and air pollution. Air pollution can be reduced through social and economic measures such as expanding healthcare access and providing incentives for sustainable energy. Enhancing urban forests and green spaces not only provides recreational areas but also improves air quality, and supports mental and physical health, particularly in urban settings.

References

1. Richard G, Izah SC, Morufu OR, Austin-Asomeji I (2023) Public and environmental health implications of artisanal petroleum refining and risk reduction strategies in The Niger Delta region of Nigeria. Bio-Research 21(1):1896–1910
2. Richard G, Izah SC, Ibrahim M (2023) Air pollution in The Niger Delta region of Nigeria: sources, health effects, and strategies for mitigation. J Environ Sci 29(1):1–15
3. Richard G, Izah SC, Ogwu MC (2022) Implications of artisanal crude oil refining on sustainable food production in The Niger Delta region of Nigeria. J Environ Bioremediat Toxicol 5(2):69–77
4. Sawyer WE, Izah SC, Richard G, Ogwu MC (2023) Microbial air contaminants: diversity and health effects. Annal Public Health Res 10(1):1120
5. Sawyer WE, Izah SC (2023) Particulate matter pollution and health status of residents in Bayelsa State, Nigeria: an assessment. Annal Commun Med Pract 8(1):1052. https://doi.org/10.47739/2475-9465/1052
6. Seiyaboh EI, Izah SC (2019) Impacts of soil pollution on air quality under Nigerian setting. J Soil Water Sci 3(1):45–53
7. Ohimain EI, Izah SC, Abah SO (2013) Air quality impacts of smallholder oil palm processing in Nigeria. J Environ Prot 4:83–98
8. Ohimain EI, Izah SC (2013) Gaseous emissions from a semi-mechanized oil palm processing mill in Bayelsa state, Nigeria. Continental J Water Air Soil Pollut 4(1):15–25
9. Izah SC, Uzoekwe SA, Aigberua AO (2021) Source, geochemical spreading and risks of trace metals in particulate matter 2.5 within a gas flaring area in Bayelsa State, Nigeria. Adv Environ Technol 7(2):101–118
10. Alvarez HAO, Kubzansky LD, Campen MJ, Slavich GM (2018) Early life stress, air pollution, inflammation, and disease: an integrative review and immunologic model of social-environmental adversity and lifespan health. Neurosci Biobehav Rev 92:226–242
11. Perera FP (2017) Multiple threats to child health from fossil fuel combustion: impacts of air pollution and climate change. Environ Health Perspect 125(2):141–148
12. Mohai P, Pellow D, Roberts JT (2009) Environmental justice. Annu Rev Environ Resour 34: 405–430
13. Laurent E (2011) Issues in environmental justice within the European Union. Ecol Econ 70(11):1846–1853
14. Méndez M (2020) Climate change from the streets: how conflict and collaboration strengthen the environmental justice movement. Yale University Press
15. Kaswan A (1997) Environmental justice: bridging the gap between environmental laws and justice. Am UL Rev 47:221
16. Uzoekwe SA, Izah SC, Aigberua AO (2021) Environmental and human health risk of heavy metals in atmospheric particulate matter (PM_{10}) around gas flaring vicinity in Bayelsa State, Nigeria. Toxicol Environ Heal Sci 13(4):323–335. https://doi.org/10.1007/s13530-021-00085-7

17. Agyeman J, Bullard RD, Evans B (2002) Exploring the nexus: bringing together sustainability, environmental justice and equity. Space and Polity 6(1):77–90
18. Li Z, Khajepour A, Song J (2019) A comprehensive review of the key technologies for pure electric vehicles. Energy 182:824–839
19. Sun X, Li Z, Wang X, Li C (2019) Technology development of electric vehicles: a review. Energies 13(1):90
20. Mehar S, Zeadally S, Remy G, Senouci SM (2014) Sustainable transportation management system for a fleet of electric vehicles. IEEE Trans Intell Transp Syst 16(3):1401–1414
21. Reynolds C, Winters M, Ries F, Gouge B (2010) Active transportation in urban areas: exploring health benefits and risks. National Collaboration Centre for Environmental Health, p 2
22. Sadeghian MM, Vardanyan Z (2013) The benefits of urban parks, a review of urban research. J Novel Appl Sci 2(8):231–237
23. Vargas-Hernández JG, Pallagst K, Zdunek-Wielgołaska J (2023) Urban green spaces as a component of an ecosystem. In: Sustainable development and environmental stewardship: global initiatives towards engaged sustainability. Springer International Publishing, Cham, pp 165–198
24. Vargas-Hernández JG, Pallagst K, Zdunek-Wielgołaska J (2017) Urban green spaces as a component of an ecosystem. Functions, services, users, community involvement, initiatives and actions. Rev Urban 37:1–26
25. Reyes-Riveros R, Altamirano A, De La Barrera F, Rozas-Vásquez D, Vieli L, Meli P (2021) Linking public urban green spaces and human well-being: a systematic review. Urban For Urban Green 61:127105
26. Wolch JR, Byrne J, Newell JP (2014) Urban green space, public health, and environmental justice: the challenge of making cities 'just green enough'. Landsc Urban Plan 125:234–244
27. Dorfman P, Leksmono N, Burnet F, Gibbs D, Longhurst J, Weitkamp E (2006) A preliminary review of local air quality management consultation practices in England. WIT Trans Ecol Environ 86:145–153
28. Tu WL (2019) Combating air pollution through data generation and reinterpretation: community air monitoring in Taiwan. East Asian Sci Technol Soc 13(2):235–255
29. Commodore A, Wilson S, Muhammad O, Svendsen E, Pearce J (2017) Community-based participatory research for the study of air pollution: a review of motivations, approaches, and outcomes. Environ Monit Assess 189:1–30
30. Gee GC, Payne-Sturges DC (2004) Environmental health disparities: a framework integrating psychosocial and environmental concepts. Environ Health Perspect 112(17):1645–1653
31. Wakefield SE, Baxter J (2010) Linking health inequality and environmental justice: articulating a precautionary framework for research and action. Environ Justice 3(3):95–102
32. Foster S (2002) Environmental justice in an era of devolved collaboration. Harv Envtl L Rev 26:459
33. Holifield R (2012) Environmental justice as recognition and participation in risk assessment: negotiating and translating health risk at a superfund site in Indian country. Ann Assoc Am Geogr 102(3):591–613
34. Bullard RD, Johnson GS (2000) Environmentalism and public policy: environmental justice: grassroots activism and its impact on public policy decision making. J Soc Issues 56(3): 555–578
35. Weber EP (2003) Bringing society back in: grassroots ecosystem management, accountability, and sustainable communities. MIT Press
36. Agyeman J, Angus B (2003) The role of civic environmentalism in the pursuit of sustainable communities. J Environ Plan Manag 46(3):345–363
37. Freudenberg N, Pastor M, Israel B (2011) Strengthening community capacity to participate in making decisions to reduce disproportionate environmental exposures. Am J Public Health 101(S1):S123–S130

38. Stacey J (2017) Preventive justice, the precautionary principle and the rule of law. In: Regulating preventive justice. Routledge, pp 23–39
39. Cameron J, Abouchar J (1991) The precautionary principle: a fundamental principle of law and policy for the protection of the global environment. BC Int'l & Comp L Rev 14:1
40. Martuzzi M, Tickner JA (2004) 1. Introduction--the precautionary principle: protecting public health, the environment and the future of our children. WHO Regional Publications, European Series
41. Ikeme J (2003) Equity, environmental justice and sustainability: incomplete approaches in climate change politics. Glob Environ Chang 13(3):195–206
42. Okereke C (2006) Global environmental sustainability: Intragenerational equity and conceptions of justice in multilateral environmental regimes. Geoforum 37(5):725–738
43. Cole BL, Fielding JE (2007) Health impact assessment: a tool to help policy makers understand health beyond health care. Annu Rev Public Health 28:393–412
44. Fakhri A, Harris P, Maleki M (2015) Proposing a framework for health impact assessment in Iran. BMC Public Health 15:1–7
45. MacDonell MM, Hertzberg RC, Rice GE, Wright JM, Teuschler LK (2018) Characterizing risk for cumulative risk assessments. Risk Anal 38(6):1183–1201
46. Tong R, Zhang B (2024) Cumulative risk assessment for combinations of environmental and psychosocial stressors: a systematic review. Integr Environ Assess Manag 20(3):602–615
47. Dean HD, Williams KM, Fenton KA (2013) From theory to action: applying social determinants of health to public health practice. Public Health Rep 128(6_suppl3):1–4
48. Shokouh SMH, Mohammad ARAB, Emamgholipour S, Rashidian A, Montazeri A, Zaboli R (2017) Conceptual models of social determinants of health: a narrative review. Iran J Public Health 46(4):435
49. Davis LF, Ramírez-Andreotta MD (2021) Participatory research for environmental justice: a critical interpretive synthesis. Environ Health Perspect 129(2):026001
50. Widener P, Rowe C (2018) Climate discourse: eluding literacy, justice & inclusion, by evading causation, privilege & diversity. Environ Sociol 4(1):162–174
51. Wan P, Zhang Z, Chen L (2024) Environmental co-benefits of climate mitigation: evidence from clean development mechanism projects in China. China Econ Rev 85:102182
52. Kriebel D, Tickner J, Epstein P, Lemons J, Levins R, Loechler EL, Quinn M, Rudel R, Schettler T, Stoto M (2001) The precautionary principle in environmental science. Environ Health Perspect 109(9):871–876
53. Levy BS (ed) (2006) Occupational and environmental health: recognizing and preventing disease and injury. Lippincott Williams & Wilkins
54. Molina MJ, Molina LT (2004) Megacities and atmospheric pollution. J Air Waste Manage Assoc 54(6):644–680
55. Houston D, Wu J, Ong P, Winer A (2004) Structural disparities of urban traffic in Southern California: implications for vehicle-related air pollution exposure in minority and high-poverty neighborhoods. J Urban Aff 26(5):565–592
56. Izzotti A, Spatera P, Khalid Z, Pulliero A (2022) Importance of punctual monitoring to evaluate the health effects of airborne particulate matter. Int J Environ Res Public Health 19(17):10587
57. Mathiarasan S, Hüls A (2021) Impact of environmental injustice on children's health—interaction between air pollution and socioeconomic status. Int J Environ Res Public Health 18(2):795
58. Tiotiu AI, Novakova P, Nedeva D, Chong-Neto HJ, Novakova S, Steiropoulos P, Kowal K (2020) Impact of air pollution on asthma outcomes. Int J Environ Res Public Health 17(17): 6212
59. Yang BY, Qian Z, Howard SW, Vaughn MG, Fan SJ, Liu KK, Dong GH (2018) Global association between ambient air pollution and blood pressure: a systematic review and meta-analysis. Environ Pollut 235:576–588

60. Münzel T, Gori T, Al-Kindi S, Deanfield J, Lelieveld J, Daiber A, Rajagopalan S (2018) Effects of gaseous and solid constituents of air pollution on endothelial function. Eur Heart J 39(38):3543–3550
61. de Bont J, Jaganathan S, Dahlquist M, Persson Å, Stafoggia M, Ljungman P (2022) Ambient air pollution and cardiovascular diseases: an umbrella review of systematic reviews and meta-analyses. J Intern Med 291(6):779–800
62. Loomis D, Guyton KZ, Grosse Y, El Ghissassi F, Bouvard V, Benbrahim-Tallaa L, Guha N, Vilahur N, Mattock H, Straif K (2017) Carcinogenicity of benzene. Lancet Oncol 18(12): 1574–1575
63. Kim KH, Jahan SA, Lee JT (2011) Exposure to formaldehyde and its potential human health hazards. J Environ Sci Health C 29(4):277–299
64. Kim KH, Jahan SA, Kabir E, Brown RJ (2013) A review of airborne polycyclic aromatic hydrocarbons (PAHs) and their human health effects. Environ Int 60:71–80
65. Kumar V, Kothiyal NC, Saruchi V, P., & Sharma, R. (2016) Sources, distribution, and health effect of carcinogenic polycyclic aromatic hydrocarbons (PAHs)–current knowledge and future directions. J Chin Adv Mater Soc 4(4):302–321
66. Dybing E, Schwarze PE, Nafstad P, Victorin K, Penning TM (2013) Polycyclic aromatic hydrocarbons in ambient air and cancer. Air Pollut Cancer 161:1
67. Ibrahimou B, Salihu H, Gasana J, Owusu H (2014) Risk of low birth weight and very low birth weight from exposure to particulate matter (PM2. 5) speciation metals during pregnancy. Gynecol Obs 4:2161–0932
68. Meyyazhagan A, Bhotla HK, Tsibizova V, Pappuswamy M, Chaudhary A, Arumugam VA, Al Qasem M, Di Renzo GC (2023) Nutrition paves the way to environmental toxicants and influences fetal development during pregnancy. Best Pract Res Clin Obstet Gynaecol 89: 102351
69. Suades-González E, Gascon M, Guxens M, Sunyer J (2015) Air pollution and neuropsychological development: a review of the latest evidence. Endocrinology 156(10):3473–3482
70. Clifford A, Lang L, Chen R, Anstey KJ, Seaton A (2016) Exposure to air pollution and cognitive functioning across the life course–a systematic literature review. Environ Res 147: 383–398
71. Sunyer J, Esnaola M, Alvarez-Pedrerol M, Forns J, Rivas I, López-Vicente M, Suades-González E, Foraster M, Garcia-Esteban R, Basagaña X, Querol X (2015) Association between traffic-related air pollution in schools and cognitive development in primary school children: a prospective cohort study. PLoS Med 12(3):e1001792
72. Kerin T, Volk H, Li W, Lurmann F, Eckel S, McConnell R, Hertz-Picciotto I (2018) Association between air pollution exposure, cognitive and adaptive function, and ASD severity among children with autism spectrum disorder. J Autism Dev Disord 48:137–150
73. Costa LG, Cole TB, Dao K, Chang YC, Coburn J, Garrick JM (2020) Effects of air pollution on the nervous system and its possible role in neurodevelopmental and neurodegenerative disorders. Pharmacol Ther 210:107523
74. Kilian J, Kitazawa M (2018) The emerging risk of exposure to air pollution on cognitive decline and Alzheimer's disease–evidence from epidemiological and animal studies. Biom J 41(3):141–162
75. Sewberath Misser VH, Hindori-Mohangoo AD, Shankar A, Wickliffe JK, Lichtveld MY, Mans DR (2022) Prenatal exposure to mercury, manganese, and Lead and adverse birth outcomes in Suriname: a population-based birth cohort study. Toxics 10(8):464
76. Izah SC, Richard G, Stanley HO, Sawyer WE, Ogwu MC, Uwaeme OR (2023) Integrating the one health approach and statistical analysis for sustainable aquatic ecosystem management and trace metal contamination mitigation. ES Food Agrofor 14:1012. https://doi.org/10.30919/esfaf1012
77. Izah SC, Chakrabarty N, Srivastav AL (2016) A review on heavy metal concentration in potable water sources in Nigeria: human health effects and mitigating measures. Expo Health 8:285–304

78. Yokoo EM, Valente JG, Grattan L, Schmidt SL, Platt I, Silbergeld EK (2003) Low level methylmercury exposure affects neuropsychological function in adults. Environ Health 2:1–11
79. Carré J, Gatimel N, Moreau J, Parinaud J, Léandri R (2017) Does air pollution play a role in infertility?: a systematic review. Environ Health 16:1–16
80. Joffe M (2003) Infertility and environmental pollutants. Br Med Bull 68(1):47–70
81. Maji S, Ahmed S, Kaur-Sidhu M, Mor S, Ravindra K (2023) Health risks of major air pollutants, their drivers and mitigation strategies: a review. Air Soil Water Res 16: 11786221231154659
82. Tanwar V, Adelstein JM, Wold LE (2021) Double trouble: combined cardiovascular effects of particulate matter exposure and coronavirus disease 2019. Cardiovasc Res 117(1):85–95
83. Ahmadi S, AlKafaas SS, Aziz SA, Ammar EE, Elsalahaty MI, Bedair H, Oroke A, Zafer MM, Pourebrahimi S, Ghosh S (2024) Effects of particulate matter on human health. In: Health and environmental effects of ambient air pollution. Academic Press, pp 35–70
84. Brazeal M, Jacobs DE (2024) The home environment. In: Textbook of children's environmental health, vol 186. Oxford Academic
85. Ponessa JT (1992) Environmental and safety issues in housing. Hous Soc 19(2):59–73
86. Ezzati M, Kammen DM (2002) The health impacts of exposure to indoor air pollution from solid fuels in developing countries: knowledge, gaps, and data needs. Environ Health Perspect 110(11):1057–1068
87. Jin Y, Ma X, Chen X, Cheng Y, Baris E, Ezzati M (2006) Exposure to indoor air pollution from household energy use in rural China: the interactions of technology, behavior, and knowledge in health risk management. Soc Sci Med 62(12):3161–3176
88. Melnick RS (2010) Regulation and the courts: the case of the clean air act. Brookings Institution Press
89. Braveman PA, Arkin E, Proctor D, Kauh T, Holm N (2022) Systemic and structural racism: definitions, examples, health damages, and approaches to dismantling: study examines definitions, examples, health damages, and dismantling systemic and structural racism. Health Aff 41(2):171–178
90. Blessett B, Littleton V (2017) Examining the impact of institutional racism in black residentially segregated communities. Ralph Bunche J Public Aff 6(1):3
91. Morey BN, Tanjasiri SP, Subica AM, Kaholokula JKA, Penaia C, Thomas K, Chang RC, Tran VD, Ponce NA, Ong P, Ong E (2020) Structural racism and its effects on native Hawaiians and Pacific islanders in the United States: issues of health equity, census undercounting, and voter disenfranchisement. AAPI Nexus Policy Pract Commun 17(1–2)
92. Sechel IC, Mariasiu F (2021) Efficiency of governmental policy and programs to stimulate the use of low-emission and electric vehicles: the case of Romania. Sustainability 14(1):45
93. Gao J, Huang J, Li X, Tian G, Wang X, Yang C, Ma C (2022) Challenges of the UK government and industries regarding emission control after ICE vehicle bans. Sci Total Environ 835:155406
94. Selmi W, Weber C, Rivière E, Blond N, Mehdi L, Nowak D (2016) Air pollution removal by trees in public green spaces in Strasbourg city, France. Urban For Urban Green 17:192–201
95. Mitchell G, Norman P, Mullin K (2015) Who benefits from environmental policy? An environmental justice analysis of air quality change in Britain, 2001–2011. Environ Res Lett 10(10):105009
96. Bhatia R, Wernham A (2009) Integrating human health into environmental impact assessment: an unrealized opportunity for environmental health and justice. Ciênc Saúde Colet 14:1159–1175
97. Feng L, Liao W (2016) Legislation, plans, and policies for prevention and control of air pollution in China: achievements, challenges, and improvements. J Clean Prod 112:1549–1558
98. Faiz A, Weaver CS, Walsh MP (1996) Air pollution from motor vehicles: standards and technologies for controlling emissions. World Bank Publications, Washington, DC

99. Richard G, Sawyer WE, Izah SC (2021) Outdoor air quality index of biomass combustion in The Niger Delta, Nigeria: a health impact perspective. J Adv Res Med Sci Technol 8(1):19–28
100. Richard G, Sawyer WE, Izah SC (2021) Health implications of air quality index of fine and coarse particulates during outdoor combustion of biomass in The Niger Delta, Nigeria. Int J World Policy Dev Stud 7(2):21–26
101. Colvile RN, Hutchinson EJ, Mindell JS, Warren RF (2001) The transport sector as a source of air pollution. Atmos Environ 35(9):1537–1565
102. Kelly FJ, Fuller GW, Walton HA, Fussell JC (2012) Monitoring air pollution: use of early warning systems for public health. Respirology 17(1):7–19
103. Kraft ME, Stephan M, Abel TD (2011) Coming clean: information disclosure and environmental performance. MIT Press
104. Behr T, Witte JM, Hoxtell W, Manzer J (2009) Towards a global carbon market. In: Potential and limits of carbon market integration, GPPI policy paper, p 5
105. Bergsager H (2013) China's state-owned enterprises as climate policy actors the power and steel sectors. Nordic Council of Ministers
106. Raiborn C, Massoud M (2010) Emissions allowances: accounting and public policy issues. Account Public Interest 10(1):105–121
107. Ross, R. L. (2011). Driving sustainable reductions in greenhouse gas emissions: more than just cap-&-trade. Available at SSRN 1863046
108. Bilgen S, Kaygusuz K, Sari A (2004) Renewable energy for a clean and sustainable future. Energy Sources 26(12):1119–1129
109. Abbasi T, Abbasi SA (2012) Is the use of renewable energy sources an answer to the problems of global warming and pollution? Crit Rev Environ Sci Technol 42(2):99–154
110. Pimentel D, Rodrigues G, Wang T, Abrams R, Goldberg K, Staecker H, Ma E, Brueckner L, Trovato L, Chow C, Boerke S (1994) Renewable energy: economic and environmental issues. Bioscience 44(8):536–547
111. Love I, Emanuel R, Echeverri DP (2024) Plugging into justice: merging environmental justice practices into the transition to electric vehicles. file:///C:/Users/hp/Downloads/Masters%20Project%20Final%20Paper-2.pdf
112. Sarfraz M (2021) Green technologies to combat air pollution. In: Air pollution and its complications: from the regional to the global scale. Springer International Publishing, Cham, pp 143–161
113. Wade J, Farrauto RJ (2012) Controlling emissions of pollutants in urban areas. In: Metropolitan Sustainability. Woodhead Publishing, pp 260–291
114. Pham TD, Lee BK, Lee CH, Nguyen MV (2015) Emission control technology. In: Nejadkoorki F (ed) Current air quality issues. IntechOpen, pp 221–249
115. Vashi D, Vashi PD, Shah DV, Kurmi KB (2015) Air pollution and its control measures. Adv Appl Sci Res 6(5):118–121
116. Brunekreef B, Holgate ST (2002) Air pollution and health. Lancet 360(9341):1233–1242
117. Manisalidis I, Stavropoulou E, Stavropoulos A, Bezirtzoglou E (2020) Environmental and health impacts of air pollution: a review. Front Public Health 8:505570
118. Nassikas NJ, McCormack MC, Ewart G, Balmes JR, Bond TC, Brigham E, Cromar K, Goldstein AH, Hicks A, Hopke PK, Kipen HM (2024) Indoor air sources of outdoor air pollution: health consequences, policy, and recommendations: an official American Thoracic Society workshop report. Ann Am Thorac Soc 21(3):365–376
119. Corburn J (2005) Street science: community knowledge and environmental health justice. MIT Press
120. Williamson DH, Yu EX, Hunter CM, Kaufman JA, Komro K, Jelks NTO, Johnson DA, Gribble MO, Kegler MC (2020) A scoping review of capacity-building efforts to address environmental justice concerns. Int J Environ Res Public Health 17(11):3765
121. McCollum DL, Krey V, Riahi K, Kolp P, Grubler A, Makowski M, Nakicenovic N (2013) Climate policies can help resolve energy security and air pollution challenges. Clim Chang 119:479–494

122. Rao S, Pachauri S, Dentener F, Kinney P, Klimont Z, Riahi K, Schoepp W (2013) Better air for better health: forging synergies in policies for energy access, climate change and air pollution. Glob Environ Chang 23(5):1122–1130
123. Jennings V, Johnson Gaither C, Gragg RS (2012) Promoting environmental justice through urban green space access: a synopsis. Environ Justice 5(1):1–7
124. Kahn ME (2007) Green cities: urban growth and the environment. Rowman & Littlefield
125. Fielding JE, Briss PA (2006) Promoting evidence-based public health policy: can we have better evidence and more action? Health Aff 25(4):969–978
126. Ncube A, Mtetwa S, Bukhari M, Fiorentino G, Passaro R (2023) Circular economy and green chemistry: the need for radical innovative approaches in the design for new products. Energies 16(4):1752
127. Mann RF (2023) Driving transformation: tax strategies for electrifying light-duty transportation. Env't L Rep 53:10298
128. Olsen L, Kemter D (2012) The International Labour Organization and the environment: the way to a socially just transition for workers. In: Trade unions in the green economy. Routledge, pp 41–57
129. Stevis D, Felli R (2015) Global labour unions and just transition to a green economy. Int Environ Agreem Politics Law Econ 15:29–43
130. Kamis A, Mustapha R, Wahab NA, Ismail BLH (2016) Green skills as an added-value element in producing competent students. Int J Eng Res Appl 6(11):12–21
131. Grey D, Sadoff CW (2007) Sink or swim? Water security for growth and development. Water Policy 9(6):545–571
132. Adanma UM, Ogunbiyi EO (2024) The public health benefits of implementing environmental policies: a comprehensive review of recent studies. Int J Appl Res Soc Sci 6(5):978–1004
133. Davies C, Lafortezza R (2017) Urban green infrastructure in Europe: is greenspace planning and policy compliant? Land Use Policy 69:93–101
134. Lafortezza R, Pauleit S, Hansen R, Sanesi G, Davies C (2017) Strategic green infrastructure planning and urban forestry. In: Routledge handbook of urban forestry. Routledge, pp 179–193
135. Nowak DJ (2019) The atmospheric system: air quality and greenhouse gases. In: Hall MHP, Balogh SB (eds) Understanding urban ecology: an interdisciplinary systems approach. Springer Nature, Cham, pp 175–199
136. Nowak DJ, Dwyer JF (2007) Understanding the benefits and costs of urban forest ecosystems. In: Urban and community forestry in the northeast. Springer Netherlands, Dordrecht, pp 25–46
137. Jamei E, Tapper N (2019) WSUD and urban heat Island effect mitigation. In: Approaches to water sensitive urban design. Woodhead Publishing, pp 381–407
138. Zhao D, Lei Q, Shi Y, Wang M, Chen S, Shah K, Ji W (2020) Role of species and planting configuration on transpiration and microclimate for urban trees. Forests 11(8):825
139. Chen CF (2013) Performance evaluation and development strategies for green roofs in Taiwan: a review. Ecol Eng 52:51–58
140. Sharma R, Malaviya P (2021) Management of stormwater pollution using green infrastructure: the role of rain gardens. Wiley Interdiscip Rev Water 8(2):e1507
141. Taguchi VJ, Weiss PT, Gulliver JS, Klein MR, Hozalski RM, Baker LA, Finlay JC, Keeler BL, Nieber JL (2020) It is not easy being green: recognizing unintended consequences of green stormwater infrastructure. Water 12(2):522
142. Chang NB, Lu JW, Chui TFM, Hartshorn N (2018) Global policy analysis of low impact development for stormwater management in urban regions. Land Use Policy 70:368–383
143. Mohammed TA, Noor MJMM, Ghazali AH (2007) Study on potential uses of rainwater harvesting in urban areas. Putrajaya Malaysia
144. Devkota J, Schlachter H, Apul D (2015) Life cycle based evaluation of harvested rainwater use in toilets and for irrigation. J Clean Prod 95:311–321

145. Izah SC, Ohimain EI (2013) The challenge of biodiesel production from oil palm feedstock in Nigeria. Greener J Biol Sci 3(1):1–12
146. Dakwale VA, Ralegaonkar RV, Mandavgane S (2011) Improving environmental performance of building through increased energy efficiency: a review. Sustain Cities Soc 1(4):211–218
147. Ottelin J, Heinonen J, Junnila S (2015) New energy efficient housing has reduced carbon footprints in outer but not in inner urban areas. Environ Sci Technol 49(16):9574–9583
148. Biswas WK (2014) Carbon footprint and embodied energy consumption assessment of building construction works in Western Australia. Int J Sustain Built Environ 3(2):179–186
149. Soni N, Soni N (2016) Benefits of pedestrianization and warrants to pedestrianize an area. Land Use Policy 57:139–150
150. Adejumo T (2010) Bikabilty in metropolitan lagos: a conceptualization of eco friendly transportation alternative. Real Corp (AT)
151. Paul N, Tiwari R (2023) Identifying potential for non-motorized transportation: a case of Siliguri. In: Annual conference on infrastructure and built environment: towards sustainable and resilient societies. Singapore, Springer Nature, pp 33–51
152. Whittinghill L, Starry O (2016) Up on the roof: considerations for food production on rooftops. In: Brown S, McIvor K, Snyder EH (eds) Sowing seeds in the City: ecosystem and municipal services. Springer, Dordrecht, pp 325–338
153. Izah SC, Aigberua AO (2023) Potential threats and possible conservation strategies of biodiversity in Niger Delta region of Nigeria. In: Izah SC, Ogwu MC (eds) Sustainable utilization and conservation of Africa's biological resources and environment. Sustainable development and biodiversity, vol 888. Springer, Singapore. https://doi.org/10.1007/978-981-19-6974-4_6
154. Ogwu MC, Iyiola AO, Izah SC (2023) Overview of African biological resources and environment. In: Izah SC, Ogwu MC (eds) Sustainable utilization and conservation of Africa's biological resources and environment. Sustainable development and biodiversity, vol 888. Springer, Singapore. https://doi.org/10.1007/978-981-19-6974-4_1
155. Ogwu MC, Izah SC, Iyiola AO (2022) An overview of the potentials, threats and conservation of biodiversity in Africa. In: Izah SC (ed) Biodiversity in Africa: potentials, threats and conservation. Sustainable development and biodiversity, vol 29. Springer Nature, Singapore, pp 1–20. https://doi.org/10.1007/978-981-19-3326
156. Inatimi SA, Popoola OM, Yarkwan B, Iyiola AO, Izah SC (2022) Therapeutic potentials of wildlife resources and options for conservation. In: Izah SC (ed) Biodiversity in Africa: potentials, threats and conservation. Sustainable development and biodiversity, vol 29. Springer Nature, Singapore, pp 143–174. https://doi.org/10.1007/978-981-19-3326-4_6
157. Shakou LM, Wybo JL, Reniers G, Boustras G (2019) Developing an innovative framework for enhancing the resilience of critical infrastructure to climate change. Saf Sci 118:364–378
158. Carter JG, Cavan G, Connelly A, Guy S, Handley J, Kazmierczak A (2015) Climate change and the city: building capacity for urban adaptation. Prog Plan 95:1–66

Green Infrastructure and Urban Planning for Sustainable Clean Air

Onah Peter Thompson, Enoch Akwasi Kosoe, and Jiajun Xu

Contents

Abstract In the face of rapid urbanization and escalating environmental concerns, the concept of green infrastructure has gained significant traction in urban planning circles. This chapter explores the integral relationship between green infrastructure and urban planning in fostering sustainable clean air. By examining the multifaceted benefits of green infrastructure, such as mitigating air pollution, enhancing urban biodiversity, and improving public health, this chapter underscores its critical role in addressing the pressing challenge of urban and suburban air quality. Moreover, it

O. P. Thompson (✉)
Government and Justice Studies Department, Appalachian State University Boone, Boone, NC, USA
e-mail: thompsonpo@appstate.edu

E. A. Kosoe
Department of Environment and Resource Studies, Faculty of Integrated Development Studies, Simon Diedong Dombo University of Business and Integrated Development Studies, Wa, Upper West Region, Ghana
e-mail: ekosoe@ubids.edu.gh

J. Xu
Institute for Advanced Studies, University of Malaya, Kuala Lumpur, Malaysia

Matthew Chidozie Ogwu and Sylvester Izah (eds.), *Sustainable Strategies for Air Pollution Mitigation: Development, Economics and Technologies*,
Hdb Env Chem (2024) 133: 343–376, DOI 10.1007/698_2024_1122,

Published online: 20 September 2024

highlights the importance of collaborative efforts among policymakers, urban planners, environmentalists, and communities in realizing the potential of green infrastructure to create healthier and more livable cities. Drawing upon principles of landscape ecology, environmental design, and urban governance, the chapter delineates various typologies of green infrastructure interventions, ranging from vegetated corridors and green roofs to urban forests and permeable surfaces. By harnessing the natural capacities of vegetation, soil, and water, these interventions not only mitigate air pollution through pollutant filtration and carbon sequestration but also confer ancillary benefits such as heat island mitigation, stormwater management, and biodiversity conservation. The chapter underscores the significance of integrated planning approaches that transcend traditional sectoral silos, fostering synergies between land use planning, transportation design, and environmental management. In conclusion, by leveraging emerging technologies for monitoring air quality to fostering interdisciplinary collaborations between planners, scientists, and policymakers, it delineates a roadmap for harnessing the full potential of green infrastructure as a cornerstone of sustainable urban development.

Keywords Bicycle infrastructure, Green corridors, Renewable energy, Sustainable communities, Sustainable urbanism, Transit-oriented development, Urban greening

Abbreviations

NO2	Nitrogen dioxide
SO2	Sulfur dioxide
SuDS	Sustainable drainage systems
UK	United Kingdom
PM	Particulate matter
NO2	Nitrogen dioxide
O3	Ozone
LID	Low impact development
GI	Green infrastructure
NOx	Nitrogen oxides
CO	Carbon monoxide
PM	Particulate matter
VOCs	Volatile organic compounds
CO2	Carbon dioxide
CH4	Methane
N2O	Nitrous oxide
EVs	Electric vehicles
BEVs	Battery electric vehicles
PHEVs	Plug-in hybrid electric vehicles
GHGs	Greenhouse gases
HC	Hydrocarbons

WHO	World Health Organization
HVAC	Heating, ventilation, and air-conditioning
M&E	Monitoring and evaluation
PM2.5	Ultrafine particles
PM10	Fine particles

1 Introduction

With the rapid expansion of urban areas worldwide, the challenges associated with air pollution have become increasingly pronounced. Poor air quality not only poses significant health risks to urban residents but also undermines the overall quality of life and environmental sustainability of cities [1, 2]. In response to these challenges, the concept of green infrastructure has emerged as a promising approach to mitigate air pollution while promoting sustainable urban development.

Green infrastructure encompasses a range of natural and engineered systems designed to provide ecological services that benefit both the environment and society [3]. Examples include urban forests, green roofs, rain gardens, permeable pavements, and constructed wetlands. These "green" elements mimic the functions of natural ecosystems, such as filtration, carbon sequestration, and temperature regulation, while also enhancing aesthetic value and biodiversity in urban areas. Green infrastructure plays a crucial role in mitigating air pollution through various mechanisms [4]. Vegetation, for instance, acts as a natural filter, removing particulate matter and absorbing harmful pollutants such as nitrogen dioxide (NO2) and sulfur dioxide (SO2) [5–7]. Trees and vegetation also reduce the urban heat island effect, which can exacerbate air pollution by increasing the formation of ground-level ozone (O3) [8]. Moreover, the implementation of urban green infrastructure development could help promote active transportation modes such as walking and cycling, thereby reducing vehicle emissions and improving air quality.

In addition to its role in mitigating air pollution, green infrastructure offers numerous benefits for urban livability. Access to green spaces has been linked to improved mental health and well-being, reduced stress levels, and increased social cohesion. Green roofs and walls can help insulate buildings, reduce energy consumption, and mitigate urban heat, creating more comfortable and resilient urban environments [9, 10]. Furthermore, green infrastructure enhances urban biodiversity, providing habitat for wildlife and supporting ecosystem services such as pollination and soil erosion control. Effective integration of green infrastructure into urban planning requires a holistic and multidisciplinary approach. Key considerations include land use planning, transportation planning, sustainable environment management, green infrastructure design, and scientific building codes and regulations [11, 12]. Planners also need to engage with stakeholders and communities to ensure that green infrastructure projects are socially inclusive and address the needs and priorities of residents [13]. Moreover, policymakers can incentivize the adoption of green infrastructure through financial incentives, tax breaks, and regulatory mechanisms.

Despite its potential benefits, the widespread adoption of green infrastructure faces several challenges and barriers. Financial constraints limited technical expertise, and lack of awareness among policymakers and the public are significant obstacles to implementation [14]. Moreover, competing land use priorities, regulatory hurdles, and institutional barriers can hinder the integration of green infrastructure into urban planning processes. Addressing these challenges requires coordinated action at the local, regional, and national levels, as well as investment in research, capacity building, and public education. Several cities around the world have successfully integrated green infrastructure into their urban planning practices to improve air quality and enhance urban livability [15, 16]. For example, Copenhagen, Denmark, has implemented extensive green roof initiatives to mitigate air pollution, reduce energy consumption, and enhance biodiversity [15]. Vancouver, Canada, has developed an urban forest strategy to increase tree canopy cover and improve air quality in urban areas. Seoul, South Korea, has transformed a former highway into the Cheonggyecheon Restoration Project, creating a linear park and restoring a natural waterway to improve air quality and enhance urban biodiversity. Other examples include Millennium Park, Chicago, USA, which is a world-renowned urban park featuring green roofs, landscaped gardens, and water features designed to improve air quality, manage stormwater, and enhance biodiversity in downtown Chicago as well as the Sustainable Drainage Systems (SuDS) retrofit project in London, UK, which aims to retrofit existing urban areas with green infrastructure features such as permeable pavements, green roofs, and rain gardens to reduce flood risk, improve water quality, and enhance air quality. The High Line Park is an elevated linear park built on a former railway line in Manhattan, featuring greenery, public art, and recreational amenities. The park has become a model for adaptive reuse and green infrastructure development, enhancing localized air quality, biodiversity, and urban livability in the heart of New York City.

This chapter aims to examine the integration of green infrastructure into urban planning practices to achieve sustainable clean air in cities. It seeks to provide a comprehensive understanding of how green infrastructure can mitigate air pollution, enhance urban livability, and contribute to ecological resilience within the context of urban planning. Green infrastructure holds tremendous potential to promote sustainable clean air in cities by mitigating air pollution, enhancing urban livability, and fostering ecological resilience [17]. However, realizing this potential requires proactive planning, interdisciplinary collaboration, and community engagement. By integrating green infrastructure into urban planning practices, policymakers and planners can create healthier, more resilient, and more sustainable urban environments for current and future generations.

2 Green Infrastructure: History, Types, Classification, and Their Benefits

Green infrastructure refers to the network of natural and engineered systems that provide ecological services and enhance environmental quality in urban areas [18]. Green infrastructure differs from traditional "gray" infrastructure, such as roads and buildings, in that it mimics natural ecosystems and functions, thereby providing multiple benefits for both people and the environment [19, 20]. Green infrastructure has its roots in urban planning and environmental conservation efforts dating back to the late nineteenth and early twentieth centuries [21, 22]. However, the modern concept of green infrastructure emerged in the late twentieth century as a response to the challenges of urbanization, climate change, and environmental degradation.

Parks and urban green spaces, such as New York City's Central Park (established in 1858), were among the earliest forms of green infrastructure, designed to provide recreational opportunities and improve public health in rapidly industrializing cities. The conservation movement of the twentieth century emphasized the importance of protecting natural resources and preserving green spaces for future generations [23]. Efforts to establish national parks, wildlife refuges, and green belts around cities contributed to the development of green infrastructure principles and practices. The term "green infrastructure" gained prominence in the late twentieth century as urban planners and environmentalists recognized the need for nature-based solutions to address urban challenges such as air and water pollution, urban heat island effects, and loss of biodiversity [21].

Green infrastructure encompasses a wide range of natural and engineered elements designed to provide ecological services and enhance environmental quality in urban and rural areas [3, 18]. These natural elements include existing natural areas such as forests, wetlands, rivers, and coastlines that provide ecosystem services such as habitat provision, flood control, water purification, and carbon sequestration whereas engineered structures include human-made elements such as green roofs, vegetated swales, rain gardens, permeable pavements, and bioswales designed to mimic the functions of natural systems and provide ecosystem services within urban and suburban environments. Some common types of green infrastructure include the following (Fig. 1):

- **Urban forests and tree canopies**: Urban forests consist of trees, woodlands, and green spaces within urban areas. Tree canopies provide shade, absorb carbon dioxide, and improve air quality by filtering pollutants from the air.
- **Green roofs and walls**: Green roofs and walls are vegetated surfaces installed on buildings to reduce stormwater runoff, insulate buildings, and mitigate urban heat island effects. They also provide habitat for wildlife and improve urban aesthetics.
- **Parks and open spaces**: Parks, street green landscapes, playgrounds, greenways, and nature reserves provide recreational opportunities, protect natural habitats, and promote biodiversity in urban and suburban areas.

Fig. 1 Common types of green infrastructures

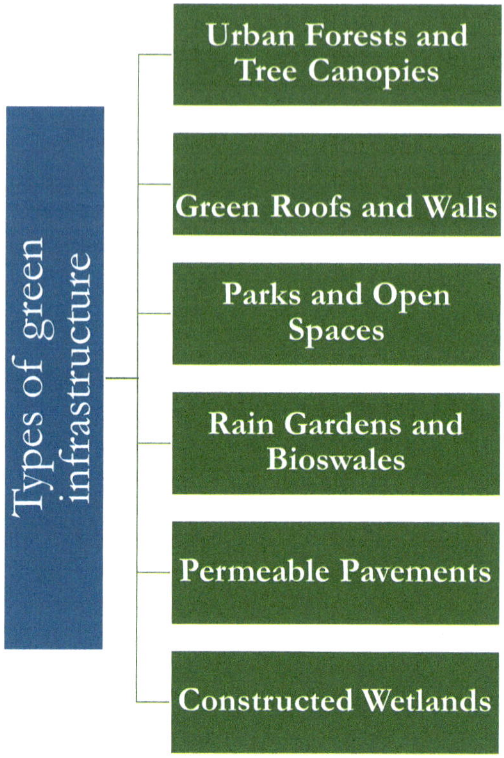

- **Rain gardens and bioswales**: Rain gardens and bioswales are vegetated depressions or channels designed to capture and filter stormwater runoff, reducing flooding and improving water quality by removing pollutants and sediment.
- **Permeable pavements**: Permeable pavements allow rainwater to infiltrate the ground, reducing runoff and preventing pollution of water bodies. They also help to recharge groundwater and mitigate flooding in urban areas.
- **Constructed wetlands**: Constructed wetlands are human-made ecosystems designed to mimic the functions of natural wetlands, such as filtering pollutants, providing habitat for wildlife, and enhancing water quality.

From the works of Hansen and Pauleit [24], Schifman et al. [25], and Adesoji and Pearce [18] due to the need for green infrastructure to function at multiple stages in concert, they may be classified based on various criteria, including scale, function, and location, where:

- **Scale**: Green infrastructure projects can be classified as microscale (e.g., individual green roofs or rain gardens), mesoscale (e.g., neighborhood parks or green corridors), or macroscale (e.g., regional green belts or watershed protection areas).

- **Function**: Green infrastructure elements can be classified based on their primary functions, such as stormwater management, air quality improvement, urban heat island mitigation, biodiversity conservation, or recreational use.
- **Location**: Green infrastructure projects can be classified based on their location within the urban or rural landscape, including roadside greenery, urban parks and squares, green corridors, riparian buffers, and green roofs or facades.

Green infrastructure offers numerous benefits for both the environment and society, including improving air quality, managing stormwater, enhancing biodiversity, mitigating urban heat effects and public health security, and increasing property values [26, 27]. Vegetation in green infrastructure elements filters pollutants from the air, reducing levels of particulate matter (PM), nitrogen dioxide (NO2), sulfur dioxide (SO2), and ozone (O3) and improving overall air quality in urban areas. Green infrastructure elements, such as rain gardens, bioswales, and permeable pavements, capture and filter stormwater runoff, reducing flooding, erosion, and pollution of water bodies [28]. According to Maes et al. [29], green infrastructure provides habitat for wildlife, promotes plant and animal diversity, and supports ecosystem services such as pollination, soil stabilization, and nutrient cycling. Vegetation in green infrastructure elements provides shade, evaporative cooling, and thermal insulation, helping to reduce temperatures in urban areas and mitigate the urban heat island effect. Access to green spaces has been linked to improved mental health, reduced stress levels, increased physical activity, and enhanced social cohesion, contributing to overall health and well-being in urban communities. Green infrastructure amenities, such as parks, greenways, and urban forests, enhance property values, attract investment, and promote economic development in urban and suburban areas [30–32] (Fig. 2).

Green infrastructure plays a vital role in enhancing environmental quality, promoting sustainability, and improving the quality of life in urban and rural

Fig. 2 Some benefits of green infrastructure

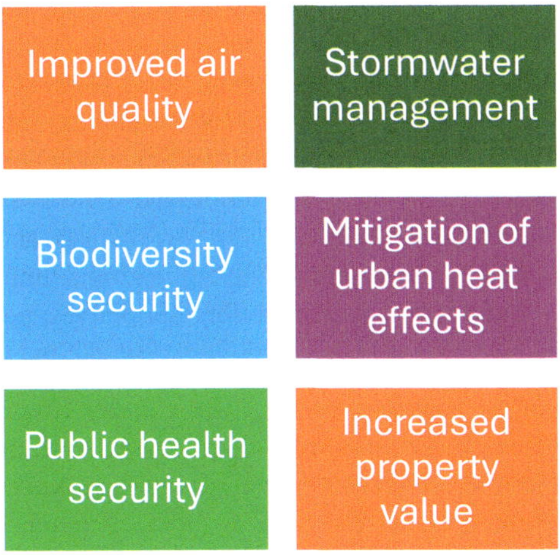

Table 1 Key differences between low impact development and green infrastructure

Aspect	Green infrastructure	Low impact development	References
Definition	Natural and engineered elements provide ecological services	Sustainable stormwater management approach	Hoover and Hopton [36], Golden and Hoghooghi [33]
Focus	Broad encompasses various environmental functions and goals	Primarily addresses stormwater management and water quality	Hoover and Hopton [36], Zellner and Massey [28]
Components	It includes urban forests, green roofs, parks, rain gardens, etc.	Features permeable pavements, bioretention areas, swales, etc.	Thiagarajan et al. [20], Lai et al. [34]
Scale	It can range from micro (e.g., green roofs) to macro (e.g., parks)	Typically implemented at the site or neighborhood scale	Eric et al. [37], Navarrete-Hernandez and Laffan [38]
Purpose	It enhances environmental quality, biodiversity, and livability	Minimizes impacts of development on hydrology and ecosystems	Golden and Hoghooghi [33], Adesoji and Pearce [18]
Approach	Nature-based solutions that mimic natural processes	Integrates engineering and landscape design principles	Yang et al. [39], Strbac et al. [19]
Regulatory framework	Often incorporated into urban planning and green space policies	Regulated by stormwater management ordinances and guidelines	Hoover and Hopton [36], Anderson and Gough [26, 27]
Benefits	Improves air quality, reduces urban heat island effect, etc.	Reduces stormwater runoff, improves water quality, etc.	Cai et al. [35] and Adesoji and Pearce [18]

communities [19]. Therefore, by integrating green infrastructure principles and practices into planning and development processes, cities and regions can create healthier, more resilient, and more sustainable environments for current and future generations. On a nearly similar note, low impact development (LID) is a sustainable approach to land development and stormwater management that aims to minimize the environmental impacts of urbanization while maintaining or restoring the natural hydrological processes of a site [33, 34]. LID principles focus on managing stormwater runoff at its source using decentralized, small-scale stormwater control measures integrated into the built environment [35]. The key differences between LID and green infrastructure are presented in Table 1. Just like green infrastructure, LID can play a significant role in mitigating air pollution in urban and suburban areas. However, LID is primarily focused on stormwater management and reducing the impact of development on hydrological systems, and its practices can indirectly contribute to improving air quality through several mechanisms [40].

Both LID and green infrastructure offer numerous benefits for urban and suburban environments, but they also present certain disadvantages and challenges. Implementing LID practices often requires additional upfront costs compared to conventional development approaches [39]. Features such as permeable pavements, green roofs, and bioswales may have higher initial installation and maintenance costs, which can deter developers and property owners from adopting LID techniques. LID practices may not be suitable for all development sites or environmental conditions. Factors such as soil type, slope, land use, and regulatory constraints can limit the feasibility and effectiveness of certain LID techniques, particularly in densely developed urban areas or regions with limited available space [41]. LID features require regular inspection, maintenance, and upkeep to ensure their continued functionality and effectiveness. Failure to properly maintain LID practices can result in clogged drainage systems, reduced infiltration capacity, and diminished stormwater management performance over time [39]. Some developers and property owners may perceive LID practices as experimental or unproven, leading to concerns about potential risks and liabilities associated with their implementation. This perception may deter investment in LID projects and hinder their widespread adoption in the development industry. Local zoning ordinances, building codes, and land use regulations may pose barriers to the implementation of LID practices by imposing restrictions or requirements that are difficult to meet or comply with [12]. Lack of coordination among regulatory agencies and unclear permitting processes can further complicate LID implementation efforts.

On the other hand, green infrastructure features such as parks, greenways, and vegetated swales require adequate land area for implementation, which may be limited in densely developed urban environments [3, 36]. Competition for land use and development pressures can restrict the availability of space for green infrastructure projects. Developing and implementing green infrastructure projects often involves complex planning, permitting, and approval processes that can be time-consuming and resource-intensive. Coordination among multiple stakeholders, including government agencies, community groups, and private landowners, may be necessary to overcome bureaucratic hurdles and regulatory challenges. While green infrastructure can provide numerous benefits for air quality, stormwater management, and urban biodiversity, its effectiveness may be limited under certain conditions or in specific contexts. Factors such as soil quality, climate variability, and maintenance requirements can affect the performance and longevity of green infrastructure features over time [19]. Green infrastructure projects require ongoing maintenance and management to ensure their continued functionality and performance. Regular upkeep, including pruning, watering, weeding, and pest control, may be necessary to preserve the health and integrity of vegetation-based features, adding to the long-term costs of green infrastructure implementation [42]. Green infrastructure projects may inadvertently exacerbate existing social inequities by disproportionately benefiting wealthier or more affluent communities while neglecting the needs of marginalized or underserved populations. Ensuring equitable access to green spaces and environmental amenities requires deliberate efforts to address issues of spatial justice and community engagement in the planning and design process.

2.1 Relationship Between Green Infrastructure and Air Quality in Urban Areas

Urban air pollution refers to the presence of harmful pollutants in the air within urban areas, typically resulting from human activities associated with transportation, industrial processes, energy production, and urbanization [43, 44]. Urban areas are often characterized by high population densities, extensive transportation networks, and concentrated sources of pollution, making them particularly vulnerable to air quality issues (Ogwu 2019). Sources of air pollution in urban areas include motor vehicles, including cars, trucks, buses, and motorcycles, which are major contributors to urban air pollution. Combustion engines of motor vehicles emit pollutants such as nitrogen oxides (NOx), carbon monoxide (CO), particulate matter (PM), volatile organic compounds (VOCs), and sulfur dioxide (SO2). Industrial facilities, such as factories, power plants, refineries, and construction sites, release pollutants into the air during manufacturing processes, energy generation, and construction activities [45]. Residential sources, such as wood-burning stoves, fireplaces, and gas stoves, emit pollutants like PM, CO, and VOCs during heating and cooking activities. Commercial establishments, restaurants, and institutions, such as schools and hospitals, may also contribute to urban air pollution through activities like food preparation, waste incineration, and chemical use [46].

Urban air pollution poses significant health risks to residents, particularly vulnerable populations such as children, the elderly, and individuals with preexisting health conditions. Short-term exposure to high levels of air pollution can cause respiratory symptoms, cardiovascular problems, exacerbation of asthma, and increased susceptibility to respiratory infections [47, 48]. Long-term exposure to urban air pollution has been linked to chronic respiratory diseases, cardiovascular diseases, lung cancer, reduced lung function, and premature death [49]. Urban air pollution can have adverse effects on ecosystems, including damage to vegetation, soil, water bodies, and wildlife habitat as well as acid rain, smog formation, and deposition of pollutants that can harm forests, crops, aquatic ecosystems, and biodiversity. Air pollution also contributes to climate change by emitting greenhouse gases, such as carbon dioxide (CO2), methane (CH4), and nitrous oxide (N2O), which trap heat in the Earth's atmosphere and lead to global warming [50].

Addressing urban air pollution requires a comprehensive approach that involves implementing policies and regulations to reduce emissions from transportation, industry, energy production, and other sources, promoting sustainable urban planning and development practices, investing in clean technologies and renewable energy sources, and raising awareness about the importance of air quality protection and public health [43, 51]. The relationship between green infrastructure and air quality is multifaceted, with green infrastructure playing a crucial role in mitigating air pollution and improving overall air quality in urban environments [52]. This relationship can be understood through various mechanisms and processes (Fig. 3):

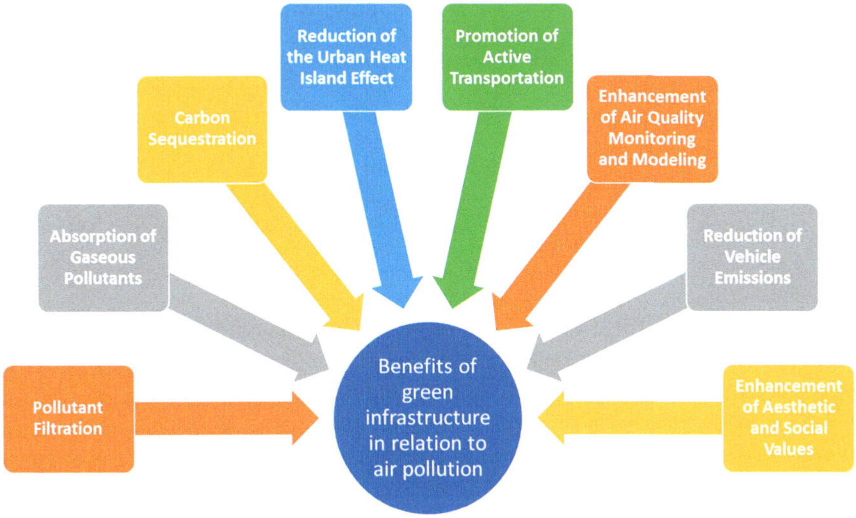

Fig. 3 Benefits of green infrastructure in relation to air pollution

- **Pollutant filtration**: Vegetation, such as trees, shrubs, and greenery in general, acts as a natural filter for airborne pollutants. Leaves and other plant surfaces capture PM suspended in the air, including dust, soot, and other harmful particles [53]. The complex surface structure of leaves and the porous nature of vegetation provide ample surface area for pollutants to adhere to, effectively removing them from the air.
- **Absorption of gaseous pollutants**: Green infrastructure, particularly vegetation, absorbs gaseous pollutants such as NO2, SO2, O3, and VOCs through stomata on leaf surfaces [5, 54]. Through the process of photosynthesis, plants convert these pollutants into oxygen and other harmless byproducts, thereby reducing their concentration in the atmosphere.
- **Carbon sequestration**: Trees and vegetation play a vital role in sequestering CO2, a major greenhouse gas and contributor to climate change [55]. By absorbing CO2 during photosynthesis and storing it in their biomass and soil, green infrastructure helps mitigate the greenhouse effect and reduce air pollution.
- **Reduction of the urban heat island effect**: Green infrastructure, such as urban forests, green roofs, and parks, helps to mitigate the urban heat island effect by providing shade, evaporative cooling, and thermal insulation. By lowering surface temperatures and reducing the energy demand for cooling, green infrastructure helps to decrease the formation of ground-level O3, a harmful pollutant formed through chemical reactions between NOx and VOCs in the presence of sunlight and heat [56].
- **Promotion of active transportation**: Green infrastructure, including pedestrian and cycling paths, greenways, and linear parks, promotes active transportation modes such as walking and cycling [57]. By reducing reliance on motor vehicles,

which are significant sources of air pollution, green infrastructure helps to decrease emissions of NOx, PM, and other pollutants associated with combustion engines [58].

- **Enhancement of air quality monitoring and modeling**: Green infrastructure can complement traditional air quality monitoring and modeling efforts by providing additional data points and insights into local air quality conditions [56]. Vegetation health, canopy density, and other green infrastructure characteristics can be monitored and integrated into air quality models to assess the effectiveness of green infrastructure in mitigating air pollution and improving overall air quality.
- **Reduction of vehicle emissions**: Green infrastructure promotes alternative modes of transportation, such as walking, cycling, and public transit, which reduce reliance on motor vehicles and decrease emissions of NOx, PM, and other pollutants associated with combustion engines [58]. By providing pedestrian and cycling paths, greenways, and landscape parks, green infrastructure encourages active transportation and helps to improve air quality in urban areas.
- **Enhancement of aesthetic and social values**: Green infrastructure enhances the aesthetic appeal and livability of urban environments, creating green spaces for recreation, relaxation, and social interaction. Access to green spaces has been linked to improved mental health and well-being, reduced stress levels, and increased social cohesion [59]. By providing opportunities for outdoor activities and community engagement, green infrastructure contributes to healthier and more vibrant urban communities.

2.2 Integrating Green Infrastructure into Urban Planning for Cleaner Air

Urban areas worldwide face the challenge of air pollution, which poses significant risks to public health and the environment. As cities continue to grow and develop, it becomes imperative to address this issue through innovative approaches that prioritize sustainability and environmental quality. One such approach is the integration of green infrastructure into urban planning, which offers multifaceted solutions for mitigating air pollution and promoting cleaner air in cities. Urban planning plays a crucial role in addressing air pollution by guiding the design, development, and management of urban spaces in ways that minimize pollution sources, mitigate exposure to harmful pollutants, and promote environmental sustainability [60]. Some key aspects of urban plannings role in addressing air pollution include the following (Fig. 4):

- **Land use planning:** Urban planners can strategically allocate land uses to minimize exposure to pollution sources and promote mixed-use development patterns that reduce the need for vehicular travel and associated emissions [43]. Zoning regulations can also be used to locate industrial facilities away from residential areas and sensitive land uses.

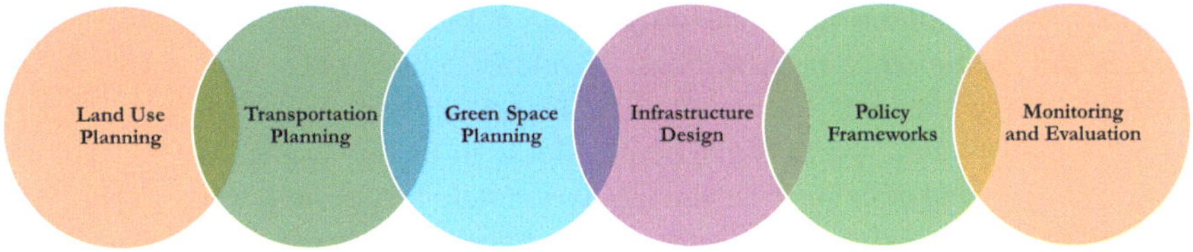

Fig. 4 Key aspects of urban planning that may be integrated for potential positive air quality implications

- **Transportation planning**: Transportation systems are major contributors to urban air pollution, emitting pollutants such as NOx, PM, and VOCs. Urban planners can promote sustainable transportation options, such as public transit, walking, cycling, and electric vehicles, through investments in infrastructure, transit-oriented development, and land use policies that prioritize nonmotorized modes of travel [61].
- **Green space planning**: Incorporating green spaces, parks, and urban forests into the urban fabric can help improve air quality by providing vegetation that absorbs pollutants, filters particulate matter, and reduces the urban heat island effect. Urban planners can prioritize the preservation and creation of green spaces through land acquisition, park design standards, and green infrastructure requirements in development projects [62].
- **Infrastructure design**: Urban planners can influence the design of infrastructure systems, such as water management, energy, and waste disposal, to minimize air pollution and environmental impacts [28]. For example, green stormwater management practices, such as rain gardens and permeable pavements, can reduce runoff and pollution from paved surfaces, while renewable energy sources can reduce emissions from fossil fuel combustion.
- **Policy frameworks**: Develop and implement policies, regulations, and incentives that support the integration of green infrastructure into urban planning processes. This may include updating zoning codes, land use plans, building regulations, and environmental ordinances to prioritize green infrastructure requirements and incentives for developers.
- **Comprehensive planning**: Integrate green infrastructure considerations into comprehensive planning processes, including master plans, transportation plans, and environmental assessments. According to Pamukcu-Albers et al. [3], this may be accomplished by ensuring that green infrastructure goals and objectives are incorporated into long-term planning documents and guiding decision-making at the local and regional levels.
- **Design guidelines**: Develop design guidelines and standards for green infrastructure features, such as vegetated roofs, rain gardens, bioswales, and urban forests, to ensure consistency, functionality, and aesthetic quality [19]. Provide technical assistance and resources to developers, designers, and practitioners to facilitate the implementation of green infrastructure projects.

- **Collaborative planning processes**: Engage stakeholders, including government agencies, community organizations, developers, businesses, and residents, in collaborative planning processes to identify priorities, assess needs, and develop green infrastructure projects that address local air quality challenges [63]. Foster partnerships and collaboration among diverse stakeholders to leverage resources, share expertise, and build consensus around green infrastructure initiatives.
- **Monitoring and evaluation**: Establish monitoring and evaluation mechanisms to assess the effectiveness of green infrastructure projects in improving air quality and achieving other environmental and social objectives [28]. Collect data on air quality, vegetation health, green infrastructure management facilities, operational efficiency, and feedback on community benefits to inform decision-making, track progress, and adjust as needed.

2.3 Designing Green Spaces for Clean Air

Green spaces play a vital role in promoting clean air and mitigating urban air pollution. By strategically planning and designing green spaces, cities can harness the natural processes of vegetation to filter pollutants, absorb carbon dioxide, and provide cooling effects, thereby improving air quality and enhancing the overall health and well-being of urban residents [17, 64]. Designing and creating a green space for cleaner air involves several approaches and steps aimed at maximizing the air quality benefits of vegetation and landscaping features including:

Location and Distribution Identify areas within the urban fabric where green spaces can have the greatest impact on air quality improvement. Consider factors such as proximity to pollution sources, population density, land use patterns, and existing vegetation cover [65]. Ensure equitable access to green spaces across diverse neighborhoods and communities, particularly in underserved areas that may experience higher levels of air pollution and environmental degradation [63].

Site Selection, Size, and Scale Identify suitable locations for the green space based on factors such as proximity to pollution sources, population density, land use patterns, and existing vegetation cover [66]. Conduct a site analysis to assess environmental conditions, including sunlight exposure, soil quality, drainage patterns, and microclimate variations, to inform the design process. Design green spaces of varying sizes and scales to accommodate diverse user needs and preferences, from pocket parks and community gardens to larger urban forests and green corridors. Integrate green spaces into the urban landscape at multiple scales, including neighborhood parks, greenways, and street-level plantings, to maximize their effectiveness in improving air quality and enhancing urban biodiversity [67].

Establishing Goals and Objectives Define clear goals and objectives for the green space project, including air quality improvement targets, ecological restoration goals, community engagement objectives, and aesthetic considerations

[66, 68]. Consider the specific air pollutants of concern in the area and establish measurable targets for pollutant reduction and air quality enhancement.

Stakeholder Engagement and Community Input Engage stakeholders, including residents, community organizations, government agencies, and businesses, in the planning and design process to solicit input, gather feedback, and build support for the project [69]. Host public meetings, workshops, and design charrettes to involve stakeholders in decision-making and ensure that their needs and preferences are reflected in the final design.

Design Concept Development Develop a conceptual design for the green space that integrates principles of green infrastructure, landscape architecture, and urban planning to maximize air quality benefits [70]. Consider factors such as vegetation selection, layout, circulation patterns, seating areas, water features, and amenities to create a functional, attractive, and sustainable green space design.

Connectivity and Accessibility Establish green networks and connectivity corridors that link individual green spaces and natural areas to create continuous pathways for wildlife movement, pedestrian mobility, and recreational opportunities [71, 72]. Ensure that green spaces are accessible to all members of the community, including pedestrians, cyclists, people with disabilities, and residents of all ages and socioeconomic backgrounds.

Multifunctionality and Cobenefits Design green spaces to serve multiple functions and provide a wide range of cobenefits beyond air quality improvement, such as stormwater management, habitat provision, biodiversity conservation, urban agriculture, and social interaction. Incorporate features such as rain gardens, bioswales, green roofs, and permeable pavements that contribute to both air quality improvement and other environmental objectives [73].

Climate Resilience and Adaptation Design green spaces to enhance climate resilience and adaptation by incorporating vegetation and landscape features that provide shade, cooling, and heat island mitigation, particularly in urban heat islands and heat-vulnerable areas [56]. Select drought-tolerant and climate-appropriate plant species that can thrive in urban environments with limited water availability and fluctuating climate conditions [18].

Maintenance and Management Develop long-term maintenance plans and management strategies to ensure the health, vitality, and sustainability of green spaces over time. Consider factors such as watering, fertilization, pruning, pest control, and litter removal to maintain optimal air quality and aesthetic appeal. Engage local communities, volunteers, and stakeholders in stewardship and monitoring efforts to foster a sense of ownership, pride, and responsibility for green spaces and their contributions to clean air and environmental quality [74].

In terms of incorporating vegetation and planting schemes for air quality improvement, the following are pertinent:

Native and Climate-Adapted Species Select native and climate-adapted plant species that are well suited to local environmental conditions and can thrive with minimal inputs of water, fertilizer, and pesticides [75]. Choose vegetation with dense foliage, broad leaves, and high surface area to maximize the capture and filtration of airborne pollutants, including particulate matter, nitrogen dioxide, sulfur dioxide, ozone, and volatile organic compounds.

Evergreen and Deciduous Trees Incorporate a mix of evergreen and deciduous trees into green space designs to provide year-round air quality benefits and seasonal variations in shade, cooling, and aesthetics [66]. Evergreen trees offer continuous pollutant filtration and carbon sequestration, while deciduous trees provide additional benefits such as shade in summer and sunlight penetration in winter.

Vertical and Horizontal Vegetation Layers Design green spaces with vertical and horizontal vegetation layers to create a diverse and multidimensional canopy that intercepts, filters, and absorbs pollutants at different heights and levels within the urban environment [4]. Incorporate climbing vines, shrubs, groundcovers, and understory plants to fill in gaps, create visual interest, and have environmental aesthetics while enhancing air quality benefits throughout the vertical profile of green spaces.

Pollinator-Friendly Plantings Include pollinator-friendly plantings such as wildflowers, herbs, and native grasses in green space designs to support biodiversity, enhance ecosystem resilience, and promote the health and abundance of pollinating insects such as bees, butterflies, and birds. Pollinator-friendly plantings contribute to air quality improvement by enhancing vegetation cover, increasing plant diversity, and attracting beneficial insects that contribute to ecosystem services such as pollination and pest control [76].

Low-Emission and Low-Allergen Species Avoid or minimize the use of high-emission and allergenic plant species that release VOCs or pollen that may exacerbate air quality problems or trigger allergic reactions in sensitive individuals [77]. Select low-emission and low-allergen species with minimal pollen production, low VOC emissions, and noninvasive growth habits.

3 Green Transportation Solutions for Air Pollution

Air pollution is a major environmental concern that affects people's health and quality of life all over the world. Vehicles are significant contributors to air pollution and greenhouse gas emissions, causing health issues such as respiratory diseases, cardiovascular diseases, and premature deaths [78, 79]. As the global population continues to grow and urban areas become more congested, it is crucial to find sustainable transportation solutions that can mitigate air pollution. Green transportation solutions have been identified as an effective strategy to address this challenge by reducing emissions and improving air quality [80]. Some green transportation solutions for the mitigation of air pollution and their benefits are provided in Table 2 and discussed below.

One of the most promising green transportation solutions for reducing air pollution is the widespread adoption of electric vehicles (EVs). EVs produce zero tailpipe emissions, which can significantly reduce air pollution and improve air quality in urban areas [78]. In addition to eliminating harmful emissions, EVs can also reduce greenhouse gas emissions and help combat climate change. According to Forsythe et al. [85], there are two main types of EVs: battery electric vehicles (BEVs) and plug-in hybrid electric vehicles (PHEVs). BEVs are powered solely by electricity and must be plugged in to charge, while PHEVs have both an electric motor and a gasoline engine, allowing them to switch between electric and gasoline power [86]. Both types of EVs offer significant environmental benefits compared to traditional gasoline- and diesel-powered vehicles.

Several countries and cities around the world have started implementing policies and incentives to promote the adoption of EVs. For example, Norway offers generous subsidies and tax incentives for EV buyers, resulting in a high market share of EVs in the country [87]. In the United States, several states offer rebates and tax credits for EV purchases, as well as incentives for installing charging infrastructure [82]. Despite the environmental benefits of EVs, there are still challenges and barriers that need to be addressed to accelerate their adoption. One of the main challenges is the high upfront cost of EVs, which can be a barrier for many consumers [88]. However, as battery technology continues to improve and production costs decrease, the cost of EVs is expected to become more competitive with traditional vehicles. Another challenge is the lack of charging infrastructure, especially in urban areas and apartment buildings [89]. To promote the adoption of EVs, governments and private companies need to invest in expanding the charging network and making it more convenient for EV owners to recharge their vehicles. Additionally, there is a need for standardization of charging infrastructure and interoperability between different charging networks to improve user experience and increase confidence in EVs.

Public transportation is another green transportation solution that can help reduce air pollution by encouraging people to use shared mobility options instead of driving their cars. Public transportation systems, such as buses, trains, and subways, produce lower emissions per passenger than individual cars, making them a more sustainable mode of transport [61, 90]. By promoting public transportation, cities can reduce traffic congestion, improve air quality, and enhance urban mobility. In addition to reducing emissions, public transportation can also provide social and economic benefits, such as improving access to jobs and services for low-income communities, reducing traffic accidents, and enhancing the overall quality of life in urban areas [91]. By investing in public transportation infrastructure and services, cities can create more sustainable and livable environments for their residents. Several cities around the world have implemented policies to promote public transportation and reduce car dependence. For example, London has introduced a congestion pricing scheme to reduce traffic congestion and air pollution in the city center, while Paris has implemented a car-free day to encourage people to use public transportation and

Table 2 Some green transportation solutions for air pollution mitigation and their benefits

Green transportation solution	Description	Air pollution benefits	References
Walking and cycling	Encouraging walking and cycling as viable transportation modes by developing pedestrian-friendly infrastructure such as sidewalks, crosswalks, and bike lanes	Reduction in vehicular emissions, particularly NOx and PM, as fewer vehicles are used as well as improved air quality due to decreased reliance on fossil fuel-powered transportation	Brand et al. [81]
Public transit	Expanding and enhancing public transit networks including buses, trains, light rail, and metro systems to provide efficient, affordable, and accessible transportation options for urban residents	Decreased emissions per passenger mile compared to single-occupancy vehicles, leading to reduced air pollution levels. And the alleviation of traffic congestion, which can further mitigate emissions from idling vehicles	Chen et al. [30, 31]
Electric vehicles (EVs)	Promoting the adoption of electric vehicles (EVs) through the development of EV charging infrastructure, incentives for EV purchases and usage, and integration of EVs into public and private fleets	Elimination of tailpipe emissions, resulting in significant reductions in greenhouse gases (GHGs) and criteria air pollutants such as CO2, NOx, and PM as well as an improvement in local air quality, especially in urban areas, by replacing combustion engine vehicles with zero-emission EVs	Hamdare et al. [82]
Carpooling and ridesharing	Encouraging carpooling and ridesharing through incentives such as preferential parking, toll discounts, and dedicated carpool lanes on highways. Facilitating the use of ridesharing platforms and apps to connect commuters with shared transportation options	Reduction in the number of single-occupancy vehicles on the road, leading to lower emissions of pollutants such as CO, hydrocarbons (HC), and NOx. Mitigation of traffic congestion and associated air pollution by promoting vehicle occupancy and reducing vehicle miles traveled	Rey-Merchan et al. [83]
Active transportation infrastructure	Designing and implementing infrastructure that supports active transportation modes such as walking and cycling, including dedicated bike lanes, pedestrian-friendly streetscapes, and bike-sharing programs	Reduction in vehicle trips and associated emissions, contributing to improved air quality and public health. Promotion of physical activity and active lifestyles, leading to additional health benefits and reduced healthcare costs	Nieuwenhuijsen [84]

alternative modes of transport [92, 93]. These initiatives have been successful in reducing emissions and improving air quality in urban areas [94]. To promote the use of public transportation, governments and city planners need to invest in expanding and improving public transportation infrastructure, including adding new bus routes, increasing the frequency of trains and buses, and upgrading existing systems to make them more efficient and reliable. Additionally, cities can implement policies such as dedicated bus lanes, electric bicycles, bike-sharing programs, and pedestrian-friendly streets to encourage people to use public transportation and alternative modes of transport.

Cycling and walking are two sustainable modes of transportation that can help reduce air pollution and promote a healthy lifestyle. By cycling or walking instead of driving, people can reduce their carbon footprint, improve air quality, and enhance their physical and mental well-being [95]. Cycling and walking are also cost-effective and convenient modes of transport, especially for short trips in urban areas. In addition to reducing emissions, cycling and walking can provide social and economic benefits, such as reducing traffic congestion, increasing physical activity levels, and improving road safety [84]. By investing in cycling infrastructure, such as bike lanes, bike racks, and bike-sharing programs, cities can encourage more people to cycle and walk instead of driving. Several cities around the world have implemented policies to promote cycling and walking as sustainable modes of transportation. For example, Copenhagen and Amsterdam have built extensive cycling infrastructure, including dedicated bike lanes and parking facilities, to encourage residents to cycle as their primary mode of transport [96]. These cities have high rates of cycling and low levels of car ownership, resulting in cleaner air and healthier urban environments. In addition, Hellobike, Mobike, and DiDi Bike, the leading Chinese companies involved in bike sharing and e-mobility, are operating and expanding their businesses in major Chinese cities, helping to spread the concept of green mobility and improve the quality of urban environments while generating economic benefits for the companies (Fig. 5).

To promote cycling and walking, cities need to invest in building safe and accessible infrastructure, such as bike lanes, pedestrian paths, and shared streets, that prioritize the needs of cyclists and pedestrians. Additionally, cities can implement policies such as traffic calming measures, reduced speed limits, and car-free zones to create a safer and more sustainable urban environment for cyclists and pedestrians.

Government policies and regulations play a crucial role in promoting green transportation solutions and reducing air pollution. By implementing supportive policies and regulations, governments can create an enabling environment for the adoption of sustainable transportation options and incentivize people to make environmentally friendly choices [97, 98]. Some key policies and regulations that governments can consider include the following:

1. Incentives for EV adoption, such as tax credits, rebates, and subsidies for EV buyers, as well as grants and loans for charging infrastructure installation.
2. Mandates for zero-emission vehicles, such as setting targets for EV market share and banning the sale of new gasoline-powered vehicles in the future.

Fig. 5 Sharing bikes and sharing electric bikes riding in major Chinese cities (Source: https://www.caupd.com/)

3. Investments in public transportation infrastructure, such as expanding bus and rail networks, improving service frequency and quality, and implementing integrated ticketing systems.
4. Support for cycling and walking, such as building bike lanes, pedestrian paths, and shared streets, as well as implementing traffic calming measures and reducing speed limits.
5. Regulations on vehicle emissions and fuel efficiency, such as setting emission standards for gasoline and diesel vehicles, as well as promoting the use of clean fuels and technologies.

By implementing these policies and regulations, governments can create a more sustainable and environmentally friendly transportation system that reduces air pollution, improves air quality, and enhances the overall quality of life for people living in urban areas. Additionally, governments can collaborate with private companies, nonprofit organizations, and other stakeholders to develop comprehensive strategies and initiatives that promote green transportation solutions and address the challenges of air pollution mitigation.

4 Green Buildings and Sustainable Development for Air Pollution Mitigation

The World Health Organization (WHO) estimates that outdoor air pollution is responsible for millions of premature deaths globally each year [99]. It is evident from the work of Nassikas et al. [46] that buildings are major contributors to air pollution, emitting pollutants such as VOCs, PM, HC, and GHGs. In response to the growing awareness of the environmental impacts of buildings, green building practices have emerged as a sustainable solution for mitigating air pollution and promoting sustainable development.

Green buildings are designed, constructed, and operated using environmentally friendly practices that reduce energy consumption, minimize waste, and improve indoor air quality [100]. These buildings incorporate sustainable design principles, energy-efficient technologies, and renewable energy sources to reduce their environmental footprint and contribute to overall air quality improvement. Green buildings play a crucial role in mitigating air pollution by reducing emissions, minimizing energy consumption, and improving indoor air quality. The design and construction of green buildings focus on energy efficiency, resource conservation, and pollution prevention to minimize the environmental impact of buildings on air quality [101]. Some ways in which green buildings help mitigate air pollution include the following:

Energy Efficiency Green buildings are designed to be energy efficient, using technologies and strategies that reduce energy consumption and greenhouse gas emissions. Energy-efficient features such as high-performance insulation, efficient HVAC systems, and energy-efficient lighting help reduce the building's energy demand and greenhouse gas emissions, thus mitigating air pollution [102].

Renewable Energy Sources Green buildings incorporate renewable energy sources such as solar panels, wind turbines, and geothermal systems to generate clean and sustainable energy. By utilizing renewable energy sources, green buildings reduce reliance on fossil fuels and contribute to the reduction of air pollution associated with energy production [103].

Indoor Air Quality Green buildings prioritize indoor air quality by using low-emission materials, adequate ventilation systems, and natural lighting to create healthy and comfortable indoor environments [101]. By improving indoor air quality, green buildings minimize the release of pollutants and enhance occupant health and well-being.

Sustainable Materials Green buildings use sustainable materials and construction practices that minimize environmental impact and reduce emissions throughout the building's life cycle. By choosing eco-friendly materials, such as recycled content, locally sourced materials, and low-VOC products, green buildings help reduce air pollution associated with the production and disposal of building materials [104].

Water Conservation Green buildings implement water-saving technologies and strategies, such as rainwater harvesting, graywater recycling, and efficient irrigation systems, to minimize water consumption and reduce the environmental impact of building operations [105]. By conserving water resources, green buildings help prevent water pollution and contamination, thereby contributing to air quality improvement.

Sustainable development aims to meet the present needs of society without compromising the ability of future generations to meet their own needs. Green buildings are integral to sustainable development as they promote environmental, social, and economic sustainability by incorporating sustainable design principles and practices. Green buildings support sustainable development in the following ways:

Environmental Sustainability Green buildings minimize environmental impact by reducing energy consumption, water use, waste generation, and emissions [101]. By promoting resource conservation, pollution prevention, and sustainable practices, green buildings contribute to environmental sustainability and help protect natural resources for future generations.

Social Sustainability Green buildings prioritize occupant health, well-being, and comfort by providing healthy indoor environments, natural daylighting, thermal comfort, and access to green spaces. By creating healthy and livable spaces, green buildings support social sustainability and enhance the quality of life for building occupants and communities [101].

Economic Sustainability Green buildings offer long-term economic benefits by reducing operating costs, increasing property value, and enhancing market competitiveness [106]. By improving energy efficiency, resource efficiency, and overall building performance, green buildings help reduce operational expenses and create economic value for building owners and stakeholders.

Community Sustainability Green buildings contribute to community sustainability by fostering social cohesion, community engagement, and environmental stewardship [73]. By implementing sustainable design features, promoting sustainable practices, and engaging with local communities, green buildings support community resilience and contribute to the overall well-being of society.

Despite the numerous benefits of green buildings, some several challenges and barriers hinder their widespread adoption and implementation. Green building practices often require upfront investment and higher initial costs compared to conventional buildings. The perceived cost barrier can deter developers, builders, and property owners from adopting green building practices, as they may prioritize short-term financial considerations over long-term environmental benefits [107]. Many stakeholders in the construction industry lack awareness and knowledge about green building practices, technologies, and certification programs [108]. Without adequate education and training, stakeholders may be hesitant to embrace green building principles and may not fully understand the benefits of sustainable design and construction. Inconsistent or inadequate regulatory frameworks, building codes, and policies can hinder the adoption of green building practices. The absence of mandatory green building requirements, incentives, and enforcement mechanisms may limit the incentives for developers and builders to invest in green building initiatives [109]. The demand for green buildings may vary depending on market conditions, consumer preferences, and regional priorities. In some regions, there may be limited demand for green building certifications or sustainable features, which can deter developers and builders from investing in green building projects. Green building technologies and strategies may present technical challenges related to integration, compatibility, and performance. From the design phase to the operation and maintenance of green buildings, stakeholders may face technical complexities and implementation barriers that require specialized expertise and resources. Also, governments play a critical role in promoting green buildings and sustainable development through policy initiatives, incentives, and regulations. By implementing supportive policies and measures, governments can create an enabling environment for green building development and encourage stakeholders to adopt sustainable design and construction practices. Some key government policies and support mechanisms for green buildings include the following:

Building Codes and Standards Governments can establish green building codes, standards, and guidelines that mandate energy efficiency, water conservation, and indoor air quality in building construction and renovation projects [100]. By setting minimum requirements for sustainable building performance, governments can encourage compliance with green building principles.

Incentives and Subsidies Governments can provide financial incentives, subsidies, grants, and tax credits to promote green building practices and encourage investment in sustainable design and construction. Incentives such as rebates for energy-efficient appliances, grants for renewable energy systems, and tax breaks for green building certification can incentivize developers and building owners to adopt green buildings [110].

5 Green Infrastructure and Urban Planning Policy and Governance for Clean Air

Green infrastructure helps improve air quality by absorbing pollutants, filtering particulate matter, and reducing emissions through natural processes. Trees, plants, and green spaces act as natural filters that capture and remove pollutants from the air, thereby reducing the concentration of harmful contaminants and improving overall air quality [8]. Green infrastructure contributes to climate change mitigation by sequestering carbon dioxide, reducing greenhouse gas emissions, and enhancing urban resilience to extreme weather events [19]. Green infrastructure elements such as green roofs, urban forests, and vegetative buffers help sequester carbon and reduce the urban heat island effect, mitigating the impacts of climate change on air quality. Green infrastructure promotes sustainable urban design principles by integrating nature into the built environment and creating healthy, livable spaces for residents. Green spaces, parks, and green corridors provide opportunities for recreation, biodiversity conservation, and community engagement, enhancing the overall quality of life in urban areas. Green infrastructure helps manage stormwater runoff and reduce flooding by absorbing rainwater, reducing runoff volumes, and improving water quality. Green roofs, rain gardens, and permeable pavements help capture and infiltrate stormwater, preventing pollutants from entering water bodies and protecting aquatic ecosystems [111]. Green infrastructure provides numerous health and well-being benefits for urban residents, including improved mental health, physical fitness, and social cohesion. Access to green spaces, trees, and parks promotes physical activity, reduces stress, and enhances overall well-being, contributing to public health and quality of life in urban areas.

Urban planning policies and governance mechanisms play a critical role in promoting clean air and sustainable development in urban areas. Effective urban planning strategies can help integrate green infrastructure, air quality management, and sustainable transportation solutions into the built environment, creating healthier and more resilient cities [26, 27]. Some key urban planning policies and governance approaches for clean air include the following:

Land Use Planning Land use planning policies can help guide sustainable development and minimize land use conflicts that contribute to air pollution. Zoning regulations, urban growth boundaries, and transit-oriented development strategies can help reduce vehicle emissions, encourage mixed land uses, and promote compact, walkable communities that support clean air goals.

Green Building Regulations Green building regulations require new construction and renovation projects to meet energy efficiency, green building standards, and sustainable design criteria. By implementing green building codes, governments can promote resource conservation, energy efficiency, and indoor air quality improvements, contributing to cleaner air and healthier buildings in urban areas.

Sustainable Transportation Planning Sustainable transportation planning focuses on promoting alternative modes of transport, reducing reliance on private vehicles, and improving public transit systems. Policies such as bike lanes, pedestrian infrastructure, car-free zones, and public transportation incentives can help reduce traffic congestion, emissions, and air pollution in urban areas.

Green Infrastructure Investment Governments can invest in green infrastructure projects, such as urban forests, green roofs, and permeable pavements, to enhance air quality, reduce heat island effects, and improve urban resilience. By allocating funding for green infrastructure initiatives, policymakers can create healthier, more sustainable cities that prioritize clean air and environmental sustainability.

Public Participation and Stakeholder Engagement Effective urban planning for clean air requires meaningful public participation, stakeholder engagement, and community involvement in decision-making processes. Engaging residents, businesses, and organizations in urban planning initiatives can help build consensus, garner support for clean air policies, and foster a sense of ownership and responsibility for creating healthier urban environments.

Some key challenges and opportunities for clean air governance include policy coordination and integration, data collection and monitoring, stakeholder engagement and collaboration, capacity building, and knowledge sharing as well as equity and social justice issues.

6 Green Infrastructure Monitoring and Evaluation for Clean Air

Green infrastructure plays a critical role in improving air quality and promoting clean air in urban areas. Monitoring and evaluating green infrastructure projects are essential to assess their effectiveness, measure their impact on air quality, and inform decision-making for future initiatives [112]. Monitoring and evaluation (M&E) are essential components of green infrastructure projects aimed at improving air quality. M&E helps project managers, policymakers, and stakeholders assess the performance of green infrastructure initiatives, track progress toward clean air goals, and identify areas for improvement [32]. The monitoring and evaluation of air quality improvements is essential for assessing the effectiveness of green infrastructure projects in mitigating air pollution and promoting clean air. By systematically collecting data, measuring outcomes, and analyzing results, M&E enables informed decision-making, evidence-based policymaking, and accountability in green infrastructure projects. Some key reasons why M&E are important in green infrastructure for clean air include:

Performance Assessment M&E helps assess the performance of green infrastructure projects in achieving air quality improvements, reducing emissions, and enhancing environmental sustainability [32, 113]. By measuring key indicators and comparing actual outcomes against expected results, project managers can evaluate the effectiveness of green infrastructure interventions and identify successful strategies for clean air.

Impact Measurement M&E enables the measurement of the impact of green infrastructure on air quality, public health, and environmental quality [84]. By quantifying changes in air pollutant concentrations, health outcomes, and ecosystem benefits, M&E provides valuable insights into the benefits of green infrastructure for clean air and supports evidence-based decision-making.

Decision-Making Support M&E provides critical information and data to support decision-making processes, prioritize investments, and allocate resources to green infrastructure projects. By identifying successful interventions, assessing costs and benefits, and evaluating the contributions of green infrastructure to clean air goals, M&E helps inform policy decisions and strategic planning [114].

Accountability and Transparency M&E promotes accountability, transparency, and stakeholder engagement in green infrastructure projects by ensuring that project outcomes are monitored, evaluated, and reported effectively [115]. By documenting progress, sharing results with the public, and soliciting feedback from stakeholders, M&E enhances the credibility and legitimacy of green infrastructure initiatives.

Learning and Knowledge Sharing M&E facilitates learning, knowledge sharing, and capacity building among project stakeholders, policymakers, and practitioners involved in green infrastructure for clean air [116]. By identifying best practices, lessons learned, and success stories from monitoring and evaluation efforts, M&E helps inform future projects, replicate successful interventions, and improve overall project performance.

7 Conclusion

Air pollution is a serious environmental problem that affects people's health and quality of life around the world. Vehicles are significant contributors to air pollution, emitting harmful pollutants and greenhouse gases that cause respiratory diseases, cardiovascular diseases, and premature deaths. To address this challenge, it is crucial to promote green transportation solutions that can reduce emissions, improve air quality, and enhance urban mobility. Electric vehicles, public transportation, cycling, and walking are all sustainable modes of transportation that can help mitigate air pollution and promote a healthier and more sustainable urban environment, by investing in EV infrastructure, expanding public transportation networks, building cycling infrastructure, and implementing supportive policies.

Integrating green infrastructure into urban planning offers a promising approach to addressing air pollution and promoting cleaner air in cities. By incorporating green infrastructure principles and practices into city planning and development processes, urban areas can reduce emissions, improve air quality, enhance environmental sustainability, and create healthier, more livable communities for residents. However, achieving these goals requires strong leadership, effective collaboration, and meaningful engagement with stakeholders to overcome challenges, build consensus, and implement innovative solutions that benefit both people and the planet. Integrating green infrastructure into urban planning offers a promising approach to addressing air pollution and promoting cleaner air in cities. By incorporating green infrastructure principles and practices into city planning and development processes, urban areas can reduce emissions, improve air quality, enhance environmental sustainability, and create healthier, more livable communities for residents. However, achieving these goals requires strong leadership, effective collaboration, and meaningful engagement with stakeholders to overcome challenges, build consensus, and implement innovative solutions that benefit both people and the planet. Monitoring and evaluation play a crucial role in assessing the effectiveness of green infrastructure projects for improving air quality and promoting clean air in urban areas. Specifically, by measuring air quality indicators, evaluating green infrastructure performance, and analyzing monitoring data, M&E provides valuable insights into the impact of green infrastructure on air quality improvements and supports evidence-based decision-making in clean air governance. Furthermore, implementing best practices for data collection, analysis, and reporting can help project managers, policymakers, and stakeholders enhance the effectiveness of green infrastructure initiatives, inform policy decisions, and contribute to sustainable urban development. As a result, by embracing green infrastructure as a key component of urban planning, cities can pave the way toward a more sustainable and resilient future for generations to come.

References

1. Li X, Lyu Y, Dong W, Xu A (2023) Exploring the relationship between air quality and health shocks to the elderly: a retrospective cross-sectional study in China. Front Public Health 11: 1087626. https://doi.org/10.3389/fpubh.2023.1087626
2. Wang Q (2018) Urbanization and global health: the role of air pollution. Iran J Public Health 47(11):1644–1652
3. Pamukcu-Albers P, Ugolini F, La Rosa D, Grădinaru SR, Azevedo JC, Wu J (2021) Building green infrastructure to enhance urban resilience to climate change and pandemics. Landsc Ecol 36(3):665–673. https://doi.org/10.1007/s10980-021-01212-y
4. Venter ZS, Hassani A, Stange E, Schneider P, Castell N (2024) Reassessing the role of urban green space in air pollution control. Proc Natl Acad Sci U S A 121(6):e2306200121. https://doi.org/10.1073/pnas.2306200121
5. Wei X, Lyu S, Yu Y, Wang Z, Liu H, Pan D, Chen J (2017) Phylloremediation of air pollutants: exploiting the potential of plant leaves and leaf-associated microbes. Front Plant Sci 8:1318. https://doi.org/10.3389/fpls.2017.01318
6. Vwioko DE, Okoekhian I, Ogwu MC (2018) Stress analysis of amaranthus hybridus L. and Lycopersicon esculentum mill. exposed to sulphur and nitrogen dioxide. Pertanika Journal of Tropical Agricultural Sciences 41(3):1169–1191
7. Ogwu MC (2023) Plants as monitors and managers of pollution. In: Srivastav AL, Grewal AS, Tiwari M, Pham TD (eds) Role of green chemistry in ecosystem restoration to achieve environmental sustainability. Elsevier, Amsterdam, Netherlands. pp. 51–60. ISBN: 978-0-443-15291-7. https://doi.org/10.1016/B978-0-443-15291-7.00022-5
8. Isaifan RJ, Baldauf RW (2020) Estimating economic and environmental benefits of urban trees in desert regions. Urban For Urban Green. https://doi.org/10.3389/fevo.2020.00016
9. Jamei E, Chau HW, Seyedmahmoudian M, Mekhilef SS, Sami FA (2023) Green roof and energy - role of climate and design elements in hot and temperate climates. Heliyon 9(5): e15917. https://doi.org/10.1016/j.heliyon.2023.e15917
10. Visvanathan G, Patil K, Suryawanshi Y, Meshram V, Jadhav S (2024) Mitigating urban heat Island and enhancing indoor thermal comfort using terrace garden. Sci Rep 14(1):9697. https://doi.org/10.1038/s41598-024-60546-0
11. Gharaibeh NG, Lee CB, Alhalbouni T, Wang F, Lee J, Newman G, Güneralp B, Zandt SV (2023) Quality of stormwater infrastructure systems in vulnerable communities: three case studies from Texas. Public Works Manag Policy 28(4):518–536. https://doi.org/10.1177/1087724x231164415
12. Lens MC (2022) Zoning, land use, and the reproduction of urban inequality. Annu Rev Sociol 48(1):421–439. https://doi.org/10.1146/annurev-soc-030420-122027
13. Solarek K, Domaradzka A, Zdunek-Wielgołaska J (2023) Changing approaches to green infrastructure design: from modernism to the future: Warsaw case study. URBAN Design Int:1–18. Advance online publication. https://doi.org/10.1057/s41289-023-00220-5

14. Dhakal KP, Chevalier LR (2017) Managing urban stormwater for urban sustainability: barriers and policy solutions for green infrastructure application. J Environ Manage 203(Pt 1): 171–181. https://doi.org/10.1016/j.jenvman.2017.07.065
15. Beery TH, Raymond CM, Kyttä M, Olafsson AS, Plieninger T, Sandberg M, Stenseke M, Tengö M, Jönsson KI (2017) Fostering incidental experiences of nature through green infrastructure planning. Ambio 46(7):717–730. https://doi.org/10.1007/s13280-017-0920-z
16. Sturiale L, Scuderi A (2019) The role of green infrastructures in urban planning for climate change adaptation. Climate 7:119. https://doi.org/10.3390/cli7100119
17. Woodward A, Hinwood A, Bennett D, Grear B, Vardoulakis S, Lalchandani N, Lyne K, Williams C (2023) Trees, climate change, and health: an urban planning, greening and implementation perspective. Int J Environ Res Public Health 20(18):6798. https://doi.org/10.3390/ijerph20186798
18. Adesoji T, Pearce A (2024) Interdisciplinary perspectives on green infrastructure: a systematic exploration of definitions and their origins. Environments 11:8. https://doi.org/10.3390/environments11010008
19. Štrbac S, Kašanin-Grubin M, Pezo L, Stojić N, Lončar B, Ćurčić L, Pucarević M (2023) Green infrastructure designed through nature-based solutions for sustainable urban development. Int J Environ Res Public Health 20(2):1102. https://doi.org/10.3390/ijerph20021102
20. Thiagarajan M, Newman G, Van Zandt S (2018) The projected impact of a neighborhood-scaled green infrastructure retrofit. Sustain For 10(10):3665. https://doi.org/10.3390/su10103665
21. Coutts C, Hahn M (2015) Green infrastructure, ecosystem services, and human health. Int J Environ Res Public Health 12(8):9768–9798. https://doi.org/10.3390/ijerph120809768
22. Shao H, Kim G, Li Q, Newman G (2021) Web of science-based green infrastructure: a bibliometric analysis in CiteSpace. Land 10(7):711. https://doi.org/10.3390/land10070711
23. Corlett RT (2020) Safeguarding our future by protecting biodiversity. Plant Diversity 42(4): 221–228. https://doi.org/10.1016/j.pld.2020.04.002
24. Hansen R, Pauleit S (2014) From multifunctionality to multiple ecosystem services? A conceptual framework for multifunctionality in green infrastructure planning for urban areas. Ambio 43(4):516–529. https://doi.org/10.1007/s13280-014-0510-2
25. Schifman LA, Herrmann DL, Shuster WD, Ossola A, Garmestani A, Hopton ME (2017) Situating green infrastructure in context: a framework for adaptive socio-hydrology in cities. Water Resour Res 53(12):10139–10154. https://doi.org/10.1002/2017WR020926
26. Anderson V, Gough WA (2022) Enabling nature-based solutions to build back better-an environmental regulatory impact analysis of green infrastructure in Ontario, Canada. Buildings 12(1):61. https://doi.org/10.3390/buildings12010061
27. Anderson V, Gough WA (2022) Nature-based cooling potential: a multi-type green infrastructure evaluation in Toronto, Ontario, Canada. Int J Biometeorol 66(2):397–410. https://doi.org/10.1007/s00484-021-02100-5
28. Zellner ML, Massey D (2024) Modeling benefits and tradeoffs of green infrastructure: evaluating and extending parsimonious models for neighborhood stormwater planning. Heliyon 10(5):e27007. https://doi.org/10.1016/j.heliyon.2024.e27007
29. Maes J, Barbosa A, Baranzelli C, Zulian G, Silva FBE, Vandecasteele I, Hiederer R, Liquete C, Paracchini ML, Mubareka S, Jacobs-Crisioni C, Castillo CP, Lavalle C (2015) More green infrastructure is required to maintain ecosystem services under current trends in land-use change in Europe. Landsc Ecol 30(3):517–534. https://doi.org/10.1007/s10980-014-0083-2
30. Chen K, Lin H, You S, Han Y (2022) Review of the impact of urban parks and green spaces on residence prices in the environmental health context. Front Public Health 10:993801. https://doi.org/10.3389/fpubh.2022.993801
31. Chen K, Zhang G, Wu H, Mao R, Chen X (2022) Uncovering the carbon emission intensity and reduction potentials of the metro operation phase: a case study in Shenzhen megacity. Int J Environ Res Public Health 20(1):206. https://doi.org/10.3390/ijerph20010206

32. Fu X, Hopton ME, Wang X (2021) Assessment of green infrastructure performance through an urban resilience lens. J Clean Prod 289. https://doi.org/10.1016/j.jclepro.2020.125146
33. Hoghooghi N, Golden HE, Bledsoe BP, Barnhart BL, Brookes AF, Djang KS, Halama JJ, McKane RB, Nietch CT, Pettus PP (2018) Cumulative effects of low impact development on watershed hydrology in a mixed land-cover system. Water 10(8):991. https://doi.org/10.3390/w10080991
34. Lai Y, Lu Y, Ding T, Sun H, Li X, Ge X (2022) Effects of low-impact development facilities (water systems of the Park) on Stormwater runoff in shallow mountainous areas based on dual-model (SWMM and MIKE21) simulations. Int J Environ Res Public Health 19(21):14349. https://doi.org/10.3390/ijerph192114349
35. Cai Z, Zhu R, Ruggier E, Newman G, Horney JA (2023) Calculating the environmental impacts of low-impact development using long-term hydrologic impact assessment: a review of model applications. Land 12(3):612. https://doi.org/10.3390/land12030612
36. Hoover FA, Hopton ME (2019) Developing a framework for stormwater management: leveraging ancillary benefits from urban greenspace. Urban Ecosyst 22(6):1139–1148. https://doi.org/10.1007/s11252-019-00890-6
37. Eric M, Fan C, Joksimovic D, Li JY (2013) Modeling low impact development potential with hydrological response units. Water Sci Technol 68(11):2382–2390. https://doi.org/10.2166/wst.2013.502
38. Navarrete-Hernandez P, Laffan K (2023) The impact of small-scale green infrastructure on the affective wellbeing associated with urban sites. Sci Rep 13(1):9687. https://doi.org/10.1038/s41598-023-35804-2
39. Yang W, Brüggemann K, Seguya KD, Ahmed E, Kaeseberg T, Dai H, Hua P, Zhang J, Krebs P (2020) Measuring performance of low impact development practices for the surface runoff management. Environ Sci Ecotechnol 1:100010. https://doi.org/10.1016/j.ese.2020.100010
40. Fu X, Goddard H, Wang X, Hopton ME (2019) Development of a scenario-based stormwater management planning support system for reducing combined sewer overflows (CSOs). J Environ Manage 236:571–580. https://doi.org/10.1016/j.jenvman.2018.12.089
41. Naeini AM, Tabesh M, Soltaninia S (2024) Modeling the effect of land use change to design a suitable low impact development (LID) system to control surface water pollutants. Sci Total Environ:172756. Advance online publication. https://doi.org/10.1016/j.scitotenv.2024.172756
42. Lõhmus M, Balbus J (2015) Making green infrastructure healthier infrastructure. Infect Ecol Epidemiol 5:30082. https://doi.org/10.3402/iee.v5.30082
43. Jafari A, Charkhloo E, Pasalari H (2021) Urban air pollution control policies and strategies: a systematic review. J Environ Health Sci Eng 19(2):1911–1940. https://doi.org/10.1007/s40201-021-00744-4
44. Manisalidis I, Stavropoulou E, Stavropoulos A, Bezirtzoglou E (2020) Environmental and health impacts of air pollution: a review. Front Public Health 8:14. https://doi.org/10.3389/fpubh.2020.00014
45. Johnston J, Cushing L (2020) Chemical exposures, health, and environmental justice in communities living on the Fenceline of industry. Curr Environ Health Rep 7(1):48–57. https://doi.org/10.1007/s40572-020-00263-8
46. Nassikas NJ, McCormack MC, Ewart G, Balmes JR, Bond TC, Brigham E, Cromar K, Goldstein AH, Hicks A, Hopke PK, Meyer B, Nazaroff WW, Paulin LM, Rice MB, Thurston GD, Turpin BJ, Vance ME, Weschler CJ, Zhang J, Kipen HM (2024) Indoor air sources of outdoor air pollution: health consequences, policy, and recommendations: an official American Thoracic Society workshop report. Ann Am Thorac Soc 21(3):365–376. https://doi.org/10.1513/AnnalsATS.202312-1067ST
47. Lee BJ, Kim B, Lee K (2014) Air pollution exposure and cardiovascular disease. Toxicol Res 30(2):71–75. https://doi.org/10.5487/TR.2014.30.2.071

48. Tiotiu AI, Novakova P, Nedeva D, Chong-Neto HJ, Novakova S, Steiropoulos P, Kowal K (2020) Impact of air pollution on asthma outcomes. Int J Environ Res Public Health 17(17): 6212. https://doi.org/10.3390/ijerph17176212
49. Guo J, Chai G, Song X, Hui X, Li Z, Feng X, Yang K (2023) Long-term exposure to particulate matter on cardiovascular and respiratory diseases in low- and middle-income countries: a systematic review and meta-analysis. Front Public Health 11:1134341. https://doi.org/10.3389/fpubh.2023.1134341
50. Cassia R, Nocioni M, Correa-Aragunde N, Lamattina L (2018) Climate change and the impact of greenhouse gasses: CO_2 and NO, friends and foes of plant oxidative stress. Front Plant Sci 9:273. https://doi.org/10.3389/fpls.2018.00273
51. Morantes G, Rincon G, Chanaba A, Jones B (2024) Addressing air quality challenges: comparative analysis of Barcelona, Venezuela, and Guayaquil, Ecuador. Heliyon 10(8): e29211. https://doi.org/10.1016/j.heliyon.2024.e29211
52. Hewitt CN, Ashworth K, MacKenzie AR (2020) Using green infrastructure to improve urban air quality (GI4AQ). Ambio 49(1):62–73. https://doi.org/10.1007/s13280-019-01164-3
53. Mandal M, Popek R, Przybysz A, Roy A, Das S, Sarkar A (2023) Breathing fresh air in the city: implementing avenue trees as a sustainable solution to reduce particulate pollution in urban agglomerations. Plants 12(7):1545. https://doi.org/10.3390/plants12071545
54. Wróblewska K, Jeong BR (2021) Effectiveness of plants and green infrastructure utilization in ambient particulate matter removal. Environ Sci Eur 33(1):110. https://doi.org/10.1186/s12302-021-00547-2
55. Domke GM, Oswalt SN, Walters BF, Morin RS (2020) Tree planting has the potential to increase carbon sequestration capacity of forests in the United States. Proc Natl Acad Sci U S A 117(40):24649–24651. https://doi.org/10.1073/pnas.2010840117
56. Kumar P, Debele SE, Khalili S, Halios CH, Sahani J, Aghamohammadi N, Andrade MF, Athanassiadou M, Bhui K, Calvillo N, Cao SJ, Coulon F, Edmondson JL, Fletcher D, Dias de Freitas E, Guo H, Hort MC, Katti M, Kjeldsen TR, Lehmann S et al (2024) Urban heat mitigation by green and blue infrastructure: drivers, effectiveness, and future needs. Innovations 5(2):100588. https://doi.org/10.1016/j.xinn.2024.100588
57. Yu J, Zhang H, Dong X, Shen J (2024) The impact of street greenery on active travel: a narrative systematic review. Front Public Health 12:1337804. https://doi.org/10.3389/fpubh.2024.1337804
58. Din AU, Ming J, Rahman IU, Han H, Yoo S, Alhrahsheh RR (2023) Green road transportation management and environmental sustainability: the impact of population density. Heliyon 9(9): e19771. https://doi.org/10.1016/j.heliyon.2023.e19771
59. Jabbar M, Yusoff MM, Shafie A (2022) Assessing the role of urban green spaces for human well-being: a systematic review. GeoJournal 87(5):4405–4423. https://doi.org/10.1007/s10708-021-10474-7
60. Slovic AD, de Oliveira MA, Biehl J, Ribeiro H (2016) How can urban policies improve air quality and help mitigate global climate change: a systematic mapping review. J Urban Health 93(1):73–95. https://doi.org/10.1007/s11524-015-0007-8
61. Xia T, Zhang Y, Crabb S, Shah P (2013) Cobenefits of replacing car trips with alternative transportation: a review of evidence and methodological issues. J Environ Public Health 2013: 797312. https://doi.org/10.1155/2013/797312
62. Chaudhuri S, Kumar A (2022) Urban greenery for air pollution control: a meta-analysis of current practice, progress, and challenges. Environ Monit Assess 194(3):235. https://doi.org/10.1007/s10661-022-09808-w
63. Ward F, Lowther-Payne HJ, Halliday EC, Dooley K, Joseph N, Livesey R, Moran P, Kirby S, Cloke J (2022) Engaging communities in addressing air quality: a scoping review. Environ Health 21(1):89. https://doi.org/10.1186/s12940-022-00896-2
64. Richard G, Izah SC, Sawyer EW, Ogwu MC (2023) Seasonal variations in the concentration of some atmospheric pollutants along a major highway in Southern Nigeria. J Environ Studies 30 (1):10–25. https://doi.org/10.21608/JESJ.2023.205673.1045

65. Junior DPM, Bueno C, da Silva CM (2022) The effect of urban green spaces on reduction of particulate matter concentration. Bull Environ Contam Toxicol 108(6):1104–1110. https://doi.org/10.1007/s00128-022-03460-3
66. Nguyen PY, Astell-Burt T, Rahimi-Ardabili H, Feng X (2021) Green space quality and health: a systematic review. Int J Environ Res Public Health 18(21):11028. https://doi.org/10.3390/ijerph182111028
67. Heo S, Bell ML (2023) Investigation on urban greenspace in relation to sociodemographic factors and health inequity based on different greenspace metrics in 3 US urban communities. J Expo Sci Environ Epidemiol 33(2):218–228. https://doi.org/10.1038/s41370-022-00468-z
68. Shanahan DF, Lin BB, Bush R, Gaston KJ, Dean JH, Barber E, Fuller RA (2015) Toward improved public health outcomes from urban nature. Am J Public Health 105(3):470–477. https://doi.org/10.2105/AJPH.2014.302324
69. Boaz A, Hanney S, Borst R, O'Shea A, Kok M (2018) How to engage stakeholders in research: design principles to support improvement. Health Res Policy Syst 16(1):60. https://doi.org/10.1186/s12961-018-0337-6
70. Veen EJ, Ekkel ED, Hansma MR, de Vrieze AGM (2020) Designing urban green space (UGS) to enhance health: a methodology. Int J Environ Res Public Health 17(14):5205. https://doi.org/10.3390/ijerph17145205
71. Dong Y, Liu H, Zheng T (2020) Does the connectivity of urban public green space promote its use? An empirical study of Wuhan. Int J Environ Res Public Health 17(1):297. https://doi.org/10.3390/ijerph17010297
72. He M, Wu Y, Liu X, Wu B, Fu H (2023) Constructing a multi-functional small urban green space network for green space equity in urban built-up areas: a case study of Harbin, China. Heliyon 9(11):e21671. https://doi.org/10.1016/j.heliyon.2023.e21671
73. Kruize H, van der Vliet N, Staatsen B, Bell R, Chiabai A, Muiños G, Higgins S, Quiroga S, Martinez-Juarez P, Aberg Yngwe M, Tsichlas F, Karnaki P, Lima ML, García de Jalón S, Khan M, Morris G, Stegeman I (2019) Urban green space: creating a triple win for environmental sustainability, health, and health equity through behavior change. Int J Environ Res Public Health 16(22):4403. https://doi.org/10.3390/ijerph16224403
74. Jennings V, Larson L, Yun J (2016) Advancing sustainability through urban green space: cultural ecosystem services, equity, and social determinants of health. Int J Environ Res Public Health 13(2):196. https://doi.org/10.3390/ijerph13020196
75. Ding N, Zhong Y, Li J, Xiao Q (2022) Study on selection of native greening plants based on eye-tracking technology. Sci Rep 12(1):1092. https://doi.org/10.1038/s41598-022-05114-0
76. Katumo DM, Liang H, Ochola AC, Lv M, Wang QF, Yang CF (2022) Pollinator diversity benefits natural and agricultural ecosystems, environmental health, and human welfare. Plant Divers 44(5):429–435. https://doi.org/10.1016/j.pld.2022.01.005
77. Reinmuth-Selzle K, Kampf CJ, Lucas K, Lang-Yona N, Fröhlich-Nowoisky J, Shiraiwa M, Lakey PSJ, Lai S, Liu F, Kunert AT, Ziegler K, Shen F, Sgarbanti R, Weber B, Bellinghausen I, Saloga J, Weller MG, Duschl A, Schuppan D, Pöschl U (2017) Air pollution and climate change effects on allergies in the anthropocene: abundance, interaction, and modification of allergens and adjuvants. Environ Sci Technol 51(8):4119–4141. https://doi.org/10.1021/acs.est.6b04908
78. Carey J (2023) The other benefit of electric vehicles. Proc Natl Acad Sci U S A 120(3):e2220923120. https://doi.org/10.1073/pnas.2220923120
79. Peters DR, Schnell JL, Kinney PL, Naik V, Horton DE (2020) Public health and climate benefits and trade-offs of U.S. Vehicle electrification. GeoHealth 4(10):e2020GH000275. https://doi.org/10.1029/2020GH000275
80. Erickson LE (2017) Reducing greenhouse gas emissions and improving air quality: two global challenges. Environ Prog Sustain Energy 36(4):982–988. https://doi.org/10.1002/ep.12665
81. Brand C, Goodman A, Ogilvie D (2014) Evaluating the impacts of new walking and cycling infrastructure on carbon dioxide emissions from motorized travel: a controlled longitudinal study. Appl Energy 128:284–295. https://doi.org/10.1016/j.apenergy.2014.04.072

82. Hamdare S, Kaiwartya O, Aljaidi M, Jugran M, Cao Y, Kumar S, Mahmud M, Brown D, Lloret J (2023) Cybersecurity risk analysis of electric vehicles charging stations. Sensors 23(15):6716. https://doi.org/10.3390/s23156716
83. Rey-Merchán MDC, López-Arquillos A, Pires Rosa M (2022) Carpooling systems for commuting among teachers: an expert panel analysis of their barriers and incentives. Int J Environ Res Public Health 19(14):8533. https://doi.org/10.3390/ijerph19148533
84. Nieuwenhuijsen MJ (2021) New urban models for more sustainable, liveable and healthier cities post covid19; reducing air pollution, noise and heat Island effects and increasing green space and physical activity. Environ Int 157:106850. https://doi.org/10.1016/j.envint.2021.106850
85. Forsythe CR, Gillingham KT, Michalek JJ, Whitefoot KS (2023) Technology advancement is driving electric vehicle adoption. Proc Natl Acad Sci U S A 120(23):e2219396120. https://doi.org/10.1073/pnas.2219396120
86. Weiss M, Zerfass A, Helmers E (2019) Fully electric and plug-in hybrid cars - an analysis of learning rates, user costs, and costs for mitigating CO_2 and air pollutant emissions. J Clean Prod 212:1478–1489. https://doi.org/10.1016/j.jclepro.2018.12.019
87. Okesanya OJ, Saclolo JMB, Mia KBP, Ntacyabukura B, Corman V, Ahmadi A, Nugraha RR, He J, Telin JM, Tshering UU, Abdullahi Y, Ogaya JB, Delos Santos FE, Pedrajas-Mendoza SA, Magramo MM, Lucero-Prisno III DE, Kouwenhoven MBN (2024) Norway's battery electric vehicles and public health: findings from the literature. Environ Environ Health Insights 18:1–5. https://doi.org/10.1177/11786302241234566
88. Alanazi F (2023) Electric vehicles: benefits, challenges, and potential solutions for widespread adaptation. Appl Sci 13:6016. https://doi.org/10.3390/app13106016
89. Mohammed A, Saif O, Abo-Adma M, Fahmy A, Elazab R (2024) Strategies and sustainability in fast charging station deployment for electric vehicles. Sci Rep 14(1):283. https://doi.org/10.1038/s41598-023-50825-7
90. Ceder AA (2021) Syncing sustainable urban mobility with public transit policy trends based on global data analysis. Sci Rep 11(1):14597. https://doi.org/10.1038/s41598-021-93741-4
91. Mahadevia D, Mukhopadhyay C, Lathia S, Gounder K (2023) The role of urban transport in delivering sustainable development goal 11: learning from two Indian cities. Heliyon 9(9): e19453. https://doi.org/10.1016/j.heliyon.2023.e19453
92. Baranzini A, Carattini S, Tesauro L (2021) Designing effective and acceptable road pricing schemes: evidence from the Geneva congestion charge. Environ Resource Econ 79(3): 417–482. https://doi.org/10.1007/s10640-021-00564-y
93. Glazener A, Wylie J, van Waas W, Khreis H (2022) The impacts of car-free days and events on the environment and human health. Curr Environ Health Rep 9(2):165–182. https://doi.org/10.1007/s40572-022-00342-y
94. Winkler L, Pearce D, Nelson J, Babacan O (2023) The effect of sustainable mobility transition policies on cumulative urban transport emissions and energy demand. Nat Commun 14(1): 2357. https://doi.org/10.1038/s41467-023-37728-x
95. Green S, Sakuls P, Levitt S (2021) Cycling for health: improving health and mitigating the climate crisis. Can Fam Physician 67(10):739–742. https://doi.org/10.46747/cfp.6710739
96. Fosgerau M, Łukawska M, Paulsen M, Rasmussen TK (2023) Bikeability and the induced demand for cycling. Proc Natl Acad Sci U S A 120(16):e2220515120. https://doi.org/10.1073/pnas.2220515120
97. Rodríguez-Rad CJ, Revilla-Camacho MÁ, Sánchez-Del-Río-Vázquez ME (2023) Exploring the intention to adopt sustainable mobility modes of transport among young university students. Int J Environ Res Public Health 20(4):3196. https://doi.org/10.3390/ijerph20043196
98. Winters M, Fuller D, Cloutier MS, Harris MA, Howard A, Kestens Y, Kirk S, Macpherson A, Moore S, Rothman L, Shareck M, Tomasone JR, Laberee K, Stephens ZP, Sones M, Ayton D, Batomen B, Bell S, Collins P, Diab E et al (2024) Building CapaCITY/É for sustainable transportation: protocol for an implementation science research program in healthy cities. BMJ Open 14(4):e085850. https://doi.org/10.1136/bmjopen-2024-085850

99. WHO (2022) Ambient (outdoor) air pollution. Available online: https://www.who.int/news-room/fact-sheets/detail/ambient-(outdoor)-air-quality-and-health#:~:text=The%20combined%20effects%20of%20ambient,premature%20deaths%20worldwide%20in%202019
100. Li Y, Rong Y, Ahmad UM, Wang X, Zuo J, Mao G (2021) A comprehensive review on green buildings research: bibliometric analysis during 1998-2018. Environ Sci Pollut Res Int 28(34): 46196–46214. https://doi.org/10.1007/s11356-021-12739-7
101. Allen JG, MacNaughton P, Laurent JG, Flanigan SS, Eitland ES, Spengler JD (2015) Green buildings and health. Curr Environ Health Rep 2(3):250–258. https://doi.org/10.1007/s40572-015-0063-y
102. Alharasees O, Kale U, Rohacs J, Rohacs D, Eva ME, Boros A (2024) Green building energy: patents analysis and analytical hierarchy process evaluation. Heliyon 10(8):e29442. https://doi.org/10.1016/j.heliyon.2024.e29442
103. Chen L, Hu Y, Wang R et al (2024) Green building practices to integrate renewable energy in the construction sector: a review. Environ Chem Lett 22:751–784. https://doi.org/10.1007/s10311-023-01675-2
104. Khoshnava SM, Rostami R, Mohamad Zin R, Štreimikienė D, Mardani A, Ismail M (2020) The role of green building materials in reducing environmental and human health impacts. Int J Environ Res Public Health 17(7):2589. https://doi.org/10.3390/ijerph17072589
105. Teston A, Ghisi E, Vaz ICM, Scolaro TP, Severis RM (2024) Modular life cycle assessment approach: environmental impact of rainwater harvesting systems in urban water systems. Sci Total Environ 908:168281. https://doi.org/10.1016/j.scitotenv.2023.168281
106. Chen S, Gou Z (2023) Spatiotemporal distribution of green-certified buildings and the influencing factors: a study of U.S. Heliyon 9(11):e21868. https://doi.org/10.1016/j.heliyon.2023.e21868
107. Ekins P, Zenghelis D (2021) The costs and benefits of environmental sustainability. Sustain Sci 16(3):949–965. https://doi.org/10.1007/s11625-021-00910-5
108. Berawi MA, Miraj P, Windrayani R, Berawi ARB (2019) Stakeholders' perspectives on green building rating: a case study in Indonesia. Heliyon 5(3):e01328. https://doi.org/10.1016/j.heliyon.2019.e01328
109. Xue S, Na J, Wang L, Wang S, Xu X (2023) The outlook of green building development in China during the "Fourteenth Five-Year Plan" period. Int J Environ Res Public Health 20(6): 5122. https://doi.org/10.3390/ijerph20065122
110. Rana A, Sadiq R, Alam MS, Karunathilake H, Hewage K (2021) Evaluation of financial incentives for green buildings in Canadian landscape. Renew Sustain Energy Rev 135:110199. https://doi.org/10.1016/j.rser.2020.110199
111. Golden HE, Hoghooghi N (2018) Green infrastructure and its catchment-scale effects: an emerging science. WIREs Water 5(1):1254. https://doi.org/10.1002/wat2.1254
112. Zhang P, Chen P, Xiao F, Sun Y, Ma S, Zhao Z (2022) The impact of information infrastructure on air pollution: empirical evidence from China. Int J Environ Res Public Health 19(21): 14351. https://doi.org/10.3390/ijerph192114351
113. Liang M, Liu L, Liang W, Mi W, Ye K, Gao J (2024) Intelligentization helps the green and energy-saving transformation of power industry-evidence from substation engineering in China. Sci Rep 14(1):8698. https://doi.org/10.1038/s41598-024-59271-5
114. Orsetti E, Tollin N, Lehmann M, Valderrama VA, Morató J (2022) Building resilient cities: climate change and health interlinkages in the planning of public spaces. Int J Environ Res Public Health 19(3):1355. https://doi.org/10.3390/ijerph19031355
115. Zikargae MH, Woldearegay AG, Skjerdal T (2022) Assessing the roles of stakeholders in community projects on environmental security and livelihood of impoverished rural society: a nongovernmental organization implementation strategy in focus. Heliyon 8(10):e10987. https://doi.org/10.1016/j.heliyon.2022.e10987
116. Parikh P, Diep L, Hofmann P, Tomei J, Campos LC, Teh TH, Mulugetta Y, Milligan B, Lakhanpaul M (2021) Synergies and trade-offs between sanitation and the sustainable development goals. UCL Open Environ 3:e016. https://doi.org/10.14324/111.444/ucloe.000016

Innovations for Cleaner Futures: Emerging Technologies in Air Pollution Mitigation

Odoligie Imarhiagbe, Miriam U. Nwodo, and Matthew Chidozie Ogwu

Contents

Abstract The escalating threat of air pollution to human health and the environment has catalyzed a pressing need for innovative solutions to mitigate its adverse effects. This chapter aims to highlight the latest advancements in emerging technologies aimed at achieving cleaner futures by effectively mitigating air pollution. Drawing upon multidisciplinary research and practical applications, the chapter delves into various innovations and their potential to revolutionize air pollution mitigation strategies. The chapter provides a comprehensive overview of the current state of

O. Imarhiagbe (✉)
Department of Health and Social Science, London School of Science and Technology, London, UK
e-mail: odoligie.imarhiagbe@lsst.ac

M. U. Nwodo
Department of Medical Laboratory Science, Faculty of Basic Medical Science, Niger Delta University, Wilberforce Island, Bayelsa, Nigeria

M. C. Ogwu
Goodnight Family Department of Sustainable Development, Appalachian State University, Boone, NC, USA
e-mail: ogwumc@appstate.edu

Matthew Chidozie Ogwu and Sylvester Izah (eds.), *Sustainable Strategies for Air Pollution Mitigation: Development, Economics and Technologies*,
Hdb Env Chem (2024) 133: 377–408, DOI 10.1007/698_2024_1124,

Published online: 21 September 2024

air pollution, which underscores the urgency of adopting sustainable and effective measures to address this multifaceted challenge. Some innovative clean air emerging technologies that hold promise in mitigating air pollution includes those related to renewable energy, transportation, industrial processes, and urban planning. For instance, advancements in renewable energy sources such as solar and wind power offer cleaner alternatives to fossil fuels and contribute to reducing greenhouse gas emissions. Also, the roles of electric vehicles and other low-emission transportation technologies in curbing air pollution from vehicular emissions are discussed as well as the potential roles of smart transportation systems and urban mobility solutions to optimize traffic flow and minimize pollutant emissions in densely populated areas. Other innovations highlighted include novel approaches to industrial processes, such as carbon capture and utilization technologies, which capture carbon dioxide emissions from industrial sources and convert them into valuable products and the application of green chemistry principles in manufacturing processes to minimize pollution and enhance resource efficiency. Robust air quality monitoring and pollution control measures using advanced sensor technologies and data analytics for real-time monitoring of air quality parameters will enable proactive interventions to mitigate pollution hotspots and safeguarding public health. In conclusion, the chapter emphasizes the transformative potential of emerging technologies in advancing air pollution mitigation efforts and fostering cleaner, healthier environments for future generations.

Keywords Air purification systems, Biofilters, Clean energy sources, Electric vehicles, Green infrastructure, Low-emission vehicles, Nanotechnology

Abbreviations

EVs	Electric vehicles
CCS	Carbon capture and storage
UNEP	United Nations Environmental Programme
UN	United Nations
IoT	Internet of Things
PV	Photovoltaic
DC	Direct current
CSP	Concentrated solar power
PHEVs	Plug-in hybrid electric vehicles
HEVs	Hybrid electric vehicles
V2G	Vehicle-to-grid
FCVs	Fuel cell vehicles
SAFs	Sustainable aviation fuels
NZEBs	Net-zero energy buildings
LEED	Leadership in energy and environmental design
BREEAM	Building research establishment environmental assessment method

HVAC	Heating, ventilation, and air conditioning
EMS	Energy management systems
VOC	Volatile organic compound
EPDs	Environmental product declarations
EPR	Extended producer responsibility
PM	Particulate matter
NOx	Nitrogen oxides
SO2	Sulfur dioxide
CO	Carbon monoxide
O3	Ozone
UAVs	Unmanned aerial vehicles
GIS	Geographic information systems

1 Introduction

As air pollution continues to be a pressing environmental issue affecting human health, ecosystems, and climate change, the development of innovative technologies and solutions is crucial for mitigating emissions, improving air quality, and promoting sustainable development [1–3]. Addressing air pollution through innovative solutions is of paramount importance due to its significant impact on human health, ecosystems, and the environment. Air pollution is a major threat to public health, contributing to respiratory diseases, cardiovascular problems, and even premature death [4]. Innovative solutions are crucial for reducing exposure to harmful pollutants and safeguarding human health. Air pollution adversely affects ecosystems, leading to biodiversity loss, acid rain, and degradation of natural habitats. Innovative approaches can help minimize the environmental damage caused by pollutants and promote ecosystem resilience [5, 6]. Many air pollutants, such as greenhouse gases and aerosols, contribute to climate change by trapping heat in the atmosphere and altering weather patterns [7, 8]. Addressing air pollution through innovative technologies can mitigate its contribution to global warming and associated climate-related impacts [9]. Vulnerable populations, including low-income communities and marginalized groups, often bear the brunt of air pollution's health effects. Innovative solutions can promote social equity by ensuring access to clean air for all members of society, regardless of socioeconomic status. The economic costs of air pollution are substantial, encompassing healthcare expenditures, lost productivity, and damage to infrastructure [10]. Innovative mitigation strategies offer opportunities for cost-effective solutions that can alleviate the financial burden associated with pollution-related expenses [11, 12].

Emerging technologies offer promising avenues for mitigating air pollution and creating cleaner, healthier, and more resilient communities for future generations. For instance, electric vehicles (EVs) and clean mobility solutions represent a promising pathway for reducing emissions from transportation, a major source of air pollution in urban areas [13]. Advances in battery technology, charging infrastructure, and electric vehicle design have enabled the widespread adoption of EVs, contributing to lower emissions, reduced noise pollution, and improved air quality in cities [14, 15]. Furthermore, the integration of EVs with renewable energy sources, smart grid systems, and energy storage technologies can enhance the sustainability and efficiency of electric transportation, ushering in a clean mobility revolution. Innovations in air quality monitoring and sensor technologies have revolutionized the way we track, assess, and respond to air pollution levels in real time [16]. Low-cost, portable air quality sensors, satellite imaging, and Internet of Things (IoT) devices provide valuable data on pollutant concentrations, sources of emissions, and pollution hotspots, enabling informed decision-making, early warning systems, and targeted interventions to improve air quality [17–20]. These technologies are capable of leveraging citizen science initiatives, community-based monitoring networks, and data visualization platforms to engage the public in environmental monitoring, foster collaboration with researchers and policymakers, and empower individuals to take action to reduce their exposure to air pollution.

Green infrastructure, nature-based solutions, and urban green spaces play a critical role in mitigating air pollution, enhancing environmental quality, and promoting human well-being in urban areas [21]. Green roofs, vertical gardens, urban forests, and permeable pavements capture pollutants, reduce heat island effects, and improve air quality through natural filtration processes. Incorporating green infrastructure into urban planning, architecture, and landscape design can enhance the resilience of cities to climate change, promote biodiversity, and create sustainable environments that benefit both people and the planet. Clean energy technologies and renewable energy sources offer sustainable alternatives to fossil fuels, coal-fired power plants, and high-emission industries, thereby reducing air pollution and combating climate change [22, 23]. Solar power, wind energy, hydropower, and geothermal energy have become increasingly cost-effective, scalable, and reliable solutions for meeting energy demand while minimizing environmental impacts. According to the results of Erickson [24], transitioning to a renewable energy economy, promoting energy efficiency measures, and investing in smart grid systems can reduce emissions, improve air quality, and create green energy jobs in a low-carbon future.

Pollution control technologies, such as scrubbers, filters, and emission reduction systems, are instrumental in reducing emissions from industrial processes, power plants, and other pollution sources [25, 26]. Advanced air pollution control devices, catalytic converters, and particulate matter filters capture pollutants before they are released into the atmosphere, mitigating the environmental impact of industrial activities and improving air quality [27]. Furthermore, carbon capture and storage (CCS) technologies capture carbon dioxide emissions from power plants, industrial facilities, and other sources, preventing greenhouse gas emissions and contributing to climate change mitigation efforts [28, 29]. Circular economy practices, resource recovery technologies, and waste-to-energy solutions offer innovative approaches to reducing waste generation, minimizing pollution, and promoting resource efficiency in a cirscular economy model [30, 31]. In the opinion of Saleem et al. [32], recycling, upcycling, and reusing materials reduce the environmental impact of waste disposal, conserve resources, and lower emissions associated with waste management processes. Implementing sustainable waste management practices, promoting circular design principles, and fostering a culture of resource conservation can reduce air pollution, protect ecosystems, and create economic opportunities in a circular economy.

Despite several promising advancements in emerging technologies and innovations for mitigating air pollution, several challenges and opportunities exist in the development and implementation of these solutions including cost and scalability, infrastructural needs, regulatory framework, collaboration, and public awareness. The initial costs of implementing new technologies and innovations for air pollution mitigation can be a barrier to widespread adoption, particularly in low-income countries and underserved communities [11]. In the same way, building the necessary infrastructure, investing in research and development, and securing funding for sustainable technologies require coordinated efforts, partnerships, and long-term commitment from governments, businesses, and stakeholders [33]. Strengthening regulatory frameworks, incentivizing investments in clean technologies, and aligning policies with environmental goals are essential for creating an enabling environment for innovation and sustainability [34]. Educating the public, raising awareness about air pollution risks, and promoting sustainable behaviors are crucial for driving behavioral change, fostering a culture of environmental stewardship, and mobilizing support for pollution mitigation efforts.

This chapter explores the latest technological advancements, trends, and innovations that are driving efforts to combat air pollution for mitigating air pollution, highlighting key advancements, challenges, and opportunities in the field of environmental protection and the promotion of sustainable development. The objectives of the chapter are to provide a comprehensive overview of emerging technologies, discuss their applications in air pollution mitigation, and analyze the potential implications for cleaner and more sustainable futures. It will contribute to the growing body of literature on environmental innovation, technology transfer, and sustainable development strategies that are essential for achieving global targets for air quality improvement, climate change mitigation, and building a more sustainable future for all.

2 Technological Innovations in Air Pollution Mitigation

Air pollution is a critical environmental issue that impacts the health and well-being of individuals, ecosystems, and communities worldwide. The combustion of fossil fuels, industrial activities, transportation, and agricultural practices are major sources of air pollutants, including particulate matter, nitrogen oxides, sulfur dioxide, and volatile organic compounds [35]. To address the challenges of air pollution and combat climate change, technological innovations play a key role in mitigating emissions, promoting cleaner energy sources, and advancing sustainable development solutions. This section examines the latest advancements in clean energy technologies for air pollution mitigation, focusing on solar energy, wind power, and hydroelectric power as sustainable alternatives to traditional fossil fuel sources as well as sustainable transportation solutions, green building technologies, waste management innovations, and air quality monitoring and data analytics (Table 1).

Table 1 Technological innovations for air pollution mitigation

Category	Technology	Example	Value to clean air
Clean energy technologies	Solar energy	Advancements in solar photovoltaic systems	Generates renewable energy, reduces reliance on fossil fuels, and decreases air pollution
		Concentrated solar power	By reducing emissions from traditional power sources
	Wind power	Innovations in wind turbine design	Harnesses clean and sustainable wind energy and reduces air pollution from fossil fuels
		Offshore wind farms	By decreasing reliance on coal and oil for electricity generation
	Hydroelectric power	Small-scale hydroelectric projects	Provides clean electricity production and reduces air pollution emissions from fossil fuels
		Low-impact development	By utilizing renewable energy sources like water
Transportation solutions	Electric vehicles	Evolution of electric cars, buses, and bikes	Reduces tailpipe emissions, decreases air pollution, and mitigates climate change impacts
	Fuel cell technology	Fuel cell vehicles	Utilizes hydrogen for cleaner mobility and reduces emissions contributing to air pollution
		Hydrogen-powered transportation	By moving away from traditional gasoline and diesel fuels
	Sustainable aviation	Biofuels	Reduces greenhouse gas emissions from aviation and contributes to cleaner air travel
		Electric aircraft	Decreases reliance on fossil fuels and lowers emissions from conventional aircraft
		Hybrid propulsion systems	Improves fuel efficiency and reduces air pollution from traditional aviation fuels
Green building technologies	Energy-efficient design	Passive solar design	Reduces energy consumption and lowers emissions from heating/cooling systems
		Insulation	Improves building efficiency and decreases energy use and associated air pollution
		Ventilation systems	Enhances indoor air quality and reduces pollutants in indoor spaces

(continued)

Table 1 (continued)

Category	Technology	Example	Value to clean air
	Smart buildings	Automation	Optimizes energy use, reduces environmental impact, and enhances indoor air quality
		Sensors	Monitors air quality, improves building efficiency, and reduces pollutants
		Energy management systems	Minimizes energy waste and lowers emissions contributing to air pollution
	Green materials	Recycled concrete	Reduces demand for new materials and lowers emissions from production processes
		Bamboo	Sustainable material choice decreases environmental impact
		Sustainable wood products	Conserves forests and minimizes air pollution from deforestation
Waste management innovations	Waste-to-energy	Anaerobic digestion	Converts waste into energy, reduces landfill emissions and mitigates air pollution
		Thermal treatment	Handles waste efficiently and reduces methane emissions contributing to air pollution
	Recycling and circular economy	Closed-loop systems	Reduces waste generation, conserves resources and minimizes air pollution from production
		Innovative recycling processes	Promotes waste reduction, lowers emissions from production and decreases air pollution
	Bioremediation	Biological agents	Cleans up contaminated environments, improves air quality, and mitigates pollution impacts
		Microorganisms	Remediate air, soil, and water pollution and reduce harmful pollutants
Air quality monitoring and data analytics	Sensor technologies	Low-cost sensors, drones, and satellite imagery	Provides real-time air quality data, maps pollution sources, and supports pollution control
			Initiatives to reduce air pollution emissions
	Big data analytics	Data analytics, machine learning, and AI	Analyzes pollution trends, develops predictive models, and informs policy for cleaner air
			Quality and improved public health outcomes
	Citizen science initiatives	Crowdsourcing platforms and participatory sensing	Engages citizens in air quality monitoring, increases public awareness, and drives action for
			Cleaner air and environmental protection

2.1 Clean Energy Technologies

Solar Energy Solar energy is a renewable and abundant source of clean power that offers significant potential for reducing greenhouse gas emissions and improving air quality. Advancements in solar photovoltaic (PV) systems, solar panels, and solar energy storage technologies have revolutionized the way we harness solar energy for electricity generation. Solar PV systems convert sunlight into electricity through the photovoltaic effect, capturing solar radiation and converting it into direct current (DC) electricity using semiconducting materials such as silicon (Fig. 1; [37, 38]). The development of thin-film solar cells, organic photovoltaics, and perovskite solar cells has led to higher efficiency, lower costs, and increased flexibility in solar panel design and installation [39, 40].

Concentrated solar power (CSP) is another innovative technology that uses mirrors or lenses to concentrate sunlight onto a receiver, generating high-temperature heat that drives a steam turbine to produce electricity [41]. CSP systems can store thermal energy in molten salt or other heat storage media, allowing for dispatchable power generation even when the sun is not shining. Hybrid solar PV-CSP systems combine the benefits of both technologies, providing a reliable and efficient source of renewable energy that can meet the electricity demand of grid-connected or off-grid applications [42, 43].

Solar energy is a versatile and scalable solution for air pollution mitigation, with utility-scale solar farms, rooftop solar installations, and community solar projects contributing to the transition to a low-carbon economy [36]. Solar power plants can be deployed in arid regions, remote areas, and urban environments, reducing reliance on traditional power sources, lowering emissions, and creating clean energy jobs. According to Umair et al. [44], the integration of solar energy with energy storage systems, smart grid technologies, and electric vehicle charging infrastructure enables

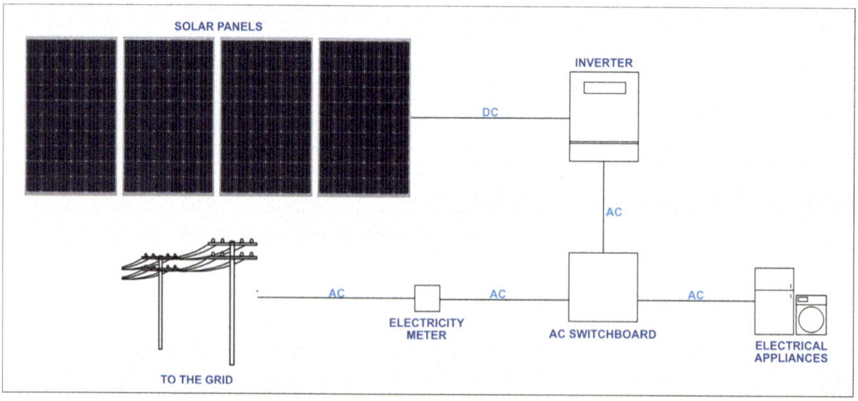

Fig. 1 Solar photovoltaic systems with solar panels and energy setup. Source: Solar Design Guard [36]

the optimization of renewable energy resources, grid stability, and energy efficiency in a sustainable energy ecosystem.

Wind Power Wind power is a mature and reliable form of renewable energy that harnesses the kinetic energy of wind to generate electricity through wind turbines. Innovations in wind turbine design, rotor blades, tower systems, and offshore wind farms have increased the efficiency, capacity, and competitiveness of wind power as a clean energy source. Horizontal-axis wind turbines are commonly used for onshore wind projects, while vertical-axis wind turbines offer a smaller footprint, quieter operation, and enhanced scalability for urban and distributed wind applications [45]. Offshore wind farms represent a growing sector in the wind energy market, with large-scale projects located in coastal waters or offshore locations to capture stronger and more consistent wind speeds [46]. Floating wind turbines, fixed-bottom foundations, and subsea cabling technologies enable the deployment of offshore wind farms in deep waters, challenging seabed conditions, and remote locations [47, 48]. Wind energy clusters, interconnected grid systems, and offshore transmission networks facilitate the integration of offshore wind power into the electricity grid, reducing curtailment, enhancing reliability, and maximizing renewable energy generation capacities [49].

Wind power is a cost-effective and environmentally friendly solution for air pollution mitigation, with wind farms displacing coal-fired power plants, reducing carbon dioxide emissions, and promoting sustainable energy transitions. Onshore wind installations can be sited in windy regions, rural areas, and agricultural lands, while offshore wind projects leverage offshore winds, coastal infrastructure, and maritime expertise to deliver clean electricity to coastal communities and coastal cities [50]. The development of hybrid wind-solar systems, energy storage solutions, and demand response technologies optimizes the integration of wind power into the energy mix, supporting grid stability, decarbonization, and energy security objectives [51, 52].

Hydroelectric Power Hydroelectric power is a well-established and sustainable form of renewable energy that utilizes the potential energy of water flow to generate electricity through hydroelectric dams, run-of-river hydropower plants, and pumped storage facilities [53]. Small-scale hydroelectric projects, microhydro systems, and low-impact run-of-river installations offer decentralized and environmentally friendly solutions for clean electricity production in rural and remote areas. Pumped storage hydropower plants store excess energy from intermittent renewable sources, such as wind and solar power, by pumping water to higher elevations and releasing it to generate electricity during peak demand periods [54].

Hydropower technologies, turbine designs, and dam construction methods have evolved to minimize environmental impacts, enhance fish passage, and improve water quality in river ecosystems [55]. According to Inam Ullah et al. [56], run-off-river hydropower plants divert a portion of river flow through a turbine without the need for large reservoirs or impoundments, preserving downstream flow regimes, aquatic habitats, and sediment transport processes. Adaptive management strategies, fish-friendly turbines, and downstream fish passage facilities mitigate the impacts of hydropower operations on fish populations, migratory species, and river biodiversity.

Hydroelectric power is a reliable and flexible energy source for air pollution mitigation, with hydropower plants providing baseload power, grid stability, and peaking capacity to meet energy demand and balance renewable energy supply. Large-scale hydroelectric projects, such as dam complexes, multipurpose reservoirs, and pumped storage facilities, offer storage capabilities, dispatchable power generation, and grid support services that complement variable renewable energy sources and reduce the need for fossil fuel backup generation [57]. The integration of hydroelectric power with smart grid systems, regional interconnections, and energy trading platforms enhances the resilience, efficiency, and sustainability of the electricity grid, reducing emissions, enhancing air quality, and promoting climate change mitigation [58].

2.2 Transportation Solutions

Transportation is a major contributor to air pollution, with emissions from vehicles, airplanes, ships, and trains impacting air quality and public health in urban areas and regions with high traffic volumes [59, 60]. Therefore, technological innovations play a crucial role in addressing the environmental impacts of transportation, reducing emissions, and promoting cleaner mobility options. Some latest advancements in transportation solutions for air pollution mitigation include electric vehicles, fuel cell technology, and sustainable aviation as key strategies to achieve zero-emission transportation and sustainable travel alternatives.

Electric Vehicles Electric vehicles (EVs) have emerged as a promising solution for reducing greenhouse gas emissions, improving air quality, and transitioning to a low-carbon transportation system [24]. The evolution of electric cars, buses, trucks, and bikes has been driven by advancements in battery technology, charging infrastructure, and vehicle design, making electric mobility more accessible, affordable, and convenient for consumers and businesses [61]. Battery electric vehicles (BEVs) use rechargeable lithium-ion batteries to store electricity, powering an electric motor to drive the vehicle without tailpipe emissions. Plug-in hybrid electric vehicles (PHEVs) combine an internal combustion engine with an electric motor and battery pack, offering extended range and flexibility for longer trips or highway driving [62, 63]. Hybrid electric vehicles (HEVs) use regenerative braking to recharge the battery, reducing fuel consumption and emissions during city driving conditions [64]. Electric motorcycles, scooters, and bicycles provide sustainable and eco-friendly options for urban mobility, reducing noise pollution, congestion, and air pollution in densely populated areas.

Electric vehicles offer numerous benefits for air pollution mitigation, including zero tailpipe emissions, lower operating costs, and reduced dependence on fossil fuels. EVs are ideal for short trips, commute distances, and urban transport needs, with fast charging stations, home charging units, and public charging networks enabling convenient recharging options for electric vehicle owners [15]. In the opinion of Ohanu et al. [65], the integration of smart grid technology, vehicle-to-grid (V2G) systems, and renewable energy sources optimizes the energy management, grid stability, and carbon footprint of electric vehicles, supporting sustainable transportation solutions for cleaner and healthier communities.

Fuel Cell Technology Fuel cell technology is a clean and efficient power generation technology that produces electricity through an electrochemical reaction between hydrogen and oxygen, with water vapor as the only byproduct [66, 67]. Fuel cell vehicles (FCVs) use hydrogen fuel cells to generate electricity, powering an electric motor to drive the vehicle with zero emissions and low noise levels. Hydrogen fueling stations, onboard hydrogen tanks, and fuel cell stacks enable FCVs to refuel quickly, achieve long-range driving capabilities, and deliver high performance and reliability for passenger cars, buses, trucks, and fleet vehicles [68].

Fuel cell technology offers several advantages for air pollution mitigation, including zero tailpipe emissions, rapid refueling, and energy efficiency compared to internal combustion engine vehicles [69]. FCVs have a longer driving range and faster refueling times than battery electric vehicles, making them suitable for long-distance travel, fleet operations, and heavy-duty applications. The development of hydrogen production, storage, and distribution infrastructure supports the expansion of fuel cell transportation systems, integrating hydrogen as a clean energy carrier for decarbonizing the transportation sector and reducing air pollution emissions [70, 71]. Fuel cell technology has the potential to facilitate the transition to a hydrogen economy, with applications in stationary power generation, portable electronics, and backup power systems in addition to transportation solutions [66, 67]. Fuel cell buses, trucks, forklifts, and drones are being deployed in various industry sectors, supporting sustainable logistics, urban freight transport, and emergency response operations. Fuel cell microgrids, remote power systems, and off-grid applications demonstrate the versatility and reliability of fuel cell technology in meeting energy needs, reducing emissions, and promoting environmental sustainability in diverse environments and contexts.

Sustainable Aviation Sustainable aviation is an emerging field of research and development that aims to reduce emissions, improve fuel efficiency, and promote environmentally friendly practices in the aviation industry [72]. Biofuels, electric aircraft, hybrid propulsion systems, and sustainable aviation fuels (SAFs) are key technologies and strategies for greener air travel, mitigating the environmental impacts of air transport and advancing sustainable aviation goals [73, 74]. Biofuels are renewable fuels derived from biomass, wastes, or algae, with lower carbon intensity and emissions compared to conventional jet fuels.

Electric aircraft are electrically powered aviation platforms that use batteries, fuel cells, or hybrid-electric systems for propulsion, reducing noise, emissions, and operating costs while enhancing performance, range, and sustainability in air transportation [75]. Short-haul regional flights, urban air mobility, and aerial taxis are target markets for electric aircraft, with advancements in aircraft design, lightweight materials, and energy storage technologies enabling the electrification of propulsion systems and the development of zero-emission aviation solutions [76]. Hybrid propulsion systems combine conventional engines with electric motors, turbofans, or alternative power sources to optimize fuel efficiency, reduce emissions, and enhance operational flexibility for commercial airlines, cargo carriers, and military applications [75].

Sustainable aviation fuels (SAFs) are drop-in substitutes for conventional jet fuels that are produced from sustainable feedstocks, such as algae, biomass, or waste oils, with lower lifecycle greenhouse gas emissions and emissions of criteria pollutants compared to petroleum-derived fuels. SAF blends, green hydrogen, and synthetic fuels offer carbon-neutral and renewable alternatives for decarbonizing the aviation sector, reducing the environmental footprint of flights, and meeting international carbon offsetting commitments [77]. The deployment of SAFs in airports, flight operations, and air traffic management systems supports low-emission aviation pathways, climate resilience, and responsible air transport practices that prioritize environmental stewardship and social responsibility in the aviation industry.

2.3 *Green Building Technologies*

Green building technologies are innovative solutions that promote energy efficiency, sustainability, and environmental stewardship in the design, construction, and operation of buildings and structures [78]. By incorporating green building practices, technologies, and materials, architects, engineers, builders, and developers can reduce resource consumption, minimize emissions, and improve indoor air quality, creating healthy, comfortable, and environmentally friendly built environments [79]. Relevant advancements in green building technologies for air pollution mitigation include the following:

Energy-Efficient Design Energy-efficient design is a fundamental principle of green building technologies that emphasizes the optimization of building envelopes, orientation, insulation, and ventilation systems to minimize energy consumption, maximize thermal comfort, and reduce greenhouse gas emissions [80]. Passive solar design integrates building orientation, window placement, shading devices, and thermal mass to capture solar heat gain, natural light, and ventilation, reducing the need for artificial heating, cooling, and lighting in buildings [81]. High-performance windows, insulation, air sealing, and daylighting strategies enhance energy efficiency, indoor comfort, and occupant well-being in residential, commercial, and institutional buildings.

Net-zero energy buildings (NZEBs) are designed to produce as much energy as they consume over a year, incorporating renewable energy systems, energy storage technologies, and energy-efficient appliances, lighting, and HVAC equipment to achieve comprehensive energy savings and carbon neutrality [82]. Passive houses, green roofs, and energy recovery ventilation systems further enhance the energy performance, indoor air quality, and thermal comfort of buildings, reducing energy costs, environmental impacts, and air pollution emissions associated with building operations [81, 83]. Building energy modeling, life-cycle assessments, and green building certifications, such as LEED, BREEAM, and Green Globes, support evidence-based decision-making and performance benchmarking for sustainable building design and construction [84, 85].

Smart Buildings Smart buildings are intelligent and connected structures that utilize automation, sensors, building management systems, and Internet of Things (IoT) technologies to optimize resource utilization, enhance occupant comfort, and improve energy efficiency in buildings [86]. Building automation systems control heating, ventilation, air conditioning (HVAC), lighting, security, and water systems based on occupancy patterns, indoor environmental conditions, and energy demand, reducing energy waste, operational costs, and carbon emissions [81, 87]. Smart sensors, meters, and controls enable real-time monitoring, fault detection, and energy management strategies to ensure building performance, occupant satisfaction, and sustainability goals are met [88].

Energy management systems (EMS) integrate building automation, smart grids, renewable energy sources, and demand response programs to optimize energy consumption, peak load management, and grid interactions in smart buildings [89]. Building energy dashboards, smart thermostats, and remote access applications empower occupants and facility managers to monitor, analyze, and control energy use, indoor air quality, and environmental impacts in buildings. The integration of smart meters, advanced metering infrastructure, and energy monitoring devices facilitate data collection, analysis, and reporting for energy audits, compliance verification, and performance measurement in smart buildings, supporting energy efficiency improvements, emissions reductions, and air pollution mitigation efforts [90, 91].

Green building technologies, including passive design strategies, energy-efficient systems, and smart building solutions, offer tangible benefits for air pollution mitigation, energy savings, and sustainable development in the built environment. By adopting green building practices, codes, and standards, stakeholders can enhance building performance, occupant health, and environmental quality, while reducing operational costs, carbon footprints, and air pollution emissions associated with building construction and operations. Through collaboration, innovation, and knowledge sharing, the green building industry can drive positive change, improve building sustainability, and contribute to global efforts to address climate change, air quality, and public health concerns in urban and rural communities.

Green Materials Green building materials play a critical role in promoting sustainability, resource efficiency, and environmental responsibility in building construction, renovation, and demolition activities. Eco-friendly building materials are characterized by their low embodied energy, renewable sourcing, recyclability, and reduced environmental impacts throughout their life cycle, from extraction and manufacturing to installation and disposal [79]. Recycled concrete, reclaimed wood, bamboo, cork, straw bales, and natural fibers are examples of green materials that minimize waste, conserve resources, and promote circular economy principles in the construction industry [92, 93].

Sustainable wood products, such as certified lumber, engineered wood, and timber composites, are harvested from responsibly managed forests, certified by forest stewardship councils, and verified for sustainable sourcing practices, protecting biodiversity, wildlife habitats, and ecosystems. Green roofs, solar panels, and rainwater harvesting systems incorporate renewable energy technologies, water conservation measures, and green infrastructure solutions to enhance building performance, reduce urban heat islands, and mitigate stormwater runoff, improving air quality, microclimate conditions, and environmental resilience in urban environments.

Low-VOC paints, adhesives, sealants, and finishes reduce indoor air pollution, volatile organic compound (VOC) emissions, and off-gassing in buildings, contributing to a healthier and more comfortable indoor environment for occupants [94]. According to Dsilva et al. [95], green building materials certification programs, such as Cradle to Cradle, Declare, and Environmental Product Declarations (EPDs), provide transparency, accountability, and credibility for sustainable building products, enabling architects, designers, and builders to specify, source, and use green materials in construction projects that prioritize environmental stewardship, occupant health, and long-term sustainability objectives.

2.4 *Waste Management Innovations*

Waste management plays a crucial role in air pollution mitigation by reducing emissions from landfills, incinerators, and waste disposal sites and promoting resource recovery, recycling, and circular economy principles to minimize waste generation and environmental impacts. Technological innovations in waste management offer sustainable solutions for converting waste into energy, recycling materials, cleaning up contaminated environments, addressing air pollution sources, protecting public health, and achieving environmental sustainability goals [96]. These innovations include waste-to-energy technologies, recycling, circular economy approaches, and bioremediation strategies as key pathways to enhance waste management practices, reduce pollution emissions, and support sustainable waste treatment and disposal options.

Waste-to-Energy Waste-to-energy technologies utilize advanced processes, such as anaerobic digestion, thermal treatment, gasification, and pyrolysis, to convert waste materials into energy products, including biogas, biofuels, electricity, heat, and steam, reducing the reliance on fossil fuels, landfill disposal, and greenhouse gas emissions [2]. Anaerobic digestion systems break down organic waste in the absence of oxygen, producing methane-rich biogas that can be used for power generation, heating, or vehicle fuel applications, capturing valuable energy resources and reducing organic waste volumes in landfills [97]. Thermal treatment technologies, including incineration, combustion, and gasification, use high temperatures to convert waste into syngas, ash, and heat, recovering energy and reducing emissions from waste disposal operations [98].

Waste-to-energy plants, bioenergy facilities, and biorefineries play an important role in waste management, energy production, and climate change mitigation, transforming waste streams into renewable energy sources, bio-based products, and sustainable fuels that support low-carbon economies, circular supply chains, and resource recovery incentives. Codigestion of organic wastes, food scraps, and agricultural residues with wastewater treatment sludge, industrial byproducts, and forestry residues enhances biogas production, feedstock flexibility, and energy efficiency in anaerobic digestion systems, optimizing waste-to-energy conversion processes and enhancing environmental performance in waste management operations [99, 100].

Recycling and Circular Economy Recycling and circular economy principles promote the reuse, repair, and recycling of materials, products, and resources, closing material loops, reducing waste streams, and conserving natural resources to promote sustainable consumption, production, and waste management practices. Innovative recycling technologies, such as mechanical sorting, chemical processing, and material recovery systems, enable the separation, recovery, and valorization of valuable materials, such as metals, plastics, paper, glass, and electronic waste, diverting them from landfill disposal, reducing pollution, and promoting resource conservation and recovery initiatives [69].

Closed-loop systems, reverse logistics, and product stewardship programs facilitate the collection, processing, and remanufacturing of end-of-life products, components, and packaging materials, promoting sustainable supply chains, waste reduction, and circular business models that prioritize material efficiency, resource optimization, and environmental responsibility [101, 102]. Waste prevention, waste minimization, and waste diversion strategies encourage the reduction of single-use plastics, packaging waste, food waste, and electronic waste through source separation, waste audits, and recycling education programs that engage consumers, businesses, and communities in waste reduction and recycling efforts.

Circular economy initiatives, such as extended producer responsibility (EPR), product take-back schemes, and eco-design requirements, shift the responsibility for waste management, recycling, and disposal from consumers to producers, manufacturers, and retailers, incentivizing eco-friendly products, packaging designs, and end-of-life management practices that promote resource recovery, material reuse, and environmental sustainability in product life cycles [103, 104]. Sustainable procurement, green labeling, and eco-certification programs empower consumers and businesses to make informed choices, support sustainable products, and drive the market demand for circular economy solutions that reduce waste, promote recycling, and mitigate air pollution emissions associated with landfills, incinerators, and waste disposal activities.

Bioremediation Bioremediation is a biological treatment technology that utilizes natural or engineered biological agents, microorganisms, enzymes, and plants to clean up contaminated environments, including air, soil, water, and waste sites, by degrading, metabolizing, or sequestering pollutants, chemicals, and toxic substances, reducing environmental risks, human exposures, and pollution impacts [105, 106]. Bioremediation processes, such as phytoremediation, bioaugmentation, composting, and biostimulation, enhance the biodegradation, detoxification, and removal of contaminants, such as heavy metals, petroleum hydrocarbons, pesticides, and industrial chemicals, from polluted sites, restoring environmental quality, ecosystem health, and natural habitats.

Bioremediation technologies provide cost-effective, sustainable, and environmentally friendly solutions for cleaning up contaminated air, soil, and water, by harnessing the power of nature, ecosystem services, and ecological processes to remediate pollution, restore ecological balance, and enhance environmental resilience in degraded landscapes, industrial sites, and brownfield areas. Microbial populations, biofilms, and consortia play key roles in metabolizing pollutants, managing waste streams, and detoxifying hazardous substances, breaking down organic compounds, oils, and toxins into harmless byproducts, such as carbon dioxide, water, and biomass, that can be recycled or reused in ecological cycles and nutrient cycles.

Biofiltration, biosorption, and biofiltration systems use living organisms, bacteria, fungi, algae, and plants to filter, absorb, and degrade pollutants, odors, and volatile organic compounds (VOCs) from air emissions, industrial effluents, and waste streams, promoting clean air, water, and land resources, while reducing emissions, pollutants, and toxic substances that contribute to air pollution, water pollution, and soil contamination. Composting, vermiculture, and microbial biodegradation processes facilitate the decomposition, mineralization, and nutrient cycling of organic matter, biowaste, and agricultural residues, supporting waste management, soil fertility, and ecosystem health; promoting circular economy principles; and minimizing environmental impacts associated with waste disposal and landfill operations.

2.5 Air Quality Monitoring and Data Analytics

Air quality monitoring and data analytics play a crucial role in air pollution mitigation by providing real-time information, insights, and solutions to address pollution sources, health risks, and environmental impacts in urban and industrial areas. Technological innovations in air quality monitoring and data analytics enable stakeholders to collect, analyze, and visualize air pollution data, trends, and patterns, leveraging sensor technologies, big data analytics, and citizen science initiatives to enhance environmental monitoring, public awareness, and policy decision-making in addressing air quality challenges [107, 108]. At the moment, air quality monitoring generates results such as those presented in Table 2.

Table 2 Air quality monitoring index results

AQI Color Code	Level of Concern	Index Value	Air Quality Description
Green	Good	0 to 50	Air quality is satisfactory, and air pollution poses little or no risk
Yellow	Moderate	51 to 100	Air quality is acceptable. However, there may be a risk for some people, particularly those who are unusually sensitive to air pollution
Orange	Unhealthy for Sensitive Groups	101 to 150	Members of sensitive groups may experience health effects. The general public is less likely to be affected
Red	Unhealthy	151 to 200	Some members of the general public may experience health effects; members of sensitive groups may experience more serious health effects
Purple	Very Unhealthy	201 to 300	Health alert: The risk of health effects is increased for everyone
Maroon	Hazardous	301 and higher	Health warning of emergency conditions: everyone is more likely to be affected

Source: Adapted from AIRNOW [109]

Source: Adapted from AIRNOW [109]

However, the latest advancements in air quality monitoring and data analytics for air pollution mitigation include sensor technologies, big data analytics, and citizen science initiatives as key tools to enhance air quality management, pollution control, and public engagement to improve air quality, protect public health, and achieve sustainable development goals.

Sensor Technologies Sensor technologies enable the real-time monitoring, measurement, and mapping of air pollutants, such as particulate matter (PM), nitrogen oxides (NOx), sulfur dioxide (SO2), volatile organic compounds (VOCs), carbon monoxide (CO), and ozone (O3), using low-cost sensors, drones, satellites, and remote sensing platforms to detect pollution hotspots, air quality violations, and emission sources in urban, industrial, and rural areas. Low-cost sensors, portable monitors, and wearable devices provide personalized air quality data, exposure assessments, and health risk information to individuals, communities, and vulnerable populations, empowering them to act, avoid pollution sources, and protect themselves from harmful pollutants in indoor and outdoor environments [110].

Drones and unmanned aerial vehicles (UAVs) equipped with air quality sensors, cameras, and data loggers enable aerial surveillance, reconnaissance, and monitoring of air pollution sources, traffic emissions, and industrial activities, enhancing data collection, visualization, and analysis capabilities in hard-to-reach or hazardous areas, such as industrial sites, urban centers, and disaster zones [111]. Satellite imagery, remote sensing, and Geographic Information Systems (GIS) applications provide spatial data, images, and maps of air quality indicators, pollution concentrations, and emission sources, facilitating monitoring, modeling, and enforcement efforts to track pollution trends, assess environmental impacts, and inform policy decisions in air quality management and pollution control initiatives [112]. Real-time air quality monitoring networks, data-sharing platforms, and open data initiatives enhance data transparency, accessibility, and accountability for air pollution monitoring and management, promoting data-driven decision-making, public engagement, and policy responses to address air quality challenges, mitigate pollution impacts, and protect public health in communities, cities, and regions around the world.

Big Data Analytics Big data analytics leverage data science, machine learning, and artificial intelligence algorithms to analyze air pollution trends, model emissions, and predict air quality changes, using multisource data, historical records, and real-time measurements to identify pollution sources, patterns, and interventions for air quality improvement, pollution prevention, and health protection. Data fusion, integration, and visualization techniques combine information from multiple sensors, models, and data streams to create comprehensive air quality maps, forecasts, and risk assessments, enabling stakeholders to understand pollution dynamics, exposure pathways, and health impacts in urban, industrial, and rural settings [113, 114].

Machine learning algorithms, neural networks, and predictive analytics models predict air pollution concentrations, dispersion patterns, and health risks based on meteorological conditions, emission sources, and pollutant interactions, supporting early warning systems, pollution alerts, and emergency responses to mitigate pollution events, reduce public exposures, and prevent adverse health effects associated with poor air quality episodes [115]. Artificial intelligence applications, deep learning algorithms, and data mining techniques analyze vast amounts of air pollution data, sensor readings, and environmental parameters to identify pollution trends, hotspots, and interventions that drive policy decisions, infrastructure investments, and pollution control measures to protect air quality, public health, and environmental resources [116].

Data analytics platforms, cloud computing services, and software applications enable data visualization, dashboarding, and sharing features that foster data transparency, citizen engagement, and stakeholder collaboration in air quality monitoring, management, and policy-making processes, promoting data-driven decision-making, public awareness, and accountability in addressing air pollution challenges, protecting environmental quality, and advancing sustainable development goals for clean air, healthy communities, and resilient ecosystems.

Citizen Science Initiatives Citizen science initiatives engage individuals, communities, and volunteers in monitoring air quality, collecting data, and participating in research projects through crowdsourcing platforms, mobile apps, and participatory sensing activities that empower citizens to become environmental stewards, data contributors, and advocates for public health, environmental protection, and pollution prevention efforts in their neighborhoods, cities, and regions. Participatory sensing, community monitoring, and citizen science campaigns enable citizens to access air quality information, sensor data, and pollution alerts in real time, fostering awareness, engagement, and action on air pollution issues, sources, and solutions that impact their daily lives, health, and well-being [117, 118].

Crowdsourced data collection, citizen reporting, and public outreach activities provide valuable insights, local knowledge, and community perspectives on air quality concerns, pollution sources, and environmental impacts in different settings, enabling researchers, policymakers, and stakeholders to collaborate, prioritize, and address air pollution challenges, public health risks, and environmental disparities that affect vulnerable populations, low-income communities, and environmental justice areas. Citizen science networks, community partnerships, and environmental education programs build capacity, engagement, and empowerment among citizens, students, and organizations to monitor air quality, advocate for pollution controls, and contribute to science-based solutions that address air pollution, climate change, and environmental health challenges in cities, neighborhoods, and rural areas [119]. Citizen science data, observations, and experiences complement official monitoring networks, regulatory agencies, and research institutions by providing diverse perspectives, local insights, and community-driven solutions to air quality issues, pollution sources, and health impacts that require collaborative action, data sharing, and public participation in air quality monitoring, management, and policy-making processes that promote environmental justice, citizen engagement, and public health protection in diverse and underserved communities.

3 Opportunities, Challenges, and Policy Perspectives of Technological Innovation for Air Pollution Mitigation

Government policies, regulations, and incentives play a crucial role in promoting the adoption and deployment of innovative air pollution mitigation technologies. Governments can enforce regulations that set emissions limits, driving industries to adopt cleaner technologies to reduce air pollution [120]. Setting standards for energy efficiency, emissions reduction, and waste management can incentivize the adoption of innovative technologies. Governments can provide financial incentives such as tax credits, grants, subsidies, and rebates to promote the use of clean technologies [121, 122]. These incentives can offset the upfront costs of adopting new technologies and encourage businesses and individuals to invest in cleaner options [77]. Governments can allocate funding for research and development in clean technologies, including air pollution mitigation. This can support the innovation and scalability of new solutions, making them more accessible and cost-effective for widespread adoption. According to Heard et al. [123] and Alami et al. [124], governments can use their purchasing power to support clean technologies by mandating the use of environmentally friendly products and services in public projects. This can create market demand for innovative solutions and drive competitiveness in the clean technology sector.

International collaborations, funding mechanisms, and partnerships are essential for supporting research, development, and implementation of clean technologies. Collaborations between countries, international organizations, research institutions, and industry can facilitate knowledge sharing, technology transfer, and capacity building for air pollution mitigation. Joint initiatives can accelerate the development and deployment of innovative solutions on a larger scale [125, 126]. International funding mechanisms, such as grants, loans, and investment funds, can support cross-border projects and initiatives for clean technology development. Multilateral agreements and initiatives like the Green Climate Fund can provide financial resources for implementing sustainable solutions. International partnerships can facilitate the transfer of clean technologies from developed countries to developing nations, addressing global environmental challenges and promoting sustainable development [127]. Capacity-building initiatives can ensure that technology transfer is effective and beneficial for all parties involved. Some ethical considerations, environmental justice, and equity implications of technological innovations in air pollution mitigation are important aspects to consider:

- **Environmental justice**: It is crucial to ensure that the benefits of clean technologies for air pollution mitigation are equitably distributed among different communities, including marginalized and vulnerable populations [128]. Policies should address environmental injustices and disparities in exposure to air pollution.
- **Equitable access:** Policies and initiatives should prioritize equitable access to clean technologies, ensuring that all individuals and communities can benefit from reduced air pollution and improved air quality [129]. And this includes addressing affordability, accessibility, and inclusivity in technology deployment.
- **Ethical standards**: The development and implementation of clean technologies should adhere to ethical standards, including transparency, accountability, and responsible governance. Considerations for social impacts, community engagement, and ethical use of data should be integrated into technological innovations for air pollution mitigation [130, 131].

Technical challenges and barriers to the widespread adoption of emerging technologies for air pollution mitigation include the following:

- **Cost**: The initial costs of implementing new technologies can be high, making it challenging for industries and governments to invest in clean solutions [132]. Cost-effective alternatives need to be developed to ensure affordability and scalability.
- **Technological maturity**: Some emerging technologies may still be in the early stages of development, lacking proven performance, reliability, and durability [133]. However, more research and testing are needed to demonstrate the effectiveness of these solutions.
- **Integration and compatibility**: Integrating new technologies with existing infrastructure and systems can be complex. Compatibility issues, interoperability, and retrofitting challenges may hinder the adoption of innovative air pollution mitigation technologies [134].

- **Regulatory hurdles**: Regulations and policies may not always be conducive to the adoption of new technologies. In the opinion of Kaur et al. [135], uncertainty around regulatory requirements, permitting processes, and compliance standards can create barriers for businesses looking to deploy clean solutions.

Opportunities for collaboration and capacity building to overcome technological, financial, and institutional constraints include the following:

- **Public-private partnerships**: Examples include collaborations between governments, industry stakeholders, research institutions, and nonprofit organizations can drive innovation, leverage expertise, and pool resources to overcome barriers to adoption.
- **Knowledge exchange platforms**: Establishing forums, networks, and platforms for sharing best practices, lessons learned, and technical expertise can facilitate capacity building and accelerate the deployment of clean technologies.
- **Capacity building programs**: Investing in training, skill development, and education for professionals in the field of air pollution mitigation can enhance technical expertise, promote innovation, and build a workforce capable of implementing new technologies [136].

Some potential future trends and emerging opportunities as well as potential breakthroughs in the field of air pollution mitigation through technological innovation include the following (Fig. 2):

- **Advancements in clean energy**: Continued advancements in renewable energy technologies, such as solar, wind, and hydropower, offer opportunities for decentralized energy production and reduced reliance on fossil fuels, contributing to air pollution reduction [137].
- **Electrification of transportation**: The widespread adoption of electric vehicles and the development of charging infrastructure present opportunities to decarbonize transportation systems and reduce emissions from vehicles, a major source of air pollution.
- **Smart cities and IoT solutions**: Integration of smart sensors, data analytics, and Internet of Things (IoT) technologies in urban infrastructure can enhance air quality monitoring, management, and decision-making, leading to more effective pollution control measures [138].
- **Circular economy practices**: The work of Di Viao et al. [139] suggests that embracing circular economy principles, such as waste reduction, recycling, and resource recovery, can minimize pollution, conserve natural resources, and promote sustainable consumption and production patterns.
- **Biotechnological innovations**: Advances in bioremediation, biofiltration, and biotechnology solutions offer opportunities for nature-based approaches to air pollution mitigation, harnessing the power of living organisms to clean the environment [140].

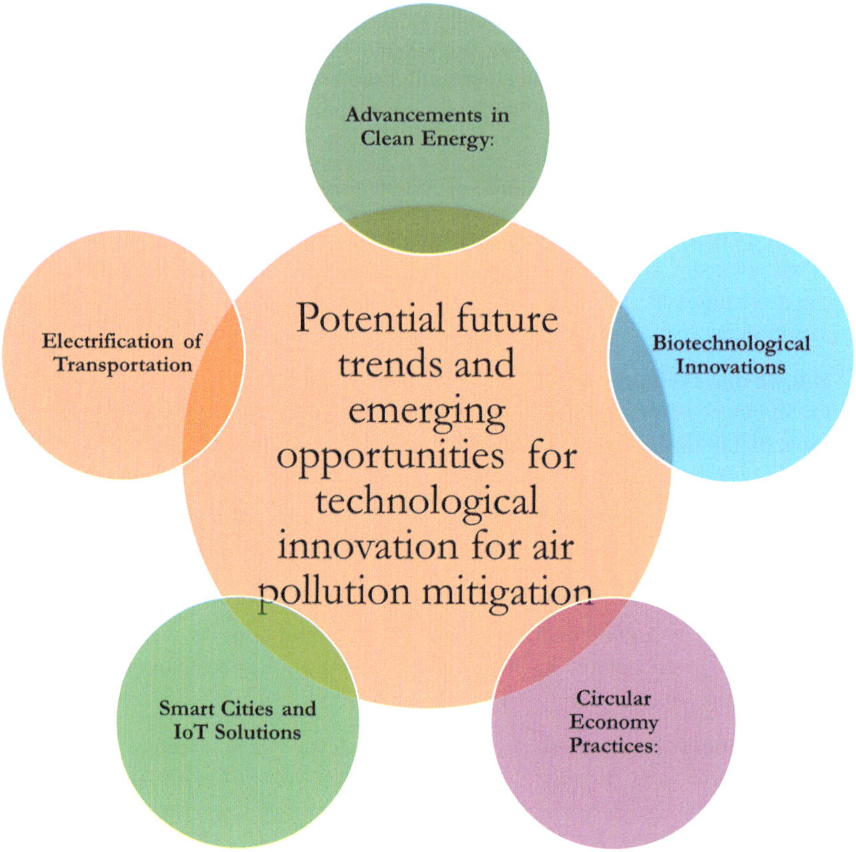

Fig. 2 Potential future trends and emerging opportunities for technological innovation for air pollution mitigation

4 Conclusion

Technological innovations in clean energy technologies, including solar energy, wind power, and hydropower, are driving advancements in air pollution mitigation, promoting sustainable development, and accelerating the transition to a low-carbon economy. Solar photovoltaic systems, concentrated solar power, wind turbines, offshore wind farms, run-of-river hydropower plants, and pumped storage facilities offer renewable and environmentally friendly alternatives to fossil fuel power generation, reducing emissions, improving air quality, and fostering cleaner and healthier futures for communities around the world.

Technological innovations in transportation solutions, including electric vehicles, fuel cell technology, and sustainable aviation, are driving advancements in air pollution mitigation, promoting sustainable mobility, and accelerating the transition to a low-carbon transportation system. Electric cars, buses, trucks, bikes, fuel cell vehicles, hydrogen-powered transportation, biofuels, electric aircraft, and hybrid propulsion systems offer clean and efficient alternatives to traditional fossil fuel vehicles, reducing emissions, enhancing air quality, and fostering sustainable travel options for individuals, communities, and industries.

Green building technologies, including energy-efficient design, smart buildings, and green materials, play a critical role in air pollution mitigation, sustainability, and environmental stewardship in the built environment. By incorporating green building practices, technologies, and materials into planning, design, construction, and operation processes, stakeholders can minimize resource consumption, reduce emissions, and improve indoor air quality, creating healthy, efficient, and environmentally friendly buildings that support human health, well-being, and productivity. Technological innovations in waste management, including waste-to-energy solutions, recycling and circular economy practices, and bioremediation strategies, play a critical role in air pollution mitigation, sustainable resource management, and environmental protection efforts. By adopting advanced waste treatment technologies, recycling programs, circular supply chain initiatives, and biological remediation processes, stakeholders can reduce waste generation, recover valuable resources, and mitigate pollution emissions, contributing to cleaner air, healthier ecosystems, and more sustainable waste management practices that support public health, environmental quality, and climate resilience in communities around the world. Through collaboration, innovation, and policy support, the waste management sector can drive positive change, promote circular economy principles, and advance sustainable waste treatment and disposal options that prioritize environmental sustainability, resource efficiency, and pollution prevention goals for a cleaner, healthier, and more sustainable future for all.

References

1. Abbass K, Qasim MZ, Song H, Murshed M, Mahmood H, Younis I (2022) A review of the global climate change impacts, adaptation, and sustainable mitigation measures. Environ Sci Pollut Res Int 29(28):42539–42559. https://doi.org/10.1007/s11356-022-19718-6
2. Ogwu MC (2019) Towards sustainable development in Africa: the challenge of urbanization and climate change adaptation. In: Cobbinah PB, Addaney M (eds) The geography of climate change adaptation in urban Africa. Springer Nature, Cham, pp 29–55. https://doi.org/10.1007/978-3-030-04873-0_2
3. Ogwu MC (2023) Plants as monitors and managers of pollution. In: Srivastav AL, Grewal AS, Tiwari M, Pham TD (eds) Role of green chemistry in ecosystem restoration to achieve environmental sustainability. Elsevier, Amsterdam, pp 51–60. ISBN: 978-0-443-15291-7. https://doi.org/10.1016/B978-0-443-15291-7.00022-5
4. Yadav R, Deora S, Yadav G (2021) Air pollution and its impact on cardiovascular health – it's time to act fast! Indian Heart J 73(1):1–6. https://doi.org/10.1016/j.ihj.2021.01.021
5. Stevens CJ, Bell JNB, Brimblecombe P, Clark CM, Dise NB, Fowler D, Lovett GM, Wolseley PA (2020) The impact of air pollution on terrestrial managed and natural vegetation. Philos Transact A Math Phys Eng Sci 378(2183):20190317. https://doi.org/10.1098/rsta.2019.0317

6. Richard G, Izah SC, Sawyer EW, Ogwu MC (2023) Seasonal variations in the concentration of some atmospheric pollutants along a major highway in Southern Nigeria. J Environ Stud 30 (1):10–25. https://doi.org/10.21608/JESJ.2023.205673.1045
7. Chang JH, Lee YL, Chang LT, Chang TY, Hsiao TC, Chung KF, Ho KF, Kuo HP, Lee KY, Chuang KJ, Chuang HC (2023) Climate change, air quality, and respiratory health: a focus on particle deposition in the lungs. Ann Med 55(2):2264881. https://doi.org/10.1080/07853890.2023.2264881
8. Wang P, Yang Y, Xue D, Ren L, Tang J, Leung LR, Liao H (2023) Aerosols overtake greenhouse gases causing a warmer climate and more weather extremes toward carbon neutrality. Nat Commun 14(1):7257. https://doi.org/10.1038/s41467-023-42891-2
9. UNEP [United Nations Environment Programme] (2019) Global environment outlook - GEO-6: healthy planet, healthy people. United Nations Environment Programme, Nairobi
10. Martinez GS, Spadaro JV, Chapizanis D, Kendrovski V, Kochubovski M, Mudu P (2018) Health impacts and economic costs of air pollution in the metropolitan area of Skopje. Int J Environ Res Public Health 15(4):626. https://doi.org/10.3390/ijerph15040626
11. Irfan M, Cameron MP, Hassan G (2021) Interventions to mitigate indoor air pollution: a cost-benefit analysis. PLoS One 16(9):e0257543. https://doi.org/10.1371/journal.pone.0257543
12. Lawler JJ, Spencer B, Olden JD, Kim SH, Lowe C, Bolton S, Beamon BM, Thompson L, Voss JG (2013) Mitigation and adaptation strategies to reduce climate vulnerabilities and maintain ecosystem services. Climate vulnerability, pp 315–335. https://doi.org/10.1016/B978-0-12-384703-4.00436-6
13. Tripp-Barba C, Barbecho P, Urquiza L, Aguilar-Calderón JA (2023) A comparison of vehicle emissions control strategies for smart cities. PeerJ Comp Sci 9:e1676. https://doi.org/10.7717/peerj-cs.1676
14. Carey J (2023) The other benefit of electric vehicles. Proc Natl Acad Sci U S A 120(3): e2220923120. https://doi.org/10.1073/pnas.2220923120
15. Mohammed A, Saif O, Abo-Adma M, Fahmy A, Elazab R (2024) Strategies and sustainability in fast charging station deployment for electric vehicles. Sci Rep 14(1):283. https://doi.org/10.1038/s41598-023-50825-7
16. Morawska L, Thai PK, Liu X, Asumadu-Sakyi A, Ayoko G, Bartonova A, Bedini A, Chai F, Christensen B, Dunbabin M, Gao J, Hagler GSW, Jayaratne R, Kumar P, Lau AKH, Louie PKK, Mazaheri M, Ning Z, Motta N, Mullins B et al (2018) Applications of low-cost sensing technologies for air quality monitoring and exposure assessment: how far have they gone? Environ Int 116:286–299. https://doi.org/10.1016/j.envint.2018.04.018
17. Alvear-Puertas VE, Burbano-Prado YA, Rosero-Montalvo PD, Tözün P, Marcillo F, Hernandez W (2022) Smart and portable air-quality monitoring IoT low-cost devices in Ibarra City, Ecuador. Sensors 22(18):7015. https://doi.org/10.3390/s22187015
18. Bobulski J, Szymoniak S, Pasternak K (2024) An IoT system for air pollution monitoring with safe data transmission. Sensors 24(2):445. https://doi.org/10.3390/s24020445
19. deSouza PN (2022) Key concerns and drivers of low-cost air quality sensor use. Sustain For 14:584. https://doi.org/10.3390/su14010584
20. Park YM (2022) A GPS-enabled portable air pollution sensor and web-mapping technologies for field-based learning in health geography. J Geogr Higher Educ 46(2). https://doi.org/10.1080/03098265.2021.1900083
21. Štrbac S, Kašanin-Grubin M, Pezo L, Stojić N, Lončar B, Ćurčić L, Pucarević M (2023) Green infrastructure designed through nature-based solutions for sustainable urban development. Int J Environ Res Public Health 20(2):1102. https://doi.org/10.3390/ijerph20021102
22. Oyebanji MO, Kirikkaleli D (2023) Green technology, green electricity, and environmental sustainability in Western European countries. Environ Sci Pollut Res Int 30(13):38525–38534. https://doi.org/10.1007/s11356-022-24793-w
23. Peng K, Feng K, Chen B, Shan Y, Zhang N, Wang P, Fang K, Bai Y, Zou X, Wei W, Geng X, Zhang Y, Li J (2023) The global power sector's low-carbon transition may enhance

sustainable development goal achievement. Nat Commun 14(1):3144. https://doi.org/10.1038/s41467-023-38987-4
24. Erickson LE (2017) Reducing greenhouse gas emissions and improving air quality: two global challenges. Environ Prog Sustain Energy 36(4):982–988. https://doi.org/10.1002/ep.12665
25. Ramaswamy K, Jule LT, Nagaprasad N, Subramanian K, Shanmugam R, Priyanka Dwarampudi L, Seenivasan V (2022) Reduction of environmental chemicals, toxicity and particulate matter in wet scrubber device to achieve zero emissions. Sci Rep 12(1):9170. https://doi.org/10.1038/s41598-022-13369-w
26. Sparks T, Chase G (2016) Air and gas filtration. Filters and filtration handbook, pp 117–198. https://doi.org/10.1016/B978-0-08-099396-6.00003-4
27. Sofia D, Gioiella F, Lotrecchiano N et al (2020) Mitigation strategies for reducing air pollution. Environ Sci Pollut Res 27:19226–19235. https://doi.org/10.1007/s11356-020-08647-x
28. Haszeldine RS, Flude S, Johnson G, Scott V (2018) Negative emissions technologies and carbon capture and storage to achieve the Paris agreement commitments. Philos Transact A Math Phys Eng Sci 376(2119):20160447. https://doi.org/10.1098/rsta.2016.0447
29. Rashid MI, Yaqoob Z, Mujtaba MA, Kalam MA, Fayaz H, Qazi A (2024) Carbon capture, utilization and storage opportunities to mitigate greenhouse gases. Heliyon 10(3):e25419. https://doi.org/10.1016/j.heliyon.2024.e25419
30. Negrete-Cardoso M, Rosano-Ortega G, Álvarez-Aros EL, Tavera-Cortés ME, Vega-Lebrún CA, Sánchez-Ruíz FJ (2022) Circular economy strategy and waste management: a bibliometric analysis in its contribution to sustainable development, toward a post-COVID-19 era. Environ Sci Pollut Res Int 29(41):61729–61746. https://doi.org/10.1007/s11356-022-18703-3
31. Sharma HB, Vanapalli KR, Samal B, Cheela VRS, Dubey BK, Bhattacharya J (2021) Circular economy approach in solid waste management system to achieve UN-SDGs: solutions for post-COVID recovery. Sci Total Environ 800:149605. https://doi.org/10.1016/j.scitotenv.2021.149605
32. Saleem J, Moghal ZKB, Shakoor RA, McKay G (2023) Sustainable solution for plastic pollution: upcycling waste polypropylene masks for effective oil-spill management. Int J Mol Sci 24(15):12368. https://doi.org/10.3390/ijms241512368
33. Leal Filho W, Vidal DG, Chen C, Petrova M, Dinis MAP, Yang P, Rogers S, Álvarez-Castañón L, Djekic I, Sharifi A, Neiva S (2022) An assessment of requirements in investments, new technologies, and infrastructures to achieve the SDGs. Environ Sci Eur 34(1):58. https://doi.org/10.1186/s12302-022-00629-9
34. United Nations (UN) (2015) Transforming our world: the 2030 agenda for sustainable development. United Nations, New York
35. Perera F (2017) Pollution from fossil-fuel combustion is the leading environmental threat to global pediatric health and equity: solutions exist. Int J Environ Res Public Health 15(1):16. https://doi.org/10.3390/ijerph15010016
36. Solar Design Guide (2024) An introduction to solar PV systems. Available online: https://solardesignguide.com/an-introduction-to-solar-pv-systems/
37. Dallaev R, Pisarenko T, Papež N, Holcman V (2023) Overview of the current state of flexible solar panels and photovoltaic materials. Materials 16(17):5839. https://doi.org/10.3390/ma16175839
38. Machín A, Márquez F (2024) Advancements in photovoltaic cell materials: silicon, organic, and perovskite solar cells. Materials 17(5):1165. https://doi.org/10.3390/ma17051165
39. IAE [International Energy Agency] (2020) Solar energy technology roadmap. International Energy Agency, Paris
40. Solak EK, Irmak E (2023) Advances in organic photovoltaic cells: a comprehensive review of materials, technologies, and performance. RSC Adv 13(18):12244–12269. https://doi.org/10.1039/d3ra01454a

41. Müller-Steinhagen H (2013) Concentrating solar thermal power. Philos Transact A Math Phys Eng Sci 371(1996):20110433. https://doi.org/10.1098/rsta.2011.0433
42. Kong L, Chen X, Gong J, Fan D, Wang B, Li S (2022) Optimization of the hybrid solar power plants comprising photovoltaic and concentrating solar power using the butterfly algorithm. Energy Conv Manag 257:115310. https://doi.org/10.1016/j.enconman.2022.115310
43. Sumayli H, El-Leathy A, Danish SN, Al-Ansary H, Almutairi Z, Al-Suhaibani Z, Saleh NS, Saeed RS, Alswaiyd A, Djajadiwinata E, Alaqel S (2023) Integrated CSP-PV hybrid solar power plant for two cities in Saudi Arabia. Case Stud Therm Eng 44:102835. https://doi.org/10.1016/j.csite.2023.102835
44. Umair M, Hidayat NM, Sukri Ahmad A, Nik Ali NH, Mawardi MIM, Abdullah E (2024) A renewable approach to electric vehicle charging through solar energy storage. PLoS One 19 (2):e0297376. https://doi.org/10.1371/journal.pone.0297376
45. Global Wind Energy Council (2021) Global wind report 2021. Global Wind Energy Council, Brussels
46. Akhtar N, Geyer B, Schrum C (2024) Larger wind turbines as a solution to reduce environmental impacts. Sci Rep 14(1):6608. https://doi.org/10.1038/s41598-024-56731-w
47. Failla G, Arena F (2015) New perspectives in offshore wind energy. Philos Transact A Math Phys Eng Sci 373(2035):20140228. https://doi.org/10.1098/rsta.2014.0228
48. Maxwell SM, Kershaw F, Locke CC, Conners MG, Dawson C, Aylesworth S, Loomis R, Johnson AF (2022) Potential impacts of floating wind turbine technology for marine species and habitats. J Environ Manage 307:114577. https://doi.org/10.1016/j.jenvman.2022.114577
49. Browning MS, Lenox CS (2020) Contribution of offshore wind to the power grid: U.S. air quality implications. Appl Energy 276:115474. https://doi.org/10.1016/j.apenergy.2020.115474
50. Castro-Santos L, deCastro M, Costoya X, Filgueira-Vizoso A, Lamas-Galdo I, Ribeiro A, Dias JM, Gómez-Gesteira M (2021) Economic feasibility of floating offshore wind farms considering near future wind resources: case study of Iberian coast and Bay of Biscay. Int J Environ Res Public Health 18(5):2553. https://doi.org/10.3390/ijerph18052553
51. Obiora SC, Bamisile O, Hu Y, Ozsahin DU, Adun H (2024) Assessing the decarbonization of electricity generation in major emitting countries by 2030 and 2050: transition to a high share renewable energy mix. Heliyon 10(8):e28770. https://doi.org/10.1016/j.heliyon.2024.e28770
52. Freeman S, Agar E (2024) The impact of energy storage on the reliability of wind and solar power in New England. Heliyon 10(6):e27652. https://doi.org/10.1016/j.heliyon.2024.e27652
53. International Hydropower Association (2018) Hydropower status report 2018. International Hydropower Association, London
54. Hunt JD, Byers E, Wada Y, Parkinson S, Gernaat DEHJ, Langan S, van Vuuren DP, Riahi K (2020) Global resource potential of seasonal pumped hydropower storage for energy and water storage. Nat Commun 11(1):947. https://doi.org/10.1038/s41467-020-14555-y
55. Bilotta GS, Burnside NG, Gray JC, Orr HG (2016) The effects of run-Of-river hydroelectric power schemes on fish community composition in temperate streams and rivers. PLoS One 11 (5):e0154271. https://doi.org/10.1371/journal.pone.0154271
56. Inam Ullah E, Ahmad S, Khokhar MF, Azmat M, Khayyam U, Qaiser FUR (2023) Hydrological and ecological impacts of run off river scheme; a case study of Ghazi Barotha hydropower project on Indus River, Pakistan. Heliyon 9(1):e12659. https://doi.org/10.1016/j.heliyon.2022.e12659
57. World Bank Group (2019) Pumped storage hydropower: benefits and markets. World Bank Group, Washington
58. Liang M, Liu L, Liang W, Mi W, Ye K, Gao J (2024) Intelligentization helps the green and energy-saving transformation of power industry-evidence from substation engineering in China. Sci Rep 14(1):8698. https://doi.org/10.1038/s41598-024-59271-5
59. Fu X (2019) In-vehicle exposures at transportation and the health concerns. Indoor environmental quality and health risk toward healthier environment for all, pp 113–126. https://doi.org/10.1007/978-981-32-9182-9_6

60. Zhang K, Batterman S (2013) Air pollution and health risks due to vehicle traffic. Sci Total Environ 450-451:307–316. https://doi.org/10.1016/j.scitotenv.2013.01.074
61. Forsythe CR, Gillingham KT, Michalek JJ, Whitefoot KS (2023) Technology advancement is driving electric vehicle adoption. Proc Natl Acad Sci U S A 120(23):e2219396120. https://doi.org/10.1073/pnas.2219396120
62. Chen X, Zhong J, Sha F, Zhong Z (2021) Research on innovative plug-in hybrid electric vehicle comprehensive energy consumption evaluation method based on statistic energy consumption. Sci Prog 1(4):368504211050284. https://doi.org/10.1177/00368504211050284
63. Goppelt G (2021) New potential for plug-in hybrids. ATZ Worldwide 123(10):8–13. https://doi.org/10.1007/s38311-021-0719-1
64. Rahmani D, Loureiro ML (2018) Why is the market for hybrid electric vehicles (HEVs) moving slowly? PLoS One 13(3):e0193777. https://doi.org/10.1371/journal.pone.0193777
65. Ohanu CP, Rufai SA, Oluchi UC (2024) A comprehensive review of recent developments in smart grid through renewable energy resources integration. Heliyon 10(3):e25705. https://doi.org/10.1016/j.heliyon.2024.e25705
66. Ahad MT, Bhuiyan MMH, Sakib AN, Becerril Corral A, Siddique Z (2023) An overview of challenges for the future of hydrogen. Materials 16(20):6680. https://doi.org/10.3390/ma16206680
67. Sazali N, Wan Salleh WN, Jamaludin AS, Mhd Razali MN (2020) New perspectives on fuel cell technology: a brief review. Membranes 10(5):99. https://doi.org/10.3390/membranes10050099
68. FCHEA [Fuel Cell and Hydrogen Energy Association] (2021) Fuel cells & hydrogen: transforming transportation. Fuel Cell and Hydrogen Energy Association, Washington
69. World Green Building Council (WorldGBC) (2018) Advancing net zero: the green building council framework. World Green Building Council, London
70. Brandon NP, Kurban Z (2017) Clean energy and the hydrogen economy. Philos Transact A Math Phys Eng Sci 375(2098):20160400. https://doi.org/10.1098/rsta.2016.0400
71. Rampai MM, Mtshali CB, Seroka NS, Khotseng L (2024) Hydrogen production, storage, and transportation: recent advances. RSC Adv 14(10):6699–6718. https://doi.org/10.1039/d3ra08305e
72. Sustainable Aviation (2021) Sustainable aviation: innovation hub. Sustainable Aviation, London
73. Amankwah-Amoah J, Debrah Y, Anang S (2023) Greening aviation in era of COVID-19: towards conceptualizing and operationalizing decarbonization. J Environ Manage 326(Pt A):116649. https://doi.org/10.1016/j.jenvman.2022.116649
74. Cardone LM, Petrone G, De Rosa S et al (2024) Review of the recent developments about the hybrid propelled aircraft. Aerotec Missili Spaz 103:17–37. https://doi.org/10.1007/s42496-023-00173-6
75. Rendón MA, Sánchez R, Carlos D, Josselyn Gallo M, Anzai AH (2021) Aircraft hybrid-electric propulsion: development trends, challenges and opportunities. J Control Autom Electr Syst 32(5):1244–1268. https://doi.org/10.1007/s40313-021-00740-x
76. Sripad S, Viswanathan V (2021) The promise of energy-efficient battery-powered urban aircraft. Proc Natl Acad Sci U S A 118(45):e2111164118. https://doi.org/10.1073/pnas.2111164118
77. International Civil Aviation Organization (ICAO) (2020) ICAO carbon offsetting and reduction scheme for international aviation (CORSIA). International Civil Aviation Organization, Montreal
78. U.S. Green Building Council (USGBC) (2021) LEED v4.1 green building rating system. U.S. Green Building Council, Washington
79. Khoshnava SM, Rostami R, Mohamad Zin R, Štreimikienė D, Mardani A, Ismail M (2020) The role of green building materials in reducing environmental and human health impacts. Int J Environ Res Public Health 17(7):2589. https://doi.org/10.3390/ijerph17072589

80. Alharasees O, Kale U, Rohacs J, Rohacs D, Eva ME, Boros A (2024) Green building energy: patents analysis and analytical hierarchy process evaluation. Heliyon 10(8):e29442. https://doi.org/10.1016/j.heliyon.2024.e29442
81. Anand V, Kadiri VL, Putcha C (2023) Passive buildings: a state-of-the-art review. J Infrastruct Preserv Resil 4(1):3. https://doi.org/10.1186/s43065-022-00068-z
82. Wu W, Skye HM (2021) Residential net-zero energy buildings: review and perspective. Renew Sustain Energy Rev:142. https://doi.org/10.1016/j.rser.2021.110859
83. Moreno-Rangel A, Sharpe T, McGill G, Musau F (2020) Indoor air quality in Passivhaus dwellings: a literature review. Int J Environ Res Public Health 17(13):4749. https://doi.org/10.3390/ijerph17134749
84. Afroz Z, Burak Gunay H, O'Brien W (2020) A review of data collection and analysis requirements for certified green buildings. Energ Buildings 226:110367. https://doi.org/10.1016/j.enbuild.2020.110367
85. Di Foggia G (2018) Energy efficiency measures in buildings for achieving sustainable development goals. Heliyon 4(11):e00953. https://doi.org/10.1016/j.heliyon.2018.e00953
86. Green Building Council of Australia (GBCA) (2019) Green star rating system. Green Building Council of Australia, Sydney
87. Luo N, Wang Z, Blum D, Weyandt C, Bourassa N, Piette MA, Hong T (2022) A three-year dataset supporting research on building energy management and occupancy analytics. Sci Data 9(1):156. https://doi.org/10.1038/s41597-022-01257-x
88. International WELL Building Institute (IWBI) (2020) WELL building standard. International WELL Building Institute, New York
89. Saeed MA, Eladl AA, Alhasnawi BN, Motahhir S, Nayyar A, Shah MA, Sedhom BE (2023) Energy management system in smart buildings based coalition game theory with fog platform and smart meter infrastructure. Sci Rep 13(1):2023. https://doi.org/10.1038/s41598-023-29209-4
90. Ezhilarasi P, Ramesh L, Sanjeevikumar P, Khan B (2023) A cost-effective smart metering approach towards affordable deployment strategy. Sci Rep 13(1):19452. https://doi.org/10.1038/s41598-023-44149-9
91. Jayachandran M, Reddy CR, Padmanaban S, Milyani AH (2021) Operational planning steps in smart electric power delivery system. Sci Rep 11(1):17250. https://doi.org/10.1038/s41598-021-96769-8
92. European Environment Agency (EEA) (2019) Air quality in Europe. European Environment Agency, Copenhagen
93. European Environment Agency (EEA) (2020) Circular economy in Europe: insights on progress and opportunities. European Environment Agency, Copenhagen
94. Kubba S (2012) Indoor environmental quality. Handbook of green building design and construction, pp 313–360. https://doi.org/10.1016/B978-0-12-385128-4.00007-X
95. Dsilva J, Zarmukhambetova S, Locke J (2023) Assessment of building materials in the construction sector: a case study using life cycle assessment approach to achieve the circular economy. Heliyon 9(10):e20404. https://doi.org/10.1016/j.heliyon.2023.e20404
96. World Health Organization (WHO) (2021) Ambient air pollution: a global assessment of exposure and burden of disease. World Health Organization, Geneva
97. Zaks DP, Winchester N, Kucharik CJ, Barford CC, Paltsev S, Reilly JM (2011) Contribution of anaerobic digesters to emissions mitigation and electricity generation under U.S. climate policy. Environ Sci Technol 45(16):6735–6742. https://doi.org/10.1021/es104227y
98. Kaushal R, Rohit, Dhaka AK (2022) A comprehensive review of the application of plasma gasification technology in circumventing the medical waste in a post-COVID-19 scenario. Biomass Convers Biorefinery:1–16. Advance online publication. https://doi.org/10.1007/s13399-022-02434-z
99. Adnane I, Taoumi H, Elouahabi K, Lahrech K, Oulmekki A (2024) Valorization of crop residues and animal wastes: anaerobic co-digestion technology. Heliyon 10(5):e26440. https://doi.org/10.1016/j.heliyon.2024.e26440

100. Kriswantoro JA, Pan KY, Chu CY (2024) Co-digestion approach for enhancement of biogas production by mixture of untreated Napier grass and industrial hydrolyzed food waste. Front Bioeng Biotechnol 11:1269727. https://doi.org/10.3389/fbioe.2023.1269727
101. Khan F, Ali Y (2022) Implementation of the circular supply chain management in the pharmaceutical industry. Environ Dev Sustain 24(12):13705–13731. https://doi.org/10.1007/s10668-021-02007-6
102. Mayanti B, Helo P (2024) Circular economy through waste reverse logistics under extended producer responsibility in Finland. Waste Manag Res 42(1):59–73. https://doi.org/10.1177/0734242X231168801
103. Dalhammar C, Wihlborg E, Milios L, Richter JL, Svensson-Höglund S, Russell J, Thidell Å (2021) Enabling reuse in extended producer responsibility schemes for white goods: legal and organisational conditions for connecting resource flows and actors. Circ Econ Sustain 1 (2):671–695. https://doi.org/10.1007/s43615-021-00053-w
104. Faibil D, Asante R, Agyemang M, Addaney M, Baah C (2023) Extended producer responsibility in developing economies: assessment of promoting factors through retail electronic firms for sustainable e-waste management. Waste Manag Res 41(1):117–142. https://doi.org/10.1177/0734242X221105433
105. National Renewable Energy Laboratory (NREL) (2021) Bioremediation: nature's cleanup crew. National Renewable Energy Laboratory, Golden
106. World Health Organization (WHO) (2016) Guidelines for the management of waste from health-care activities. World Health Organization, Geneva
107. United States Environmental Protection Agency (US EPA) (2020) Air quality monitoring. US EPA, Washington
108. Citizen Science [International Society for Citizen Science] (2021) Citizen science: engaging the public in research. International Society for Citizen Science, Geneva
109. AIRNOW (2023) Air quality index (AQI) basics. Available online: AQI Basics | AirNow.gov
110. Park YM, Chavez D, Sousan S, Figueroa-Bernal N, Alvarez JR, Rocha-Peralta J (2023) Personal exposure monitoring using GPS-enabled portable air pollution sensors: a strategy to promote citizen awareness and behavioral changes regarding indoor and outdoor air pollution. J Expo Sci Environ Epidemiol 33(3):347–357. https://doi.org/10.1038/s41370-022-00515-9
111. Mohsan SAH, Othman NQH, Li Y, Alsharif MH, Khan MA (2023) Unmanned aerial vehicles (UAVs): practical aspects, applications, open challenges, security issues, and future trends. Intell Serv Robot 16(1):109–137. https://doi.org/10.1007/s11370-022-00452-4
112. Anenberg SC, Bindl M, Brauer M, Castillo JJ, Cavalieri S, Duncan BN, Fiore AM, Fuller R, Goldberg DL, Henze DK, Hess J, Holloway T, James P, Jin X, Kheirbek I, Kinney PL, Liu Y, Mohegh A, Patz J, Jimenez MP et al (2020) Using satellites to track indicators of global air pollution and climate change impacts: lessons learned from a NASA-supported science-stakeholder collaborative. GeoHealth 4(7):e2020GH000270. https://doi.org/10.1029/2020GH000270
113. Gressent A, Malherbe L, Colette A, Rollin H, Scimia R (2020) Data fusion for air quality mapping using low-cost sensor observations: feasibility and added-value. Environ Int 143:105965. https://doi.org/10.1016/j.envint.2020.105965
114. Novak R, Petridis I, Kocman D, Robinson JA, Kanduč T, Chapizanis D, Karakitsios S, Flückiger B, Vienneau D, Mikeš O, Degrendele C, Sáňka O, García Dos Santos-Alves S, Maggos T, Pardali D, Stamatelopoulou A, Saraga D, Persico MG, Visave J, Gotti A et al (2021) Harmonization and visualization of data from a transnational multi-sensor personal exposure campaign. Int J Environ Res Public Health 18(21):11614. https://doi.org/10.3390/ijerph182111614
115. Dong J, Goodman N, Rajagopalan P (2023) A review of artificial neural network models applied to predict indoor air quality in schools. Int J Environ Res Public Health 20(15):6441. https://doi.org/10.3390/ijerph20156441

116. Guo Q, Ren M, Wu S, Sun Y, Wang J, Wang Q, Ma Y, Song X, Chen Y (2022) Applications of artificial intelligence in the field of air pollution: A bibliometric analysis. Front Public Health 10:933665. https://doi.org/10.3389/fpubh.2022.933665
117. Hubbell BJ, Kaufman A, Rivers L, Schulte K, Hagler G, Clougherty J, Cascio W, Costa D (2018) Understanding social and behavioral drivers and impacts of air quality sensor use. Sci Total Environ 621:886–894. https://doi.org/10.1016/j.scitotenv.2017.11.275
118. Sîrbu A, Becker M, Caminiti S, De Baets B, Elen B, Francis L, Gravino P, Hotho A, Ingarra S, Loreto V, Molino A, Mueller J, Peters J, Ricchiuti F, Saracino F, Servedio VD, Stumme G, Theunis J, Tria F, Van den Bossche J (2015) Participatory patterns in an international air quality monitoring initiative. PLoS One 10(8):e0136763. https://doi.org/10.1371/journal.pone.0136763
119. Ward F, Lowther-Payne HJ, Halliday EC, Dooley K, Joseph N, Livesey R, Moran P, Kirby S, Cloke J (2022) Engaging communities in addressing air quality: a scoping review. Environ Health 21(1):89. https://doi.org/10.1186/s12940-022-00896-2
120. Singh S, Kulshrestha MJ, Rani N, Kumar K, Sharma C, Aswal DK (2023) An overview of vehicular emission standards. Mapan 38(1):241–263. https://doi.org/10.1007/s12647-022-00555-4
121. Du C, Zhang Q, Huang D (2023) Environmental protection subsidies, green technology innovation and environmental performance: evidence from China's heavy-polluting listed firms. PLoS One 18(2):e0278629. https://doi.org/10.1371/journal.pone.0278629
122. Bölük G, Kaplan R (2022) Effectiveness of renewable energy incentives on sustainability: evidence from dynamic panel data analysis for the EU countries and Turkey. Environ Sci Pollut Res Int 29(18):26613–26630. https://doi.org/10.1007/s11356-021-17801-y
123. Heard BR, Kerxhalli-Kleinfield M, Holmes KJ (2022) Beyond A single stimulus: how to leverage the federal government to advance clean energy innovation? iScience 25(6):104366. https://doi.org/10.1016/j.isci.2022.104366
124. Alami H, Rivard L, Lehoux P, Ag Ahmed MA, Fortin JP, Fleet R (2023) Integrating environmental considerations in digital health technology assessment and procurement: stakeholders' perspectives. Digital Health 9:20552076231219113. https://doi.org/10.1177/20552076231219113
125. King JS, Manning J, Woodward A (2023) In this together: international collaborations for environmental and human health. J Law Med Ethics 51(2):271–286. https://doi.org/10.1017/jme.2023.82
126. Pulford J, El Hajj T, Tancred T, Ding Y, Crossman S, Silvester L, Savio M, Bevan N, Tagoe N, Bates I (2023) How international research consortia can strengthen organisations' research systems and promote a conducive environment and culture. BMJ Glob Health 8(4):e011419. https://doi.org/10.1136/bmjgh-2022-011419
127. Kalfagianni A, Young OR (2022) The politics of multilateral environmental agreements lessons from 20 years of *INEA*. Int Environ Agreements 22(2):245–262. https://doi.org/10.1007/s10784-022-09567-6
128. Goforth T, Nock D (2022) Air pollution disparities and equality assessments of US national decarbonization strategies. Nat Commun 13(1):7488. https://doi.org/10.1038/s41467-022-35098-4
129. Richmond-Bryant J, Wilson A (2024) Invited perspective: increasing disparities in air pollution health burdens in the United States - a concerning trend for PM2.5 and a path forward. Environ Health Perspect 132(3):31302. https://doi.org/10.1289/EHP14233
130. Ao XY, Ong TS, Aprile R, Di Vaio A (2023) Environmental uncertainty and digital technologies corporate in shaping corporate green behavior and tax avoidance. Sci Rep 13(1):22170. https://doi.org/10.1038/s41598-023-49687-w
131. Singhal A, Neveditsin N, Tanveer H, Mago V (2024) Toward fairness, accountability, transparency, and ethics in AI for social media and health care: scoping review. JMIR Med Inform 12:e50048. https://doi.org/10.2196/50048

132. Ekins P, Zenghelis D (2021) The costs and benefits of environmental sustainability. Sustain Sci 16(3):949–965. https://doi.org/10.1007/s11625-021-00910-5
133. Junaid SB, Imam AA, Balogun AO, De Silva LC, Surakat YA, Kumar G, Abdulkarim M, Shuaibu AN, Garba A, Sahalu Y, Mohammed A, Mohammed TY, Abdulkadir BA, Abba AA, Kakumi NAI, Mahamad S (2022) Recent advancements in emerging technologies for healthcare management systems: a survey. Healthcare 10(10):1940. https://doi.org/10.3390/healthcare10101940
134. Yang Z, Fu L, Chen Y (2023) Digital economy and pollution reduction-mechanism and regional heterogeneity. PLoS One 18(2):e0277852. https://doi.org/10.1371/journal.pone.0277852
135. Kaur J, Kumar S, Narkhede BE, Dabić M, Rathore APS, Joshi R (2022) Barriers to blockchain adoption for supply chain finance: the case of Indian SMEs. Electron Commer Res 24:1–38. Advance online publication. https://doi.org/10.1007/s10660-022-09566-4
136. Li L (2022) Reskilling and upskilling the future-ready workforce for industry 4.0 and beyond. Inform Syst Front:1–16. Advance online publication. https://doi.org/10.1007/s10796-022-10308-y
137. Akash FA, Shovon SM, Rahman W, Rahman MA, Chakraborty P, Monir MU (2024) Greening the grid: A comprehensive review of renewable energy in Bangladesh. Heliyon 10(5):e27477. https://doi.org/10.1016/j.heliyon.2024.e27477
138. Toma C, Alexandru A, Popa M, Zamfiroiu A (2019) IoT solution for smart cities' pollution monitoring and the security challenges. Sensors 19(15):3401. https://doi.org/10.3390/s19153401
139. Di Vaio A, Hasan S, Palladino R, Hassan R (2023) The transition towards circular economy and waste within accounting and accountability models: a systematic literature review and conceptual framework. Environ Dev Sustain 25(1):734–810. https://doi.org/10.1007/s10668-021-02078-5
140. Ezeonu CS, Tagbo R, Anike EN, Oje OA, Onwurah IN (2012) Biotechnological tools for environmental sustainability: prospects and challenges for environments in Nigeria – a standard review. Biotechnol Res Int 2012:450802. https://doi.org/10.1155/2012/450802

Policies and Regulations for Sustainable Clean Air: An Overview

Hanadi G. Al-Thani and Rima J. Isaifan

Contents

H. G. Al-Thani (✉) and R. J. Isaifan
College of Science and Engineering, Hamad Bin Khalifa University, Doha, Qatar
e-mail: hgalthani@qf.org.qa

Matthew Chidozie Ogwu and Sylvester Izah (eds.), *Sustainable Strategies for Air Pollution Mitigation: Development, Economics and Technologies*,
Hdb Env Chem (2024) 133: 409–438, DOI 10.1007/698_2024_1093,

Published online: 20 July 2024

Abstract This chapter aims to give a thorough analysis of the laws and policies that are essential to maintaining sustainable clean air as the globe struggles with the growing problems caused by air pollution. Setting the scene, the introduction emphasizes the urgent need for practical solutions to address the growing threat through which air pollution impacts ecosystems, human health, and the planet's overall well-being. To comprehend the intricate world of air quality management, the chapter presents itself as a vital resource, stressing the connections between industrialization, urbanization, and economic development. The examination of national and regional policies, which provides illuminating case studies from top nations and regions dedicated to clean air standards, forms the central focus of the conversation. These case studies offer proactive insights from various socioeconomic circumstances in addition to highlighting effective strategies. Policies that incorporate sustainability principles are given particular attention, acknowledging the critical balance between environmental protection and economic advancement. The chapter looks at how data-driven methods, emerging technology, and public involvement can help shape and implement air quality regulations to stay up-to-date with the always-changing environment. Case studies highlight creative approaches and effective technological implementations, highlighting the need for a multidisciplinary and cooperative approach. The chapter also highlights the difficulties in enforcing policies, such as cross-border pollution complexity, political and economic obstacles, and enforcement concerns, as the story progresses. The chapter concludes by synthesizing policy evaluations, research findings, and real-world case studies to provide a comprehensive overview of sustainable clean air policies and regulations. To address the changing nature of air quality challenges, it calls for further innovation, knowledge sharing, and reinvigorated international cooperation in its conclusion. The scope and complexity of this investigation supports a lively discussion aimed at promoting a better, more sustainable future for the quality of the air worldwide.

Keywords Air Quality Regulations, Clean Air, Climate Change Mitigation, Environmental Policies, Policy Implementation, Pollution Control, Public Health

Abbreviations

CER	Certified emission reduction
CH_4	Methane
CDM	Clean development mechanism
CO_2	Carbon dioxide
EU	European Union
IRENA	International Renewable Energy Agency
NDC	Nationally determined contributions
N_2O	Nitrous oxide
PM	Particulate matter

$PM_{2.5}$	Particulate matter of diameter less than or equal to 2.5 μm
PM_{10}	Particulate matter of diameter less than or equal to 10 μm
SDG	Sustainable development goal
UN	United Nations
UNEP	United Nations Environment Program
UNFCCC	United Nations Framework Convention on Climate Change
WHO	World Health Organization

1 Introduction

1.1 Background and Rationale

Air pollution remains a persistent and pervasive challenge in contemporary society, posing grave threats to ecosystems, human health, and environmental sustainability on a global scale [1]. Rooted in the rapid pace of industrialization, urbanization, and economic expansion, air pollutants infiltrate the atmosphere, compromising air quality and endangering both current and future generations' living conditions. The historical trajectory of air pollution traces back centuries, evolving alongside human activities from the early stages of industrialization to the present era characterized by unprecedented levels of globalization and urban sprawl. Throughout this trajectory, emissions from industrial processes, transportation, energy production, and other anthropogenic sources have continually tainted the air with a diverse array of gaseous and particulate pollutants [2].

The consequences of air pollution are diverse and multifaceted, encompassing damaging impacts on public health, ecosystems, and the global climate system [3]. Firstly, clean air is a fundamental prerequisite for safeguarding public health and well-being, as exposure to air pollutants has been linked to a myriad of adverse health effects, including respiratory diseases, cardiovascular disorders, and premature mortality. Sensitive population groups are particularly susceptible to the harmful effects of air pollution, exacerbating health disparities and inequalities within society. These vulnerable categories include individuals with pre-existing health conditions, children, and the elderly [4]. Secondly, air pollution poses significant threats to ecosystems and biodiversity, disrupting ecosystem processes, impairing nutrient cycling, and degrading habitat quality for wildlife [1]. Acid deposition resulting from the emission of sulfur and nitrogen oxides contributes to soil acidification, forest decline, and aquatic ecosystem degradation, with far-reaching implications for ecosystem resilience and functioning [5, 6]. Additionally, ground-level ozone, a key component of smog, damages vegetation, reduces crop yields, and undermines agricultural productivity, further exacerbating the impacts of air pollution on ecosystems and food security.

Moreover, air pollution exacerbates climate change by emitting greenhouse gases such as carbon dioxide and methane, contributing to global warming and its associated impacts on weather patterns, sea levels, and natural habitats [7]. The interconnection between air pollution and climate change underscores the urgency of adopting integrated approaches to address both issues simultaneously, recognizing their shared sources, impacts, and mitigation strategies [8].

In light of these challenges, addressing air pollution is not only a matter of environmental stewardship but also a moral imperative to safeguard human health, protect ecosystems, and mitigate climate change. The urgency of action is underscored by the staggering toll of air pollution on public health and the economy, with millions of premature deaths attributed to air pollution-related illnesses annually and trillions of dollars in economic costs incurred worldwide.

The imperative of preserving clean air has never been more pressing, as nations grapple with the complex interplay between environmental degradation, economic development, and societal well-being. In this context, the formulation and implementation of comprehensive policies and regulations are paramount to addressing the root causes of air pollution and fostering sustainable development. By adopting a holistic and multi-sectoral approach that integrates environmental, social, and economic considerations, governments can chart a course toward cleaner air, healthier communities, and a more resilient future for all.

1.2 Importance of Clean Air for Sustainable Development

Clean air serves as a cornerstone of sustainable development, underpinning environmental integrity, societal well-being, and economic prosperity. Beyond its intrinsic value for human health and ecosystem functioning, clean air is essential for supporting a wide range of socioeconomic activities, from agriculture and tourism to industry and innovation. Moreover, clean air is a prerequisite for achieving the Sustainable Development Goals (SDGs) outlined by the United Nations, spanning areas such as health, education, poverty reduction, and environmental sustainability [9].

From an environmental perspective, clean air is indispensable for maintaining the health and resilience of ecosystems, sustaining biodiversity, and preserving natural habitats. Air pollutants such as nitrogen oxides, sulfur dioxide, and ozone can disrupt ecosystem processes, impairing nutrient cycling, plant growth, and wildlife habitat quality. Moreover, airborne toxins can bioaccumulate in food chains, posing risks to human health and wildlife populations [10]. By preserving clean air, policymakers can safeguard ecosystem services such as clean water, fertile soils, and pollination, which are essential for supporting agriculture, food security, and biodiversity conservation.

From a societal standpoint, having clean air is crucial to preserving public health and well-being since it lowers the likelihood of respiratory conditions, heart problems, and other harmful health effects linked to air pollution. The World Health Organization (WHO) states that exposure to air pollutants including $PM_{2.5}$ and ozone is a major global cause of illness and mortality [11]. These disorders include heart disease, lung cancer, and asthma. Health inequities and disparities are further exacerbated when air pollution affects vulnerable populations, including children, the elderly, and people with pre-existing medical illnesses [12]. Moreover, clean air contributes to social cohesion, quality of life, and community resilience, fostering a conducive environment for human flourishing and well-being.

From an economic perspective, clean air is essential for supporting sustainable economic development and prosperity, as it enhances the productivity and competitiveness of economies, attracts investment, and creates jobs. Industries such as tourism, agriculture, and outdoor recreation depend on clean air and natural landscapes to thrive, generating revenue, employment, and economic growth. Moreover, clean air reduces healthcare costs, improves labor productivity, and enhances human capital development, creating a conducive environment for sustainable economic development. By prioritizing public health and environmental protection, policymakers can build trust, foster social cohesion, and promote equitable development for present and future generations [13].

In light of these considerations, the pursuit of clean air is not only a moral imperative but also an economic opportunity to build resilient, inclusive, and sustainable societies. By investing in clean energy, green technologies, and pollution control measures, governments can stimulate economic growth, create green jobs, and enhance competitiveness in the global marketplace [14]. Moreover, by prioritizing public health and environmental protection, policymakers can build trust, foster social cohesion, and promote equitable development for present and future generations.

1.3 Scope and Objectives of the Chapter

This chapter seeks to provide a comprehensive elucidation of policies and regulations geared toward attaining sustainable clean air. By examining national and regional approaches to air quality management, the chapter aims to discern pivotal strategies, confront inherent challenges, and delineate best practices in the realm of air pollution control and environmental conservation. Emphasizing the importance of sustainability principles, the chapter delves into the intricate interplay between environmental preservation and economic progress, highlighting the need for integrated and holistic approaches to address the multifaceted challenges of pollution mitigation and sustainable development.

The chapter is divided into five main sections. The first section is an introduction that sheds light on the importance of clean air and the rationale of such research. The second section paves the way for the historical content of the international cooperation to address environmental challenges with focus on Kyoto Protocol and Paris Agreement. The third chapter focuses on sustainable policies that depend on the integration with sustainable technologies and lessons learned from successful implementations. The fourth chapter discusses the role of community engagement and citizen science. The fifth chapter provides details on challenges in implementation to be accounted for.

By addressing these objectives, the chapter aims to provide policymakers, researchers, practitioners, and stakeholders with valuable insights and guidance for developing evidence-based strategies and interventions to combat air pollution and promote sustainable development. Through a nuanced analysis of policy frameworks, case studies, and empirical evidence, the chapter seeks to foster informed discourse, facilitate knowledge exchange, and catalyze collective action toward achieving clean air goals at the local, national, and global levels.

2 International Cooperation and Environmental Challenges

2.1 Historical Context of International Cooperation to Address Environmental Challenges

The quest for international cooperation to address environmental challenges, including air pollution, dates back to the mid-twentieth century. As industrialization accelerated and pollutants began to accumulate in the atmosphere, concerns about the transboundary nature of air pollution prompted nations to seek collaborative solutions. The post-World War II era witnessed the emergence of multilateral efforts to address environmental issues through diplomacy and international cooperation.

One of the earliest manifestations of international environmental cooperation was the establishment of the United Nations (UN) and its specialized agencies, including the United Nations Environment Program (UNEP), in the 1970s. These institutions provided a platform for countries to exchange information, coordinate policies, and negotiate agreements to protect the environment and promote sustainable development [15].

The 1972 United Nations Conference on the Human Environment, held in Stockholm, Sweden, marked a watershed moment in global environmental governance. The conference highlighted the interconnectedness of environmental issues and emphasized the need for collective action to address common challenges. The Stockholm Declaration, adopted at the conference, outlined principles for environmental protection and laid the groundwork for future international agreements on air pollution and other environmental issues [16].

Subsequent decades saw the negotiation and adoption of several landmark treaties and agreements aimed at addressing air pollution and its impacts on human health and the environment. These agreements reflected growing recognition of the urgency of the problem and the need for concerted action at the international level to mitigate its adverse effects.

2.2 *Key Global Agreements*

2.2.1 Kyoto Protocol

Under the auspices of the United Nations Framework Convention on Climate Change (UNFCCC), the Kyoto Protocol was enacted in 1997 and marked a turning point in global efforts to tackle climate change, especially air pollution brought on by greenhouse gas emissions. For developed nations, also referred to as Annex I parties, the protocol set legally binding emissions reduction targets. The countries were required to cut their emissions of six greenhouse gases, such as carbon dioxide (CO_2), methane (CH_4), and nitrous oxide (N_2O), by an average of 5.2% below 1990 levels between 2008 and 2012 [17].

Moreover, the Kyoto Protocol introduced innovative mechanisms to facilitate emissions reductions and promote sustainable development, including emissions trading, joint implementation, and the Clean Development Mechanism (CDM) [18]. Emissions trading allowed countries to buy and sell emission allowances, providing flexibility and cost-effectiveness in meeting their targets. While the CDM allowed Annex I parties to invest in emissions reduction projects in developing nations and earn certified emission reductions (CERs) as credits toward their own targets, joint implementation allowed Annex I parties to carry out emissions reduction projects in other Annex I countries.

Despite its ambitious goals and innovative mechanisms, the Kyoto Protocol faced several challenges in implementation and effectiveness. One of the main criticisms was the lack of binding emissions reduction targets for developing countries, which were exempt from obligations under the protocol due to their lower historical emissions and development needs. This asymmetry in obligations raised concerns about fairness and equity and hindered global efforts to address climate change comprehensively [19]. Furthermore, the withdrawal of key countries, including the United States, from the Kyoto Protocol undermined its effectiveness and credibility. The absence of the world's largest emitter of greenhouse gases from the agreement weakened its impact and limited its ability to achieve meaningful emissions reductions on a global scale.

Despite these challenges, the Kyoto Protocol laid the groundwork for subsequent international climate agreements and played a crucial role in raising awareness about the need for collective action to address climate change. Its legacy lives on in the ongoing efforts of the international community to strengthen and enhance global climate governance through agreements such as the Paris Agreement [19].

2.2.2 Paris Agreement

A major advancement in the fight against climate change and air pollution, the Paris Agreement was reached in 2015 under the auspices of the United Nations Framework Convention on Climate Change. By acknowledging the serious risks associated with rising above these thresholds, the Paris Agreement seeks to build on the goals and principles of the Kyoto Protocol by limiting global warming to well below 2°C above pre-industrial levels and pursuing efforts to keep the increase in temperature to 1.5°C [20].

One of the key features of the Paris Agreement is its bottom-up approach to climate action, whereby countries submit nationally determined contributions (NDCs) outlining their individual emissions reduction targets and mitigation measures [21]. This approach reflects the principle of common but differentiated responsibilities and respective capabilities, recognizing the varying historical emissions and development needs of different countries. The Paris Agreement also emphasizes the importance of adaptation to the impacts of climate change, particularly for vulnerable countries and communities that are most at risk. It includes provisions for enhancing climate resilience, mobilizing financial resources, and facilitating technology transfer to support adaptation efforts in developing countries [20].

Moreover, the Paris Agreement recognizes the crucial role of non-state actors, including cities, regions, businesses, and civil society organizations, in advancing climate action and driving emission reductions. It encourages collaboration and partnership between governments and non-state actors to enhance ambition, mobilize resources, and scale up climate action at all levels [20].

One of the defining features of the Paris Agreement is its emphasis on transparency, accountability, and international cooperation in the implementation of climate commitments. The agreement establishes a robust framework for monitoring, reporting, and verifying emissions reduction efforts, ensuring transparency and comparability of action among countries. It also includes provisions for regular review and updating of NDCs to reflect evolving circumstances and new scientific findings [20].

Despite its widespread support and ambitious goals, the Paris Agreement faces challenges in implementation and enforcement. One of the main challenges is the lack of binding emissions reduction targets and enforcement mechanisms, which rely on voluntary compliance and peer pressure to ensure compliance. Moreover, similar to Kyoto Protocol case, the withdrawal of the United States, the world's second-largest emitter of greenhouse gases, from the agreement dealt a significant blow to its credibility and effectiveness [20].

However, despite these challenges, the Paris Agreement represents a historic achievement in international climate governance and a beacon of hope for global efforts to address climate change and air pollution. Its adoption signaled a renewed commitment by the international community to work together to safeguard the planet for present and future generations, underscoring the urgency of collective action to tackle one of the greatest challenges of our time [20].

3 Integration of Sustainability in Policies

3.1 Balancing Economic Development and Environmental Protection

The intersection between economic development and environmental protection represents a critical juncture where policies must be carefully crafted to ensure both objectives are met harmoniously. Traditionally, economic growth has often been pursued at the expense of environmental degradation, including air pollution. However, modern perspectives on sustainable development emphasize the interdependence of economic, social, and environmental dimensions, highlighting the need for integrated policies that prioritize long-term sustainability.

A cornerstone of integrating sustainability into policies is recognizing that economic activities are dependent on environmental resources and services. Clean air, for instance, is not only essential for human health but also for supporting economic sectors such as agriculture, tourism, and outdoor recreation [13]. Therefore, policies that protect air quality contribute not only to public well-being but also to economic productivity and resilience.

Governments worldwide are increasingly adopting strategies to decouple economic growth from environmental degradation, a concept often referred to as "green growth" or "sustainable development." These strategies emphasize investments in clean technologies, renewable energy, energy efficiency, and sustainable infrastructure to drive economic progress while minimizing environmental impacts [22]. By fostering innovation, promoting resource efficiency, and enhancing environmental quality, green growth strategies aim to achieve economic prosperity without compromising the ability of future generations to meet their needs.

Moreover, sustainable development frameworks, such as the United Nations Sustainable Development Goals (SDGs), provide a comprehensive blueprint for integrating economic, social, and environmental objectives (Fig. 1). The SDGs encompass a wide range of targets and indicators, including those related to air quality (SDG 3 – Good Health and Well-being), sustainable cities (SDG 11 – Sustainable Cities and Communities), climate action (SDG 13 – Climate Action), and responsible consumption and production (SDG 12 – Responsible Consumption and Production). By aligning policies with the SDGs, governments can ensure coherence and synergy across different policy domains, maximizing positive impacts and minimizing trade-offs [23].

One key challenge in balancing economic development and environmental protection is addressing market failures and externalities that lead to environmental degradation. For instance, industries that pollute the air often do not bear the full costs of their actions, as the negative impacts on public health and the environment are externalized. Therefore, policies such as pollution taxes, emissions trading schemes, and environmental regulations are essential for internalizing these externalities and ensuring that polluters bear the true costs of their activities [23]. By incorporating the "polluter pays" principle into economic decision-making,

Fig. 1 The global goals for sustainable development, source (UNICEF)

governments can create incentives for businesses to adopt cleaner technologies and practices while generating revenues to fund environmental conservation and restoration efforts [23].

Another consideration in integrating sustainability into policies is the need to address social equity and justice concerns. Environmental degradation often disproportionately affects marginalized and vulnerable communities, exacerbating inequalities and disparities [24]. Therefore, policies that promote inclusive growth, social justice, and community participation are essential for ensuring that the benefits of economic development are equitably distributed and that no one is left behind. By engaging stakeholders in decision-making processes, governments can foster trust, transparency, and accountability, building support for sustainability initiatives and creating pathways for meaningful participation in the transition to a more sustainable future [24].

In brief, integrating sustainability into policies requires a holistic approach that balances economic, social, and environmental objectives. By prioritizing long-term sustainability over short-term gains, governments can foster inclusive and resilient economies that benefit people and the planet.

3.2 *Sustainable Technologies and Practices*

Advancements in technology play a pivotal role in driving progress toward clean air and sustainable development. From renewable energy sources to innovative pollution control technologies and development of nanomaterials [25], a wide array of sustainable technologies and practices are available to mitigate air pollution and promote environmental sustainability [26–28].

Renewable energy sources such as solar, wind, and hydropower offer clean and abundant alternatives to fossil fuels, reducing emissions of greenhouse gases and conventional air pollutants [29]. The declining costs of renewable energy technologies, coupled with policy support and market incentives, have led to rapid deployment and adoption worldwide. In many regions, renewable energy now competes favorably with conventional energy sources, offering a viable pathway to decarbonization and improved air quality.

Energy efficiency measures also play a crucial role in reducing energy consumption and emissions. By improving the efficiency of appliances, buildings, and industrial processes, governments can achieve significant reductions in energy use and greenhouse gas emissions while lowering energy bills for consumers and businesses [30]. Moreover, energy efficiency measures often have co-benefits for air quality, as they reduce the demand for fossil fuels and the associated emissions of air pollutants.

In addition to technological solutions, nature-based approaches such as afforestation, reforestation, and ecosystem restoration can contribute to air quality improvement and carbon sequestration [31]. Green infrastructure solutions such as green roofs, permeable pavements, and urban forests can help absorb pollutants, mitigate urban heat island effects, and enhance biodiversity and ecosystem services. By integrating nature into urban planning and development, cities can create healthier and more resilient environments for their residents [32].

Furthermore, sustainable agriculture practices such as organic farming, agroforestry, and precision agriculture can reduce emissions of air pollutants such as methane and nitrous oxide from agricultural activities [33]. By promoting soil health, biodiversity, and resource efficiency, sustainable agriculture not only mitigates air pollution but also enhances resilience to climate change and improves food security and livelihoods for farmers.

The adoption of sustainable technologies and practices is essential not only for mitigating air pollution but also for achieving broader environmental and development goals [27, 28]. By harnessing the power of innovation, collaboration, and investment, governments, businesses, and communities can accelerate the transition to a more sustainable and equitable future for all.

3.3 Lessons from Successful Implementations

Several countries and regions have demonstrated successful approaches to integrating sustainability principles into policies and practices for clean air and sustainable development. These success stories offer valuable insights and lessons learned for policymakers seeking to replicate and scale up effective interventions.

One such example is the European Union's experience with air quality management and environmental regulation. Through a combination of legislative measures, technological innovation, and public engagement, the EU has made significant strides in improving air quality and reducing emissions of air pollutants. The EU's Air Quality Directive sets standards for ambient air quality and emissions from key sources such as vehicles, industry, and agriculture, while the National Emission Ceilings Directive establishes national emission reduction targets for member states. Moreover, the EU's Clean Air Program for Europe provides a framework for coordinated action to address transboundary air pollution and achieve sustainable development objectives [34].

Another example is the city of Copenhagen, Denmark, which has emerged as a global leader in sustainable urban development and transportation planning. Through investments in cycling infrastructure, public transit, and green spaces, Copenhagen has significantly reduced reliance on private vehicles, improved air quality, and enhanced quality of life for its residents. The city's ambitious climate goals, including becoming carbon-neutral by 2025, have spurred innovation and collaboration across sectors to achieve sustainable development objectives [35].

Similarly, the Republic of Korea has implemented a comprehensive set of policies and programs to address air pollution and promote sustainable development [36]. The Korean government's Green Growth Strategy aims to decouple economic growth from environmental degradation by investing in clean energy, green technologies, and sustainable infrastructure. Through regulatory measures, incentives for clean technologies, and public awareness campaigns, Korea has made significant progress in reducing emissions of air pollutants and greenhouse gases while fostering economic growth and innovation.

These successful initiatives underscore the importance of political leadership, stakeholder engagement, and institutional capacity in driving progress toward clean air and sustainable development. By adopting a holistic and integrated approach that balances economic, social, and environmental considerations, governments can achieve lasting results and build a healthier, more resilient future for current and future generations.

4 Community Engagement and Citizen Science

4.1 Importance of Public Awareness

Public awareness stands as a cornerstone in addressing air pollution, serving as a catalyst for informed decision-making and collective action [37]. Understanding the sources, impacts, and potential solutions to air pollution empowers individuals and communities to advocate for change, driving policy initiatives and behavior modifications that contribute to cleaner air and healthier environments. Furthermore, heightened public awareness fosters accountability among policymakers, industries, and individuals, creating a demand for transparent governance and responsible practices [37].

Communication of air pollution and its effects is facilitated by media campaigns, outreach programs, and educational activities. The goal of these initiatives is to increase public knowledge of the health risks – such as respiratory disorders, cardiovascular diseases, and unfavorable pregnancy outcomes – that come with being exposed to air pollution. These programs encourage people to take proactive steps to limit their exposure to pollution, such as reducing outdoor activities during peak pollution hours and utilizing air purifiers indoors, by emphasizing the link between air quality and public health. Moreover, public awareness campaigns play a crucial role in dispelling misconceptions and myths about air pollution, promoting an accurate understanding of the causes and effects of poor air quality. By debunking misinformation and providing evidence-based information, these campaigns empower individuals to make informed decisions about their health and well-being, as well as advocate for policies and initiatives that promote cleaner air and environmental sustainability.

Public awareness campaigns draw attention to the wider environmental and societal ramifications of poor air quality in addition to the health effects of air pollution. For instance, the emission of greenhouse gases like CO_2 and CH_4 by air pollution increases global warming and its consequences, which include extreme weather, sea level rise, and ecological damage. Public awareness campaigns encourage a comprehensive understanding of the interconnection of environmental issues and the significance of taking collective action to address them by tying air quality to more general environmental challenges. Furthermore, public awareness campaigns can empower individuals to take meaningful action to reduce their contribution to air pollution and mitigate its impacts on their communities. For example, campaigns promoting energy conservation and efficiency, sustainable transportation options, and waste reduction strategies encourage individuals to adopt environmentally friendly behaviors that reduce emissions and improve air quality. By providing practical tips and resources, these campaigns empower individuals to make positive changes in their daily lives and contribute to collective efforts to combat air pollution [38].

Hence, public awareness plays a vital role in addressing air pollution by raising awareness about its causes, consequences, and solutions. By educating individuals and communities about the health, environmental, and societal impacts of poor air quality, public awareness campaigns motivate action, foster accountability, and promote collective efforts to achieve cleaner air and healthier environments.

5 Citizen Science Initiatives

5.1 Air Quality Monitoring by Citizens

Citizen science initiatives harness the power of ordinary citizens to collect, analyze, and interpret air quality data, providing valuable insights into local pollution levels and trends. Advances in sensor technology, data visualization, and community engagement platforms have democratized access to air quality monitoring tools, enabling individuals and communities to participate actively in environmental monitoring and research [38].

One of the key advantages of citizen science initiatives is their ability to complement official monitoring networks operated by government agencies. While government monitoring stations provide valuable data on regional air quality trends, they may not always capture localized pollution hotspots or variations in pollution levels within communities. Citizen science projects fill this gap by deploying low-cost, portable air quality sensors in neighborhoods, schools, workplaces, and public spaces, generating high-resolution data that reflect real-time conditions at the local level [38].

By collecting and sharing air quality data, citizen scientists can raise awareness about local pollution issues, identify sources of pollution, and advocate for policy changes and interventions to improve air quality. For example, community-based monitoring projects have been instrumental in identifying pollution hotspots near industrial facilities, highways, and other sources of emissions, leading to targeted regulatory action and pollution control measures.

Additionally, citizen science data can offer insightful information about how air pollution affects the health of vulnerable groups, including children, the elderly, and people with pre-existing medical disorders. Citizen scientists can monitor the quality of the air in residential areas, schools, and playgrounds in order to evaluate the efficacy of pollution mitigation strategies and to support laws that safeguard public health and welfare.

5.2 Community-Driven Initiatives

Community-driven initiatives empower residents to take ownership of air quality management and advocate for cleaner air in their neighborhoods [39]. These initiatives often emerge in response to perceived environmental injustices, such as disproportionate exposure to pollution in low-income and minority communities, where residents may lack access to resources and information to address environmental concerns.

One example of a community-driven initiative is the development of air quality action plans by residents in collaboration with environmental organizations, academic institutions, and government agencies. These action plans typically involve community members in identifying air quality priorities, setting goals and targets, and implementing strategies to reduce emissions and exposure to air pollutants. By empowering communities to take ownership of air quality management, these initiatives can foster collaboration, build social cohesion, and create lasting impacts on public health and well-being [39].

Another example is the establishment of community-based monitoring networks that engage residents in collecting and analyzing air quality data. These networks often involve partnerships between community organizations, academic institutions, and government agencies to provide training, technical support, and quality assurance for citizen scientists. By empowering citizens to monitor their own air quality, these initiatives can enhance public awareness, accountability, and advocacy for clean air policies and regulations [39].

Furthermore, community-driven initiatives often leverage social networks, grassroots organizing, and advocacy campaigns to mobilize support for environmental justice and equity. By amplifying the voices of affected communities and advocating for policies that address the root causes of environmental inequalities, these initiatives empower marginalized populations to advocate for their rights and demand solutions to environmental injustices [39].

6 Collaborative Approaches for Effective Engagement

Collaborative approaches that engage multiple stakeholders, including government agencies, community organizations, businesses, and academic institutions, are essential for effective community engagement in air quality management [40]. By bringing together diverse perspectives, expertise, and resources, collaborative initiatives can generate innovative solutions, build consensus, and mobilize support for action. One example of a collaborative approach is the establishment of multi-stakeholder partnerships or coalitions that work together to address air pollution at the local, regional, or national level. These partnerships may involve joint planning, resource sharing, and implementation of strategies to reduce emissions, improve monitoring and enforcement, and promote public awareness and education. By fostering collaboration and coordination among stakeholders, these partnerships can leverage collective action and maximize impact on air quality improvement.

Another example is the integration of community input and feedback into decision-making processes through participatory approaches such as public consultations, stakeholder forums, and citizen advisory boards. By soliciting input from affected communities, governments can ensure that air quality policies and regulations reflect the needs, priorities, and concerns of local residents. Moreover, involving stakeholders in decision-making processes can enhance transparency, accountability, and legitimacy, building trust and credibility in air quality management efforts [40].

Hence, community engagement and citizen science are essential components of effective air quality management and sustainable development. By raising public awareness, empowering citizens to monitor their own air quality, and fostering collaboration and partnership among stakeholders, governments and organizations can build support for clean air policies and initiatives, mobilize resources, and create healthier and more resilient communities for current and future generations.

Collaborative approaches that engage multiple stakeholders, including government agencies, community organizations, businesses, and academic institutions, are essential for effective community engagement in air quality management [40]. By bringing together diverse perspectives, expertise, and resources, collaborative initiatives can generate innovative solutions, build consensus, and mobilize support for action. One example of a collaborative approach is the establishment of multi-stakeholder partnerships or coalitions that work together to address air pollution at the local, regional, or national level. These partnerships may involve joint planning, resource sharing, and implementation of strategies to reduce emissions, improve monitoring and enforcement, and promote public awareness and education. By fostering collaboration and coordination among stakeholders, these partnerships can leverage collective action and maximize impact on air quality improvement.

Another example is the integration of community input and feedback into decision-making processes through participatory approaches such as public consultations, stakeholder forums, and citizen advisory boards. By soliciting input from affected communities, governments can ensure that air quality policies and regulations reflect the needs, priorities, and concerns of local residents. Moreover, involving stakeholders in decision-making processes can enhance transparency, accountability, and legitimacy, building trust and credibility in air quality management efforts [40].

Hence, community engagement and citizen science are essential components of effective air quality management and sustainable development. By raising public awareness, empowering citizens to monitor their own air quality, and fostering collaboration and partnership among stakeholders, governments and organizations can build support for clean air policies and initiatives, mobilize resources, and create healthier and more resilient communities for current and future generations.

7 Challenges in Implementation

Implementing policies and regulations aimed at achieving sustainable clean air faces a multitude of challenges that span political, economic, enforcement, and cross-border dimensions. Overcoming these challenges requires a comprehensive understanding of the complexities involved and the development of innovative strategies to address them effectively. This section delves into each of these challenges in detail, exploring their underlying causes and potential solutions.

7.1 Political and Economic Barriers

Political and economic interests often intersect with environmental policies, posing significant barriers to their effective implementation. In many cases, policymakers must navigate competing priorities and conflicting agendas, balancing the interests of various stakeholders, including industry, government, and the public.

One of the primary political challenges is resistance from vested interests, particularly industries that may oppose stringent regulations due to concerns about competitiveness and profitability [41]. Lobbying efforts by industry groups can influence policymaking processes, leading to the watering down or delay of air quality regulations. Moreover, political polarization and ideological differences may hinder consensus-building and bipartisan support for environmental initiatives, further complicating efforts to address air pollution.

Economic considerations also play a significant role in shaping air quality policies, as policymakers must weigh the costs and benefits of regulatory interventions. Industries may argue that compliance with environmental regulations imposes undue financial burdens, leading to job losses, reduced competitiveness, and economic downturns. Consequently, policymakers may face pressure to prioritize short-term economic interests over long-term environmental sustainability, undermining efforts to implement effective air quality regulations [42].

Furthermore, fiscal constraints and budgetary limitations may constrain the resources available for air quality management, limiting investment in monitoring, enforcement, and pollution control measures. Inadequate funding for environmental agencies and regulatory bodies can hinder their ability to effectively enforce existing regulations and develop new policies to address emerging air quality challenges.

Addressing political and economic barriers requires proactive engagement with stakeholders, including industry, government, civil society, and the public. Building consensus around the importance of clean air for public health, environmental protection, and economic prosperity is essential for overcoming resistance to regulatory interventions [43]. Moreover, policymakers must demonstrate the economic benefits of investing in air quality improvement, such as reduced healthcare costs, increased productivity, and job creation in clean technology industries.

In addition to stakeholder engagement, policymakers can employ a range of policy instruments to address political and economic barriers effectively. For example, market-based mechanisms such as emissions trading schemes and pollution taxes can provide economic incentives for industries to reduce their emissions while generating revenue to fund air quality improvement initiatives. Similarly, targeted subsidies and incentives for clean technologies and renewable energy sources can encourage investment in low-carbon alternatives and stimulate innovation in the clean energy sector.

7.2 *Enforcement and Compliance Challenges*

Effective enforcement and compliance mechanisms are critical for ensuring the success of air quality policies and regulations. However, enforcement challenges, including limited resources, capacity constraints, and regulatory loopholes, can undermine the effectiveness of regulatory interventions.

One of the key enforcement challenges is the lack of adequate monitoring and enforcement resources, including staff, equipment, and funding [44]. Environmental agencies may struggle to conduct regular inspections, monitor emissions, and enforce compliance with air quality standards due to resource constraints. Moreover, the complexity of air pollution sources and pathways, including mobile sources such as vehicles and stationary sources such as industrial facilities, further complicates enforcement efforts.

Another challenge is the prevalence of non-compliance and regulatory violations, as some industries may flout environmental regulations to cut costs or maximize profits [45]. Weak penalties, lenient enforcement, and inadequate deterrence mechanisms may embolden polluters to disregard environmental regulations, undermining the credibility of regulatory authorities and eroding public trust in the regulatory process. Additionally, regulatory loopholes and exemptions may undermine the effectiveness of air quality regulations, allowing certain industries or activities to evade compliance requirements. For example, exemptions for small businesses or grandfather clauses for existing facilities may create disparities in regulatory burdens and incentivize non-compliance [45].

Overcoming enforcement and compliance challenges requires a multifaceted approach that strengthens regulatory capacity, improves transparency and accountability, and enhances public participation in monitoring and enforcement efforts. Investing in training and capacity-building for environmental enforcement agencies can enhance their ability to detect and deter violations, while improving coordination and collaboration with other government agencies and law enforcement authorities [45].

Moreover, strengthening penalties and sanctions for regulatory violations can increase the deterrence effect and send a clear signal that non-compliance will not be tolerated. Implementing innovative enforcement strategies, such as using remote sensing technologies, satellite monitoring, and data analytics, can enhance the efficiency and effectiveness of enforcement efforts, allowing regulatory authorities to target high-risk areas and prioritize resources for maximum impact.

Furthermore, promoting public awareness and engagement in monitoring and reporting environmental violations can complement regulatory enforcement efforts, empowering communities to hold polluters accountable and advocate for stronger regulatory action. By fostering a culture of compliance and accountability, policymakers can create a regulatory environment that incentivizes businesses to adopt cleaner technologies and practices, leading to improved air quality and environmental protection [5, 6].

7.3 Cross-Border Pollution Issues

Air pollution knows no borders, and transboundary pollution sources pose significant challenges for air quality management. Pollution generated in one region can travel long distances, crossing national boundaries and affecting air quality in neighboring countries. Consequently, addressing cross-border pollution requires international cooperation and coordination to develop shared solutions and mitigate transboundary impacts [46].

One of the key challenges in addressing cross-border pollution is the lack of harmonization and consistency in air quality standards and regulations across countries. Differences in regulatory frameworks, monitoring methodologies, and emission standards can create disparities in air quality management practices, leading to regulatory gaps and enforcement challenges [47].

Moreover, jurisdictional issues and legal complexities may hinder efforts to hold polluters accountable for transboundary pollution. In many cases, identifying the source of pollution and attributing responsibility for its impacts can be challenging, particularly when multiple countries are involved. Disputes over liability, compensation, and responsibility for mitigating transboundary pollution can further complicate efforts to address cross-border air quality issues [47].

Furthermore, political tensions and diplomatic disputes may impede cooperation and collaboration on transboundary pollution issues, as countries may prioritize national interests over regional or global environmental concerns. Competing economic interests, geopolitical considerations, and historical grievances may exacerbate tensions and hinder efforts to develop shared strategies for addressing cross-border pollution [41].

Overcoming cross-border pollution challenges requires enhanced international cooperation, information sharing, and collaboration among countries, as well as the development of multilateral agreements and mechanisms for addressing transboundary air quality issues. Regional air quality management initiatives, such as the European Union's Air Quality Directive and the United Nations Economic Commission for Europe's Convention on Long-Range Transboundary Air Pollution, provide frameworks for coordinating action and sharing best practices to address cross-border pollution in Europe and beyond [47].

Moreover, promoting dialogue, trust-building, and diplomacy among countries can help overcome political barriers and facilitate cooperation on transboundary pollution issues. By fostering mutual understanding and consensus on the importance of clean air for public health, environmental protection, and sustainable development, policymakers can create a conducive environment for international collaboration and collective action to address cross-border air quality challenges [47].

Furthermore, leveraging international organizations, such as the World Health Organization (WHO) and the United Nations Environment Program (UNEP), can provide technical expertise, capacity-building support, and policy guidance to countries seeking to address cross-border air pollution. By facilitating dialogue, knowledge exchange, and capacity-building initiatives, these organizations can strengthen regional and global cooperation on air quality management and contribute to improved public health and environmental outcomes.

7.4 Strategies for Overcoming Implementation Challenges

Overcoming the challenges associated with implementing policies and regulations for sustainable clean air requires a comprehensive and integrated approach that addresses political, economic, enforcement, and cross-border issues. Several strategies can be employed to overcome these challenges and achieve tangible progress in air quality management [5, 6].

First and foremost, political leadership and commitment are essential for driving action and mobilizing resources to address air pollution. Governments must prioritize air quality improvement as a public health and environmental priority, enacting robust regulatory frameworks, allocating adequate funding, and fostering interagency coordination to achieve clean air goals [5, 6].

Moreover, engaging stakeholders, including industry, government, civil society, and the public, in the policymaking process can build consensus, enhance accountability, and promote ownership of air quality management efforts. By involving stakeholders in decision-making processes, policymakers can ensure that policies are informed by diverse perspectives and reflect the needs and priorities of affected communities [40].

Furthermore, investing in monitoring and enforcement infrastructure, including the deployment of advanced sensor networks, satellite monitoring systems, and data analytics platforms, can enhance the capacity of environmental agencies to detect and deter violations, as well as prioritize resources for maximum impact. By leveraging technology and innovation, policymakers can improve the efficiency and effectiveness of enforcement efforts, allowing regulatory authorities to target high-risk areas and hold polluters accountable.

Additionally, strengthening international cooperation and collaboration is essential for addressing cross-border air pollution and mitigating transboundary impacts. Governments must engage in dialogue, diplomacy, and multilateral negotiations to develop shared strategies and mechanisms for addressing cross-border air quality issues, such as harmonizing air quality standards, sharing data and best practices, and establishing cooperative monitoring and enforcement initiatives [21].

Furthermore, promoting public awareness, education, and engagement is crucial for fostering a culture of environmental stewardship and accountability. By empowering citizens to monitor their own air quality, report violations, and advocate for policy changes, policymakers can harness the collective power of communities to drive positive change and create healthier and more sustainable environments for all.

Hence, addressing the challenges associated with implementing policies and regulations for sustainable clean air requires a multifaceted approach that combines political leadership, stakeholder engagement, technological innovation, and international cooperation. By adopting a comprehensive and integrated strategy, policymakers can overcome barriers to implementation and achieve tangible progress in air quality management, leading to improved public health, environmental protection, and sustainable development outcomes.

8 Future Directions and Recommendations

As the world grapples with the complex challenges of air pollution and its detrimental impacts on public health, ecosystems, and the global climate, it is imperative to chart a course toward sustainable clean air. This section explores future directions and provides recommendations for policymakers, stakeholders, and researchers to advance the agenda of air quality management and environmental sustainability.

8.1 Evolving Strategies for Sustainable Clean Air

The landscape of air quality management is continuously evolving in response to emerging technologies, shifting socioeconomic dynamics, and evolving environmental priorities. To achieve sustainable clean air, it is essential to embrace innovative strategies that integrate environmental protection with economic development and social equity.

One promising approach is the adoption of integrated air quality management frameworks that recognize the interconnectedness of environmental issues and address air pollution in conjunction with other sustainability challenges, such as climate change, biodiversity loss, and resource depletion. By adopting a holistic approach to environmental governance, policymakers can develop synergistic solutions that maximize co-benefits and minimize trade-offs across different policy domains [34].

Furthermore, advancing technological innovations hold immense potential for improving air quality monitoring, modeling, and mitigation efforts. The deployment of low-cost sensor networks, satellite imaging technologies, and data analytics platforms can enhance our understanding of air pollution sources, trends, and impacts, enabling more targeted and effective interventions. Additionally, the development of clean energy technologies, such as renewable energy sources, energy-efficient transportation systems, and sustainable urban planning strategies, can reduce reliance on fossil fuels and mitigate emissions from key sources of air pollution [25].

Moreover, promoting nature-based solutions, such as urban green spaces, afforestation initiatives, and sustainable land management practices, can enhance air quality, sequester carbon, and enhance resilience to climate change. By harnessing the power of nature to purify the air and regulate microclimates, cities can create healthier and more livable environments for their residents while mitigating the impacts of urbanization on air quality.

In addition to technological and nature-based solutions, fostering behavioral change and sustainable consumption patterns is crucial for reducing individual and collective contributions to air pollution. Public awareness campaigns, education programs, and community engagement initiatives can empower individuals to make informed choices about their transportation habits, energy consumption, and lifestyle practices, reducing their environmental footprint and promoting sustainable living.

Overall, evolving strategies for sustainable clean air require a multifaceted approach that leverages technological innovation, nature-based solutions, and behavioral change strategies to address air pollution in a holistic and integrated manner. By embracing innovative approaches and fostering collaboration across sectors and disciplines, policymakers can pave the way toward a cleaner, healthier, and more sustainable future for all.

8.2 Key Recommendations for Policymakers

In light of the multifaceted challenges posed by air pollution, policymakers must prioritize the development and implementation of evidence-based policies and regulations that promote sustainable clean air. Key recommendations for policymakers include:

- Strengthening regulatory frameworks: Policymakers should enact robust air quality regulations that set stringent emission standards, promote pollution prevention, and incentivize the adoption of clean technologies and practices [34].
- Enhancing enforcement and compliance: Policymakers should invest in monitoring and enforcement infrastructure, strengthen penalties for regulatory violations, and promote transparency and accountability in air quality management efforts [39].
- Promoting intersectoral collaboration: Policymakers should foster collaboration across government agencies, industry sectors, civil society organizations, and research institutions to develop integrated approaches to air quality management that address the root causes of pollution and promote sustainable development [36].
- Prioritizing public health and equity: Policymakers should prioritize public health considerations in air quality management efforts, particularly in vulnerable and marginalized communities disproportionately affected by air pollution. Additionally, policies should promote environmental justice and equitable access to clean air for all populations (Public Health England 2019).
- Investing in research and innovation: Policymakers should allocate resources for research and development initiatives aimed at advancing technologies, monitoring methodologies, and mitigation strategies for air quality improvement. Moreover, policymakers should support interdisciplinary research collaborations that integrate scientific expertise from diverse fields to address complex air quality challenges.
- Promoting public awareness and engagement: Policymakers should invest in public awareness campaigns, education programs, and community engagement initiatives to raise awareness about air pollution, its impacts, and potential solutions. By empowering citizens to take action and advocate for clean air policies, policymakers can harness the collective power of communities to drive positive change [37].

Overall, policymakers play a critical role in shaping the trajectory of air quality management efforts and must prioritize sustainable clean air as a fundamental pillar of environmental governance and public health policy.

8.3 Opportunities for International Collaboration

Air pollution is a global challenge that transcends national borders, requiring concerted efforts and international cooperation to address effectively. Opportunities for international collaboration in air quality management include:

- Sharing best practices and lessons learned: Countries can learn from each other's experiences in air quality management, sharing best practices, innovative technologies, and successful policy interventions to accelerate progress toward sustainable clean air.
- Harmonizing standards and regulations: Harmonizing air quality standards, monitoring methodologies, and regulatory frameworks can facilitate cross-border cooperation and coordination, ensuring consistency and comparability in air quality management practices across regions [34].
- Collaborative research and data sharing: International collaboration in research and data sharing can enhance our understanding of air pollution sources, trends, and impacts, informing evidence-based policymaking and fostering innovation in air quality management.
- Building capacity and providing technical assistance: Developed nations can help developing nations boost their institutional capacity, acquire technical know-how, and improve their ability to successfully handle air quality issues by offering them support in building capacity.
- Multilateral initiatives and agreements: Participation in multilateral initiatives and agreements, such as the Paris Agreement and the Sustainable Development Goals, can provide a framework for collective action on air quality management and promote shared goals and objectives for environmental sustainability [20].

By leveraging international collaboration and cooperation, countries can amplify their efforts to address air pollution and achieve sustainable clean air for present and future generations.

8.4 Research and Development Needs

Despite significant progress in air quality management in recent years, several research and development needs remain to be addressed to advance the agenda of sustainable clean air. Key areas for future research include:

- Understanding air pollution dynamics: Research is needed to improve our understanding of the sources, transport pathways, and chemical transformations of air pollutants, particularly in urban areas and industrial hotspots.
- Developing advanced monitoring technologies: Research on the development of advanced monitoring technologies, such as low-cost sensor networks, satellite imaging, and remote sensing, can enhance our ability to track air pollution in real-time and at high spatial resolutions [39].
- Assessing health impacts and vulnerability: Research on the health impacts of air pollution, particularly on vulnerable populations such as children, the elderly, and individuals with pre-existing health conditions, can inform targeted interventions and public health policies [48].
- Evaluating policy effectiveness: Research is needed to evaluate the effectiveness of air quality policies and regulations in achieving their intended goals, identifying barriers to implementation, and informing evidence-based policymaking.
- Advancing clean energy technologies: Research on the development of clean energy technologies, such as renewable energy sources, energy storage systems, and fuel-efficient vehicles, can accelerate the transition to a low-carbon economy and mitigate emissions from key sources of air pollution.
- Promoting interdisciplinary collaboration: Research initiatives that foster interdisciplinary collaboration across fields such as environmental science, public health, engineering, economics, and social sciences can generate innovative solutions to complex air quality challenges.

By prioritizing research and development in these key areas, policymakers, researchers, and stakeholders can advance the agenda of sustainable clean air and pave the way toward a healthier, more resilient, and more sustainable future for all. Hence, addressing the challenges of air pollution and achieving sustainable clean air requires a concerted effort from policymakers, stakeholders, and researchers at the local, national, and international levels. By embracing innovative strategies, fostering collaboration, and prioritizing research and development, we can create a world where clean air is a fundamental right and a cornerstone of environmental sustainability and public health.

9 Conclusions

In the pursuit of sustainable clean air, this comprehensive examination has illuminated critical insights, challenges, and recommendations for policymakers, stakeholders, and researchers. The convergence of environmental, social, and economic imperatives underscores the urgency of addressing air pollution through multifaceted approaches that integrate regulatory frameworks, technological innovations, public engagement, and international collaboration. This section synthesizes the key findings and offers concluding remarks on the path forward toward achieving sustainable clean air.

Air pollution is a global challenge that transcends geographical boundaries and affects populations worldwide. The World Health Organization (WHO) estimates that air pollution contributes to millions of premature deaths each year and is a leading environmental health risk factor globally. Moreover, air pollution exacerbates climate change, degrades ecosystems, and undermines sustainable development goals, highlighting the urgent need for coordinated action at the local, national, and international levels. Hence, clean air is not merely a commodity; it is a fundamental human right and a prerequisite for healthy ecosystems, thriving communities, and sustainable development.

The journey toward sustainable clean air is fraught with challenges, ranging from political and economic barriers to enforcement and compliance issues, and cross-border pollution complexities. However, amidst these challenges lie opportunities for innovation, collaboration, and transformative change. *Technological innovations* hold immense potential for improving air quality monitoring, modeling, and mitigation efforts. The deployment of low-cost sensor networks, satellite imaging technologies, and data analytic platforms can enhance our understanding of air pollution sources, trends, and impacts, enabling more targeted and effective interventions. Additionally, the development of *clean energy technologies*, such as renewable energy sources, energy-efficient transportation systems, and sustainable urban planning strategies, can reduce reliance on fossil fuels and mitigate emissions from key sources of air pollution. Furthermore, promoting nature-based solutions, such as urban green spaces, afforestation initiatives, and sustainable land management practices, can enhance air quality, sequester carbon, and enhance resilience to climate change.

In addition to technological and nature-based solutions, fostering *behavioral change and sustainable consumption* patterns is crucial for reducing individual and collective contributions to air pollution. Public awareness campaigns, education programs, and community engagement initiatives can empower individuals to make informed choices about their transportation habits, energy consumption, and lifestyle practices, reducing their environmental footprint and promoting sustainable living.

On the other hand, *policymakers* play a pivotal role in shaping the trajectory of air quality management efforts and must prioritize evidence-based policies and regulations that promote sustainable clean air. *Key recommendations* include strengthening regulatory frameworks, enhancing enforcement and compliance mechanisms, promoting intersectoral collaboration, prioritizing public health and equity, and investing in research and innovation. By adopting a holistic and integrated approach to air quality management, policymakers can pave the way toward a cleaner, healthier, and more sustainable future for all.

Strengthening regulatory frameworks is essential for ensuring the effectiveness of air quality management efforts. For example, emission trading schemes and pollution taxes can provide economic incentives for industries to reduce their emissions while generating revenue to fund air quality improvement initiatives. Similarly, targeted subsidies and incentives for clean technologies and renewable energy sources can encourage investment in low-carbon alternatives and stimulate innovation in the clean energy sector.

Moreover, promoting intersectoral collaboration is crucial for addressing the complex and interconnected drivers of air pollution. Policymakers should foster collaboration across government agencies, industry sectors, civil society organizations, and research institutions to develop integrated approaches to air quality management that address the root causes of pollution and promote sustainable development. By breaking down silos and fostering synergies across different sectors, policymakers can maximize the effectiveness and efficiency of air quality management efforts.

Furthermore, prioritizing public health and equity considerations is essential for ensuring that air quality management efforts benefit all segments of society, particularly vulnerable and marginalized populations. Policymakers should prioritize public health considerations in air quality management efforts, particularly in communities disproportionately affected by air pollution. Additionally, policies should promote environmental justice and equitable access to clean air for all populations.

Investing in research and innovation is critical for advancing the science and technology of air quality management. Policymakers should allocate resources for research and development initiatives aimed at advancing technologies, monitoring methodologies, and mitigation strategies for air quality improvement. Moreover, policymakers should support interdisciplinary research collaborations that integrate scientific expertise from diverse fields to address complex air quality challenges.

At the international level, addressing air pollution requires coordinated action between countries as air pollutants know no borders and transboundary pollution sources pose significant challenges for air quality management. Opportunities for international collaboration include sharing best practices and lessons learned, harmonizing standards and regulations, collaborative research and data sharing, capacity-building and technical assistance, and participation in multilateral initiatives and agreements. By leveraging international cooperation and collaboration, countries can amplify their efforts to address air pollution and achieve sustainable clean air for present and future generations.

Sharing best practices and lessons learned is essential for accelerating progress toward sustainable clean air. Countries can learn from each other's experiences in air quality management, sharing best practices, innovative technologies, and successful policy interventions to accelerate progress toward sustainable clean air. Moreover, collaborative research and data sharing are essential for advancing our understanding of air pollution sources, trends, and impacts. International collaboration in research and data sharing can enhance our understanding of air pollution sources, trends, and impacts, informing evidence-based policymaking and fostering innovation in air quality management. By sharing data, research findings, and best practices, countries can collaborate more effectively to address air pollution and achieve sustainable clean air.

In conclusion, achieving sustainable clean air is a complex but achievable goal that requires collective action, political will, and sustained commitment from all stakeholders. By prioritizing evidence-based policies, investing in clean technologies, fostering public awareness, and strengthening international collaboration, we can overcome the challenges of air pollution and create a world where clean air is a fundamental right and a cornerstone of environmental sustainability and public health. As we embark on this journey toward a cleaner, healthier, and more sustainable future, let us remember that the air we breathe knows no boundaries, and our collective efforts today will shape the air quality of tomorrow for generations to come.

References

1. De Marco A, Proietti C, Anav A, Ciancarella L, D'Elia I, Fares S, Fornasier MF, Fusaro L, Gualtieri M, Manes F, Marchetto A, Mircea M, Paoletti E, Piersanti A, Rogora M, Salvati L, Salvatori E, Screpanti A, Vialetto G et al (2019) Impacts of air pollution on human and ecosystem health, and implications for the National Emission Ceilings Directive: insights from Italy. Environ Int 125:320–333. https://doi.org/10.1016/j.envint.2019.01.064
2. Davy P, Gunchin G, Markwitz A, Trompetter W, Barry B, Shagjjamba D, Lodoysamba S (2011) Air particulate matter pollution; determination of composition, source contribution and source locations in Ulaanbaatar, Mongolia. Atmos Pollut Res 2:126–137
3. Mahmoud N, Al-Shahwani D, Al-Thani H, Isaifan RJ (2023) Risk assessment of the impact of heavy metals in urban traffic dust on human health. Atmos 14(6). https://doi.org/10.3390/atmos14061049
4. Chen S, Guo C, Huang X (2018) Air pollution, student health, and school absences: evidence from China. J Environ Econ Manag 92:465–497. https://doi.org/10.1016/j.jeem.2018.10.002
5. Al-Thani H, Koç M, Isaifan RJ (2018) A review on the direct effect of particulate atmospheric pollution on materials and its mitigation for sustainable cities and societies. Environ Sci Pollut Res 25(August):27839–27857. https://doi.org/10.1007/s11356-018-2952-8
6. Al-Thani H, Koc M, Isaifan RJ (2018) Investigations on deposited dust fallout in Urban Doha: characterization, source apportionment and mitigation. Environ Ecol Res 6(5):493–506. https://doi.org/10.13189/eer.2018.060510
7. Khoder MI (2009) Diurnal, seasonal and weekdays-weekends variations of ground level ozone concentrations in an urban area in greater Cairo. Environ Monit Assess 149:349–362
8. Ramanathan V (2014) The two worlds approach for mitigating air pollution and climate change. Sustainable humanity, sustainable nature, pp 1–16
9. WHO (2018) Air quality and health. In: Health and sustainable development goals. https://www.who.int/sustainable-development/SDG-Factsheets-AQ-med-res-120718b.pdf
10. Bell ML, Samet J (2016) Air pollution. In: Frumkin H (ed) Environmental health: from global to local3rd edn. Wiley
11. World Health Organisation (2020) Air pollution. http://www9.who.int/airpollution/en/
12. Li M, Mallat L (2018) Health impacts of air pollution. Art Sci Risk:42
13. Khoshnevis Yazdi S, Khanalizadeh B (2017) Air pollution, economic growth and health care expenditure. Econ Res-Ekonomska Istraživanja 30(1):1181–1190. https://doi.org/10.1080/1331677X.2017.1314823
14. Schaltegger S, Synnestvedt T (2002) The link between green and economic success: environmental management as the crucial trigger between environmental and economic performance. J Environ Manage 65:339–346
15. United Nations (2019) Mitigation | UNEP – UN Environment Programme. https://www.unenvironment.org/explore-topics/climate-change/what-we-do/mitigation
16. Sohn LB (1973) The Stockholm declaration on the human environment. Harv Int Law J 14
17. Bonzanini D, Giudici G, Patrucco A (2016) The crowdfunding of renewable energy projects. In: Handbook of environmental and sustainable finance. Elsevier, pp 429–444. https://doi.org/10.1016/B978-0-12-803615-0.00021-2

18. Sutter C, Parreño JC (2007) Does the current clean development mechanism (CDM) deliver its sustainable development claim? An analysis of officially registered CDM projects. Clim Change 84(1):75–90. https://doi.org/10.1007/s10584-007-9269-9
19. Rosen A (2012) The wrong solution at the right time: the failure of the Kyoto protocol on climate change. Policy Polit 40(3):444–470. https://doi.org/10.1111/j.17471346.2007.00080.x/abstract
20. Schleussner CF, Rogelj J, Schaeffer M, Lissner T, Licker R, Fischer EM, Knutti R, Levermann A, Frieler K, Hare W (2016) Science and policy characteristics of the Paris agreement temperature goal. In: Nature climate change, vol 6. Nature Publishing Group, pp 827–835. https://doi.org/10.1038/nclimate3096
21. Ministry of Environment (2011) State of Qatar Initial National Communication to the United Nations Framework Convention on Climate Change Strengthen technical and institutional capacities. pp 1–62. http://unfccc.int/resource/docs/natc/qatnc1.pdf
22. IRENA – International Renewable Energy Agency (2016) The renewable route to sustainable transport. In: The renewable route to sustainable transport: a working paper based on REmap (Issue August). http://www.irena.org
23. Luppi B, Parisi F, Rajagopalan S (2012) The rise and fall of the polluter-pays principle in developing countries. Int Rev Law Econ 32(1):135–144. https://doi.org/10.1016/j.irle.2011.10.002
24. Minde H (2001) Sami land rights in Norway: a test case for indigenous peoples. Int J Minority Group Rights 8(2 and 3):67–126
25. Isaifan RJ, Dole HAE, Obeid E, Lizarraga L, Baranova EA, Vernoux P (2011) Catalytic CO oxidation over Pt nanoparticles prepared from the polyol reduction method supported on yttria-stabilized zirconia. ECS Trans 35(28):43. https://doi.org/10.1149/1.3641818
26. Fares E, Buffiere M, Figgis B, Haik Y, Isaifan RJ (2021) Soiling of photovoltaic panels in the Gulf Cooperation Council countries and mitigation strategies. Sol Energy Mater Sol Cells 231:111303. https://doi.org/10.1016/j.solmat.2021.111303
27. Isaifan RJRJ, Couillard M, Baranova EAEA (2017) Low temperature-high selectivity carbon monoxide methanation over yttria-stabilized zirconia-supported Pt nanoparticles. Int J Hydrogen Energy 42(19):13754–13762. https://doi.org/10.1016/j.ijhydene.2017.01.049
28. Isaifan RJ, Samara A, Suwaileh W, Johnson D, Yiming W, Abdallah AA, Aïssa B (2017) Improved self-cleaning properties of an efficient and easy to scale up TiO2 thin films prepared by adsorptive self-assembly. Sci Rep 7(1):9466. https://doi.org/10.1038/s41598-017-07826-0
29. Aïssa B, Nedil M, Kroeger J, Ali A, Isaifan RJ, Essehli R, Mahmoud KA (2018) Graphene nanoplatelet doping of P3HT:PCBM photoactive layer of bulk heterojunction organic solar cells for enhancing performance. Nanotechnology 29(10). https://doi.org/10.1088/13616528/aaa62d
30. Al-Qahtani S, Koç M, Isaifan RJ (2023) Mycelium-based thermal insulation for domestic cooling footprint reduction: a review. In: Sustainability (Switzerland), vol 15. Multidisciplinary Digital Publishing Institute (MDPI). https://doi.org/10.3390/su151713217
31. Wang R, Helbich M, Yao Y, Zhang J, Liu P, Yuan Y, Liu Y (2019) Urban greenery and mental wellbeing in adults: cross-sectional mediation analyses on multiple pathways across different greenery measures. Environ Res 176:108535. https://doi.org/10.1016/j.envres.2019.108535
32. Isaifan RJ, Baldauf RW (2020) Estimating economic and environmental benefits of urban trees in desert regions. Front Ecol Evol 8:1–14. https://doi.org/10.3389/fevo.2020.00016
33. Atef N, Gamal F (2019) Food security via improving crop water productivity in some Arab countries. World J Agric Soil Sci 2(2):10.33552/wjass.2019.02.000532
34. Wettestad J, Farmer A (2003) The EU air quality framework directive: shaped and saved by interaction?
35. Glazener A, Khreis H (2019) Transforming our cities: best practices towards clean air and active transportation. In: Current environmental health reports, vol 6. Springer, pp 22–37. https://doi.org/10.1007/s40572-019-0228-1

36. Hoon Moon T (2009) Korea's sustainable development strategy. In: Korea OBSERVER, vol 40. https://www.researchgate.net/publication/298670979
37. Ubachukwu NN, Phil-Eze PO, Emeribe CN (2014) Analysis of household hazardous wastes awareness level in Enugu Metropolis. AJIS. https://doi.org/10.5901/ajis.2014.v3n1p369
38. Mahajan S, Kumar P, Pinto JA, Riccetti A, Schaaf K, Camprodon G, Smári V, Passani A, Forino G (2020) A citizen science approach for enhancing public understanding of air pollution. Sustain Cities
39. Conrad CC, Hilchey KG (2011) A review of citizen science and community-based environmental monitoring: issues and opportunities. Environ Monit Assess 176(1–4):273–291. https://doi.org/10.1007/s10661-010-1582-5
40. Kaginalkar A, Kumar S, Gargava P, Niyogi D (2022) Stakeholder analysis for designing an urban air quality data governance ecosystem in 3 smart cities 4. https://www.elsevier.com/open-access/userlicense/1.0/
41. Selin H, Vandeveer SD (2003) Mapping institutional linkages in European air pollution politics. Glob Environ Pol 3(3)
42. Millock K, Nauges C, Sterner T (2004) Environmental taxes: a comparison of French and Swedish experience from taxes on industrial air pollution. Econstor 2(1):30–34. http://hdl.handle.net/10419/166797
43. Acemoglu D, Robinson J (2000) Political losers as a barrier to economic development. Am Econ Rev
44. Hamburg FC (2012) Some basic consideration in the design of an air pollution monitoring system. J Air Pollut Control Assoc 21(10):609–613. https://doi.org/10.1080/00022470.1971.10469574
45. Shapiro J, Walker R (2020) Is air pollution regulation too stringent? NBER working paper series. http://www.nber.org/papers/w28199
46. Roshan DR, Koc M, Isaifan R, Shahid MZ, Fountoukis C (2019) Aerosol optical thickness over large urban environments of the Arabian Peninsula-speciation, variability, and distributions. Atmos 10(5). https://doi.org/10.3390/atmos10050228
47. Byrne A (2017) Trouble in the air: recent developments under the 1979 convention on long-range transboundary air pollution. Rev Eur Comp Int Environ Law 26(3):210–219. https://doi.org/10.1111/reel.12219
48. Roshan D, Isaifan RJ (2018) Household air pollution from burning biomass for cooking and heating - a review of health hazards and intervention programs. J Environ Sci Pollut Res 4:289–296

Action Taken to Reduce Air Pollution and Its One Health Impacts in MENA Countries

Rima J. Isaifan and Hanadi G. Al-Thani

Contents

R. J. Isaifan (✉) and H. G. Al-Thani
College of Science and Engineering, Hamad Bin Khalifa University, Doha, Qatar
e-mail: hgalthani@qf.org.qa

Matthew Chidozie Ogwu and Sylvester Izah (eds.), *Sustainable Strategies for Air Pollution Mitigation: Development, Economics and Technologies*,
Hdb Env Chem (2024) 133: 439–474, DOI 10.1007/698_2024_1094,

Published online: 20 July 2024

Abstract This chapter explores the extensive plans and programs put in place throughout the Middle East and North Africa (MENA) region to reduce air pollution and deal with its many effects on One Health. Due to their distinct environmental problems, the MENA countries have realized how critical it is to address air pollution as an avenue to protect both human populations and the larger ecosystem. The first section of the chapter gives a general review of the problems with the MENA region's air quality, highlighting the main pollutants and their sources. The many effects of air pollution on ecosystems and human health are then examined, along with how these effects are related to one another within the context of One Health. The chapter then goes into the variety of initiatives and laws that the MENA region has implemented to tackle air pollution. This covers policy initiatives, technology developments, and cooperative initiatives on a regional and global scale. The usefulness and difficulties of these initiatives are evaluated critically, illuminating the difficulties in tackling a global problem like air pollution in a heterogeneous and dynamic geopolitical environment. The chapter also includes case studies and success stories from MENA nations that have made notable strides toward lowering air pollution and lessening its negative effects on one's health. For other countries dealing with comparable issues, these instances offer insightful lessons and effective methods. The chapter concludes by highlighting the significance of ongoing initiatives, creative solutions, and global cooperation to attain long-term gains in air quality throughout the MENA area. To address the intricate web of problems presented by air pollution, it emphasizes the interdependence of environmental, human, and animal health and advocates for an integrated and holistic approach.

Keywords Air Pollution, Arab countries, Emission Reduction, Environmental Policy, One Health, Pollution Control Measures, Sustainable Development

Abbreviations

AOT	Aerosol optical thickness
AP	Attributable proportion
CNG	Compressed natural gas
CO	Carbon monoxide
COPD	Chronic obstructive pulmonary disease
CSP	Concentrated solar power
EEAA	Egyptian Environmental Affairs Agency
EPA	Environmental Protection Agency
EV	Electric vehicles
GCC	Gulf Cooperation Council
GDP	Gross domestic product
MOBILE	Model for measuring mobile source emissions
MoE	Ministry of Energy
NGO	Non-governmental Organization
NOx	Nitrogen oxides
PM	Particulate matter
$PM_{2.5}$	Particulate matter of diameter less than or equal to 2.5 μm
PM_{10}	Particulate matter of diameter less than or equal to 10 μm
PV	Photovoltaic
RR	Relative risk
SO2	Sulfur dioxide
UAE	United Arab Emirates
VOC	Volatile organic carbon
WHO	World Health Organization

1 Introduction

1.1 Background and Context

The Middle East and North Africa (MENA) region, known for its varied landscapes and rich cultural legacy, is confronted with a serious environmental problem in the form of air pollution [1]. This region includes countries such as Algeria, Bahrain, Egypt, Iran, Iraq, Israel, Jordan, Kuwait, Lebanon, Libya, Morocco, Oman, Qatar, Saudi Arabia, Syria, Tunisia, United Arab Emirates, Palestine, and Yemen. Due to the region's increasing urbanization brought on by population growth and economic development, industrial activity has surged, vehicular traffic has expanded, and agricultural practices have intensified. Consequently, air pollutants such as particulate matter (PM), nitrogen dioxide (NO_2), sulfur dioxide (SO_2), volatile organic compounds (VOCs), and ozone (O_3) have all seen a substantial increase in emissions, endangering both human health and the sustainability of the environment.

MENA's urban centers serve as hubs of economic activity and innovation, attracting millions of inhabitants seeking employment opportunities and improved living standards. However, the concentration of industries, power plants, and transportation networks in these urban areas have resulted in localized air pollution hotspots [2]. As a result, industrial emissions, vehicular exhaust, construction dust, and agricultural activities contribute to the emission of harmful pollutants, exacerbating air quality concerns and compromising public health.

With a focus on the main anthropogenic sources for air emissions, industrial activities, ranging from manufacturing and power generation to petrochemical processing, are significant contributors to air pollution in the MENA region. The expansion of industrial zones and the proliferation of manufacturing facilities have led to elevated levels of air pollutants in the atmosphere [3]. Similarly, power plants fueled by fossil fuels emit large quantities of greenhouse gases, further exacerbating air quality concerns. The transportation sector is another major source of air pollution in the MENA region, driven by the increasing number of vehicles on the roads. Urban congestion, inadequate public transportation infrastructure, and a reliance on outdated vehicles contribute to elevated levels of vehicular emissions, including carbon monoxide, nitrogen dioxide, and volatile organic compounds [4]. Additionally, dust emissions from unpaved roads and construction sites worsen air quality issues, particularly in arid regions prone to dust storms. In addition, agricultural practices, while essential for food production and livelihoods, also contribute to air pollution by fertilizers, pesticides, and agricultural burning. Inefficient irrigation practices and land degradation further speed up soil erosion and increase dust emissions, leading to air quality degradation in rural areas.

Moreover, the MENA region's unique geographical features, including its arid deserts, rugged mountains, and coastal plains, influence air pollution patterns and dispersion mechanisms. Dust storms, prevalent in desert regions, transport fine particulate matter over vast distances, affecting air quality in urban and rural areas alike [5]. Similarly, sea salt aerosols generated by coastal winds contribute to atmospheric pollution, posing challenges for coastal communities and marine ecosystems [5].

1.2 Significance of Addressing Air Pollution in MENA

Research has proven that air pollution has such detrimental effects on human health, ecological integrity, and socioeconomic growth, MENA countries must address this issue. Numerous health impacts, such as respiratory conditions, heart problems, and different types of cancers are associated with poor air quality [6]. Children, the elderly, and people with pre-existing medical issues are among the vulnerable groups that are disproportionately affected, which worsens health disparities and places a heavy load on healthcare systems.

Furthermore, air pollution poses a grave threat to the region's ecosystems and biodiversity. Elevated levels of pollutants can degrade soil quality, contaminate water bodies, and disrupt delicate ecological balances [7]. Agricultural productivity suffers as crops are exposed to harmful pollutants, leading to yield losses and compromising food security. Additionally, air pollution can harm wildlife populations, with implications for biodiversity conservation and ecosystem resilience.

Another impact of air pollution is that it poses a serious threat to the region's ecosystems and biodiversity. Elevated levels of pollutants can harm plant and animal species, disrupt ecological processes, and degrade natural habitats. Soil and water contamination, resulting from air pollution deposition, further accelerate environmental degradation and compromise ecosystem resilience [8].

The MENA region's vulnerability to climate change adds to the challenges posed by air pollution, amplifying risks such as water scarcity, desertification, and extreme weather events. Hence, addressing air pollution in the context of climate change adaptation and mitigation is essential for building resilience and promoting sustainable development in the region.

From an economic perspective, the costs of air pollution are significant, encompassing healthcare expenditures, productivity losses, and environmental cleanup efforts. The socioeconomic impacts of air pollution extend beyond health and environmental concerns, affecting livelihoods, economic productivity, and quality of life. Moreover, air pollution undermines the region's efforts toward sustainable development by hindering progress toward environmental stewardship and resilience, social equity, and economic prosperity [9].

1.3 Scope and Objectives of the Chapter

This chapter seeks to provide a comprehensive overview of the measures undertaken by MENA countries to address air pollution and mitigate its impacts on One Health. By examining the root causes of air pollution, its effects on ecosystems and human health, and the initiatives implemented to combat it, this chapter aims to identify successful strategies and areas for improvement.

Moreover, through an in-depth analysis of existing literature, case studies, and empirical evidence, this chapter endeavors to elucidate the complexities of air pollution management in the MENA region. By critically evaluating the effectiveness of policy interventions, technological innovations, and regional collaborations, it aims to distill key insights and best practices for addressing air pollution in diverse geopolitical contexts.

Furthermore, this chapter emphasizes the importance of interdisciplinary approaches and cross-sectoral cooperation in tackling air pollution holistically. By advocating for integrated solutions that consider the interconnectedness of environmental, human, and animal health, it seeks to catalyze transformative change and promote resilience in the face of emerging environmental challenges. The chapter aims to contribute to the ongoing discourse on air pollution mitigation in the MENA region, offering nuanced insights, practical recommendations, and pathways for sustainable development.

2 Air Pollution in MENA Region: A Comprehensive Overview

Air pollution is a complicated environmental concern that the Middle East and North Africa (MENA) region faces. It is caused by a combination of natural events and human-caused activity. The sources of air pollution in the MENA region are thoroughly examined in this section, including dust storms, vehicle exhaust, and industrial emissions. Furthermore, it explores regional variations and pinpoints pollution hotspots while taking into account the various topographies and climates that affect the dynamics of air quality [5].

2.1 Sources of Air Pollution in MENA Region

2.1.1 Industrial Emissions

Several MENA nations, including Saudi Arabia, the United Arab Emirates, Kuwait, Qatar, Bahrain, and Oman, have developed into important hubs for international trade between East and West. Ninety percent of all oil tanker traffic travels through the Suez Canal, making it the most significant oil trade route. Large volumes of oil, chemicals, and other industrial exports from the major oil production and industrial centers on the east coast of the Arabian Peninsula, as well as trade into and out of the free-trade zones of the Arabian Gulf are served by them [10]. Therefore, it is equally crucial to comprehend the characteristics of the emissions from all air and marine transportation sources in order to undertake studies on the consequences on health [11].

Industrial activities represent a significant source of air pollution in the MENA region, fueled by the expansion of manufacturing, energy production, and petrochemical industries. These industries emit a range of pollutants, including gaseous and particulate matter, in addition to heavy metals and volatile organic compounds. Emissions occur during various stages of production, such as combustion processes, chemical reactions, and material handling operations [8].

Major industrial sources of air pollution include power plants, refineries, cement kilns, steel mills, and chemical manufacturing facilities. Power generation, primarily from fossil fuels such as coal and oil, contributes significantly to air pollution levels, releasing large quantities of sulfur dioxide and nitrogen oxides. Similarly, refineries emit sulfur dioxide and volatile organic compounds during crude oil processing and refining operations [12].

Industrialization, driven by the extraction of natural resources, manufacturing processes, and energy production, has transformed the MENA region's economic landscape. With vast reserves of oil, gas, and minerals, countries in the region have capitalized on these resources to fuel economic growth and modernization. However, the expansion of industrial activities has come at a cost, as emissions from power plants, refineries, and manufacturing facilities pollute the air and degrade environmental quality [13].

Power generation is a significant contributor to air pollution in the MENA region, with fossil fuel-fired power plants emitting large quantities of sulfur dioxide, nitrogen oxides, and particulate matter. The burning of coal, oil, and natural gas releases greenhouse gases and air pollutants, which contribute to regional air quality degradation and global climate change. Similarly, refineries, petrochemical plants, and chemical manufacturing facilities emit a variety of pollutants, including volatile organic compounds and hazardous air pollutants, which pose risks to human health and the environment [14].

Air pollution from industrial sources does not only impact air emissions directly, but it also pollutes water and adds stress to water resources in the region. Sakhel et al. [15] reported on the quantification of water consumption, wastewater generation, and virtual water export for eight relevant MENA businesses. Their study demonstrated that approximately 793 million m^3 of wastewater is produced annually from items intended for both local and international consumption which also shows that industrial processes impact water resources as well [15].

Furthermore, the production of cement and steel involves high-temperature processes that release significant amounts of carbon dioxide and particulate matter. Chemical manufacturing facilities emit a variety of pollutants, including volatile organic compounds and hazardous air pollutants, which pose risks to human health and the environment. In a case study that investigated the gaseous and dust emissions in one of the main cement industrial nations in the MENA, Jordan, the authors found that although both particulate and gaseous emissions are high at the location of the project, however, the concentrations of the emitted air pollutants are found to be significantly below the acceptable Jordanian Standards for ambient air quality following the execution of the suggested project [8]. As a result, there is little chance that the planned activity will significantly worsen the air quality in the area around the proposed project [3].

2.1.2 Vehicular Exhaust

The transportation sector is another major source of air pollution in the MENA region, driven by the proliferation of motor vehicles and inadequate public transportation infrastructure. Urban centers are characterized by congested roads, outdated vehicles, and inefficient traffic management systems, leading to high levels of vehicular emissions and poor air quality. Diesel-powered vehicles, in particular, are significant contributors to air pollution, emitting nitrogen oxides and particulate matter that pose health risks to urban populations. Rapid urbanization and motorization in the MENA area have increased vehicle traffic and related pollutants. Internal combustion engines in cars, trucks, buses, and motorbikes release various pollutants into the air as they burn fossil fuels, including carbon monoxide (CO), nitrogen oxides (NOx), hydrocarbons, and particulate matter (PM) [16].

Older vehicles, particularly those equipped with outdated engine technologies and inadequate emission control systems, are major contributors to vehicular emissions. There are two reasons why vehicle age and air pollution emission levels are related. First, federal regulations lowered new car emissions of air pollutants beginning in 1968. More advanced pollution control technology, including catalytic converters, is found in newer cars. Second, equipment used to manage pollution ages. According to the U.S. Environmental Protection Agency's (EPA) model for measuring mobile source emissions (MOBILE), for instance, emissions of carbon monoxide are estimated to be ten times higher after four years or fifty thousand miles, emissions of volatile organic compounds to be approximately four times higher, and emissions of nitrogen oxide to be approximately twice as high as when the car was new [17].

Diesel-powered vehicles, in particular, emit high levels of nitrogen oxides and particulate matter, posing significant health risks to urban populations. Traffic congestion exacerbates emissions, as vehicles idle in traffic jams, leading to increased fuel consumption and pollutant emissions. Al-Thani et al. reported on the findings of the first assessment of the particulate matter emissions from all heavy-duty diesel vehicles that are registered in Qatar [4]. The investigation was conducted in a busy traffic area of Doha. Dust samples were gathered, and their size distribution and form were noted. Particle sizes were observed to vary from a few to 600 μm, with a tiny size fraction (less than 100 μm) predominating. In-situ elemental composition study of traffic dust from side and main roads was also done, and the results were compared with PM from non-traffic sources. The enrichment factor and preliminary source apportionment were assessed using the results. Anthropogenic elements had an enrichment factor of 350. Compared to samples from non-traffic areas, the traffic source based on sulfur element fingerprint was nearly five times greater in main roadways. Additionally, road dust resuspension, brake wear, tire wear, and PM exhaust and non-exhaust emissions were assessed [4]. It was discovered that the main sources of dust were 33% from tire wear, 31% from road dust resuspension, 19% from brake wear, and 17% from exhaust emissions. The majority of registered vehicle types were newer models with effective exhaust PM reduction systems, which explains the minimal contribution of exhaust coarse particulate (PM_{10}) emissions [4].

Moreover, the lack of efficient public transportation infrastructure and the prevalence of personal vehicle ownership increases air pollution levels in urban centers. Inadequate urban planning, characterized by sprawling development patterns and insufficient investment in public transit, further exacerbates traffic congestion and air pollution. Addressing vehicular emissions requires a multifaceted approach, including promoting alternative transportation modes, improving fuel efficiency standards, and implementing stringent emission testing and vehicle inspection programs [4].

2.1.3 Dust Storms and Natural Sources

The MENA region's arid and semi-arid landscapes are susceptible to natural phenomena such as dust storms, sandstorms, and wildfires, which contribute to air pollution. Dust storms, driven by strong winds and arid conditions, mobilize fine particles of dust and sand into the atmosphere, significantly elevating particulate matter concentrations [5].

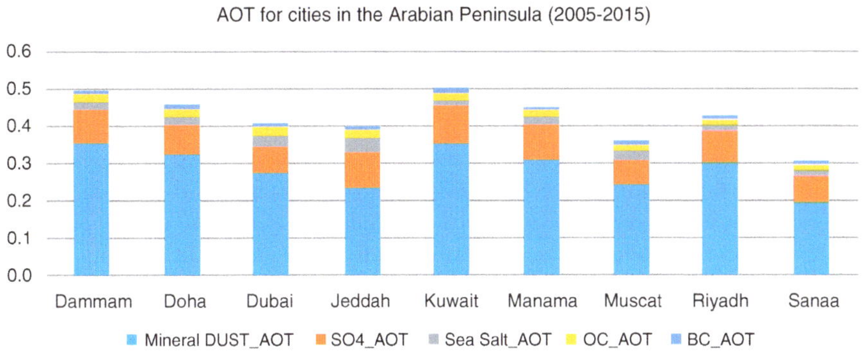

Fig. 1 Total and speciated AOT among the nine cities in the Arabian Peninsula (average from 2005 to 2015 [5])

Dust particles can travel vast distances, affecting air quality in urban, rural, and coastal areas. Dust storms can lead to respiratory problems and visibility impairments, posing health risks for vulnerable populations. In addition to dust storms, natural sources such as volcanic eruptions and wildfires can release large quantities of pollutants into the atmosphere. While volcanic activity is relatively rare in the MENA region, wildfires can occur during periods of drought and high temperatures, releasing smoke, ash, and particulate matter into the air. These natural events contribute to episodic spikes in air pollution levels, posing challenges for air quality management and public health protection [18].

Furthermore, coastal areas are subject to sea salt aerosols generated by marine processes, such as wave action and sea spray [19]. Sea salt aerosols can contribute to atmospheric pollution, particularly in coastal cities and port areas, where maritime activities are prevalent. While natural sources of air pollution may not be as significant as anthropogenic sources in the MENA region, they can still impact air quality and pose challenges for environmental management and public health protection. Roshan et al. focused on nine major cities in the region (Dammam, Doha, Dubai, Jeddah, Kuwait, Manama, Muscat, Riyadh, and Sanaa) and examined the speciation, variability, and distribution of aerosol optical depth over the Arabian Peninsula between 2005 and 2015 [5]. They found that the mean annual aerosol optical thickness (AOT) values over the last ten years have been between 0.3 and 0.5 as shown in Fig. 1. Moreover, anthropogenic activities accounted for 20–30%, while mineral dust for 60–70% with both sources considered the two main causes of this variation.

More importantly, the study found that the increased dust generation from the Fertile Crescent (Syria, Iraq, and Jordan) and the Sahel region of Northern Africa is blamed for an increase in AOT values between 2005 and 2009. This is because of an extended dry season. Even though desert dust contributes significantly to the region's overall particulate matter levels, reducing local urban emissions is still thought to be an effective way to enhance the quality of the air in major population centers [5].

2.2 Geographical Variances and Hotspots

The MENA region exhibits significant geographical variances in air pollution levels, influenced by topographical features, climatic conditions, and human activities. Urban centers, industrial zones, and transportation hubs often experience elevated pollution levels due to localized emissions and population density. Coastal cities, especially those adjacent to major ports and industrial facilities, may experience higher levels of air pollution due to maritime activities and sea salt aerosols [19].

Moreover, desert regions are prone to dust storms and sandstorms, which can result in episodic spikes in particulate matter concentrations. Dust particles can travel long distances, affecting air quality in neighboring countries and regions. Mountainous areas may experience localized pollution due to temperature inversions and restricted air circulation, trapping pollutants in valleys and basins [20].

Identifying pollution hotspots is essential for implementing targeted interventions and mitigating the adverse effects of air pollution on public health and the environment. Satellite data, air quality monitoring networks, and atmospheric modeling techniques can aid in mapping pollution sources and understanding pollutant transport pathways [21]. By focusing resources and efforts on areas with the highest pollution levels, policymakers can maximize the effectiveness of pollution control measures and safeguard the health and well-being of communities across the MENA region [14].

Geographical variances further complicate air pollution dynamics in the MENA region, as topographical features, climatic conditions, and human activities influence pollution levels and dispersion mechanisms. Urban centers, industrial zones, and transportation hubs often experience elevated pollution levels due to localized emissions and population density, while coastal cities may face additional pollution from maritime activities and sea salt aerosols. Desert regions are prone to dust storms and sandstorms, which can result in episodic spikes in particulate matter concentrations, while mountainous areas may experience localized pollution due to temperature inversions and restricted air circulation [22].

Identifying pollution hotspots and understanding the sources and spatial distribution of air pollution are essential for developing effective mitigation strategies and safeguarding human health and the environment. Remote sensing technologies, air quality monitoring networks, and atmospheric modeling techniques play a crucial role in mapping pollution sources, tracking pollutant transport pathways, and assessing the impact of air pollution on public health and environmental quality. By leveraging these tools and implementing targeted interventions, policymakers can mitigate the adverse effects of air pollution and promote sustainable development in the MENA region [1].

Hence, air pollution in the MENA region is a complex and multifaceted challenge, driven by a combination of industrial activities, vehicular emissions, and natural phenomena. Understanding the sources, spatial distribution, and impacts of air pollution is essential for developing effective policies and interventions to safeguard public health, protect the environment, and promote sustainable development in the region. By adopting a holistic approach to air quality management and fostering collaboration between governments, industries, and communities, the MENA region can address the challenges posed by air pollution and build a cleaner, healthier, and more sustainable future for all [1].

3 One Health Framework: Understanding Interconnected Impacts

Air pollution in the Middle East and North Africa (MENA) region is a pervasive issue with profound implications for human health, ecosystem integrity, and biodiversity. This section delves into the intricate web of interconnected impacts within the One Health framework, providing a comprehensive examination of how air pollution affects various facets of life in the region.

3.1 Human Health Impacts

3.1.1 Respiratory Diseases

The consequences of air pollution on the respiratory system are some of the most obvious and direct health repercussions. Particulate matter, sulfur dioxide, and nitrogen dioxide are among the airborne pollutants that can irritate the respiratory tract and aggravate pre-existing illnesses like bronchitis, asthma, and chronic obstructive pulmonary disease (COPD) [9, 23, 24].

Particulate matter, and more specifically, fine particles known as $PM_{2.5}$ (particles with diameters of 2.5 μm or less) have the ability to enter the lungs deeply and, over time, cause inflammation and deteriorate lung function. This may result in a higher risk of respiratory infections, such as pneumonia and bronchiolitis, especially in susceptible groups like children, the elderly, and people with underlying respiratory disorders [25, 26].

Research studies conducted across the MENA region have consistently highlighted the detrimental effects of air pollution on respiratory health. For example, a study conducted in urban areas of Cairo, Egypt, found that high levels of coarse particulate matter were associated with an increased prevalence of respiratory symptoms among both children and adults [27]. More specifically, the authors evaluated the impact of coarse particulate matter (PM_{10}) in the air on respiratory hospital admissions in Egypt during the year 2016. Data on daily hospital admissions for chronic bronchitis and COPD were gathered from specialty chest hospitals and correlated with air pollution data. The findings imply that one major predictor of respiratory illness is the amount of PM_{10} in the air. We discover that there is a 2% increase in hospital admission rates for COPD and chronic bronchitis for every

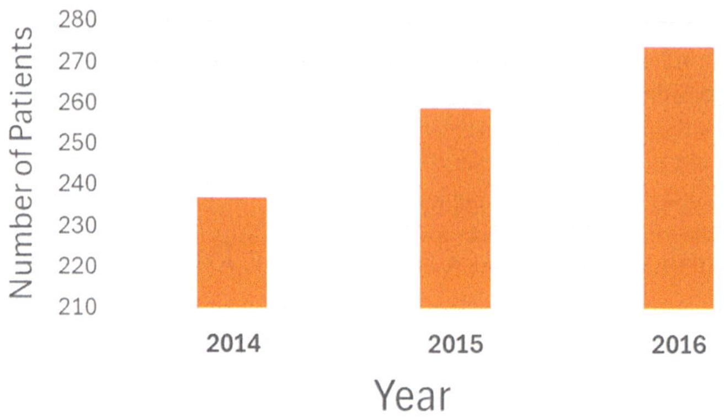

Fig. 2 Number of individuals exposed to PM 10 with cardiovascular disorders, data from [28]

10 μg/m^3 increase in PM_{10} concentration. These more admissions are linked to higher health costs and a decline in Egypt's Gross Domestic Product (GDP) [27].

Moreover, several studies conducted in Iran revealed a significant association between exposure to nitrogen dioxide and hospital admissions for respiratory diseases. Momtazan et al. [28] highlighted the detrimental effects on health caused by dust storms and air pollution in the southwest of Iran. Their study examined the correlation between dust storms and hospital admissions for cardiovascular disorders in Abadan and Khorramshahr, two southwest Iranian cities, between 2014 and 2016. They estimated the number of individuals with cardiovascular disorders in 2014, 2015, and 2016 who were exposed to PM_{10}, and the results showed that these numbers were 237, 259, and 274, respectively, as shown in Fig. 2 [28].

Dastoorpoor et al. found a significant increase in hospital admissions and the proportion of women in the population when they looked at a 10-year data set to examine the relationship between air pollution and cardiovascular hospital admissions in Ahwaz, Iran. This was especially true when looking at ozone exposure. Additionally, there was a noteworthy rise in hospital admissions related to cardiovascular disorders across all age and gender categories in relation to ambient levels of nitrogen dioxide and nitrogen monoxide [29].

Moreover, in a recent study, Isaifan [26] used the World Health Organization approach and software (AirQ+) to assess the effects of ambient $PM_{2.5}$ on human health in terms of overall mortality and the estimated attributable proportion (AP) resulting from prolonged exposure to $PM_{2.5}$ in the nine Middle Eastern nations that have the highest population density – roughly 363 million people as of 2012 [6]. The study reported that, in Saudi Arabia, the annual median $PM_{2.5}$ concentrations in 2012 were 108 μg/m^3, while in Turkey and Syria they were 34 μg/m^3. An estimated 152,925 people are thought to have died overall in the nine most populous Middle Eastern nations as a result of prolonged exposure to fine particulate matter (with half of those deaths occurring in Egypt). Furthermore, the study found that Turkey and Syria had the lowest relative risk (RR), at 1.1553, while Saudi Arabia

Table 1 The relative risk (RR) of total mortality due to long-term exposure to $PM_{2.5}$ in the highest populated ME countries [6]

Country	Lower	Central	Upper
Egypt	1.3847	1.6475	1.9382
Iran	1.1337	1.2122	1.2906
Turkey	1.0987	1.1553	1.2109
Iraq	1.1698	1.2720	1.3756
Saudi Arabia	1.4686	1.8031	2.1845
Yemen	1.1381	1.2195	1.3009
Syria	1.0987	1.1553	1.2109
UAE	1.2310	1.3754	1.5259
Jordan	1.1073	1.1692	1.2303

had the highest, at 1.8031. Saudi Arabia had the greatest AP (central value), at 44.5%, while Turkey and Syria had the lowest, at 13.4% (Table 1) [6].

3.1.2 Cardiovascular Effects

In addition to respiratory diseases, air pollution has significant impacts on cardiovascular health. Exposure to pollutants such as PM, NO_2, and CO can trigger systemic inflammation, oxidative stress, and endothelial dysfunction, contributing to the development and progression of cardiovascular diseases such as hypertension, atherosclerosis, and heart attacks. Moreover, particulate matter has been implicated in cardiovascular disorders, with fine particles capable of entering the bloodstream and causing systemic damage. Long-term exposure to air pollution is also found to be associated with an increased risk of cardiovascular mortality, highlighting the importance of mitigating air pollution to protect public health [6].

Research evidence from the MENA region underscores the cardiovascular effects of air pollution. For example, a study conducted in Jeddah, Saudi Arabia, reported a significant association between the morbidity risk associated with $PM_{2.5}$ exposure for respiratory and cardiovascular disorders [30]. Similarly, research conducted in urban areas of Lebanon found that higher levels of NO_2 were associated with an increased risk of cardiovascular mortality [31].

3.1.3 Brain Damage Effects

According to recent research, $PM_{2.5}$ is also closely linked to brain damage, specifically cerebral vascular injury (stroke) and neurological brain injury (dementia, mental problems, altered cognitive function, etc.). $PM_{2.5}$ can directly enter brain tissue through the olfactory nerve, eventually causing damage to the cerebral blood vessels and brain nerves, or it can travel across the lung-gas-blood barrier and the "gut-microbial-brain" axis to induce systemic oxidative stress and inflammation. It is important to note that brain damage caused by $PM_{2.5}$ has a window of opportunity for self-healing. But our understanding of the precise pathophysiological mechanisms behind brain injury and brain repair is still lacking [32]. The processes of $PM_{2.5}$-induced brain damage and self-healing following injury are gathered and discussed in this article, which may offer fresh perspectives on prevention and treatment [33].

3.1.4 Adverse Impact on Pregnancy and Low Birth Weight

Furthermore, negative pregnancy outcomes like preterm birth, low birth weight, and congenital abnormalities have all been related to air pollution. Elevated levels of air pollution during pregnancy can cause problems for expectant mothers and their unborn children, which can harm their health. Research conducted in the MENA region has revealed links between exposure to air pollution during pregnancy and heightened risks of preterm delivery and low birth weight. These findings underscore the need to safeguard maternal and fetal health against the harmful effects of air pollution [34].

3.2 Ecosystem Health Impacts

3.2.1 Biodiversity Loss

Air pollution not only affects human health but also poses significant threats to ecosystem health and biodiversity in the MENA region. Elevated levels of pollutants can disrupt ecological processes, degrade habitats, and threaten the survival of plant and animal species. Acid deposition resulting from SO_2 and NOx emissions can acidify soil and water bodies, affecting nutrient availability and harming sensitive plant species [18].

Moreover, air pollution can directly damage vegetation through foliar injury, reduce photosynthesis, and alter plant physiology. High concentrations of ozone, for example, can cause leaf necrosis and reduce crop yields, affecting agricultural productivity and food security. Changes in vegetation composition and structure can have cascading effects on wildlife populations, leading to shifts in species distributions and ecosystem dynamics. Furthermore, air pollution can disrupt ecological interactions and biodiversity networks, leading to population declines and ecosystem degradation. Vulnerable species, such as amphibians, pollinators, and migratory birds, are particularly susceptible to the impacts of air pollution, exacerbating global biodiversity loss and conservation challenges [8].

Research conducted in the MENA region has documented the impacts of air pollution on biodiversity and ecosystem health. For example, studies conducted in urban areas of Jordan have shown that air pollution from vehicular emissions and industrial activities can reduce plant diversity and alter community structure in urban parks and green spaces [8]. Oran et al. sampled the leaves of five distinct species from different locations to assess the impact of cement dust in Jordan [8]. The findings show that plant species cultivated close to the cement factory have greater levels of dust accumulations than species that are planted farther away from the facility. Scanning Electron Microscopy analysis reveals a wide range of elements, including calcium, potassium, aluminum, and other elements, deposited on the leaves of all plants examined in the research region. Among the mineral salts, calcium was determined to have the highest concentration [8].

Similarly, research conducted in coastal regions of Egypt has revealed that air pollution can impair coral reef health and decrease fish biodiversity, threatening marine ecosystems and fisheries [35].

Furthermore, air pollution has been linked to declines in pollinator populations, such as bees and butterflies, which play a crucial role in pollinating flowering plants and maintaining ecosystem biodiversity. Pollinator declines can have far-reaching ecological consequences, affecting plant reproduction, food production, and ecosystem stability. Studies in the MENA region have found associations between air pollution exposure and declines in pollinator populations, highlighting the need for conservation efforts to protect pollinator species from the impacts of air pollution [36].

3.2.2 Soil and Water Contamination

Airborne pollutants can also contaminate soil and water bodies, posing risks to ecosystem health and human well-being. Deposition of particulate matter and heavy metals can accumulate in soil, altering soil chemistry and nutrient cycling processes. Contaminated soils may impair plant growth, reduce crop yields, and compromise soil fertility, with implications for agricultural sustainability and land use management [37].

Furthermore, atmospheric deposition of pollutants can contaminate surface water and groundwater resources, affecting aquatic ecosystems and drinking water supplies. Heavy metals, such as lead, mercury, and cadmium, can bioaccumulate in aquatic organisms, posing risks to human health through the consumption of contaminated fish and seafood. Moreover, nutrient enrichment resulting from nitrogen deposition can contribute to eutrophication, algal blooms, and water quality impairments, further exacerbating ecological degradation [38].

Research studies in the MENA region have highlighted the impacts of air pollution on soil and water contamination. For example, investigations conducted in agricultural areas of Morocco have shown that atmospheric deposition of heavy metals, such as lead and cadmium, can accumulate in soils and affect crop quality [39]. Based on the agronomic parameters of the soils analyzed, it can be inferred that the pH and organic matter levels are higher in the soils supporting wheat growth than in those supporting vines. Compared to vines, wheat had a greater total nitrogen rate. Both farm manure contributions and residue-type agricultural soil contributions have the potential to influence this tendency [39]. Similarly, research conducted in urban areas of Kuwait has revealed that air pollution from industrial activities can contaminate groundwater resources and pose risks to human health [40].

3.3 Interconnected Impacts and the One Health Approach

The intricate relationships between air pollution and biodiversity, ecological integrity, and human health highlight the significance of implementing a comprehensive One Health strategy. This approach advocates for integrated measures to enhance sustainability and resilience, acknowledging the interdependence of animal, environmental, and human health. Air pollution's complex interconnections with human health, ecosystem integrity, and biodiversity underscore the importance of adopting a holistic One Health approach. This framework recognizes the interconnectedness of human health, animal health, and environmental health, advocating for integrated strategies to promote sustainability and resilience [41].

A One Health approach facilitates interdisciplinary collaboration, stakeholder engagement, and community participation in addressing air pollution challenges. By fostering partnerships between governments, industries, and communities, policymakers can develop comprehensive mitigation strategies that consider the diverse impacts of air pollution on public health and ecosystem integrity [41].

Public awareness and education play a crucial role in promoting environmental stewardship and sustainability. By empowering individuals and communities to take proactive measures to reduce their environmental footprint, the One Health approach fosters a culture of environmental responsibility and resilience [42].

Hence, air pollution in the MENA region poses significant challenges to human health, ecosystem integrity, and biodiversity. By adopting a holistic One Health approach, policymakers can develop comprehensive strategies to mitigate air pollution's impacts and promote sustainability. Through collaborative efforts and community engagement, we can build a healthier and more resilient future for all.

4 Policy and Regulatory Approaches

Addressing air pollution in the Middle East and North Africa (MENA) region necessitates comprehensive policy and regulatory approaches. This section explores legislative frameworks, emission control policies, technological innovations, policy implementation challenges, and future directions to achieve sustainable development goals and safeguard public health and the environment.

4.1 Legislative Frameworks

Legislative frameworks serve as the foundation for combating air pollution by providing a legal framework for monitoring, regulating, and enforcing pollution control measures. In the MENA region, countries have developed national air quality standards based on international guidelines and scientific evidence to protect public health and the environment [14]. These standards establish permissible limits for pollutants such as particulate matter (PM_{10} and $PM_{2.5}$), nitrogen dioxide, sulfur dioxide, ozone, and carbon monoxide.

4.1.1 Air Quality Standards

Robust air quality standards are essential for ensuring that pollution levels remain within acceptable limits and do not pose risks to human health and ecosystems. MENA countries have adopted various sets of air quality standards, often tailored to their specific environmental conditions and pollution sources. For instance, Qatar has established ambient air quality standards that set limits for pollutants such as sulfur dioxide, nitrogen dioxide, and particulate matter, based on recommendations from international organizations like the World Health Organization (WHO) and the United States Environmental Protection Agency (EPA) [43, 44]. These standards serve as benchmarks for assessing air quality, guiding policy decisions, and prioritizing pollution control measures. Compliance with air quality standards is monitored through a network of strategically located air quality monitoring stations. Real-time data from these stations inform policymakers and the public about pollution levels, enabling timely interventions and public health advisories during episodes of poor air quality. By aligning national standards with international best practices, countries enhance regional cooperation and coordination in addressing transboundary air pollution issues [13].

4.1.2 Emission Control Policies

In addition to setting air quality standards, MENA countries have implemented emission control policies aimed at reducing pollution from key sources such as industries, transportation, and energy production. These policies encompass regulatory measures, fiscal incentives, and voluntary agreements designed to incentivize industries to adopt cleaner technologies and practices. For example, countries like the United Arab Emirates (UAE) have introduced emission standards and regulations for industrial facilities, mandating the installation of pollution control equipment and adherence to strict emission limits [12]. Similarly, regulations governing vehicle emissions and fuel quality standards have been implemented to reduce pollution from transportation sources. Incentive-based approaches such as emissions trading schemes and eco-labeling programs have also been explored to encourage businesses to reduce their environmental footprint.

Furthermore, multilateral agreements such as the Paris Agreement and the Kyoto Protocol [45] provide a framework for international cooperation on climate change mitigation and emissions reduction [46, 47]. MENA countries have pledged to contribute to global efforts to reduce greenhouse gas emissions and transition to low-carbon economies, aligning their emission control policies with international commitments.

5 Technological Innovations

Technological innovations play a pivotal role in addressing air pollution by offering cleaner and more sustainable alternatives to conventional energy sources and transportation systems. In the MENA region, advancements in clean energy technologies and sustainable transportation solutions are crucial for reducing emissions and promoting environmental sustainability.

5.1 Clean Energy Technologies

Transitioning to clean energy technologies is essential for reducing air pollution and mitigating climate change. MENA countries, endowed with abundant renewable energy resources such as solar and wind, have made significant investments in renewable energy infrastructure and technology development. Solar power projects, in particular, have gained traction in the region, with countries like Saudi Arabia and the UAE leading the way in solar energy deployment [48, 49].

Advancements in solar photovoltaic (PV) technology and concentrated solar power (CSP) systems have made solar energy increasingly cost-effective and scalable, offering a viable alternative to fossil fuels. Large-scale solar projects such as solar parks and concentrated solar power plants have been commissioned across the MENA region, contributing to the diversification of the energy mix and reducing dependence on conventional power sources.

Moreover, investments in energy efficiency measures and smart grid technologies are essential for optimizing energy consumption and reducing emissions from power generation and consumption. Smart meters, demand response programs, and energy-efficient appliances are being deployed to enhance energy efficiency and promote sustainable consumption patterns.

5.2 Sustainable Transportation Solutions

Transportation is a major source of air pollution in urban areas, emitting pollutants such as nitrogen oxides, particulate matter, and volatile organic compounds [4]. MENA countries are exploring sustainable transportation solutions to reduce emissions and alleviate congestion in cities. Electric vehicles (EVs), hybrid vehicles, and public transportation systems are being promoted as alternatives to traditional gasoline and diesel vehicles [50].

The adoption of electric vehicles is gaining momentum, supported by government incentives, subsidies, and infrastructure development. Charging infrastructure networks are expanding to accommodate the growing fleet of electric vehicles, with charging stations installed in urban centers, residential areas, and along major transportation corridors [4].

Moreover, investments in public transportation infrastructure such as metro systems, bus rapid transit networks, and cycling lanes aim to reduce reliance on private vehicles and promote sustainable mobility options. Integrated urban planning and land use policies are essential for creating pedestrian-friendly cities and reducing car dependency [4].

Additionally, promoting non-motorized transportation modes such as walking and cycling can improve air quality, public health, and urban livability. Investments in pedestrian infrastructure, bike-sharing programs, and green corridors encourage active transportation and reduce emissions from motorized vehicles [51].

6 Policy Implementation and Challenges

While significant progress has been made in developing and implementing policy and regulatory approaches to address air pollution in the MENA region, several challenges persist which are summarized in the following sub-sections.

6.1 Capacity Building and Institutional Strengthening

Many countries in the MENA region face challenges in building institutional capacity and strengthening regulatory frameworks for effective air pollution management. Technical expertise, financial resources, and institutional coordination are essential for implementing and enforcing pollution control measures.

6.2 Data Availability and Monitoring

Limited availability of reliable air quality data and monitoring infrastructure poses challenges for assessing pollution levels, tracking progress, and informing policy decisions. Investing in robust monitoring networks and data management systems is crucial for evidence-based policymaking and public awareness.

6.3 Technological Barriers

Despite advancements in clean energy technologies, barriers such as high upfront costs, technological complexity, and grid integration challenges hinder widespread adoption and deployment. Supporting research and development, promoting innovative ecosystems, and providing financial incentives can overcome these barriers and accelerate the transition to clean energy solutions.

6.4 Policy Coherence and Coordination

Ensuring policy coherence and coordination across sectors and levels of government is essential for addressing air pollution comprehensively. Fragmentation, overlapping mandates, and competing priorities can hinder the effective implementation and coordination of pollution control measures.

6.5 Behavioral Change and Public Engagement

Encouraging behavioral change and fostering public engagement is critical for achieving sustainable solutions to air pollution. Raising awareness, promoting sustainable lifestyles, and fostering community participation empower individuals and communities to take proactive measures and support policy implementation efforts.

6.6 Enforcement Issues

Enforcing air quality regulations and emission control measures presents significant challenges in the MENA region. Weak enforcement mechanisms, limited regulatory capacity, and inadequate monitoring infrastructure contribute to gaps in compliance, undermining the effectiveness of pollution control efforts.

6.6.1 Current State of Regulations

The Ministry of the Environment (MoE) is typically in charge of creating air quality standards and putting solutions in place to reduce air pollution in each MEA nation. A detailed understanding of local dose–effect interactions is crucial for achieving targeted PM levels and safeguarding public health. As a result, national risk management and environmental policies depend on air quality regulations. Nevertheless, there are currently no local PM air quality guidelines available in certain MENA nations, like Kuwait and Oman. Only a small number of MENA nations, including Jordan and Saudi Arabia, have unique laws [11]. Tight air quality rules are taken from well-established laws elsewhere in the world into the nations without local PM norms. For ambient air quality, for instance, the Ministry of Environment and Climate Affairs (MECA) in Oman suggests using the U.S. Environmental Protection Agency (USEPA) guidelines (HMR 2010). Kuwait's air quality regulations are also based on WHO $PM_{2.5}$ norms and the US Clean Air Act for PM_{10}. Every two years, these limitations are examined to make sure they still meet the requirements necessary to protect the public's health and welfare. But occasionally, because these standards are frequently established following the practical norms used in highly industrialized nations, such adoption could be prone to a vulnerability [11].

Similarly, Kuwait's air quality regulations are derived from the WHO's $PM_{2.5}$ guidelines and the US Clean Air Act's PM_{10} regulations. Every two years, these limitations are examined to make sure they still meet the requirements necessary to protect the public's health and welfare. Unfortunately, there is occasionally a risk that such an adoption will be vulnerable because these criteria are frequently established per the applicable norms used in rich industrialized nations and may not accurately reflect the unique environmental conditions. Because of this, before directly adopting the rules, regulatory agencies should carefully assess their unique local situation [11].

6.6.2 Regulatory Compliance

Achieving regulatory compliance among industries, transportation sectors, and other pollution sources is a primary challenge. Despite the existence of air quality standards and emission limits, some facilities may flout regulations due to lax enforcement, inadequate penalties, or limited oversight. Non-compliance can result in elevated pollution levels, endangering public health and the environment.

Improving regulatory compliance necessitates strengthening enforcement mechanisms, enhancing monitoring capabilities, and imposing stricter penalties for violations. Investing in surveillance technologies, remote sensing, and real-time monitoring systems can bolster regulatory oversight and facilitate timely enforcement actions against polluting entities. Additionally, streamlining permit processes, conducting regular inspections, and imposing fines for non-compliance can incentivize businesses to adhere to environmental regulations.

6.6.3 Legal and Institutional Challenges

Navigating complex legal frameworks, bureaucratic processes, and jurisdictional ambiguities poses challenges for effective enforcement of air quality regulations. Overlapping mandates, fragmented responsibilities, and institutional barriers may hinder coordination and collaboration among government agencies responsible for environmental management.

Addressing these challenges requires streamlining regulatory frameworks, enhancing institutional coordination, and strengthening the capacity of regulatory agencies. Establishing clear guidelines, standard operating procedures, and interagency coordination mechanisms can facilitate seamless enforcement and compliance monitoring. Furthermore, investing in training programs, capacity-building initiatives, and technical assistance can empower regulatory authorities to effectively enforce air quality regulations.

In some instances, political interference and corruption may undermine enforcement efforts, leading to inconsistent application of regulations and lenient penalties for violators. Addressing these issues requires promoting transparency, accountability, and integrity in regulatory processes, as well as strengthening anti-corruption measures to ensure that enforcement actions are impartial and effective.

7 Future Directions and Opportunities

Despite these challenges, significant opportunities exist to strengthen policy and regulatory approaches to address air pollution in the MENA region which are listed briefly below.

7.1 Investing in Clean Energy Infrastructure

Expanding investments in renewable energy infrastructure, energy efficiency measures, and clean transportation technologies can reduce emissions, improve air quality, and promote economic growth and job creation.

7.2 Enhancing Regional Cooperation

Strengthening regional cooperation and collaboration on air quality management can facilitate knowledge exchange, capacity building, and joint action to address transboundary air pollution issues.

7.3 Promoting Innovation and Technology Transfer

Supporting research and development, technology transfer, and innovation partnerships can accelerate the adoption and deployment of clean energy technologies and sustainable transportation solutions.

7.4 Strengthening Governance and Regulatory Frameworks

Enhancing governance structures, strengthening regulatory frameworks, and improving enforcement mechanisms are essential for ensuring effective implementation and enforcement of pollution control measures.

7.5 Engaging Stakeholders and Fostering Partnerships

Engaging stakeholders, fostering partnerships, and promoting multi-stakeholder collaboration can mobilize resources, build political support, and catalyze action on air pollution mitigation efforts.

Hence, it can be seen that addressing air pollution in the MENA region requires a multifaceted approach encompassing legislative frameworks, emission control policies, technological innovations, and effective policy implementation. By overcoming challenges and seizing opportunities, countries in the MENA region can achieve sustainable development goals, improve public health, and protect the environment for current and future generations.

8 Resource Constraints

Resource constraints, including limited funding, technical expertise, and infrastructure, present significant barriers to implementing pollution control measures in the MENA region. Insufficient financial resources, competing priorities, and economic challenges may impede investments in clean technologies, pollution monitoring, and enforcement efforts.

8.1 Financial Challenges

A primary obstacle to addressing air pollution is the lack of adequate funding for pollution control initiatives and environmental management. Limited budget allocations, fiscal constraints, and competing demands for public expenditure may hinder investments in clean energy infrastructure, emission reduction projects, and pollution monitoring systems [52].

To overcome financial challenges, governments can explore innovative financing mechanisms, such as public–private partnerships, green bonds, and carbon pricing mechanisms. Mobilizing private sector investments, leveraging international funding sources, and prioritizing environmental spending can help bridge the funding gap and accelerate the implementation of pollution control measures [52].

International donors and development agencies can also play a crucial role in supporting pollution control efforts through grants, concessional loans, and technical assistance programs. By providing financial resources and capacity-building support, these organizations can complement domestic efforts and help countries in the MENA region overcome resource constraints to address air pollution effectively [52].

8.2 *Technological and Infrastructure Constraints*

In addition to financial constraints, technological and infrastructure limitations pose challenges to implementing pollution control measures. Outdated infrastructure, inadequate monitoring equipment, and limited access to clean technologies may hinder efforts to monitor, mitigate, and regulate air pollution effectively [53].

Investing in technological upgrades, modernizing monitoring infrastructure, and promoting technology transfer can address these constraints and enhance pollution control capabilities. Adopting innovative solutions, such as satellite imagery, remote sensing technologies, and mobile monitoring units, can expand monitoring coverage and provide real-time data on pollution levels. Moreover, investing in clean energy infrastructure, renewable energy projects, and energy efficiency initiatives can reduce emissions and improve air quality in the long term [54].

Public–private partnerships can also play a crucial role in overcoming technological and infrastructure constraints by leveraging private sector expertise, resources, and innovation. Collaborative initiatives between government agencies, research institutions, and private companies can facilitate the development and deployment of cutting-edge technologies for pollution monitoring, emissions reduction, and clean energy generation [55].

9 Transboundary Cooperation

Air pollution transcends geopolitical borders, necessitating cooperation and collaboration among neighboring countries to address shared pollution challenges. Transboundary air pollution, resulting from cross-border transport of pollutants, industrial activities, and natural sources, poses complex environmental, health, and diplomatic issues that require coordinated action and mutual cooperation.

9.1 *Regional Air Quality Management*

Regional cooperation mechanisms play a crucial role in facilitating dialogue, information exchange, and joint action on air quality management issues. These platforms provide a forum for sharing best practices, harmonizing standards, and developing collaborative initiatives to address transboundary air pollution challenges [56].

Enhancing regional cooperation requires strengthening institutional mechanisms, fostering trust and mutual understanding, and promoting transparency and data sharing among participating countries. Establishing regional air quality monitoring networks, conducting joint research projects, and coordinating emergency response efforts can enhance preparedness and resilience to transboundary pollution events.

At the same time, addressing transboundary air pollution requires diplomatic engagement and political commitment to overcome barriers such as sovereignty concerns, political tensions, and conflicting interests. By promoting dialogue, building consensus, and fostering goodwill among neighboring countries, regional cooperation mechanisms can facilitate joint action to address shared pollution [57] challenges and promote sustainable development in the MENA region [20].

9.2 Diplomatic and Political Challenges

Despite the importance of transboundary cooperation, diplomatic tensions, political disputes, and sovereignty concerns may hinder collaboration on air pollution issues among neighboring countries. Disparities in regulatory frameworks, enforcement capacities, and policy priorities may further complicate efforts to harmonize standards and coordinate actions to address shared pollution challenges.

Overcoming diplomatic and political challenges requires diplomatic engagement, conflict resolution mechanisms, and confidence-building measures to foster cooperation and trust among neighboring countries. Building consensus on common environmental goals, promoting dialogue and negotiation, and leveraging international mediation mechanisms can facilitate cooperation and collaboration on transboundary air pollution issues.

International organizations and multilateral initiatives can also play a crucial role in facilitating diplomatic and political dialogue on air pollution issues and promoting regional cooperation among countries in the MENA region. By providing a neutral platform for discussion, technical assistance, and capacity-building support, these organizations can help overcome diplomatic obstacles and foster collaboration on shared environmental challenges [58].

10 Future Directions and Opportunities

Despite these challenges, there are significant opportunities to strengthen policy and regulatory approaches to address air pollution in the MENA region. These include:

- Strengthening regulatory enforcement and compliance monitoring mechanisms to ensure adherence to air quality standards and emission limits.
- Mobilizing financial resources, promoting innovative financing mechanisms, and leveraging international funding to support pollution control initiatives.
- Investing in technological innovation, upgrading monitoring infrastructure, and adopting clean energy technologies to reduce emissions and improve air quality.
- Enhancing regional cooperation, fostering diplomatic engagement, and promoting dialogue among neighboring countries to address transboundary air pollution challenges.

- Raising public awareness, fostering community engagement, and empowering stakeholders to advocate for clean air policies and sustainable development practices.

By addressing enforcement issues, overcoming resource constraints, and promoting transboundary cooperation, countries in the MENA region can achieve significant progress in combating air pollution, protecting public health, and preserving the environment for future generations.

11 Case Studies: Exemplary Initiatives

In this section, we delve into three case studies exemplifying outstanding initiatives from different countries in the MENA region aimed at tackling air pollution. These case studies illuminate innovative approaches to combatting pollution through regional collaboration, technological advancements, and community engagement.

11.1 Legislative and Regulatory Efforts

The majority of Arab states have enacted environmental protection laws that comply with international WHO and EPA rules. However, according to [59], over half of MENA nations do not adhere to WHO criteria. This suggests that to guarantee adherence to the established requirements, capabilities must be advanced. The majority of Arab states have already enacted environmental protection laws. As guidelines or standards, certain air quality restrictions have also been established in accordance with worldwide WHO or EPA standards [43, 60]. Nevertheless, several Arab nations have set up monitoring networks to keep an eye on and gather data on the state of the air in large cities and urban areas [61].

A few MENA nations have made initiatives to prevent air pollution [62]. Algeria, Jordan, and Syria provide incentives to boost investment in energy efficiency, clean technologies, and renewable energy since energy production is the primary cause of air pollution [59].

In Bahrain, the Bahrain National Oil Company (BANOCO) and the Directorate of Traffic and Licensing are working together to execute an Action Plan aimed at lowering vehicle emissions and introducing unleaded gasoline. In 1994, Bahrain launched a program named "Fume watch" to report cars that were producing smoke. There has been a discernible improvement as a result (League of Arab State (Al-Raw.

According to a recent study conducted in Egypt on behalf of the Egyptian Environmental Affairs Agency (EEAA), the lowest bound of the annual cost of air pollution in Cairo is estimated to be between US$ 1–2 billion, or between 3–6% of GDP. As a result, the Egyptian government has acknowledged that compressed natural gas (CNG) will have positive effects on the environment, society, and economy. It has also endorsed and promoted CNG as the fuel of choice for vehicles. Benefits to the environment can be seen in the fact that CNG vehicle exhaust normally contains 85% less pollutants than gasoline-powered vehicle exhaust. Reductions in health issues linked to pollution follow from improved air quality. The cost is less than half that of gasoline, at 0.45 pounds per cubic meter, which is an economic advantage [61].

On the other hand, in the Arab world, air pollution, toxic air and water emissions, and organic waste are largely caused by industrial activity and the burning of fossil fuels. Industrial pollution is the main cause of harmful pollution and human and industrial waste accounts for the majority of organic waste released into waterways. The Arab countries exhibit varying levels of toxic severity. Chemical, refining, and occasionally metallurgical industries are the most polluting sectors when it comes to harmful compounds. This is putting pressure on the Arab world to implement additional programs for pollution prevention and remediation [61].

In Syria for instance, the government oversees the majority of industrial activity (heavy industries) while the private sector manages a smaller portion of it (craft and medium-sized enterprises). The craft industries are scattered across the urban areas. It makes use of antiquated, extremely polluting technology (textiles, tanneries). An inherent feature of all industrial activity is the lack of resources dedicated to pollution control. The facilities that process industrial waste are not well maintained or managed. In general, the production processes are incompatible with the emission control equipment [61].

11.2 Technological Innovations

Embraced technological innovations as a cornerstone of its strategy to combat air pollution and advance sustainable development is a vital action for each country in the MENA region. Through strategic investments in clean energy technologies, smart infrastructure, and digital solutions, these countries can pioneer innovative approaches to reduce emissions and enhance air quality.

Saudi Arabia has implemented a comprehensive suite of technological innovations to address pollution from key sources such as energy production, transportation, and industry. This includes widespread adoption of renewable energy technologies, deployment of electric vehicles (EVs), and implementation of smart city solutions to promote environmental sustainability. One of the largest green hydrogen factories in the world is being constructed in Saudi Arabia. It is expected to be operational by 2025 and run on 4 gigawatts of solar and wind energy. It is anticipated that the NEOM project's plant will produce 650 tons of green hydrogen daily. Moreover, in the energy sector, Saudi Arabia has made significant investments in solar and wind power projects, leveraging its abundant renewable energy resources to diversify the energy mix and reduce reliance on fossil fuels. By promoting the transition to clean energy sources, the kingdom has succeeded in reducing emissions from power generation and improving air quality in urban areas. Wind farms are being constructed by the Kingdom in Yanbu, Waad Al-Shamal, and Al-Ghat. The first wind farm in Saudi Arabia and the largest in the Middle East, Dumat Al-Jandal, started generating 400 MW of carbon-free energy in August 2021 [63].

In the transportation sector, in the United Arab Emirates (UAE), the number of electric vehicles (EVs) has increased and is projected to represent 1.3% of all passenger cars by 2022 [64]. UAE has introduced incentives for the adoption of electric vehicles and invested in charging infrastructure to support the growing fleet of EVs. By encouraging zero-emission transportation options, the UAE has successfully reduced emissions from vehicles and mitigated air pollution in densely populated urban areas.

Furthermore, some countries have implemented smart city solutions such as intelligent traffic management systems, remote sensing technologies, and air quality monitoring networks to optimize urban mobility and minimize pollution hotspots. These digital solutions enable real-time monitoring of pollution levels, facilitate data-driven decision-making, and empower citizens to take proactive measures to reduce their environmental footprint.

Such technological innovations have yielded significant environmental, economic, and social benefits. By transitioning to clean energy sources and promoting electric vehicles, these countries have reduced emissions of greenhouse gases and criteria air pollutants, leading to improvements in air quality and public health outcomes.

Moreover, investments in smart infrastructure and digital solutions have enhanced urban livability, promoted sustainable transportation options, and facilitated the transition to low-carbon, resilient cities. By embracing technological innovations, MENA countries can position themselves as global leaders in sustainable development and set a precedent for other countries seeking to address air pollution through technological solutions.

11.3 Community Engagement

The adoption of a community-centered approach to address air pollution, recognizing the importance of grassroots participation and community empowerment in driving environmental change has proved to play a great role in involving citizens in the right action toward the environment in several parts of the world.

In the MENA and Arab country region, initiatives that promote community engagement and civic participation to take proactive measures to combat air pollution and promote sustainable lifestyles are still under development. Several countries in the MENA region have implemented a range of community engagement initiatives aimed at raising awareness, building capacity, and fostering partnerships to address air pollution at the grassroots level. These initiatives include environmental education programs, citizen science projects, and community-led pollution monitoring efforts.

Through environmental education programs in schools, universities, and community centers, these countries can raise awareness about the causes and consequences of air pollution, empowering citizens to take informed action to reduce their environmental footprint. By promoting environmental literacy and behavior change, these programs have instilled a sense of environmental stewardship among future generations.

In Egypt, influential allies have been engaged in several protests such as "Egyptian against coal" [65]. Moreover, the facilitation of citizen science projects engages local communities in monitoring air quality, collecting data, and identifying pollution hotspots. By providing citizens with access to low-cost air quality sensors and training on data collection methods, several countries have empowered communities to generate local air quality data and advocate for pollution control measures.

A recent study discovered that during the past ten years, programs about the environment and sustainable development have rapidly expanded in Arab universities. The 57 colleges that were evaluated provided 221 degree programs related to environmental issues. Most postgraduate courses focus on research. The countries that produce the most environmental research include Saudi Arabia and Egypt. The report observed specific weaknesses in the areas of environmental law and green economy and emphasized that Arab universities can further accelerate their commitment to sustainability by integrating all of the SDGs into their research and community projects, learning and teaching activities, and research [66].

The most popular environmental subjects in Arab schools, based on the assessment of textbooks and curriculum, are ecosystems, natural resources, pollution, and sustainable development. The poll revealed that 40% of the countries did not have a proper discussion on climate change, in contrast to the inclusion of the relatively new concept of sustainable development. Ironically, several nations, like Egypt and Lebanon, which had the most extensive initiatives to incorporate the environment into school curricula, implemented them slowly. Offering an introduction course on environment and sustainability to all new university students is one of the report's key recommendations [66].

Additionally, it was found that forged partnerships with civil society organizations, non-governmental organizations (NGOs), and community groups mobilize resources, leverage expertise, and amplify the impact of community-led initiatives. By fostering collaboration and cooperation among diverse stakeholders, community engagement initiatives can have a transformative impact on pollution control efforts and environmental sustainability.

The community engagement initiatives in Morocco, for example, have also contributed to social cohesion, community resilience, and environmental justice by empowering marginalized communities, raising awareness about environmental inequalities, and advocating for inclusive and equitable pollution control measures.

Hence, the case studies presented in this section highlight the diverse approaches adopted by countries in the MENA region to address air pollution. From regional collaboration and technological innovations to community engagement and grassroots initiatives, these exemplary initiatives showcase the potential for transformative change and sustainable development.

By learning from these case studies, policymakers, stakeholders, and communities can identify strategies and best practices to replicate and scale up efforts to combat air pollution and promote environmental sustainability. Through collaboration, innovation, and community engagement, countries in the MENA region can achieve significant progress in addressing air pollution, protecting public health, and preserving the environment for future generations.

12 Conclusions

In our exploration of air pollution mitigation efforts in the Middle East and North Africa (MENA) region, several key findings have emerged. From an overview of air pollution challenges to exemplary initiatives and recommendations for future action, this section provides a comprehensive summary of our research.

The MENA region faces unique challenges related to air pollution due to a combination of factors such as rapid urbanization, industrial growth, and increasing vehicular emissions. These factors contribute to heightened levels of pollutants such as particulate matter and other gaseous pollutants. Urban centers, industrial zones, and transportation hubs are particularly affected, leading to adverse impacts on human health, ecosystems, and economic development.

Our analysis underscores the importance of adopting a One Health framework to understand the interconnected impacts of air pollution. While the focus is often on human health effects, air pollution also poses significant risks to ecosystems and animal populations. Respiratory diseases, cardiovascular effects, biodiversity loss, soil, and water contamination are among the key impacts that highlight the complex web of interactions between environmental, human, and animal health.

Effective policy and regulatory frameworks are crucial for addressing air pollution in the MENA region. Legislative measures such as air quality standards and emission control policies provide a foundation for regulating pollution sources and setting targets for pollution reduction. However, effective implementation and enforcement of these regulations remain key challenges. Technological innovations, including clean energy technologies and sustainable transportation solutions, play a vital role in reducing emissions and improving air quality.

Despite progress in policy development, several challenges persist in the implementation of pollution control measures. Enforcement issues, including inadequate monitoring and enforcement capacities, weak penalties for non-compliance, and bureaucratic hurdles, hinder the effective implementation of regulations. Resource constraints, such as limited funding, technical expertise, and infrastructure, pose additional barriers to pollution control efforts. Transboundary pollution presents a complex challenge that requires coordinated action and cooperation among neighboring countries.

Moving forward, several key recommendations can guide efforts to address air pollution in the MENA region:

- *Strengthening regulatory enforcement and compliance monitoring mechanisms*: Ensuring adherence to air quality standards and emission limits is crucial. Strengthening enforcement mechanisms and investing in monitoring technologies can enhance compliance.
- *Mobilizing financial resources*: Innovative financing mechanisms and international funding can support pollution control initiatives. Public–private partnerships and green bonds can help bridge funding gaps.
- *Investing in technological innovation*: Upgrading monitoring infrastructure and adopting clean energy technologies are essential. Research and development in pollution control technologies should be prioritized.
- *Enhancing regional cooperation*: Collaborating with neighboring countries is vital for addressing transboundary air pollution. Sharing best practices, data, and resources can improve pollution management efforts.
- *Raising public awareness*: Educating communities about the impacts of air pollution and promoting sustainable behaviors are essential. Public engagement and advocacy can drive support for clean air policies and initiatives.

By implementing these recommendations, countries in the MENA region can achieve significant progress in combating air pollution, protecting public health, and preserving the environment for future generations.

In conclusions, the MENA region faces significant challenges related to air pollution, driven by rapid urbanization, industrial growth, and increasing vehicular emissions. Adopting a One Health framework is essential for understanding the interconnected impacts of air pollution on environmental, human, and animal health.

Moreover, effective policy and regulatory frameworks, technological innovations, and regional collaboration are crucial for addressing air pollution and promoting sustainable development.

However, challenges in implementing pollution control measures include enforcement issues, resource constraints, and the need for transboundary cooperation. Future efforts should focus on strengthening regulatory enforcement, mobilizing financial resources, and promoting technological innovation to achieve sustainable development goals and improve environmental quality in the MENA region.

References

1. Al-Rawashdeh R, Jaradat AQ, Al-Shboul M (2015) Air pollution and economic growth in MENA countries: testing EKC hypothesis. Environ Res Eng Manag 70(4). https://doi.org/10.5755/j01.erem.70.4.7743
2. Kanakidou M, Mihalopoulos N, Kindap T, Im U, Vrekoussis M, Gerasopoulos E, Dermitzaki E, Unal A, Koçak M, Markakis K, Melas D, Kouvarakis G, Youssef AF, Richter A, Hatzianastassiou N, Hilboll A, Ebojie F, Wittrock F, von Savigny C et al (2011) Megacities as hot spots of air pollution in the East Mediterranean. Atmos Environ 45(6):1223–1235. https://doi.org/10.1016/j.atmosenv.2010.11.048

3. Abu-Allaban M, Abu-Qdais H (2011) Impact assessment of ambient air quality by cement industry: a case study in Jordan. Aerosol Air Qual Res 11(7):802–810. https://doi.org/10.4209/aaqr.2011.07.0090
4. Al-Thani H, Koç M, Fountoukis C, Isaifan RJ (2020) Evaluation of particulate matter emissions from non-passenger diesel vehicles in Qatar. J Air Waste Manage Assoc 70(2):228–242. https://doi.org/10.1080/10962247.2019.1704939
5. Roshan DR, Koc M, Isaifan R, Shahid MZ, Fountoukis C (2019) Aerosol optical thickness over large urban environments of the Arabian Peninsula-speciation, variability, and distributions. Atmosphere 10(5):228. https://doi.org/10.3390/atmos10050228
6. Isaifan RJ (2023) Air pollution burden of disease over highly populated states in the Middle East. Front Public Health 10:2296–2565
7. Rabl A, Curtiss P, Pons P (1998) Air pollution and buildings: estimation of damage costs in France. Environ Impact Assess Rev 19(4):361–385
8. Oran S, Abu Zahra H (2014) Impact of the cement dust emitted from the south cement factory in Tafila/Jordan on plant diversity of the surrounding area. Int J Biodivers Conserv 6(5):400–414. https://doi.org/10.5897/ijbc2014.0694
9. Karimzadegan H, Rahmatian M, Farhud D, Yunesian M (2008) Economic valuation of air pollution health impacts in the Tehran Area, Iran. Iran J Public Health 37(1):20–30
10. Rodrigue JP (2004) Straits, passages and chokepoints. A maritime geostrategy of petroleum distribution. Cahiers de Geographie Du Quebec 48(135):357–374. https://doi.org/10.7202/011797ar
11. Tsiouri V, Kakosimos KE, Kumar P (2015) Concentrations, sources and exposure risks associated with particulate matter in the Middle East area – a review. Air Qual Atmos Health 8(1):67–80. https://doi.org/10.1007/s11869-014-0277-4
12. Samara F, Ibrahim S, Yousuf ME, Armour R (2022) Carbon footprint at a United Arab Emirates University: GHG protocol. Sustainability (Switzerland) 14(5):2522. https://doi.org/10.3390/su14052522
13. Gelil IA, Dean V (2010) Arab climate resilience initiative climate change: economic challenges and opportunities in the Arab regions. United Nations Development Programme Regional Bureau for Arab States
14. Abou-Ali H (2023) Environment and regulations in MENA. In: Creating incentives for more effective wastewater reuse in the Middle East and North Africa. Edward Elgar Publishing, pp 1–14. https://doi.org/10.4337/9781035305780.00007
15. Sakhel SR, Geissen S-U, Vogelpohl A (2013) Hydrology and earth system sciences discussions virtual industrial water usage and wastewater generation in the Middle East/North African region. Hydrol Earth Syst Sci Discuss 10:999–1039. https://doi.org/10.5194/hessd-10-999-2013
16. Al-Bassam E, Khan A (2004) Air pollution and road traffic in Kuwait. In: Brebbia CA, Wadhwa LC (eds) Urban transport. WIT Press, Southampton, pp 741–751
17. Dill J (2001) Older vehicles and air pollution insights from the 1995 NPTS. National Academies, pp 491–514
18. Al-Thani H, Koç M, Isaifan RJ (2018) A review on the direct effect of particulate atmospheric pollution on materials and its mitigation for sustainable cities and societies. Environ Sci Pollut Res 25(August):27839–27857. https://doi.org/10.1007/s11356-018-2952-8
19. Aïssa B, Isaifan RJ, Madhavan VE, Abdallah AA (2016) Structural and physical properties of the dust particles in Qatar and their influence on the PV panel performance. Sci Rep 6:31467. https://doi.org/10.1038/srep31467
20. Byrne A (2017) Trouble in the air: recent developments under the 1979 convention on long-range transboundary air pollution. Rev Eur Comp Int Environ Law 26(3):210–219. https://doi.org/10.1111/reel.12219
21. Roshan DR, Koc M, Isaifan R, Shahid MZ, Fountoukis C (2019) Aerosol optical thickness over large urban environments of the Arabian Peninsula – speciation, variability, and distributions. Atmosphere 10:228–249

22. Al-Dabbas M, Abbas M, Al-Khafaji R (2012) Dust storms loads analysis-Iraq. Arab J Geosci 5: 121–131
23. De Marco A, Proietti C, Anav A, Ciancarella L, D'Elia I, Fares S, Fornasier MF, Fusaro L, Gualtieri M, Manes F, Marchetto A, Mircea M, Paoletti E, Piersanti A, Rogora M, Salvati L, Salvatori E, Screpanti A, Vialetto G et al (2019) Impacts of air pollution on human and ecosystem health, and implications for the National Emission Ceilings Directive: insights from Italy. Environ Int 125(February):320–333. https://doi.org/10.1016/j.envint.2019.01.064
24. Li M, Mallat L (2018) Health impacts of air pollution. The Art & Science of Risk, 42
25. Amoatey P, Sicard P, De Marco A, Khaniabadi YO (2019) Long-term exposure to ambient PM2.5 and health impacts in megacity of Rome, Italy. Clin Epidemiol Glob Health 8:531. https://doi.org/10.1016/j.cegh.2019.11.009
26. Isaifan D, Al-Nesf M, Soubra L, Mobayed H, Alfarra MR, Crovella S (2023) The rate of exacerbations in patients with asthma in Qatar: a retrospective study during 2019-2021. Qatar Med J 2023(2):22. https://doi.org/10.5339/qmj.2023.sqac.22
27. Heger M, Zens G, Meisner C (2019) Particulate matter ambient air pollution and respiratory disease in Egypt. www.worldbank.org
28. Momtazan M, Geravandi S, Rastegarimehr B, Valipour A, Ranjbarzadeh A, Yari AR, Dobaradaran S, Bostan H, Farhadi M, Darabi F, Khaniabadi YO (2018) An investigation of particulate matter and relevant cardiovascular risks in Abadan and Khorramshahr in 2014–2016. Toxin Rev 38:290
29. Dastoorpoor M, Sekhavatpour Z, Masoumi K, Mohammadi MJ, Aghababaeian H, Khanjani N, Hashemzadeh B, Vahedian M (2019) Air pollution and hospital admissions for cardiovascular diseases in Ahvaz, Iran. Sci Total Environ 652:1318–1330
30. Nayebare SR, Aburizaiza OS, Siddique A, Carpenter DO, Arden Pope C, Mirza HM, Zeb J, Aburiziza AJ, Khwaja HA (2019) Fine particles exposure and cardiopulmonary morbidity in Jeddah: a time-series analysis. Sci Total Environ 647:1314–1322. https://doi.org/10.1016/j.scitotenv.2018.08.094
31. Nasser Z, Salameh P, Dakik H, Elias E, Abou Abbas L, Levêque A (2015) Outdoor air pollution and cardiovascular diseases in Lebanon: a case-control study. J Environ Public Health 2015:1. https://doi.org/10.1155/2015/810846
32. Bray F, Ferlay J, Soerjomataram I, Siegel R, Torre L, Jemal A (2020) Global cancer statistics 2018: GLOBOCAN estimates of incidence and mortality worldwide for 36 cancers in 185 countries. Erratum in: CA Cancer J Clin 70:313
33. Li W, Lin G, Xiao Z, Zhang Y, Li B, Zhou Y, Ma Y, Chai E (2022) A review of respirable fine particulate matter (PM ..)-induced brain damage. Front Mol Neurosci 15:967174
34. Aghaei M et al (2021) Long-term exposure to air pollution and risk of preterm birth and low birth weight in Tehran, Iran. J Air Pollut Health 6(4):233–242
35. Khaled M (2013) Modeling of climate changes effect on coral reefs using remote sensing data at the Red Sea Egypt. https://doi.org/10.13140/RG.2.2.10787.73760
36. Kluser S et al (2010) Global honey bee colony disorders and other threats to insect pollinators. UNEP
37. Adesuyi A, Njoku K, Akinola M (2015) Assessment of heavy metals pollution in soils and vegetation around selected industries in Lagos State, Nigeria. J Geosci Environ Protect 9:11–19
38. Ozaki H, Watanabe I, Kuno K (2004) Investigation of the heavy metal sources in relation to automobiles. Water Air Soil Pollut 157:209–223
39. Zaakour F, Kholaiq M, Khouchlaa A, El Mjiri I, Rahimi A, Saber N (2022) Heavy metal contamination in agricultural soils: a case study in Mohammedia Benslimane region (Morocco). J Ecol Eng 23(5):1. https://doi.org/10.12911/22998993/146409
40. Hamoda MF (2006) Air pollutants emissions from waste treatment and disposal facilities. J Environ Sci Health A Tox Hazard Subst Environ Eng 41(1):77–85. https://doi.org/10.1080/10934520500298895

41. Rabl A, Spadaro JV (2000) Public health impact of air pollution and implications for the energy system. Annu Rev Energy Environ 25:601–627. https://doi.org/10.1146/annurev.energy.25.1.601
42. Akasha H, Ghaffarpasand O, Pope FD (2023) Climate change, air pollution and the associated burden of disease in the Arabian Peninsula and neighbouring regions: a critical review of the literature. Sustainability (Switzerland) 15(4):3766. https://doi.org/10.3390/su15043766
43. EPA (2024) National Ambient Air Quality Standards (NAAQS). https://www.epa.gov/criteria-air-pollutants/naaqs-table
44. WHO (2005) Air quality guidelines - global update 2005
45. Qatar S (2011) Matters relating to articles 3.14 and 2.3 of the Kyoto Protocol (Vol. 8, Issue February)
46. Poulakis E, Theodosi C, Bressi M, Sciare J, Ghersi V, Mihalopoulos N (2015) Airborne mineral components and trace metals in Paris region: spatial and temporal variability. Environ Sci Pollut Res 22:14663–14672
47. Schleussner, C. F., Rogelj, J., Schaeffer, M., Lissner, T., Licker, R., Fischer, E. M., Knutti, R., Levermann, A., Frieler, K., & Hare, W. (2016). Science and policy characteristics of the Paris agreement temperature goal. In Nature climate change, 6, 9, pp. 827–835). Nature Publishing Group. doi:https://doi.org/10.1038/nclimate3096
48. Aïssa B, Nedil M, Kroeger J, Ali A, Isaifan RJ, Essehli R, Mahmoud KA (2018) Graphene nanoplatelet doping of P3HT:PCBM photoactive layer of bulk heterojunction organic solar cells for enhancing performance. Nanotechnology 29(10):105405. https://doi.org/10.1088/1361-6528/aaa62d
49. Fares E, Buffiere M, Figgis B, Haik Y, Isaifan RJ (2021) Soiling of photovoltaic panels in the Gulf Cooperation Council countries and mitigation strategies. Sol Energy Mater Sol Cells 231:111303. https://doi.org/10.1016/j.solmat.2021.111303
50. Al-Thani SK, Amato A, Koç M, Al-Ghamdi SG (2019) Urban sustainability and livability: an analysis of Doha's urban-form and possible mitigation strategies. Sustainability (Switzerland) 11(3):786. https://doi.org/10.3390/su11030786
51. Litman T (2013) Transportation and public health. Annu Rev Public Health 34:217–233. https://doi.org/10.1146/annurev-publhealth-031912-114502
52. Pinder R et al (2019) Opportunities and challenges for filling the air quality data gap in low- and middle-income countries. Atmos Environ 215:116794
53. Acampa G, Contino F, Grasso M, Ticali D (2019) Evaluation of infrastructure: application of TOD to Catania underground metro station. Paper presented at the AIP conference proceedings. AIP Conference Proceedings
54. Kazmerski L (2011) Photovoltaic: history, technology, markets, manufacturing, applications, and outlook. In: 84th international seminar in Brighton, renewable energy policy, security, electricity, sustainable transport, water resources/management and environment, 3–9
55. Van Song N, Cong Tiep N, van Tien D, Van Ha T, Phuong TM, N., & Thi Hoang Mai, T. (2022) The role of public-private partnership investment and eco-innovation in environmental abatement in USA: evidence from quantile ARDL approach. Environ Sci Pollut Res 29:12164–12175. https://doi.org/10.1007/s11356-021-16520-8/Published
56. Gong K, Li L, Li J, Qin M, Wang X, Ying Q, Liao H, Guo S, Hu M, Zhang Y, Hu J (2021) Quantifying the impacts of inter-city transport on air quality in the Yangtze River Delta urban agglomeration, China: implications for regional cooperative controls of PM2.5 and O3. Sci Total Environ 779:146619. https://doi.org/10.1016/j.scitotenv.2021.146619
57. Saaida MBE (n.d.) Understanding global politics and diplomacy within the international relations context. SFAP 230711(1):1–18. https://doi.org/10.5281/zenodo.10841899
58. Martinez-Fernandez C, Audirac I, Fol S, Cunningham-Sabot E (2012) Shrinking cities: urban challenges of globalization. Int J Urban Reg Res 36:213–225
59. UNEP (2020) Actions taken by governments to improve air quality
60. WHO (2016) Ambient air pollution: a global assessment of exposure and burden of disease. https://www.who.int/phe/publications/air-pollution-global-assessment/en/

61. League of Arab States. (2022). Air quality and atmospheric pollution in the Arab region
62. US EPA. (2014). Air quality index: a guide to air quality and your health. https://www3.epa.gov/airnow/aqi_brochure_02_14.pdf
63. Khamis, J. (2023, October 2). How Arab countries can address pollution and improve urban air quality
64. Bridi RM, Ben Jabra M, Al Hosani N, Almurshidi AH (2024) The propensity to adopt electric vehicles in The United Arab Emirates: an analysis of economic and geographic factors. Sustainability (Switzerland) 16(2):770. https://doi.org/10.3390/su16020770
65. Orrnert A (2020) Drivers and barriers to environmental engagement in the MENA region. http://www.ifees.org.uk/declaration/
66. Saab N, Badran N, Sadik AK (2019) Environmental education for sustainable development in Arab countries. AFED

Promoting Behavioral Change and Public Awareness for Sustainable Clean Air

Onah Peter Thompson, Reagan S. Breitenstein, Brendon R. Barnes, and Richard Glory

Contents

O. P. Thompson (✉)
Government and Justice Studies Department, Appalachian State University Boone, Boone, NC, USA
e-mail: thompsonpo@appstate.edu

R. S. Breitenstein
Department of Psychology, Appalachian State University, Boone, NC, USA
e-mail: breitensteinrs@appstate.edu

B. R. Barnes
Department of Psychology, University of Johannesburg, Johannesburg, South Africa
e-mail: bbarnes@uj.ac.za

R. Glory
School of Community Health and Policy, Morgan State University, Baltimore, MD, USA

Matthew Chidozie Ogwu and Sylvester Izah (eds.), *Sustainable Strategies for Air Pollution Mitigation: Development, Economics and Technologies*,
Hdb Env Chem (2024) 133: 475–512, DOI 10.1007/698_2024_1171,

Published online: 17 October 2024

Abstract Psychological factors play a crucial role in shaping individual behaviors related to air pollution. This chapter focuses on avenues for promoting behavioral changes and public awareness for sustainable clean air. It discusses how the nexus of behavioral change and public awareness forms the cornerstone of efforts to achieve sustainable clean air. By synergistically leveraging individual actions, community engagement, policy support, and continuous improvement, this dynamic interplay can drive positive environmental outcomes and a culture of environmental responsibility and stewardship for the benefit of present and future generations. Behavioral science offers valuable insights into the complex interactions between human behavior and air quality. Therefore, by integrating behavioral principles into air quality management strategies, policymakers, researchers, and practitioners can develop more effective interventions to promote sustainable behaviors, reduce air pollution, and protect public health and the environment. Behavioral science concepts and theories can be tailored to contribute to the development of more effective interventions and strategies to address air quality management challenges, promote sustainable behaviors, and improve environmental outcomes. For instance, by combining public awareness campaigns, behavioral nudges, incentive programs, social marketing campaigns, community-based initiatives, and policy advocacy, stakeholders can create a comprehensive approach to address air quality challenges and foster a culture of environmental stewardship and sustainability. The nexus of behavioral changes and public awareness creates a feedback loop wherein successful behavioral interventions inform future awareness campaigns and policy decisions. In conclusion, the chapter offers examples to demonstrate the effectiveness of evidence-based behavioral interventions and strategies in promoting sustainable clean air through the incorporation of behavior change techniques such as goal setting, feedback, incentives, and social norms into intervention programs and stakeholders to effectively motivate individuals and communities to adopt pollution-reducing behaviors and contribute to air quality improvement efforts. This will also require the integration of behavioral insights into regulatory and enforcement mechanisms that leverage behavioral science principles to demonstrate the effectiveness of using behavior change techniques to incentivize sustainable practices, enhance compliance, and achieve sustainable clean air goals. The monitoring and evaluation of behavioral changes and public awareness initiatives for mitigating air pollution is necessary for assessing their effectiveness, improvement, and execution of target interventions where needed.

Keywords Behavioral interventions, Clean air, Pollution-reducing behaviors, Social norms, Sustainable behaviors, Sustainable future

Abbreviations

SLT	Social learning theory
PM	Particulate matter

NOx	Nitrogen oxides
SO2	Sulfur dioxide
CO	Carbon monoxide
VOCs	Volatile organic compounds
TPB	Theory of planned behavior
HBM	Health belief model
DoI	Diffusion of innovations
CBSM	Community-based social marketing
VMT	Vehicle miles traveled
UNEP	United Nations Environment Programme
WHO	World Health Organization
ETS	Emissions trading schemes
EU ETS	The European Union Emissions Trading System
EPA	Environmental Protection Agency
PBC	Perceived behavioral control

1 Introduction

Air pollution refers to the presence of harmful or excessive concentrations of substances in the air that can have adverse effects on human health, the environment, and ecosystems. These substances, known as pollutants, can be emitted from natural sources or human activities and can vary widely in composition and origin [1]. Air pollution encompasses a wide range of pollutants emitted from various sources, including transportation, industry, agriculture, and household activities [2]. These pollutants can have a range of adverse effects on human health, including respiratory diseases, cardiovascular problems, neurological disorders, and cancer [3, 4]. Additionally, air pollution can harm ecosystems; damage crops, forests, and bodies of water; and contribute to climate change by altering the Earth's energy balance and exacerbating global warming [5]. Addressing air pollution requires coordinated efforts at the local, national, and international levels, including implementing regulations and policies to limit emissions, promoting cleaner technologies and renewable energy sources, improving public transportation and urban planning, and raising public awareness about the importance of air quality and individual actions to reduce pollution [6–8]. By taking these actions, we can work toward achieving cleaner air and protecting the health and well-being of present and future generations.

As noted, there are a number of air pollutants, all of which have been shown to have a negative impact on physical health. Particulate matter (PM) consists of tiny particles suspended in the air, varying in size and composition. PM can originate from natural sources such as wildfires and volcanic eruptions, as well as human activities such as vehicle emissions, industrial processes, and construction activities [9]. Fine particles, known as PM2.5 (particles with a diameter of 2.5 μm or smaller), can penetrate deep into the lungs and pose significant health risks [10]. Nitrogen oxides (NOx) are a group of gases composed of nitrogen and oxygen, primarily produced during combustion processes, such as those occurring in vehicles, power plants, and industrial facilities [11]. Household air pollution in low-income countries caused by the indoor burning of solid fuels such as wood, coal, and animal dung is also a major contributor to the global burden of disease even surpassing ambient air pollution. NOx can contribute to the formation of ground-level ozone (smog) and particulate matter, leading to respiratory problems, cardiovascular diseases, and environmental degradation [12, 13]. Sulfur dioxide (SO2) is a gas produced from the combustion of sulfur-containing fuels, such as coal and oil, in power plants and industrial processes. SO2 can cause respiratory issues, particularly in individuals with asthma or other pre-existing respiratory conditions, and contribute to the formation of acid rain, which can harm ecosystems and corrode buildings and infrastructure [14]. Carbon monoxide (CO) is a colorless, odorless gas produced by the incomplete combustion of fossil fuels, biomass, and other organic materials. CO can interfere with the body's ability to transport oxygen, leading to symptoms such as headaches, dizziness, and fatigue, and in high concentrations, it can be fatal [15, 16]. Volatile organic compounds (VOCs) are a diverse group of chemicals that can evaporate into the air at room temperature. VOCs are emitted from various sources, including vehicle exhaust, industrial processes, and household products such as paints and solvents [15]. Some VOCs can react with other pollutants to form ground-level ozone and contribute to the formation of smog.

While some pollutants may have more natural or organic origins, many pollutants and subsequent air quality results have human behavior impacts, whether directly or indirectly. Thus, behavior science may offer valuable insights into understanding human behavior and decision-making processes related to air quality [17]. Psychological factors such as perception, motivation, and social norms related to air pollution influence individual behaviors, such as vehicle use, energy consumption, and waste management practices, which contribute to air pollution [18, 19]. Therefore, by applying behavioral science principles, policymakers and practitioners can design effective interventions to promote sustainable behaviors and reduce air pollution. Behavioral interventions should encourage individuals and communities to adopt behaviors that reduce air pollution and improve air quality and by the very least protect themselves from air pollution. These interventions target behavior change techniques such as goal setting, feedback, incentives, and social norms to motivate and empower individuals to make sustainable choices [20, 21]. Examples of behavioral interventions include promoting energy-efficient practices, encouraging active transportation, and fostering waste reduction and recycling behaviors. Behavioral change has also been used to promote the uptake of technologies such as improved cookstoves in the Global South [22].

Public awareness and education campaigns play a crucial role in raising awareness about air pollution and its impacts on health and the environment [23, 24]. These campaigns use various communication strategies, including traditional media, social media, community events, and educational materials, to inform and engage the public. By increasing knowledge and understanding of air quality issues, these campaigns empower individuals to take action to reduce their environmental footprint and advocate for cleaner air [25]. Community engagement is essential for addressing air pollution effectively, as local communities are often disproportionately affected by poor air quality [26]. Engaging communities in decision-making processes, citizen science initiatives, and environmental monitoring programs empowers them to take ownership of air quality issues and collaborate with policymakers and stakeholders to implement solutions. By involving communities in air quality management efforts, policymakers can ensure that interventions are culturally appropriate, socially inclusive, and locally relevant. Policy and governance frameworks play a critical role in promoting behavioral changes for air quality improvement. Governments can implement regulations, incentives, and public policies that encourage sustainable behaviors and discourage activities that contribute to air pollution [27]. By integrating behavioral insights into policy design and implementation, policymakers can enhance the effectiveness and acceptability of air quality management measures [28, 29]. Advancements in technology and innovation offer new opportunities to promote behavioral changes for sustainable clean air. Mobile apps, sensor technologies, and online platforms provide real-time information on air quality and empower individuals to make informed decisions about their daily activities [30]. By harnessing the power of technology, policymakers and practitioners can create innovative solutions that engage and motivate individuals to adopt sustainable behaviors and reduce air pollution [31, 32].

This chapter focuses on avenues for promoting behavioral changes and public awareness for sustainable clean air. Promoting behavioral changes and public awareness is essential for achieving sustainable clean air. By understanding human behavior, designing effective interventions, raising public awareness, engaging communities, implementing supportive policies, and leveraging technology and innovation, we can empower individuals and communities to take action to reduce air pollution and protect public health and the environment.

2 Nexus of Behavioral Change and Public Awareness in Achieving Sustainable Clean Air

Behavioral changes and public awareness are essential components of efforts to achieve sustainable clean air. Behavioral changes and public awareness initiatives empower individuals and communities to take action to reduce air pollution and create healthier environments for all [24]. The nexus of promoting behavioral changes and public awareness for sustainable clean air in no particular order includes the following (Fig. 1):

Individual Responsibility Behavioral changes promote individual responsibility for reducing air pollution. By raising awareness about the sources and impacts of air pollution, individuals become more conscious of their daily behaviors that contribute to pollution, such as transportation choices, energy consumption, and waste management practices [3]. Empowering individuals to make sustainable choices, such as

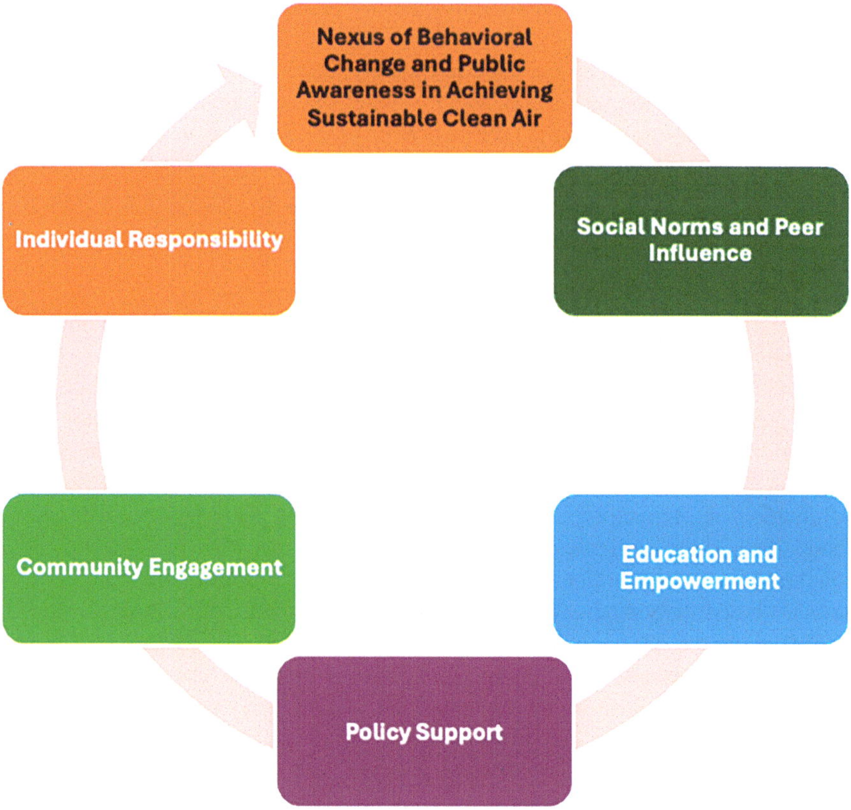

Fig. 1 Nexus of behavioral change and public awareness in achieving sustainable clean air

using public transportation, conserving energy, and reducing waste, can collectively lead to significant reductions in air pollution [33].

Community Engagement Public awareness initiatives foster community engagement and participation in air quality management efforts. Through informing and mobilizing local communities, public awareness campaigns empower residents to take ownership of air quality issues in their neighborhoods and advocate for cleaner air [25]. Community-based initiatives, such as citizen science projects and environmental monitoring programs, enable residents to actively contribute to monitoring air quality, identifying pollution sources, and implementing solutions tailored to local needs [34, 35].

Policy Support Public awareness and public pressure can drive policymakers to implement stronger regulations and policies to address air pollution. Raising public awareness about the health and environmental impacts of air pollution and advocacy efforts can generate political will and support for policies that prioritize clean air, such as emissions standards for vehicles and industrial facilities, investments in renewable energy and clean technologies, and measures to promote sustainable transportation and urban planning [36, 37]. Behavioral changes can complement regulatory approaches by reinforcing compliance with air quality regulations and promoting voluntary initiatives to reduce pollution [38]. At the policy level, various stakeholders create the correct enabling environments to foster behavioral change, for example, subsidies for certain technologies and enabling market conditions to foster the uptake of those technologies.

Education and Empowerment Public awareness campaigns provide education and empower individuals with knowledge and tools to take action to improve air quality. Increasing the knowledge and understanding of the links between human activities and air pollution, awareness initiatives empower individuals to make informed decisions and adopt behaviors that contribute to cleaner air [39]. Education programs in schools, workplaces, and communities can instill lifelong habits of environmental stewardship and sustainability, fostering a culture of responsibility and accountability for air quality among future generations.

Social Norms and Peer Influence Behavioral changes can be influenced by social norms and peer behavior. Public awareness campaigns that highlight sustainable behaviors and showcase positive role models can help shift social norms toward more environmentally friendly practices. By promoting peer-to-peer learning and social support networks, public awareness initiatives can encourage collective action and motivate individuals to adopt sustainable behaviors through social influence and reinforcement [40].

Put together, the nexus of behavioral change and public awareness is instrumental in achieving sustainable clean air by fostering individual and collective action to reduce air pollution and improve air quality. This interconnected relationship between behavioral changes and public awareness is pivotal for driving positive environmental outcomes and creating a culture of sustainability. Public awareness campaigns raise consciousness about the causes and consequences of air pollution, enlightening individuals about their roles in contributing to or mitigating pollution [24]. Through education and dissemination of information, public awareness initiatives empower individuals to make informed choices and adopt behaviors that minimize their environmental footprint [41]. Awareness of the health impacts of air pollution and the benefit of clean air motivates people to modify their habits and lifestyle practices to reduce pollution emissions.

As individuals adopt sustainable behaviors to reduce air pollution, they become living examples of the positive impacts of behavioral change, reinforcing the messaging conveyed through public awareness campaigns [42, 43]. Actions such as carpooling, using public transportation, conserving energy, and reducing waste not only contribute directly to air quality improvement but also serve as tangible demonstrations of environmental stewardship. The visible adoption of sustainable behaviors by individuals influences social norms and perceptions, encouraging others to follow suit and amplifying the effects of public awareness efforts [44]. Behavioral changes catalyze collective action and community engagement, as individuals unite around a shared goal of improving air quality [45]. Community-led initiatives, supported by public awareness campaigns, mobilize residents to participate in environmental monitoring, advocacy, and grassroots efforts to address local air pollution sources [46]. By fostering collaboration and social cohesion, these initiatives strengthen community resilience and capacity to effect meaningful change in air quality management. Informed and engaged citizens, driven by public awareness, advocate for policies and regulations that prioritize clean air and promote sustainable practices. Public pressure generated through awareness campaigns influences policymakers to enact legislation, enforce regulations, and allocate resources toward air quality improvement initiatives. Behavioral changes, backed by public support, reinforce the implementation and enforcement of policies aimed at reducing pollution emissions from various sources, including transportation, industry, and energy production [47].

It is important to note that within the nexus of behavioral changes and public awareness and feedback loop, change might start from any one of the components and have positive "trickle down" effects on the other components of the loop. Therefore, components of the loops should be perceived as successful behavioral interventions to inform current and future awareness campaigns and policy decisions for clean air. Also, the monitoring and evaluation of behavioral changes and public awareness initiatives provide valuable insights into their effectiveness, enabling stakeholders to refine strategies and target interventions where they are most needed. Continuous engagement with communities and stakeholders ensures ongoing dialogue, responsiveness to emerging challenges, and adaptation of approaches to evolving air quality concerns.

3 Behavioral Science and Air Quality

Behavioral science plays a crucial role in understanding human behavior and decision-making processes related to air quality [29]. By applying principles from psychology, sociology, economics, and other disciplines, behavioral science can help elucidate the underlying factors influencing individual and collective actions that contribute to air pollution. Behavioral science examines how individuals perceive and assess risks associated with air pollution [48]. Factors such as cognitive biases, heuristics, and social cues can influence people's perceptions of air quality and their susceptibility to its health effects [49]. Understanding these perceptual processes can inform the design of public awareness campaigns and risk communication strategies to accurately convey the severity of air pollution and motivate individuals to take protective actions. Behavioral science explores the motivations and decision-making processes underlying individual behaviors related to air quality, such as transportation choices, energy consumption, and waste management practices [50, 51]. Concepts such as self-efficacy, social norms, and temporal discounting influence people's decisions to adopt environmentally friendly behaviors or engage in activities that contribute to air pollution [52]. By identifying barriers to behavior change and leveraging behavioral insights, interventions can be designed to promote sustainable actions and reduce pollution emissions. Behavioral science informs the development and implementation of interventions aimed at promoting proenvironmental behaviors and improving air quality [53]. Behavior change techniques, such as goal setting, feedback, incentives, and social norms, are used to motivate individuals to adopt sustainable practices, such as using public transportation, conserving energy, and reducing waste. These interventions leverage psychological principles to nudge people toward more environmentally friendly choices and reinforce positive behaviors over time. This enables behavioral science to recognize the importance of social and environmental factors in shaping individual behaviors and influencing air quality outcomes [28]. Socioeconomic status, cultural norms, peer influences, and built environment characteristics can all impact people's exposure to air pollution and their ability to adopt sustainable behaviors [54, 55]. By considering these contextual factors, interventions can be tailored to specific populations and settings to address disparities in air quality and promote equitable access to clean air. Behavioral science provides methods for evaluating the effectiveness of interventions and gathering feedback to refine strategies over time. Through rigorous evaluation studies, researchers can assess the impact of behavioral interventions on air quality outcomes, such as pollution emissions, ambient pollutant concentrations, and public health indicators. By incorporating feedback from stakeholders and target audiences, interventions can be iteratively improved to maximize their effectiveness and sustainability.

3.1 Behavioral Science Concepts and Theories Relevant to Air Quality Management

Behavioral science provides a rich framework of concepts and theories that are highly relevant to understanding and addressing air quality management challenges. Some behavioral science concepts and theories that are applicable in the context of air quality management include the following (Fig. 2):

Theory of Planned Behavior The theory of planned behavior (TPB) posits that individual behavior is influenced by three factors: attitudes, subjective norms, and perceived behavioral control [56, 57]. In the context of air quality management, TPB can help explain how individuals' attitudes toward environmental protection, social norms (which may be influences by others around them) regarding sustainable behaviors, and perceived ability to enact change influence their decisions to adopt pollution-reducing actions, such as using public transportation or reducing energy consumption.

Health Belief Model The Health Belief Model (HBM) suggests that individuals are more likely to take preventive actions if they perceive themselves to be at risk of negative health outcomes, believe that a proposed action will reduce the risk, and

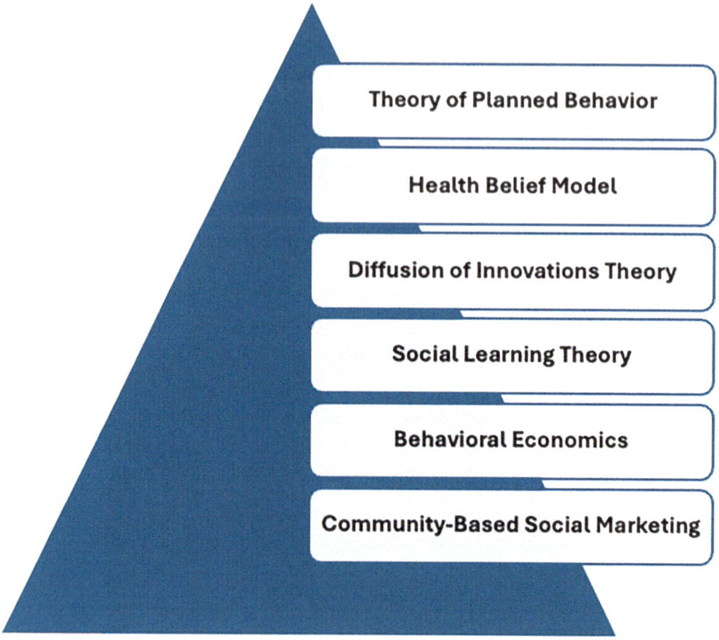

Fig. 2 Some behavioral science concepts and theories relevant to air quality management

perceive the benefits of the action to outweigh the costs [58, 59]. Concerning air quality management, HBM can be applied to understand how individuals perceive the health risks associated with air pollution, their awareness of mitigation strategies, and their willingness to adopt protective behaviors to reduce exposure to pollutants.

Diffusion of Innovations Theory The diffusion of innovations (DoI) theory explores how new ideas, technologies, or behaviors spread within a population over time [60]. The DoI theory may help understand the adoption and diffusion of clean technologies, renewable energy sources, and sustainable practices that reduce emissions and improve air quality. Understanding the factors that influence the adoption and spread of these innovations can inform strategies to accelerate their uptake.

Social Learning Theory Social learning theory (SLT) suggests that people learn by observing the behaviors of others and the consequences of those behaviors [61–64]. Applied to air quality management, this theory emphasizes the importance of social norms, peer influences, and role models in shaping individuals' behaviors related to pollution reduction and environmental conservation. Encouraging positive social modeling and providing opportunities for vicarious learning can promote the adoption of sustainable practices.

Behavioral Economics Behavioral economics integrates insights from psychology into economic theory to understand how individuals make decisions in real-world contexts [65]. In the context of air quality management, behavioral economics helps identify cognitive biases, heuristics, and decision-making shortcuts that may lead individuals to undervalue the long-term benefits of pollution-reducing actions or to prioritize immediate gratification over environmental considerations. Strategies such as framing, default options, and incentives can be designed to nudge individuals toward more sustainable choices.

Community-Based Social Marketing Community-based social marketing (CBSM) applies principles from social psychology and marketing to promote sustainable behaviors within communities [66]. In air quality management, CBSM strategies involve identifying barriers to behavior change, segmenting target audiences, designing tailored interventions, and leveraging social norms and social networks to encourage the adoption of pollution-reducing actions at the community level.

3.2 Other Psychological Factors Influencing Individual Behaviors Related to Air Pollution

Psychological factors play a significant role in shaping individual behaviors related to air pollution. Understanding these factors – such as perception, motivation, and decision-making – is crucial for developing effective interventions to mitigate pollution and improve air quality [67, 68]. Perception refers to the way individuals interpret and make sense of information about air pollution [69]. People's perceptions of air quality are shaped by various factors, including sensory experiences, media coverage, and social influences. Several research such as the works of [7, 70, 71] suggest that individuals may underestimate the severity of air pollution due to several cognitive biases and perceptual distortions:

Optimism Bias Many people tend to believe that they are less susceptible to the health risks of air pollution than others, leading to an optimism bias. This bias may result in individuals downplaying the health effects of pollution and underestimating their vulnerability.

Availability Heuristic Individuals often rely on readily available information when assessing risks, particularly information that is more recent, more negative, and more extreme. Thus, media coverage of acute pollution events, such as smog episodes or wildfires, may lead people to overestimate the immediate risks of air pollution while overlooking the chronic, long-term health impacts.

Anchoring Effect People's perceptions of air quality may be influenced by their initial impressions or reference points. For example, individuals living in areas with consistently poor air quality may become desensitized to pollution and perceive it as normal, leading to a normalization bias.

3.2.1 Motivation to Address Air Pollution

Motivation refers to the internal processes that energize and direct individuals' behavior toward achieving specific goals [72, 73]. Several motivational factors influence people's willingness to address air pollution and engage in pollution-reducing behaviors:

Intrinsic Motivation Individuals may be intrinsically motivated to protect the environment and improve air quality due to personal values, beliefs, and concerns for future generations [74, 75]. Intrinsic motivations, such as a desire for environmental stewardship or a sense of moral obligation, can drive individuals to adopt sustainable behaviors voluntarily. People may also want to protect their own and their children's health. In the same way, people may want a better life particularly in the Global South. For example, in one study, mothers of young children expressed the fact that smelling of smoke was symbolic of poverty stigmatizing and that they were motivated to reduce the air pollution in their homes and reduce that stigma [76].

Extrinsic Motivation External incentives, rewards, or penalties can also influence people's motivation to address air pollution [77]. Policies, regulations, and financial incentives, such as pollution taxes or subsidies for clean technologies, can provide extrinsic motivations that encourage individuals to reduce pollution emissions and adopt environmentally friendly practices.

Social Norms and Social Comparison Social norms – implicit or explicit rules governing acceptable behaviors within a group – play a crucial role in shaping individuals' motivations and in turn their likelihood of actually completing a behavior [56] such as sustainability efforts. People are more likely to engage in pollution-reducing behaviors if they perceive them as socially desirable or normative within their social circles [44]. Social comparison processes, whereby individuals compare their behaviors to those of others, can further influence motivation by providing benchmarks for acceptable conduct [78].

3.2.2 Decision-Making Processes

Decision-making involves evaluating alternatives and choosing courses of action based on perceived costs, benefits, and trade-offs [79]. Several cognitive processes and decision-making heuristics influence individuals' decisions related to air pollution:

Cost-Benefit Analysis People weigh the expected costs and benefits of pollution-reducing behaviors when making decisions. Perceived costs, such as inconvenience, time, or financial expenses associated with adopting sustainable practices, may deter individuals from taking action unless the anticipated benefits, such as improved health outcomes or environmental protection, outweigh these costs [80].

Temporal Discounting Individuals often discount future consequences of their actions in favor of immediate rewards or gratification. This temporal discounting bias may lead people to prioritize short-term benefits over long-term gains, such as choosing convenience or comfort over pollution-reducing behaviors with delayed or uncertain benefits [81]. Another example would be an example of this; the ease (or lack thereof) of recycling is a good example of this.

Cognitive Dissonance Cognitive dissonance arises when individuals experience inconsistency between their beliefs, attitudes, and/or behaviors. People may engage in psychological mechanisms, such as rationalization or denial, to reduce cognitive dissonance and maintain consistency between their actions and environmental values [82, 83]. For example, individuals who drive "gas-guzzling" vehicles may justify their behavior by downplaying the environmental impacts, emphasizing other virtues of car ownership, or attributing phenomenon like global warming to factors other than driving.

Overall, psychological factors play a crucial role in shaping individual behaviors related to air pollution. Perception, motivation, and decision-making processes influence people's perceptions of air quality, their willingness to address pollution, and their choices regarding pollution-reducing behaviors [48, 84]. Therefore, understanding these psychological drivers is essential for designing effective interventions and policies to mitigate pollution, promote sustainable behaviors, and improve air quality. By addressing cognitive biases, leveraging intrinsic and extrinsic motivations, and providing information and incentives that align with individuals' values and goals, policymakers, researchers, and practitioners can empower people to take proactive steps toward cleaner air and a healthier environment.

3.3 Roles of Social Norms, Attitudes, Beliefs, and Perceived Control in Shaping Collective Behaviors Toward Sustainable Clean Air

Social norms, attitudes, and beliefs play pivotal roles in shaping collective behaviors toward sustainable clean air. These sociopsychological factors influence how individuals perceive, interpret, and respond to air quality issues, ultimately shaping societal norms and driving collective actions to mitigate air pollution [29, 85]. Social norms, attitudes, and beliefs likely play crucial roles in shaping collective behaviors toward sustainable clean air. By fostering positive norms, promoting favorable attitudes, and providing accurate information and support for pollution-reducing behaviors, communities can create environments that encourage individuals to take action to protect air quality and promote a healthier, more sustainable future.

3.3.1 Social Norms

Social norms are unwritten rules or standards of behavior that guide individuals' actions within a society or group [56, 86]. They encompass both descriptive norms (perceptions of what others commonly do) and injunctive norms (perceptions of what others approve or disapprove of). Social norms influence behavior by providing cues for what is considered acceptable or appropriate within a given context [87]. In the context of air quality, social norms can influence behaviors such as transportation choices, energy consumption habits, and waste management practices. For example, if carpooling or using public transportation is perceived as common and socially acceptable within a community, individuals may be more likely to adopt these sustainable transportation options. Norms can also shape attitudes toward environmental stewardship and sustainability. Communities that prioritize clean air and environmental protection may establish norms that promote pollution-reducing behaviors and discourage actions that contribute to air pollution.

3.3.2 Attitudes

Attitudes are individual evaluations or judgments about objects, people, or issues, which can influence behavior by guiding decisions and actions [56, 88]. Attitudes toward air quality and environmental conservation reflect people's beliefs, values, and emotional responses to pollution-related issues. Positive attitudes toward sustainable clean air are associated with beliefs in the importance of environmental protection, concern for public health, and a sense of responsibility for future generations [89]. Individuals with favorable attitudes toward clean air are more likely to engage in pollution-reducing behaviors and support initiatives aimed at improving air quality [48]. Attitudes can be influenced by various factors, including personal experiences, education, media exposure, and social influences. Notably, social norms are often associated with attitudes, suggesting that they may influence one another as well [56]. Positive experiences with clean air, such as enjoying outdoor activities in unpolluted environments, can reinforce favorable attitudes and motivate individuals to take action to preserve air quality.

3.3.3 Beliefs

Beliefs are cognitive representations of reality that shape perceptions, attitudes, and behaviors. Beliefs about air pollution encompass factual knowledge about pollution sources, health effects, and mitigation strategies, as well as subjective interpretations and interpretations of pollution-related issues. Accurate beliefs about air pollution, its causes, and its consequences are essential for informed decision-making and behavior change. Education and communication efforts that provide reliable information about the sources and impacts of air pollution can help correct misconceptions and promote awareness of the importance of clean air [24, 90]. In addition to factual knowledge, beliefs about the efficacy of pollution-reducing behaviors and the feasibility of collective action can influence behavior. Individuals who believe that their actions can make a difference in improving air quality are more likely to adopt sustainable behaviors and support community-wide initiatives aimed at reducing pollution emissions.

3.3.4 Perceived Control

Perceived behavioral control (PBC) as a component of the theory of planned behavior (TPB) plays a crucial role in shaping collective behaviors toward sustainable clean air. PBC directly influences individuals' intentions to engage in a particular behavior. In the context of clean air initiatives, individuals' perceptions of their ability to contribute to cleaner air through their actions (such as reducing car emissions, conserving energy, or supporting environmental policies) significantly affect their willingness to adopt sustainable behaviors. PBC can help individuals identify barriers to engaging in sustainable actions and empowers them to overcome these barriers [91]. For instance, if individuals perceive that they lack the necessary resources, knowledge, or social support to contribute to clean air initiatives, efforts can be made to provide them with the required resources, education, or community support, thus enhancing their perceived control and enabling them to act [26, 92]. PBC is closely related to self-efficacy, which refers to individuals' beliefs in their ability to successfully perform a behavior. Efforts to enhance individuals' self-efficacy regarding clean air actions can be instrumental in promoting collective behaviors toward sustainability. This can be achieved through various means, such as educational campaigns, skills training, and providing opportunities for successful experiences in contributing to clean air. Also, efforts to promote positive social norms surrounding clean air behaviors can amplify the impact of PBC on collective actions.

4 Behavioral Interventions for Air Quality Improvement

Behavioral interventions play a vital role in promoting air quality improvement by influencing individual and collective behaviors that contribute to pollution reduction. These interventions leverage insights from behavioral science to encourage people to adopt sustainable practices, reduce emissions, and support policies and initiatives aimed at improving air quality. Some key behavioral interventions for air quality improvement include the following (Fig. 3):

Public Awareness Campaigns Public awareness campaigns raise awareness about air quality issues, pollution sources, and the health impacts of poor air quality [39]. These campaigns use various communication channels, such as media

Fig. 3 Some behavioral interventions for improving air quality

advertisements, public service announcements, educational materials, and community events, to disseminate information and motivate behavior change [93]. Messaging in awareness campaigns often emphasizes the importance of individual actions, such as reducing vehicle idling, conserving energy, and using alternative transportation modes, in improving air quality and protecting public health.

Behavioral Nudges Behavioral nudges are subtle interventions designed to influence people's choices and behaviors without restricting their freedom of choice [94, 95]. These nudges capitalize on cognitive biases and decision-making heuristics to steer individuals toward more environmentally friendly actions. Examples of behavioral nudges for air quality improvement include default options that promote sustainable choices (e.g., defaulting to energy-saving settings on appliances – as cases of opt out), feedback mechanisms that provide real-time information on pollution levels and encourage pollution-reducing behaviors, and social norms messaging that highlights the prevalence of sustainable behaviors within a community.

Incentive Programs Incentive programs provide tangible rewards or incentives to individuals who adopt pollution-reducing behaviors or use cleaner technologies [96]. These incentives can take various forms, such as financial incentives, rebates, discounts, or reward points. For example, cash rebates or tax incentives for purchasing low-emission vehicles, energy-efficient appliances, or solar panels can encourage individuals to invest in cleaner technologies and reduce their environmental impact. Similarly, rewards programs that incentivize sustainable transportation options, such as biking, walking, or carpooling, can motivate commuters to choose alternatives to driving alone.

Social Marketing Campaigns Social marketing campaigns use marketing principles and techniques to promote behavior change for social and environmental causes, including air quality improvement [97]. These campaigns employ persuasive messaging, storytelling, and social influence strategies to encourage the adoption of pollution-reducing behaviors. Social marketing campaigns often leverage social norms, peer influence, and positive role models to inspire behavior change [98]. By highlighting the collective benefits of clean air and the actions of environmentally conscious individuals and communities, these campaigns can shift social norms toward more sustainable behaviors. It might be worth mentioning that social marketing techniques often accompany technologies. So behavior change is not only about individuals engaging in behaviors that will reduce air pollution but also to adopt technologies that might reduce air pollution.

Community-Based Initiatives Community-based initiatives engage residents, businesses, and organizations in collaborative efforts to address air quality issues at the community level. These initiatives involve grassroots organizing, community outreach, and participatory decision-making processes. Examples of community-based interventions for air quality improvement include neighborhood clean-up events, tree-planting campaigns, car-free days, and community gardening projects. By fostering a sense of collective responsibility and ownership of air quality issues, these initiatives empower communities to act and advocate for policies and practices that promote cleaner air.

Policy Support and Advocacy Behavioral interventions can also target policymakers and stakeholders to advocate for policies and regulations that support air quality improvement. By mobilizing public support and engaging decision-makers, advocacy efforts can influence the adoption of measures to reduce pollution emissions, promote renewable energy, and enhance environmental regulations. Grassroots organizing, petitions, letter-writing campaigns, and public demonstrations are common advocacy tactics used to raise awareness of air quality issues, build political momentum, and hold policymakers accountable for addressing pollution sources and protecting public health [25, 99].

Promoting sustainable clean air requires the implementation of evidence-based behavioral interventions and strategies that encourage individuals and communities to adopt pollution-reducing behaviors. Leveraging behavior change techniques such as goal setting, feedback, incentives, and social norms can effectively motivate

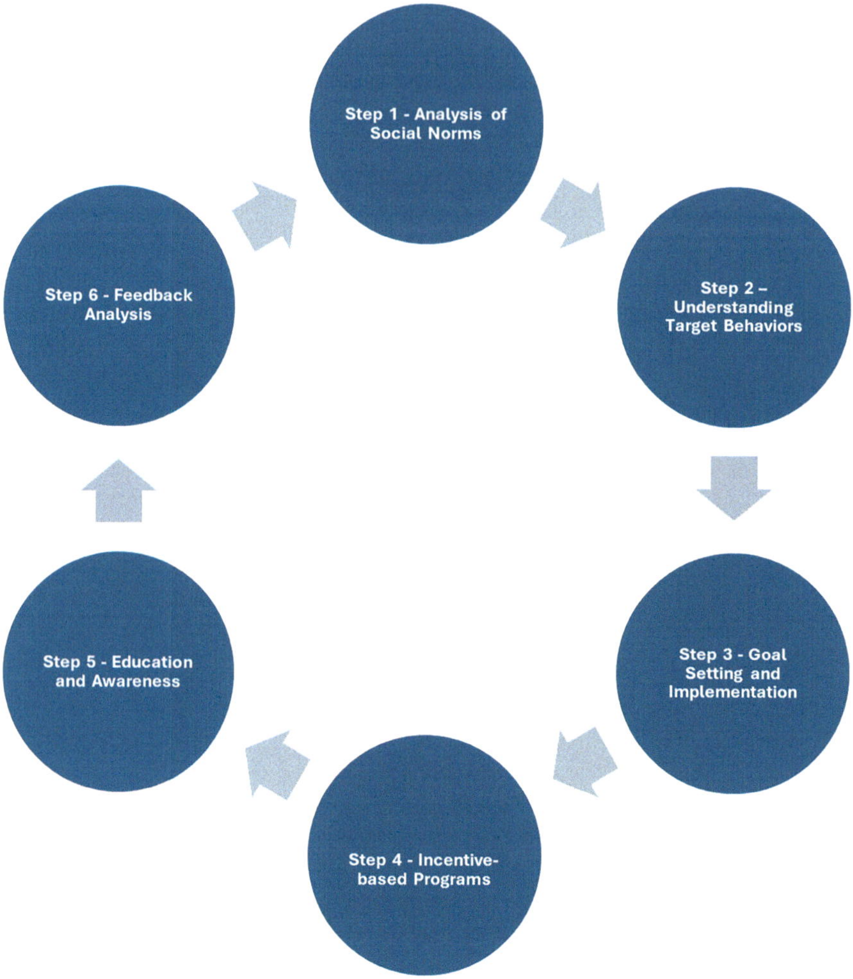

Fig. 4 Steps for developing and implementing typical behavioral interventions and strategies for mitigating air pollution

people to make environmentally friendly choices [40, 100]. Steps for typical behavioral interventions and strategies for mitigating air pollution should include the following (Fig. 4):

Step 1: Analysis of Social Norms Social norms influence behavior by establishing expectations for what is considered acceptable or appropriate within a social group or community. Leveraging social norms can encourage individuals to adopt pollution-reducing behaviors by highlighting the prevalence of these behaviors among their peers. For instance, the "Cool California Challenge" engaged households and communities in California to compete in reducing their carbon footprint and promoting sustainability [101]. By creating a social competition framework and highlighting community-wide progress toward sustainability goals, the program leveraged social norms to motivate behavior change and foster a culture of environmental stewardship.

Step 2: Understanding Target Behaviors (and Their Motivation and Barriers) This step is vital for understanding the target behaviors (and their motivations and barriers). Often, there are likely to be several possible target behaviors. However, a core principle of behavior change is to limit the number of target behaviors so that the program is as focused as possible. Often this involves an analysis of what is doable and what is effective. Sometimes a target behavior may be doable but have very little impact [102]. On the other hand, a target behavior might be difficult but have a greater impact. Programmers need to decide on the balance. It is advisable to pilot the target behaviors through participatory techniques in a small sample [103].

Step 3: Goal Setting and Implementation Goal setting involves establishing specific, measurable objectives for behavior change and providing individuals with a clear target to work toward. Setting achievable goals related to air quality improvement can motivate people to act and track their progress over time. An example is the "Drive Less, Breathe Better" campaign in Salt Lake City, Utah, which encouraged residents to set goals for reducing vehicle miles traveled (VMT) by using alternative transportation modes such as biking, walking, carpooling, or using public transit [104, 105]. Participants were provided with personalized tracking tools to monitor their VMT reduction progress and received incentives for meeting their goals.

Step 4: Incentive-Based Programs Incentive-based programs offer rewards, incentives, or tangible benefits to individuals who engage in pollution-reducing behaviors or use cleaner technologies. Incentives can motivate behavior change by providing immediate or tangible rewards for environmentally friendly actions. An example is the "Cash for Clunkers" program implemented in various countries providing financial incentives or rebates to individuals who traded in old, high-emission vehicles for new, fuel-efficient models [106]. By offering financial incentives for upgrading to cleaner vehicles, the program aimed to reduce emissions and improve air quality.

Step 5: Education and Awareness Education and awareness campaigns provide information about air quality issues, pollution sources, health impacts, and mitigation strategies to raise public awareness and encourage behavior change [24, 26]. These campaigns aim to empower individuals with knowledge and resources to act and make informed choices. For example, the "Clean Air Carolina" organization in North Carolina conducts educational outreach programs in schools, communities, and workplaces to raise awareness of air quality issues and promote clean air initiatives [107]. By providing educational resources, workshops, and outreach events, the organization empowers individuals and communities to take proactive steps toward improving air quality.

Step 6: Feedback Analysis Feedback mechanisms provide individuals with information about their actions, performance, or progress toward a goal. Feedback can help raise awareness of air quality issues, highlight the impact of individual behaviors on pollution levels, and reinforce positive changes. For instance, the "AirVisual" mobile app provides real-time air quality data and personalized air quality forecasts to users based on their location [108]. Users receive notifications and alerts about pollution levels, allowing them to adjust their activities and behaviors accordingly to reduce exposure to harmful pollutants.

5 Policy and Governance for Behavior Change for Sustainable Clean Air

Policy and governance frameworks play a critical role in promoting behavioral changes for air quality improvement by providing the necessary regulatory and institutional support to incentivize and enforce pollution-reducing behaviors [109, 110]. These frameworks integrate behavioral insights into air quality regulations, standards, and enforcement mechanisms to achieve sustainable clean air goals effectively [111]. Policy frameworks establish legal requirements and standards for air quality management, emission control, and pollution reduction. These regulations set targets for pollutant emissions, mandate compliance with air quality standards, and outline enforcement mechanisms to ensure accountability [112]. Behavioral insights can inform the design of regulatory frameworks by considering factors such as human decision-making, motivation, and social norms. Policies may incorporate behavior change techniques such as defaults, incentives, and feedback mechanisms to encourage compliance and promote sustainable practices [21, 113].

Policy incentives provide tangible rewards or benefits to individuals, businesses, and organizations that adopt pollution-reducing behaviors or invest in clean technologies [96]. Incentive programs aim to overcome barriers to behavior change by offering financial incentives, tax credits, rebates, or subsidies for environmentally friendly actions. Examples of incentive programs include grants for renewable energy projects, tax incentives for energy-efficient appliances, and subsidies for electric vehicles. These incentives encourage individuals and businesses to transition to cleaner technologies and practices, contributing to improved air quality [114]. Policy initiatives support public awareness campaigns and educational programs to raise awareness of air quality issues, health impacts of pollution, and opportunities for behavior change. These campaigns use various communication channels to disseminate information, engage communities, and promote sustainable practices. Governments, nonprofit organizations, and community groups collaborate to develop educational materials, host workshops, and conduct outreach activities that empower individuals to take action to improve air quality. By increasing knowledge and understanding, these initiatives encourage behavior change at the individual and community levels. Understanding human behavior, decision-making processes, and social influences helps policymakers identify barriers to action and develop strategies to overcome them [115]. Behavioral science techniques such as nudges, defaults, social norms messaging, and personalized feedback are incorporated into policies and programs to encourage sustainable behaviors [116]. For example, opt-out enrolment in clean energy programs or default settings for energy-saving appliances can promote energy efficiency and reduce emissions.

Policy mechanisms include monitoring and enforcement measures to ensure compliance with air quality regulations and standards [111, 117]. Monitoring systems track pollution levels, emission sources, and environmental indicators to assess progress toward clean air goals. Enforcement mechanisms, such as fines, penalties, and regulatory sanctions, deter noncompliance and hold polluters accountable for their actions. International cooperation and collaboration are essential for addressing transboundary air pollution and global environmental challenges [118]. Policy frameworks facilitate multilateral agreements, partnerships, and initiatives to coordinate efforts, share best practices, and mobilize resources for air quality improvement. International organizations, such as the United Nations Environment Programme (UNEP) and the World Health Organization (WHO), provide platforms for dialogue, capacity-building, and knowledge exchange on air quality management and sustainable development. By working together, countries can achieve collective goals and protect air quality on a global scale.

Policy and governance frameworks establish the legal and regulatory framework for addressing air quality issues, setting emissions targets, and implementing pollution control measures. These frameworks provide the necessary infrastructure and resources to support behavior change initiatives and promote sustainable practices. Effective policy and governance structures facilitate collaboration between government agencies, industry stakeholders, nonprofit organizations, and the public to develop and implement strategies for air quality improvement [26, 119, 120]. Behavioral insights are increasingly being integrated into air quality regulations, standards, and enforcement mechanisms to encourage compliance and promote behavior change. By understanding the psychological factors influencing individual behaviors, policymakers can design more effective interventions that address barriers to action and incentivize pollution-reducing behaviors. Behavioral insights can inform the design of policy interventions that leverage behavior change techniques such as defaults, incentives, social norms, and feedback mechanisms to encourage sustainable practices and reduce pollution emissions [20, 121].

Examples of policy interventions include the following:

Behavioral-Based Emission Trading Schemes Emissions trading schemes (ETS) use market-based mechanisms to regulate pollution emissions by assigning a cap on total emissions and allowing companies to buy and sell emission permits [122]. Behavioral-based ETS incorporates behavioral insights into the design of trading platforms to encourage participation and compliance [123]. The European Union Emissions Trading System (EU ETS) introduced behavioral nudges, such as simplified registration processes, personalized feedback on emissions performance, and default options for emissions trading, to increase participation and improve compliance among regulated entities [123, 124].

Smart Regulation and Compliance Assistance Smart regulation approaches leverage behavioral insights to streamline regulatory processes, provide targeted compliance assistance, and improve enforcement effectiveness [125]. These approaches aim to reduce administrative burdens, enhance transparency, and facilitate voluntary compliance with air quality regulations. The Environmental Protection Agency (EPA) in the United States has implemented the SmartWay program, which provides resources, tools, and incentives to help freight companies reduce emissions and improve fuel efficiency voluntarily [126]. By offering technical assistance, benchmarking tools, and recognition for top performers, the program encourages companies to adopt sustainable transportation practices and reduce their environmental impact.

Public Engagement and Behavioral Change Campaigns Public engagement and behavioral change campaigns use mass media, social marketing, and community outreach strategies to raise awareness of air quality issues, promote behavior change, and mobilize public support for pollution reduction efforts. The "Every Breath Matters" campaign launched by the Indian government aims to raise public awareness of air pollution and its health impacts, encourage the adoption of clean air practices, and mobilize community action for air quality improvement [127]. The campaign utilizes multimedia platforms, public events, and educational materials to engage citizens and promote behavior change at the grassroots level.

Policy and governance frameworks play a crucial role in promoting behavioral changes for air quality improvement by integrating behavioral insights into regulatory and enforcement mechanisms [128]. Examples of policy interventions that leverage behavioral science principles demonstrate the effectiveness of using behavior change techniques to incentivize sustainable practices, enhance compliance, and achieve sustainable clean air goals. By incorporating behavioral considerations into air quality policies and regulations, policymakers can create more effective and impactful interventions that empower individuals and communities to contribute to air quality improvement efforts.

6 Technology and Innovation for Behavioral Change for Sustainable Clean Air

Technology and innovation play a crucial role in promoting behavioral change and increasing public awareness for sustainable clean air by providing tools, platforms, and solutions that facilitate and incentivize pollution-reducing behaviors (Table 1). These technological advancements leverage digital platforms, data analytics, and communication technologies to raise awareness, provide real-time feedback, and promote engagement with air quality issues. Technology enables the dissemination of real-time air quality information, pollution forecasts, and health advisories to the public through various digital platforms, including websites, mobile applications,

Table 1 Technologies with behavioral consequences for mitigating air pollution

Technology/ innovation	Description	Examples	Benefits	References
Air Quality Monitoring	Sensors, mobile apps/software, and devices to measure air pollution (in real time)	Low-cost sensors and remote monitoring systems	Real-time data on air quality, early detection of pollution sources, and public awareness and education	Kumar and Morawska [129] and Karagulian et al. [130]
Clean Energy Technologies	Renewable energy sources for reducing emissions	Solar panels, wind turbines, and electric vehicles	Reduction of greenhouse gas emissions as well as promotion of energy independence and sustainable development	Jacobson and Delucchi [131] and Turrentine et al. [132]
Smart Grid Systems	Intelligent electricity distribution networks	Demand-response systems and energy-efficient appliances	Optimization of energy consumption, integration of renewable energy sources, and reduction of energy waste	Mohsenian-Rad et al. [133] and Han et al. [134]
Transportation Solutions	Innovative transportation methods to reduce emissions	Public transit systems, carpooling and ride-sharing apps, and bicycle-sharing programs	Decreased traffic congestion, reduced air pollution, and greenhouse gas emissions improved public health and quality of life	Noland and Cowart [135] and Shaheen et al. [136]
Green Building Design	Sustainable construction practices for cleaner indoor air	Energy-efficient heating, ventilation, and air condition systems, green roofs and walls, and low-emission building materials	Improved indoor air quality, energy efficiency cost savings, and reduction of environmental impact	Santamouris and Asimakopoulos [137] and Rock and Singh [138]
Public Awareness Campaigns	Educational initiatives to promote clean air behaviors	Social media campaigns, community workshops, events, and school technology-assisted programs for air pollution control	Increased awareness and knowledge, behavior change toward sustainable practices, and collaboration and community engagement	Abrahamse et al. [139]
Green Transportation Technologies	Vehicles with lower emissions and higher fuel efficiency	Electric vehicles, bicycles, green trains, carpooling, etc.	Reduction of air pollution and greenhouse gas emissions, improved fuel economy and cost savings, and enhanced public health and environmental quality	Shahin and El-Khatib [140]

(continued)

Table 1 (continued)

Technology/innovation	Description	Examples	Benefits	References
Urban Planning Strategies	Design and development of cities for sustainable air quality	Green roofs, green walls, public gardens, etc.	Promotion of active transportation (walking, cycling), green space planning and preservation, mixed land use, and transit-oriented development. Reduction of car dependency and traffic congestion, improved public health and quality of life, and enhanced environmental sustainability	Vyas and Reddy [141]
Industrial Pollution Control	Technologies and practices to reduce industrial emissions	Use of renewable energy, pollution capture technologies, etc.	Air pollution abatement systems, clean production processes, and emission trading schemes, reduction of industrial emissions and air pollutants, compliance with environmental regulations, and protection of public health and environmental quality	Moser [142]
Indoor Air Quality Technologies	Solutions to improve indoor air quality	Low-emitting stoves and heaters, scrubbers, catalytic converters, use of VOC-free products, etc.	Ventilation systems with air filters, indoor plants for air purification, low-emission building materials, enhanced indoor air quality and health, and improved occupant comfort and productivity	Mata et al. [143]

and social media channels. Access to accurate and timely information empowers individuals to make informed decisions about their activities and behaviors based on current air quality conditions. The AirVisual app provides users with real-time air quality data, pollution forecasts, and personalized health recommendations based on their location. By delivering actionable information to users' smartphones, the app raises awareness of air quality issues and encourages behavior change to reduce exposure to harmful pollutants.

Technology allows for the collection and analysis of data on individual behaviors and environmental exposures, enabling personalized feedback and recommendations tailored to users' preferences and circumstances. Personalized feedback mechanisms provide users with insights into their pollution-reducing behaviors, identify areas for improvement, and motivate continued engagement. Smart home energy monitoring systems track energy usage patterns and provide homeowners with personalized feedback on their energy consumption habits. By visualizing energy usage data and offering suggestions for energy-saving actions, these systems encourage users to adopt more sustainable behaviors and reduce their environmental footprint. Gamification techniques integrate elements of game design, such as rewards, challenges, and competitions, into behavioral change interventions to make them more engaging and motivating. Gamified platforms leverage intrinsic and extrinsic rewards to incentivize pollution-reducing behaviors and foster a sense of achievement and progress. The "Smogathon" initiative organizes hackathons and innovation challenges to develop technology-based solutions for air quality monitoring, pollution reduction, and public engagement. By gamifying the innovation process and offering prizes for winning solutions, Smogathon encourages participation from diverse stakeholders and fosters creativity in addressing air quality challenges. Technology facilitates community engagement and social networking platforms that connect individuals, organizations, and communities interested in air quality improvement. Online forums, social media groups, and virtual communities provide opportunities for information sharing, collaboration, and collective action to address air quality issues at the grassroots level. The "MyAirVisual" platform enables users to join local air quality communities, share pollution data, and participate in citizen science projects to monitor and report air quality in their neighborhoods. By fostering community engagement and peer support, MyAirVisual empowers users to take collective action for cleaner air and advocate for policy change.

Sensor technologies and Internet of Things (IoT) devices enable real-time monitoring of air quality parameters, such as particulate matter, ozone, and nitrogen dioxide, at a granular level. These sensors can be deployed in homes, schools, workplaces, and public spaces to provide localized air quality data and empower individuals to take proactive measures to reduce exposure to pollution. Air quality monitoring networks, such as "PurpleAir" and "Aircasting," utilize low-cost sensors and crowdsourcing techniques to collect and share air quality data in communities worldwide. By democratizing access to air quality information and engaging citizen scientists in data collection efforts, these platforms promote public awareness and empower communities to advocate for cleaner air. Mobile apps offer accessible and user-friendly platforms for accessing real-time air quality information, pollution forecasts, and health advisories. These apps provide users with personalized insights into local air quality conditions, empowering them to make informed decisions to protect their health. Examples of mobile apps for air quality monitoring include AirVisual, Plume Labs' Air Report, and the Environmental Protection Agency's (EPA) AirNow app. These apps use data from monitoring stations, satellite observations, and crowd-sourced sources to provide users with accurate and up-to-date information on pollution levels in their area. Mobile apps also incorporate features such as personalized health recommendations, pollution alerts, and air quality maps to help users understand the health impacts of air pollution and take appropriate actions to reduce exposure. Sensor technologies enable the deployment of low-cost air quality sensors for real-time monitoring of pollution levels at a granular level. These sensors can be installed in homes, schools, workplaces, and public spaces, providing localized air quality data and empowering individuals to take proactive measures to reduce exposure to harmful pollutants. Examples of sensor-based air quality monitoring systems include PurpleAir, Aircasting, and AirBeam. These systems use compact, portable sensors to measure particulate matter (PM), volatile organic compounds (VOCs), nitrogen dioxide (NO2), and other pollutants, allowing users to track pollution levels in real time and make informed decisions about their activities. Sensor networks also support citizen science initiatives, enabling communities to collect and share air quality data, identify pollution sources, and advocate for policy change. By democratizing access to air quality information, sensor technologies empower individuals to take action to improve air quality in their communities.

Online platforms provide virtual spaces for community engagement, information sharing, and collaboration on air quality issues. These platforms connect individuals, organizations, and communities interested in air quality improvement, fostering dialogue, raising awareness, and mobilizing collective action. Examples of online platforms for air quality engagement include MyAirVisual, OpenAQ, and the World Air Quality Index (WAQI). These platforms offer interactive maps, data visualization tools, and discussion forums where users can access air quality data, share experiences, and collaborate on initiatives to address pollution sources. Online platforms also support public awareness campaigns, educational outreach, and advocacy efforts, providing resources, tools, and information to empower individuals and communities to advocate for clean air policies and practices. Innovative technological solutions have a profound impact on public awareness and engagement with air quality issues. By providing accessible, real-time air quality information and personalized feedback, these solutions raise awareness of the health risks associated with air pollution and motivate behavior change. Technology-enabled engagement platforms facilitate community mobilization and collective action for air quality improvement. Citizens can use these platforms to report pollution incidents, organize clean-up events, and advocate for policy change, amplifying their voices and driving positive change at the grassroots level. Through data visualization, storytelling, and interactive features, technology enhances the communication of complex air quality concepts and promotes understanding among diverse audiences.

By making air quality data accessible and engaging, technological innovations empower individuals to take ownership of their environment and work together to create cleaner, healthier communities.

7 Challenges and Opportunities in Promoting Behavioral Changes and Increased Public Awareness for Sustainable Clean Air

Promoting behavioral changes and public awareness for sustainable clean air presents both challenges and opportunities that require collaborative efforts, innovation, and policy support. Below are some key challenges and opportunities in this area, along with recommendations for addressing them:

7.1 Challenges

Some challenges associated with promoting behavioral changes and increased public awareness for better air quality include the following:

- **Lack of awareness**: Many people are unaware of the health impacts of air pollution and the actions they can take to improve air quality. Limited access to information and low levels of environmental literacy contribute to this lack of awareness.
- **Behavioral barriers**: Individuals may face barriers to adopting pollution-reducing behaviors due to factors such as convenience, cost, and social norms. Ingrained habits, competing priorities, and perceived barriers to action can hinder behavior change efforts.
- **Information overload**: Despite the availability of air quality information through various channels, individuals may feel overwhelmed by the volume of data and struggle to interpret its relevance to their daily lives. Information overload can lead to disengagement and inaction.
- **Equity and access**: Vulnerable populations, including low-income communities and marginalized groups, may face disproportionate exposure to air pollution and barriers to accessing clean air resources and information. Addressing equity issues is essential for promoting inclusive and equitable solutions. Sometimes behavior change may inadvertently blame the victim by putting the onus to address air pollution on individuals who are least responsible for those emissions [144].

7.2 Opportunities

Some opportunities associated with promoting behavioral changes and increased public awareness for better air quality include the following:

- **Collaboration and partnerships:** Collaborative efforts between governments, nonprofit organizations, businesses, academia, and communities can amplify impact and reach diverse audiences. Partnerships foster knowledge sharing, resource mobilization, and collective action for air quality improvement.
- **Innovative technologies:** Advances in technology, such as mobile apps, sensor networks, and online platforms, offer innovative solutions for raising awareness, monitoring air quality, and promoting behavior change. These technologies enhance accessibility, engagement, and data-driven decision-making.
- **Policy support:** Supportive policy frameworks and regulations can incentivize pollution-reducing behaviors, invest in clean technologies, and prioritize public health. Policies that integrate behavioral insights provide incentives for sustainable practices and promote community engagement are essential for driving change.
- **Education and outreach:** Educational programs, outreach campaigns, and community-based initiatives play a crucial role in raising awareness, building environmental literacy, and empowering individuals to act. Targeted messaging, culturally relevant materials, and interactive activities can increase engagement and motivation.

7.3 Recommendations

Some recommendations for promoting behavioral changes and increased public awareness for better air quality include the following:

- **Invest in research:** There is a need to conduct research to better understand the drivers of behavior change related to air quality, including psychological factors, social dynamics, and environmental influences. Explore effective communication strategies, behavior change techniques, and interventions tailored to different audiences and contexts.
- **Empower communities**: Provide opportunities for empowering communities to take ownership of air quality issues through participatory approaches, citizen science initiatives, and grassroots advocacy. Provide resources, training, and support for community-led monitoring, action planning, and policy advocacy efforts. It is important to involve communities in the design of the interventions.
- **Foster innovation:** Encourage innovation in technology, data analytics, and communication tools for air quality monitoring, public engagement, and behavior change. Support entrepreneurs, researchers, and developers in creating user-

friendly, accessible, and scalable solutions that empower individuals and communities to improve air quality.
- **Strengthen policy action:** Advocate for policies and regulations that prioritize air quality improvement, promote sustainable practices, and address equity and social justice considerations. Engage policymakers, stakeholders, and the public in policy dialogue, decision-making processes, and implementation efforts.

8 Conclusion

Harnessing behavioral changes and public awareness is paramount in achieving sustainable clean air and safeguarding public health and the environment. This chapter has highlighted the critical role of behavioral science, community engagement, technology, and policy support in driving positive change toward cleaner air. Behavioral changes, driven by informed decision-making and motivation, are essential for individuals to adopt pollution-reducing behaviors in their daily lives. Public awareness plays a crucial role in raising understanding of air pollution issues, its health impacts, and the actions individuals can take to mitigate pollution. By empowering individuals with knowledge and motivation, we can foster a collective sense of responsibility and action toward cleaner air. Additionally, technological innovations offer powerful tools for enhancing public awareness, facilitating behavior change, and monitoring air quality in real time. Mobile apps, sensor technologies, and online platforms provide accessible and user-friendly platforms for accessing air quality information, receiving personalized feedback, and engaging with communities.

Furthermore, policy support is essential for creating an enabling environment for behavior change and public awareness initiatives. Supportive policy frameworks can incentivize sustainable practices, invest in clean technologies, and prioritize public health in air quality management efforts. There is a need to continue harnessing the power of behavioral science, community engagement, technology, and policy support to address air pollution and promote a healthier environment for all. By working together, leveraging innovative solutions, and advocating for supportive policies, humanity can create cleaner air and healthier communities for current and future generations. A future where clean air is a reality for everyone is a possibility.

References

1. Abbass K, Qasim MZ, Song H, Murshed M, Mahmood H, Younis I (2022) A review of the global climate change impacts, adaptation, and sustainable mitigation measures. Environ Sci Pollut Res Int 29(28):42539–42559. https://doi.org/10.1007/s11356-022-19718-6
2. Nishida C, Yatera K (2022) The impact of ambient environmental and occupational pollution on respiratory diseases. Int J Environ Res Public Health 19(5):2788. https://doi.org/10.3390/ijerph19052788

3. Manisalidis I, Stavropoulou E, Stavropoulos A, Bezirtzoglou E (2020) Environmental and health impacts of air pollution: a review. Front Public Health 8:14. https://doi.org/10.3389/fpubh.2020.00014
4. Wu IP, Liao SL, Lai SH, Wong KS (2022) The respiratory impacts of air pollution in children: global and domestic (Taiwan) situation. Biom J 45(1):88–94. https://doi.org/10.1016/j.bj.2021.12.004
5. von Schneidemesser E, Driscoll C, Rieder HE, Schiferl LD (2020) How will air quality effects on human health, crops and ecosystems change in the future? Philos Transact A Math Phys Eng Sci 378(2183):20190330. https://doi.org/10.1098/rsta.2019.0330
6. Ghorani-Azam A, Riahi-Zanjani B, Balali-Mood M (2016) Effects of air pollution on human health and practical measures for prevention in Iran. J Res Med Sci 21:65. https://doi.org/10.4103/1735-1995.189646
7. Kalair AR, Seyedmahmoudian M, Stojcevski A, Abas N, Khan N (2021) Waste to energy conversion for a sustainable future. Heliyon 7(10):e08155. https://doi.org/10.1016/j.heliyon.2021.e08155
8. Shivanna KR (2022) Climate change and its impact on biodiversity and human welfare. Proc Indian Nat Sci Acad A Phys Sci 88(2):160–171. https://doi.org/10.1007/s43538-022-00073-6
9. Zeb B, Alam K, Sorooshian A, Blaschke T, Ahmad I, Shahid I (2018) On the morphology and composition of particulate matter in an urban environment. Aerosol Air Qual Res 18(6):1431–1447. https://doi.org/10.4209/aaqr.2017.09.0340
10. Thangavel P, Park D, Lee YC (2022) Recent insights into particulate matter ($PM_{2.5}$)-mediated toxicity in humans: an overview. Int J Environ Res Public Health 19(12):7511. https://doi.org/10.3390/ijerph19127511
11. Jarvis DJ, Adamkiewicz G, Heroux ME et al (2010) Nitrogen dioxide. In: WHO guidelines for indoor air quality: selected pollutants. World Health Organization, Geneva, p 5. Available from: https://www.ncbi.nlm.nih.gov/books/NBK138707/
12. César AC, Carvalho Jr JA, Nascimento LF (2015) Association between NOx exposure and deaths caused by respiratory diseases in a medium-sized Brazilian city. Braz J Med Biol Res 48(12):1130–1135. https://doi.org/10.1590/1414-431X20154396
13. Lee BJ, Kim B, Lee K (2014) Air pollution exposure and cardiovascular disease. Toxicol Res 30(2):71–75. https://doi.org/10.5487/TR.2014.30.2.071
14. Reno AL, Brooks EG, Ameredes BT (2015) Mechanisms of heightened airway sensitivity and responses to inhaled SO2 in asthmatics. Environ Health Insights 9(Suppl 1):13–25. https://doi.org/10.4137/EHI.S15671
15. Kinoshita H, Türkan H, Vucinic S, Naqvi S, Bedair R, Rezaee R, Tsatsakis A (2020) Carbon monoxide poisoning. Toxicol Rep 7:169–173. https://doi.org/10.1016/j.toxrep.2020.01.005
16. Olson K, Smollin C (2008) Carbon monoxide poisoning (acute). BMJ Clin Evid 2008:2103
17. Torres G, Mourad M, Leheste JR (2022) Indoor air pollution and decision-making behavior: an interdisciplinary review. Cureus 14(6):e26247. https://doi.org/10.7759/cureus.26247
18. Bai G, Bai Y (2020) Voluntary or forced: different effects of personal and social norms on urban residents' environmental protection behavior. Int J Environ Res Public Health 17(10):3525. https://doi.org/10.3390/ijerph17103525
19. Niu N, Fan W, Ren M, Li M, Zhong Y (2023) The role of social norms and personal costs on pro-environmental behavior: the mediating role of personal norms. Psychol Res Behav Manag 16:2059–2069. https://doi.org/10.2147/PRBM.S411640
20. Bergquist M, Thiel M, Goldberg MH, van der Linden S (2023) Field interventions for climate change mitigation behaviors: a second-order meta-analysis. Proc Natl Acad Sci U S A 120(13):e2214851120. https://doi.org/10.1073/pnas.2214851120
21. Duckworth AL, Gross JJ (2020) Behavior change. Organ Behav Hum Decis Process 161 (Suppl):39–49. https://doi.org/10.1016/j.obhdp.2020.09.002
22. Barnes B, Rosenbaum J, Mehta S, Williams KN, Jagoe K, Graham J (2015) Behavior change communication: a key ingredient for advancing clean cooking. J Health Commun 20(sup1):3–5. https://doi.org/10.1080/10810730.2014.996305

23. Alzayani S, Alsabbagh M (2022) Public awareness and practices towards health impacts of PM2.5 in the Kingdom of Bahrain: identifying areas for intervention. Environ Anal Health Toxicol 37(2):e2022014. https://doi.org/10.5620/eaht.2022014
24. Ramírez AS, Ramondt S, Van Bogart K, Perez-Zuniga R (2019) Public awareness of air pollution and health threats: challenges and opportunities for communication strategies to improve environmental health literacy. J Health Commun 24(1):75–83. https://doi.org/10.1080/10810730.2019.1574320
25. Ward F, Lowther-Payne HJ, Halliday EC, Dooley K, Joseph N, Livesey R, Moran P, Kirby S, Cloke J (2022) Engaging communities in addressing air quality: a scoping review. Environ Health 21(1):89. https://doi.org/10.1186/s12940-022-00896-2
26. Tan T, Junghans C, Varaden D (2023) Empowering community health professionals for effective air pollution information communication. BMC Public Health 23(1):2547. https://doi.org/10.1186/s12889-023-17462-1
27. Puiu S, Udriştioiu MT, Velea L (2022) Air pollution management: a multivariate analysis of Citizens' perspectives and their willingness to use greener forms of transportation. Int J Environ Res Public Health 19(21):14613. https://doi.org/10.3390/ijerph192114613
28. Hubbell BJ, Kaufman A, Rivers L, Schulte K, Hagler G, Clougherty J, Cascio W, Costa D (2018) Understanding social and behavioral drivers and impacts of air quality sensor use. Sci Total Environ 621:886–894. https://doi.org/10.1016/j.scitotenv.2017.11.275
29. McCarron A, Semple S, Braban CF, Swanson V, Gillespie C, Price HD (2023) Public engagement with air quality data: using health behaviour change theory to support exposure-minimising behaviours. J Expo Sci Environ Epidemiol 33(3):321–331. https://doi.org/10.1038/s41370-022-00449-2
30. Bobulski J, Szymoniak S, Pasternak K (2024) An IoT system for air pollution monitoring with safe data transmission. Sensors 24(2):445. https://doi.org/10.3390/s24020445
31. Anadon LD, Chan G, Harley AG, Matus K, Moon S, Murthy SL, Clark WC (2016) Making technological innovation work for sustainable development. Proc Natl Acad Sci U S A 113(35):9682–9690. https://doi.org/10.1073/pnas.1525004113
32. Mosca O, Manunza A, Manca S, Vivanet G, Fornara F (2024) Digital technologies for behavioral change in sustainability domains: a systematic mapping review. Front Psychol 14:1234349. https://doi.org/10.3389/fpsyg.2023.1234349
33. Debrah JK, Vidal DG, Dinis MAP (2021) Raising awareness on solid waste management through formal education for sustainability: a developing countries evidence review. Recycling 6:6. https://doi.org/10.3390/recycling6010006
34. Fraisl D, Hager G, Bedessem B et al (2022) Citizen science in environmental and ecological sciences. Nat Rev Methods Primers 2:64. https://doi.org/10.1038/s43586-022-00144-4
35. Gignac F, Righi V, Toran R, Errandonea LP, Ortiz R, Nieuwenhuijsen M, Creus J, Basagaña X, Balestrini M (2022) Co-creating a local environmental epidemiology study: the case of citizen science for investigating air pollution and related health risks in Barcelona, Spain. Environ Health 21(1):11. https://doi.org/10.1186/s12940-021-00826-8
36. Academy of Science of South Africa, Brazilian Academy of Sciences, German National Academy of Sciences Leopoldina, U. S. National Academy of Medicine, & U. S. National Academy of Sciences (2019) Air pollution and health – a science-policy initiative. Ann Glob Health 85(1):140. https://doi.org/10.5334/aogh.2656
37. Saadeh R, Khader Y, Malkawi M, Allouh MZ (2022) Communicating the risks of air pollution to the public: a perspective from Jordan and Lebanon. Environmental Health Insights 16: 11786302221127851. https://doi.org/10.1177/11786302221127851
38. Yang X, Zhang Z, Rao S, Liu B, Li Y (2022) How does environmental information disclosure affect pollution emissions: firm-level evidence from China. Int J Environ Res Public Health 19(19):12763. https://doi.org/10.3390/ijerph191912763
39. Ngobeni FC, Kekana MP (2024) Community health risk awareness and knowledge of air pollution in Annadale, Polokwane local municipality, Limpopo Province, South Africa.

Environmental Health Insights 18:11786302241226775. https://doi.org/10.1177/11786302241226775
40. Nguyen-Van P, Stenger A, Tiet T (2021) Social incentive factors in interventions promoting sustainable behaviors: a meta-analysis. PLoS One 16(12):e0260932. https://doi.org/10.1371/journal.pone.0260932
41. Stegeman I, Godfrey A, Romeo-Velilla M, Bell R, Staatsen B, van der Vliet N, Kruize H, Morris G, Taylor T, Strube R, Anthun K, Lillefjell M, Zvěřinová I, Ščasný M, Máca V, Costongs C (2020) Encouraging and enabling lifestyles and Behaviours to simultaneously promote environmental sustainability, health and equity: key policy messages from INHERIT. Int J Environ Res Public Health 17(19):7166. https://doi.org/10.3390/ijerph17197166
42. Rau H, Nicolai S, Stoll-Kleemann S (2022) A systematic review to assess the evidence-based effectiveness, content, and success factors of behavior change interventions for enhancing pro-environmental behavior in individuals. Front Psychol 13:901927. https://doi.org/10.3389/fpsyg.2022.901927
43. Riley R, de Preux L, Capella P et al (2021) How do we effectively communicate air pollution to change public attitudes and behaviours? A review. Sustain Sci 16:2027–2047. https://doi.org/10.1007/s11625-021-01038-2
44. Kinzig AP, Ehrlich PR, Alston LJ, Arrow K, Barrett S, Buchman TG, Daily GC, Levin B, Levin S, Oppenheimer M, Ostrom E, Saari D (2013) Social norms and global environmental challenges: the complex interaction of behaviors, values, and policy. Bioscience 63(3):164–175. https://doi.org/10.1525/bio.2013.63.3.5
45. Sprague Martinez L, Ginzburg SL, Ron S, Brinkerhoff CA, Haque S, England SA, Khimani K, Zamore W, Reisner E, Lowe L, Brugge D (2024) Communities catalyzing change with data to mitigate an invisible menace, traffic-related air pollution. BMC Public Health 24(1):411. https://doi.org/10.1186/s12889-024-17864-9
46. Aubourg MA, Sawtell G, Deanes L, Fabricant N, Thomas M, Spicer K, Wagar C, Campbell S, Ulman A, Heaney CD (2023) Community-driven research and capacity building to address environmental justice concerns with industrial air pollution in Curtis Bay, South Baltimore. Front Epidemiol 3:1198321. https://doi.org/10.3389/fepid.2023.1198321
47. Amann M, Kiesewetter G, Schöpp W, Klimont Z, Winiwarter W, Cofala J, Rafaj P, Höglund-Isaksson L, Gomez-Sabriana A, Heyes C, Purohit P, Borken-Kleefeld J, Wagner F, Sander R, Fagerli H, Nyiri A, Cozzi L, Pavarini C (2020) Reducing global air pollution: the scope for further policy interventions. Philos Transact A Math Phys Eng Sci 378(2183):20190331. https://doi.org/10.1098/rsta.2019.0331
48. Cori L, Donzelli G, Gorini F, Bianchi F, Curzio O (2020) Risk perception of air pollution: a systematic review focused on particulate matter exposure. Int J Environ Res Public Health 17(17):6424. https://doi.org/10.3390/ijerph17176424
49. Engler JO, Abson DJ, von Wehrden H (2019) Navigating cognition biases in the search of sustainability. Ambio 48(6):605–618. https://doi.org/10.1007/s13280-018-1100-5
50. Perret JK, Udalov V, Fabisch N (2022) Motivations behind individuals' energy efficiency investments and daily energy-saving behavior: the case of China. IEEP 19(1):129–155. https://doi.org/10.1007/s10368-021-00521-6
51. Vieira J, Castro SL, Souza AS (2023) Psychological barriers moderate the attitude-behavior gap for climate change. PLoS One 18(7):e0287404. https://doi.org/10.1371/journal.pone.0287404
52. Tian J, Zheng X, Sun Y (2023) Fostering public climate change discussions from a social interaction perspective. Front Psychol 14:1258150. https://doi.org/10.3389/fpsyg.2023.1258150
53. Vlasceanu M, Doell KC, Bak-Coleman JB, Todorova B, Berkebile-Weinberg MM, Grayson SJ, Patel Y, Goldwert D, Pei Y, Chakroff A, Pronizius E, van den Broek KL, Vlasceanu D, Constantino S, Morais MJ, Schumann P, Rathje S, Fang K, Aglioti SM, Alfano M et al (2024) Addressing climate change with behavioral science: a global intervention tournament in 63 countries. Sci Adv 10(6):eadj5778. https://doi.org/10.1126/sciadv.adj5778

54. Ferguson L, Taylor J, Davies M, Shrubsole C, Symonds P, Dimitroulopoulou S (2020) Exposure to indoor air pollution across socio-economic groups in high-income countries: a scoping review of the literature and a modelling methodology. Environ Int 143:105748. https://doi.org/10.1016/j.envint.2020.105748
55. Hajat A, Hsia C, O'Neill MS (2015) Socioeconomic disparities and air pollution exposure: a global review. Curr Environ Health Rep 2(4):440–450. https://doi.org/10.1007/s40572-015-0069-5
56. Ajzen I (1985) From intentions to actions: a theory of planned behavior. From cognition to behavior, Action control, pp 11–39
57. Bosnjak M, Ajzen I, Schmidt P (2020) The theory of planned behavior: selected recent advances and applications. Eur J Psychol 16(3):352–356. https://doi.org/10.5964/ejop.v16i3.3107
58. Jones CL, Jensen JD, Scherr CL, Brown NR, Christy K, Weaver J (2015) The health belief model as an explanatory framework in communication research: exploring parallel, serial, and moderated mediation. Health Commun 30(6):566–576. https://doi.org/10.1080/10410236.2013.873363
59. Kim S, Kim S (2020) Analysis of the impact of health beliefs and resource factors on preventive behaviors against the COVID-19 pandemic. Int J Environ Res Public Health 17(22):8666. https://doi.org/10.3390/ijerph17228666
60. Dearing JW (2009) Applying diffusion of innovation theory to intervention development. Res Soc Work Pract 19(5):503–518. https://doi.org/10.1177/1049731509335569
61. Bandura A (1971) Social learning theory. General Learning Press, New York
62. Bandura A (1972) Modeling theory: some traditions, trends, and disputes. In: Parke RD (ed) Recent trends in social learning theory. Academic Press, New York, pp 35–61
63. Bandura A (1977) Social learning theory. Prentice Hall, Englewood Cliffs, NJ
64. Mukhalalati B, Elshami S, Eljaam M, Hussain FN, Bishawi AH (2022) Applications of social theories of learning in health professions education programs: a scoping review. Front Med 9:912751. https://doi.org/10.3389/fmed.2022.912751
65. Hallsworth M (2023) A manifesto for applying behavioural science. Nat Hum Behav 7:310–322. https://doi.org/10.1038/s41562-023-01555-3
66. Fries S, Cook J, Lynes JK (2020) Community-based social marketing in theory and practice: five case studies of water efficiency programs in Canada. Soc Mark Q 26(4):325–344. https://doi.org/10.1177/1524500420971170
67. D'Antoni D, Smith L, Auyeung V et al (2017) Psychosocial and demographic predictors of adherence and non-adherence to health advice accompanying air quality warning systems: a systematic review. Environ Health 16:100. https://doi.org/10.1186/s12940-017-0307-4
68. Kureshi RR, Thakker D, Mishra BK, Barnes J (2023) From raising awareness to a Behavioural change: a case study of indoor air quality improvement using IoT and COM-B model. Sensors 23(7):3613. https://doi.org/10.3390/s23073613
69. Noël C, Van Landschoot L, Vanroelen C, Gadeyne S (2022) The Public's perceptions of air pollution. What's in a name? Environmental Health Insights 16:11786302221123563. https://doi.org/10.1177/11786302221123563
70. Chandra M, Rai CB, Kumari N, Sandhu VK, Chandra K, Krishna M, Kota SH, Anand KS, Oudin A (2022) Air pollution and cognitive impairment across the life course in humans: a systematic review with specific focus on income level of study area. Int J Environ Res Public Health 19(3):1405. https://doi.org/10.3390/ijerph19031405
71. Korteling JEH, Paradies GL, Sassen-van Meer JP (2023) Cognitive bias and how to improve sustainable decision making. Front Psychol 14:1129835. https://doi.org/10.3389/fpsyg.2023.1129835
72. Cook DA, Artino Jr AR (2016) Motivation to learn: an overview of contemporary theories. Med Educ 50(10):997–1014. https://doi.org/10.1111/medu.13074

73. Simpson EH, Balsam PD (2016) The behavioral neuroscience of motivation: an overview of concepts, measures, and translational applications. Curr Top Behav Neurosci 27:1–12. https://doi.org/10.1007/7854_2015_402
74. De Dominicis S, Schultz PW, Bonaiuto M (2017) Protecting the environment for self-interested reasons: altruism is not the only pathway to sustainability. Front Psychol 8:1065. https://doi.org/10.3389/fpsyg.2017.01065
75. Si W, Jiang C, Meng L (2022) The relationship between environmental awareness, habitat quality, and community Residents' pro-environmental behavior-mediated effects model analysis based on social capital. Int J Environ Res Public Health 19(20):13253. https://doi.org/10.3390/ijerph192013253
76. Barnes BR (2010) Why do caregivers protect their children from indoor air pollution exposure? MILA 11:1–16
77. Nakagawa M, Lefebvre M, Stenger A (2022) Long-lasting effects of incentives and social preference: a public goods experiment. PLoS One 17(8):e0273014. https://doi.org/10.1371/journal.pone.0273014
78. Tigges BB (2009) Psychometric properties of the social comparison motives scale. J Nurs Meas 17(1):29–44. https://doi.org/10.1891/1061-3749.17.1.29
79. Sidarus N, Palminteri S, Chambon V (2019) Cost-benefit trade-offs in decision-making and learning. PLoS Comput Biol 15(9):e1007326. https://doi.org/10.1371/journal.pcbi.1007326
80. Ekins P, Zenghelis D (2021) The costs and benefits of environmental sustainability. Sustain Sci 16(3):949–965. https://doi.org/10.1007/s11625-021-00910-5
81. Duan J, Wu SJ, Sun L (2017) Do the powerful discount the future less? The effects of power on temporal discounting. Front Psychol 8:1007. https://doi.org/10.3389/fpsyg.2017.01007
82. Cancino-Montecinos S, Björklund F, Lindholm T (2020) A general model of dissonance reduction: unifying past accounts via an emotion regulation perspective. Front Psychol 11:540081. https://doi.org/10.3389/fpsyg.2020.540081
83. Bosone L, Chevrier M, Zenasni F (2022) Consistent or inconsistent? The effects of inducing cognitive dissonance vs. cognitive consonance on the intention to engage in pro-environmental behaviors. Front Psychol 13:902703. https://doi.org/10.3389/fpsyg.2022.902703
84. Cobbold AT, Crane MA, Knibbs LD, Hanigan IC, Greaves SP, Rissel CE (2022) Perceptions of air quality and concern for health in relation to long-term air pollution exposure, bushfires, and COVID-19 lockdown: a before-and-after study. J Clim Change Health 6:100137. https://doi.org/10.1016/j.joclim.2022.100137
85. Cao J, Chen J (2021) The impact of an authoritarian personality on pro-environmental behaviour for air pollution mitigation through interactions with social norms. Int J Environ Res Public Health 18(17):9301. https://doi.org/10.3390/ijerph18179301
86. Zhang W, Liu Y, Dong Y, He W, Yao S, Xu Z, Mu Y (2023) How we learn social norms: a three-stage model for social norm learning. Front Psychol 14:1153809. https://doi.org/10.3389/fpsyg.2023.1153809
87. Wanders F, Homan AC, van Vianen AEM, Rahal RM, van Kleef GA (2021) How norm violators rise and fall in the eyes of others: the role of sanctions. PLoS One 16(7):e0254574. https://doi.org/10.1371/journal.pone.0254574
88. Fazio RH (2007) Attitudes as object-evaluation associations of varying strength. Soc Cogn 25(5):603–637. https://doi.org/10.1521/soco.2007.25.5.603
89. Begum A, Liu J, Qayum H, Mamdouh A (2022) Environmental and moral education for effective environmentalism: an ideological and philosophical approach. Int J Environ Res Public Health 19(23):15549. https://doi.org/10.3390/ijerph192315549
90. Ramondt S, Ramírez AS (2020) Media reporting on air pollution: health risk and precautionary measures in national and regional newspapers. Int J Environ Res Public Health 17(18):6516. https://doi.org/10.3390/ijerph17186516

91. La Barbera F, Ajzen I (2020) Control interactions in the theory of planned behavior: rethinking the role of subjective norm. Eur J Psychol 16(3):401–417. https://doi.org/10.5964/ejop.v16i3.2056
92. Suntornsan S, Chudech S, Janmaimool P (2022) The role of the theory of planned behavior in explaining the energy-saving behaviors of high school students with physical impairments. Behav Sci 12(9):334. https://doi.org/10.3390/bs12090334
93. Wakefield MA, Loken B, Hornik RC (2010) Use of mass media campaigns to change health behaviour. Lancet 376(9748):1261–1271. https://doi.org/10.1016/S0140-6736(10)60809-4
94. Rouyard T, Engelen B, Papanikitas A, Nakamura R (2022) Boosting healthier choices. BMJ 376:e064225. https://doi.org/10.1136/bmj-2021-064225
95. Vande Velde F, Overgaard HJ, Bastien S (2021) Nudge strategies for behavior-based prevention and control of neglected tropical diseases: a scoping review and ethical assessment. PLoS Negl Trop Dis 15(11):e0009239. https://doi.org/10.1371/journal.pntd.0009239
96. Li B, Yin S, Zhang B (2024) Incentive mechanism of multiple green innovation behaviors of equipment manufacturing enterprises: a managers, green coordination groups and employees perspective. PLoS One 19(3):e0300533. https://doi.org/10.1371/journal.pone.0300533
97. Petrescu DG, Tribus LC, Raducu R, Purcarea VL (2021) Social marketing and behavioral change. Rom J Ophthalmol 65(2):101–103. https://doi.org/10.22336/rjo.2021.21
98. Cislaghi B, Heise L (2018) Theory and practice of social norms interventions: eight common pitfalls. Glob Health 14(1):83. https://doi.org/10.1186/s12992-018-0398-x
99. Flores W, Samuel J (2019) Grassroots organisations and the sustainable development goals: no one left behind? BMJ:365. https://doi.org/10.1136/bmj.l2269
100. Michaelsen MM, Esch T (2023) Understanding health behavior change by motivation and reward mechanisms: a review of the literature. Front Behav Neurosci 17:1151918. https://doi.org/10.3389/fnbeh.2023.1151918
101. Drehobl A, Chiokumbo M, Tanabe K (2018) Reducing energy waste through municipally led behavior change programs. November 2018 Report U1810. American Council for an Energy Efficiency Economy. 63 p
102. Martin SL, Arney JK, Mueller LM, Kumakech E, Walugembe F, Mugisha E (2013) Using formative research to design a behavior change strategy to increase the use of improved cookstoves in peri-urban Kampala, Uganda. Int J Environ Res Public Health 10(12):6920–6938
103. Glanz K (1997) Theory at a glance: a guide for health promotion practice (no. 97). US Department of Health and Human Services, Public Health Service, National Institutes of Health, National Cancer Institute
104. Liu H, Hao P, Liao Y, Boriboonsomsin K, Barth M (2023) Model-based vehicle-miles traveled and emission evaluation of on-demand food delivery considering the impact of COVID-19 pandemic. Transp Res Rec:03611981231169276. https://doi.org/10.1177/03611981231169276
105. EPA (2024) Salt Lake City, Utah adapts to improve air quality through smart growth. Available online: https://www.epa.gov/arc-x/salt-lake-city-utah-adapts-improve-air-quality-through-smart-growth
106. Naumov S, Keith DR, Sterman JD (2023) Accelerating vehicle fleet turnover to achieve sustainable mobility goals. J Oper Manag 69(1):36–66. https://doi.org/10.1002/joom.1173
107. NC DEQ (2015) NC air awareness. Annual Report State Fiscal Year 2014–2015. North Carolina Department of Environmental Quality Division of Air Quality Planning Section. 32 p
108. Kim S, Park Y, Ackerman MK (2021) Designing an indoor air quality monitoring app for asthma management in children: user-centered design approach. JMIR Form Res 5(9):e27447. https://doi.org/10.2196/27447
109. Jonidi Jafari A, Charkhloo E, Pasalari H (2021) Urban air pollution control policies and strategies: a systematic review. J Environ Health Sci Eng 19(2):1911–1940. https://doi.org/10.1007/s40201-021-00744-4

110. Ruan H, Qiu L, Chen J, Liu S, Ma Z (2022) Government trust, environmental pollution perception, and environmental governance satisfaction. Int J Environ Res Public Health 19(16):9929. https://doi.org/10.3390/ijerph19169929
111. World Health Organization (2021) WHO global air quality guidelines: particulate matter (PM2.5 and PM10), ozone, nitrogen dioxide, sulfur dioxide and carbon monoxide [internet]. World Health Organization, Geneva, p 6. Implementation of the guidelines.Available from: https://www.ncbi.nlm.nih.gov/books/NBK574587/
112. Russell AG, Tolbert P, Henneman L, Abrams J, Liu C, Klein M, Mulholland J, Sarnat SE, Hu Y, Chang HH, Odman T, Strickland MJ, Shen H, Lawal A (2018) Impacts of regulations on air quality and emergency department visits in the Atlanta metropolitan area, 1999-2013. Res Rep Health Eff Inst 2018(195):1–93
113. Carey RN, Connell LE, Johnston M, Rothman AJ, de Bruin M, Kelly MP, Michie S (2019) Behavior change techniques and their mechanisms of action: a synthesis of links described in published intervention literature. Ann Behav Med 53(8):693–707. https://doi.org/10.1093/abm/kay078
114. Sarpong KA, Xu W, Gyamfi BA, Ofori EK (2023) Can environmental taxes and green-energy offer carbon-free E7 economies? An empirical analysis in the framework of COP-26. Environ Sci Pollut Res Int 30(18):51726–51739. https://doi.org/10.1007/s11356-023-25904-x
115. Murayama H, Takagi Y, Tsuda H, Kato Y (2023) Applying nudge to public health policy: practical examples and tips for designing nudge interventions. Int J Environ Res Public Health 20(5):3962. https://doi.org/10.3390/ijerph20053962
116. Trujillo CA, Estrada-Mejia C, Rosa JA (2021) Norm-focused nudges influence pro-environmental choices and moderate post-choice emotional responses. PLoS One 16(3): e0247519. https://doi.org/10.1371/journal.pone.0247519
117. Nazarenko Y, Pal D, Ariya PA (2021) Air quality standards for the concentration of particulate matter 2.5, global descriptive analysis. Bull World Health Organ 99(2):125D–137D. https://doi.org/10.2471/BLT.19.245704
118. King JS, Manning J, Woodward A (2023) In this together: international collaborations for environmental and human health. J Law Med Ethics 51(2):271–286. https://doi.org/10.1017/jme.2023.82
119. Grootjans SJM, Stijnen MMN, Kroese MEAL, Ruwaard D, Jansen MWJ (2022) Collaborative governance at the start of an integrated community approach: a case study. BMC Public Health 22(1):1013. https://doi.org/10.1186/s12889-022-13354-y
120. Taylor LA, Aveling EL, Roberts J, Bhuiya N, Edmondson A, Singer S (2023) Building resilient partnerships: how businesses and nonprofits create the capacity for responsiveness. Front Health Serv 3:1155941. https://doi.org/10.3389/frhs.2023.1155941
121. Naito R, Zhao J, Chan KMA (2022) An integrative framework for transformative social change: a case in global wildlife trade. Sustain Sci 17(1):171–189. https://doi.org/10.1007/s11625-021-01081-z
122. Yu W, Luo J (2022) Impact on carbon intensity of carbon emission trading-evidence from a pilot program in 281 cities in China. Int J Environ Res Public Health 19(19):12483. https://doi.org/10.3390/ijerph191912483
123. Mirzaee Ghazani M, Jafari MA (2021) The efficiency of CO_2 market in the phase III EU ETS: analyzing in the context of a dynamic approach. Environ Sci Pollut Res Int 28(43): 61080–61095. https://doi.org/10.1007/s11356-021-15044-5
124. Bayer P, Aklin M (2020) The European Union Emissions Trading System reduced CO_2 emissions despite low prices. Proc Natl Acad Sci U S A 117(16):8804–8812. https://doi.org/10.1073/pnas.1918128117
125. Last BS, Schriger SH, Timon CE, Frank HE, Buttenheim AM, Rudd BN, Fernandez-Marcote-S, Comeau C, Shoyinka S, Beidas RS (2021) Using behavioral insights to design implementation strategies in public mental health settings: a qualitative study of clinical decision-making. Implement Sci Commun 2(1):6. https://doi.org/10.1186/s43058-020-00105-6

126. Bynum C, Sze C, Kearns D, Polovick B, Simon K (2018) An examination of a voluntary policy model to effect behavioral change and influence interactions and decision making in the freight sector. Transp Res D Transp Environ 61(A):19–32. https://doi.org/10.1016/j.trd.2016.11.018
127. Atwoli L, Baqui AH, Benfield T, Bosurgi R, Godlee F, Hancocks S, Horton R, Laybourn-Langton L, Monteiro CA, Norman I, Patrick K, Praities N, Rikkert MGO, Rubin EJ, Sahni P, Smith R, Talley NJ, Turale S, Vázquez D (2021) Call for emergency action to limit global temperature increases, restore biodiversity, and protect health - wealthy nations must do much more, much faster. Rev Saude Publica 55:1ed. 10.11606/s1518-8787.2021055000 1ed
128. McCarthy G, Burke TA (2017) We need a strong Environmental Protection Agency: it's about public health! Am J Public Health 107(5):649–651. https://doi.org/10.2105/AJPH.2017.303728
129. Kumar P, Morawska L (2014) Could fighting air pollution be made smarter? Environ Sci Technol 48(5):2367–2368
130. Karagulian F, Gerboles M, Barbiere M, Kotsev A, Lagler F, Borowiak A (2019) Review of sensors for air quality monitoring, EUR 29826 EN. Publications Office of the European Union, Luxembourg, 2019, ISBN 978-92-76-09255-1. https://doi.org/10.2760/568261, JRC116534
131. Jacobson MZ, Delucchi MA (2011) Providing all global energy with wind, water, and solar power, part I: technologies, energy resources, quantities and areas of infrastructure, and materials. Energy Policy 39(3):1154–1169
132. Turrentine T, Kurani KS, Heffner RR (2011) Why consumers buy plug-in hybrid electric vehicles, and implications for industry policy. Institute of Transportation Studies, University of California, Davis
133. Mohsenian-Rad H, Leon-Garcia A, Wong VW (2010) Autonomous demand-side management based on game-theoretic energy consumption scheduling for the future smart grid. IEEE Trans Smart Grid 1(3):320–331
134. Han S, Han J, Seidel D (2012) A survey of smart grid technologies for sustainable development. IEEE Commun Surv Tutor 14(4):944–980
135. Noland RB, Cowart WA (2000) Analysis of metropolitan bus transit energy consumption. Transp Res D Transp Environ 5(1):1–18
136. Shaheen SA, Cohen AP, Zohdy IA (2016) Shared mobility: current practices and guiding principles. UC Berkeley, Transportation Sustainability Research Center
137. Santamouris M, Asimakopoulos D (2016) Passive cooling of buildings. Springer
138. Rock BA, Singh BR (2013) Determination of VOCs in indoor air of a newly constructed building and evaluation of their impact on human health. Environ Monit Assess 185(3):2341–2351
139. Abrahamse W, Steg L, Vlek C, Rothengatter T (2005) A review of intervention studies aimed at household energy conservation. J Environ Psychol 25(3):273–291
140. Shahin MA, El-Khatib SS (2008) A comprehensive review of hybrid electric vehicles. Renew Sustain Energy Rev 12(1):116–140
141. Vyas A, Reddy A (2014) Fuel economy improvements in India's passenger vehicle fleet: comparison of technological options and policy scenarios. Energy Policy 66:405–417
142. Moser SC (2007) More bad news: the risk of neglecting emotional responses to climate change information. Int J Global Warming 1(1–2):75–79
143. Mata TM, Oliveira GM, Monteiro H, Silva GV, Caetano NS, Martins AA (2021) Indoor air quality improvement using nature-based solutions: design proposals to greener cities. Int J Environ Res Public Health 18(16):8472. https://doi.org/10.3390/ijerph18168472
144. Lee KK, Bing R, Kiang J, Bashir S, Spath N, Stelzle D et al (2020) Adverse health effects associated with household air pollution: a systematic review, meta-analysis, and burden estimation study. Lancet Glob Health 8(11):e1427–e1434

International Partnerships and Agreements for Addressing Air Pollution

Zouina Sarfraz, Azza Sarfraz, and Muzna Sarfraz

Contents

Abstract This chapter critically examines the suite of international partnerships and multilateral agreements fundamental to combating global air pollution. As industrialization, urban expansion, and increased vehicular use continue to compromise air quality, the resultant health implications – such as respiratory diseases and cardiovascular problems – and environmental degradation pose urgent challenges worldwide. Addressing these issues necessitates robust international cooperation, given the transboundary nature of air pollutants. This analysis places significant emphasis on evaluating the efficacy of global and regional regulatory frameworks, such as the World Health Organization's air quality guidelines, which target key pollutants, and the United Nations Environment Programme's global standards. The chapter also discusses key national legislative measures and how they align with and support international objectives. Strategic international agreements such as the Paris Climate

Z. Sarfraz (✉)
Department of Medicine, Fatima Jinnah Medical University, Lahore, Pakistan

A. Sarfraz
Department of Pediatrics, The Aga Khan University, Karachi, Pakistan
e-mail: azza.sarfraz@aku.edu

M. Sarfraz
Department of ORIC, King Edward Medical University, Lahore, Pakistan

Matthew Chidozie Ogwu and Sylvester Izah (eds.), *Sustainable Strategies for Air Pollution Mitigation: Development, Economics and Technologies*, Hdb Env Chem (2024) 133: 513–548, DOI 10.1007/698_2024_1109,

Published online: 28 August 2024

Agreement and the European Union's stringent emission targets play critical roles in integrating air quality management with broader climate change mitigation efforts. The analysis in this chapter emphasizes the importance of technological advancements in air quality monitoring and pollution control technologies, which are disseminated and implemented through these international agreements. The synergy between international partnerships and technological innovation can catalyze the development and adoption of sophisticated air quality management strategies and drive collaboration. The collaboration, in turn, can enhance compliance with established standards but also facilitate shared innovation, capacity building, and technical support across nations. In conclusion, this chapter elucidates how international partnerships and agreements form an indispensable framework for effectively addressing air pollution. Moving forward, there is a need to foster international cooperation and continual technological progression on a global scale.

Keywords Air pollution, Climate change, Environmental policies, Health outcomes, Regional partnerships, Sustainable development, Technological innovations

Abbreviations

CCAC	Climate and clean air coalition
CBD	Convention on biological diversity
EJScreen	Environmental Justice Screening Method
EPA	Environmental Protection Agency
EVs	Electric vehicles
GAW	Global Atmosphere Watch Programme
GDP	Gross domestic product
HEPA	Health and Energy Platform of Action
HAPs	Hazardous air pollutants
IAQ	Indoor air quality
IGP-HF	Indo-Gangetic Plain and Himalayan Foothills
LMICs	Lower- and middle-income countries
LRTAP	Long-range transboundary air pollution
MEA	Multilateral environmental agreement
NDCs	Nationally determined contributions
NEACAP	North-East Asia Clean Air Partnership
NEASPEC	North-East Asian Subregional Programme for Environmental Cooperation
OECD	Organisation for Economic Co-operation and Development
PAHs	Polycyclic aromatic hydrocarbons
$PM_{2.5}$	Particulate matter less than 2.5 μm in diameter
SCP	Sustainable consumption and production
SDG	Sustainable development goal

UNEP	United Nations Environment Programme
UNFCCC	United Nations Framework Convention on Climate Change
US	United States
VOC	Volatile organic compound
WASH	Water, sanitation, and hygiene
WHO	World Health Organization
WMO	World Meteorological Organization

1 Introduction

In recent centuries, the composition of the Earth's atmosphere has been undergoing a significant shift, a trend that dramatically accelerated with the advent of the Industrial Revolution in Europe and North America in the last few centuries [1]. This period introduced a substantial change as human activities began to alter atmospheric conditions considerably. The last 50 years have seen an even more rapid evolution as a multitude of countries have engaged in swift economic growth.

Air pollution is characterized by the contamination of both indoor and outdoor environments with chemical, physical, or biological agents, which has altered the natural state of the atmosphere. This degradation is largely due to emissions from domestic combustion devices, vehicles, industrial activities, and wildfires [2]. Such pollution is a source of serious health concerns, releasing particulate matter, carbon monoxide, ozone, nitrogen dioxide, and sulfur dioxide into the air [3]. These pollutants are known to provoke severe respiratory disorders and other health problems, significantly contributing to morbidity and mortality rates [4, 5].

The link between air quality and the global climate and ecosystems is important. The overlap between sources of air pollution and greenhouse gas emissions indicates that efforts to mitigate air pollution could simultaneously address climate change, thereby benefiting public health and reducing the impacts of climate change in both the immediate and distant future [6].

The impact of air pollution on health and the environment cannot be overstated, with nearly seven million deaths globally attributed annually to this crisis [7]. The primary driver of air pollution is the combustion of fossil fuels, releasing a slew of harmful chemicals into the atmosphere [8]. This issue has necessitated significant legislative actions, like the US Clean Air Act (CAA), which grants the Environmental Protection Agency (EPA) the authority to regulate these pollutants to safeguard public health [9].

The generation and consumption of energy stand as the principal contributors to air pollution. Despite advancements made possible by regulations such as the CAA, climate change presents new obstacles in achieving air quality standards that protect health [10]. Air pollution ranks as a leading cause of premature death worldwide, imposing a heavy disease burden, especially in densely populated countries like China and India [11]. Climate change is expected to worsen these challenges, with increased heat and ultraviolet radiation intensifying smog and climate-related extreme weather conditions prolonging the pollen seasons, thus boosting allergenic pollutants like mold and pollen [12].

The study by Wu et al. [13] revealed a correlation between COVID-19 mortality rates and prolonged exposure to particulate matter, highlighting the environmental justice issue of disproportionate pollution burdens on marginalized communities. Numerous pollutants, including mercury, lead, dioxins, benzene, and polycyclic aromatic hydrocarbons (PAHs) – primarily released through the combustion of fuels and vehicle exhausts – pose severe health risks ranging from respiratory irritation to life-threatening conditions like cancer [14].

While greenhouse gases do not have the immediate health impacts of other pollutants, their contribution to climate change poses significant long-term health risks through the induction of warmer temperatures and more frequent extreme weather events [15]. Addressing air pollution through the adoption of renewable energy sources and enhanced industrial processes is crucial for public health and combating climate change [16].

In the USA, the enactment of the Clean Air Act has played a pivotal role in reducing air pollution [17]. However, the battle against air pollution faces new challenges due to climate change-induced phenomena such as wildfires and extreme heat [18]. The exposure disparities to air pollution underscore a critical environmental justice challenge, with marginalized groups at greater risk. The use of tools like the Environmental Justice Screening Method and the EPA's EJScreen is essential for pinpointing and addressing these inequalities [19].

To effectively manage and reduce air pollution, various global and regional partnerships have been implemented by different bodies. The UNEP's Global Assessment of Air Pollution Legislation (GAAPL) provides recommendations to enhance air quality governance [20]. This initiative serves as a resource aimed at supporting effective air quality legislation across numerous countries worldwide. The UNECE Convention on Long-Range Transboundary Air Pollution focuses on pollutants like sulfur dioxide and nitrogen oxides. It is specifically targeted at countries in Europe and North America, aiming to mitigate the effects of air pollution that transcend national borders [21]. The WHO Global Air Quality Guidelines, updated in 2021, set international standards for safe pollutant levels to protect public health. These guidelines are intended for global application, urging countries everywhere to adopt and implement stricter air quality measures [22]. The International Union of Air Pollution Prevention and Environmental Protection Associations (IUAPPA) coordinates the efforts of civil society organizations globally to combat air pollution, influencing a wide range of countries through advocacy and the sharing of best practices [23]. At the national level, various countries have implemented specific regulatory measures. For example, China has introduced air pollution control policies that include tradable pollutant permits in pilot provinces and cities [24]. Finland operates under the National Air Pollution Control Programme 2030, which is specifically designed to address and manage air pollution within the country [25]. In the Asia and Pacific region, air pollution measures are actively implemented in 23 countries, as outlined in regional cooperative agreements and national policies

[26]. These policies and initiatives are enacted by a combination of global organizations, regional bodies, and national governments, all aiming to collectively enhance air quality and public health on an international scale.

This chapter discusses various international policies, partnerships, and agreements that have been established to address air pollution. These global initiatives represent collaborative efforts to not only curb emissions of harmful pollutants but also inculcate a unified approach toward sustainable development and climate change mitigation. Through a detailed exploration of these international frameworks, we aim to understand their efficacy, challenges, and the role they play in shaping a cleaner, healthier future for our planet.

2 International Partnerships for Air Pollution Mitigation

Air pollution in ambient air is identified as a significant global health risk, with the World Health Organization (WHO) estimating that approximately 90% of urban dwellers are exposed to $PM_{2.5}$ levels surpassing the recommended guideline of 10 μg m [27]. Annually, between three and nine million premature deaths are attributed to exposure to ambient air pollution globally [28]. The UN Environment Programme (UNEP) has documented policies and programs from 195 countries targeting key sectors contributing to air pollution, including power generation, transportation, industrial emissions, household air pollution, solid waste management, and agriculture [29].

In particular, efforts have been made to enhance air quality monitoring, management, and adherence to air quality standards as primary policy instruments to mitigate air pollution impacts. As of 2020, 124 countries have established national ambient air quality standards, representing about two-thirds of the global community, an increase of 17 countries since 2016 [30]. However, merely 9% of these nations comply with the limits outlined by the WHO guidelines [29]. Notably, actions to reduce air pollution, although disproportionately affecting low-income countries, offer multiple development benefits encompassing climate mitigation, agricultural productivity, energy security, and economic growth [31].

Significant progress has been made since 2016 in various policy domains (Fig. 1).

This bar chart illustrates the number of countries that have implemented various environmental policies:

- **Vehicle emission standards**: A total of 71 countries have adopted new vehicle emission standards equivalent to Euro 4/IV or higher, with a growing focus on regulating emissions from imported used vehicles [32].
- **Cleaner production policies**: Cleaner production policies have been adopted by 108 countries, promoting environmentally sustainable industrial practices [33].
- **Clean cooking and heating programs**: Programs promoting clean cooking and heating have expanded to 95 countries, resulting in reduced rates of diseases

Fig. 1 Implementation of environmental policies by number of countries

linked to household air pollution, particularly in South and East Asia and the Pacific [34].

- **Solid waste burning regulations**: Regulations governing the burning of solid waste have been strengthened in 26 additional countries, totaling 38 nations, focusing on landfill gas capture, improved waste collection, segregation, and disposal methods [35].
- **Sustainable agriculture incentives**: Incentives for sustainable agriculture have been implemented by 58 countries, emphasizing methane emission mitigation strategies such as alternatives to open burning of agricultural residues, improved livestock manure management, composting, and methane capture for energy [36].
- Air quality monitoring: Despite these advancements, monitoring infrastructure remains inadequate, with only 57 out of 124 countries continuously monitoring air quality, leaving significant data gaps and capacity constraints impeding global progress on air quality management [29].

Furthermore, regional blocs such as the European Union (EU) and individual countries like the UK have committed to ambitious emissions reduction targets to combat air pollution and mitigate climate change [37]. The EU aims to achieve a net-zero economy by 2050, with Member States mandated to reduce emissions by at least 55% by 2030 as part of the European Green Deal [38]. Similarly, the UK targets a reduction in greenhouse gas emissions by more than 68% by 2030 compared to 1990 levels, ultimately striving for net-zero emissions by 2050 [39]. These objectives necessitate substantial reductions in emissions from various sectors including electricity generation, transportation, and residential energy consumption, alongside encouraging sustainable lifestyle choices such as reduced meat consumption and increased use of active transportation modes.

Research conducted in England and Wales demonstrates the health benefits associated with policy interventions targeting air pollution reduction [40]. By quantifying the impacts of different policy actions on mortality using a life expectancy model, researchers observed significant reductions in $PM_{2.5}$ concentrations and associated mortality rates under various decarbonization pathways. These findings highlight the importance of implementing comprehensive policy measures and fostering behavioral changes to achieve substantial improvements in air quality and public health outcomes.

The global community has recognized the urgent need to address the health impacts of air pollution. In response, significant efforts have been undertaken to develop comprehensive strategies and partnerships aimed at mitigating air pollution and its adverse effects on public health and the environment. In 2015, WHO Member States adopted a resolution, World Health Assembly (WHA) 68.8, focusing on addressing the health impact of air pollution (Lode and Toussaint). This resolution was followed by the adoption of a roadmap for an enhanced global response to air pollution-related health effects at the 69th World Health Assembly in 2016. The roadmap outlined key priority areas, including expanding the knowledge base through Sustainable Development Goal (SDG) monitoring of air pollution and its health impacts, as well as building the capacity of the health sector to raise awareness and collaborate with other sectors to address air pollution-related health impacts.

The WHO Manifesto for a healthy recovery from COVID-19 provides six prescriptions for a healthy and green recovery from the pandemic, emphasizing measures such as protecting nature, investing in WASH and clean energy for healthcare facilities, promoting sustainable food systems, and advocating for financial reform to stop funding pollution. Global and regional engagement initiatives such as the SDGs, Health and Energy Platform of Action (HEPA), Climate and Clean Air Coalition (CCAC), and UN-Energy have been instrumental in driving coordinated action to address air pollution and its health impacts [41]. Furthermore, the WHO, in collaboration with the UNEP and the World Meteorological Organization (WMO), has established the Health, Environment, and Climate Change Coalition to reduce deaths from air pollution and other environmental risks [42]. Moreover, efforts such as the Global Atmosphere Watch Programme (GAW), Actions on Air Quality Report, and the Geneva Convention on Long-range Transboundary Air Pollution (LRTAP) have contributed to building a global understanding of air pollution and its impacts, setting targets, and developing technical standards for emission reductions.

These initiatives highlight the importance of collective action in tackling air pollution and its far-reaching consequences. By fostering collaboration among stakeholders and utilizing international partnerships, the global community can drive progress toward cleaner air and healthier environments. Such efforts align with broader agendas for sustainable development and climate resilience, emphasizing the interconnectedness of environmental health and human well-being on a global scale.

3 Economic Implications of Air Pollution Partnerships

The economic ramifications of air pollution are multilayered, encompassing various costs ranging from healthcare expenditures associated with pollution-induced illnesses to the expenses incurred due to environmental degradation and diminished ecosystem services. Poor air quality adversely affects workplace productivity and tourism, taking a significant toll on global economies. Despite these challenges, research suggests that the economic advantages linked with air pollution mitigation far outweigh the associated costs, highlighting the importance of transitioning toward cleaner technologies and fostering the development of sustainable industries.

In 2018, the global economy incurred an alarming cost of USD 2.9 trillion due to air pollution, equivalent to 3.3% of the world's GDP, with unhealthy air quality leading to 1.8 billion days of work absences worldwide [43]. Estimates suggest that air pollution imposes welfare costs of USD 5 trillion annually on the global economy [44]. Notably, the economic toll of air pollution is evident across various regions, as evidenced by the WHO and the Organisation for Economic Co-operation and Development's (OECD) estimation of the economic cost of premature death and disability from air pollution in Europe, reaching close to USD 1.6 trillion in 2015 [45].

A report by Greenpeace Southeast Asia and the Centre for Research on Energy and Clean Air sheds light on the extensive human and economic costs associated with air pollution from fossil fuels [46]. The report highlights that burning gas, coal, and oil contributes to three times as many deaths as road traffic accidents worldwide, with air pollution imposing a staggering economic cost of $2.9 trillion, equivalent to 3.3% of the global GDP [46, 47]. $PM_{2.5}$ pollution alone is accountable for 1.8 billion days of work absence, four million new cases of child asthma, and two million preterm births, illustrating the profound impact of air pollution on both public health and economic well-being [48].

Air pollution's economic impacts also include increased rates of asthma, diabetes, and chronic respiratory diseases, resulting in reduced workforce productivity and labor force participation rates. Children vulnerable to asthma attacks often miss school days, hindering their educational attainment, while the healthcare expenses associated with pollution-induced illnesses impose additional burdens on households and healthcare systems. For example, disability from chronic diseases incurred a cost of $200 billion to the global economy in 2018, with sick leave and preterm births amounting to $100 billion and $90 billion, respectively [48, 49]. The economic toll extends beyond healthcare costs, encompassing welfare costs associated with premature deaths and illness-related pain and suffering. Projections indicate a substantial rise in global welfare costs, reaching USD 18–25 trillion annually by 2060, emphasizing the societal implications of outdoor air pollution [50].

The economic toll of air pollution extends across various regions, with excessive costs incurred annually. In China, the total annual cost of air pollution is estimated at $900 billion, while the USA faces an economic burden of $600 billion annually [51]. Similarly, air pollution in Indian cities, persistently rated unfavorably on air pollution indexes, imposes a significant cost of approximately $150 billion per year on average [52]. In 2018, the economic impact of poor air quality amounted to 6.6% of China's GDP, 5.4% of India's GDP, and 3% of the GDP in the USA [53].

Efforts to mitigate air pollution often involve the development and implementation of new technologies aimed at reducing emissions. Setting emission limit values for air pollutants, as demonstrated by various Protocols of the Convention, serves as an effective strategy to incentivize investment in cleaner technologies. Economic analyses have quantified the benefits of adopting improved technologies to curb air pollution, revealing potential job losses in certain sectors, such as fossil fuels, offset by job gains in others, like the building and equipment sectors. Environmental policies promoting the adoption of cleaner technologies not only foster more efficient resource utilization but also yield substantial health benefits, potentially increasing GDP by up to 10% [54].

Studies also report the efficacy of air pollution control measures, highlighting significant returns on investment. Research examining the impacts of the Clean Air Act in the USA revealed a remarkable 30:1 ratio between the economic benefits and the costs of air pollution mitigation [55]. A substantial portion of the economic benefits associated with mitigating air pollution stems from reductions in premature mortality linked to particulate matter pollution exposure. In the USA alone, cardiovascular disease and respiratory conditions resulting from air pollution contribute to 107,000 premature deaths annually, along with $820 billion in healthcare costs [56, 57]. Studies focusing on nitrogen dioxide exposure demonstrate notable increases in emergency room costs, outpatient costs, and annual direct healthcare costs in areas with heightened pollutant concentrations [58].

Forecasts suggest an increase in air pollution-related healthcare costs, from USD 21 billion in 2015 to a staggering USD 176 billion by 2060 [59]. Concurrently, the annual number of lost working days due to air pollution-induced illnesses is projected to surge to 3.7 billion globally by 2060 [60]. Premature deaths attributed to outdoor air pollution are expected to escalate dramatically, from approximately three million in 2010 to 6–9 million annually by 2060, with densely populated regions like China and India bearing a disproportionate burden [61].

The combustion of fossil fuels, a significant contributor to poor air quality and climate change, imposes considerable economic costs as well. Air pollution exposure from fossil fuels translates to approximately $2,500 in additional medical expenses in the USA [62]. When coupled with rising temperatures, fossil fuel emissions worsen ozone pollution, leading to an increase in annual health costs in the USA to $7.9 billion due to aggravated asthma and various other health outcomes. Additionally, wildfire smoke, another source of air pollution, contributes significantly to health damage, with associated costs estimated at $16 billion annually in the USA, resulting in numerous respiratory hospital visits and $PM_{2.5}$-related deaths [62].

The economic implications of air pollution are vast and far-reaching, exerting a significant burden on both national economies and individual households. From healthcare expenditures to lost productivity and environmental cleanup costs, the economic toll of air pollution spans various sectors and regions globally. Despite the substantial challenges posed by air pollution, there is optimism surrounding the potential economic benefits of mitigation efforts, particularly through the adoption of cleaner technologies and the implementation of effective policy measures.

4 Technological Innovations for Air Pollution Control

Technological innovations, including renewable energy sources, electric vehicles, air quality sensors, AI, and wearable air quality sensors, play an important role in combating air pollution. They offer solutions for pollution control and mitigation by reducing reliance on fossil fuels, providing real-time pollution data, enhancing measurement accuracy, and empowering individuals. Other innovations, such as filtration units on buses, air purification units at intersections, ionization technology, smog towers, emission control devices, photonic systems, and advancements in electric vehicle autonomous technology, further contribute to pollution reduction efforts. Together, these innovations offer a comprehensive approach to air pollution control, promoting cleaner and healthier environments. Refer to Table 1 for a summary of key technological innovations for air pollution control ([77–82]; Corning).

In 2021, the Ministry of Ecology and Environment in China announced a catalog of advanced technologies designed to combat air pollution, noise, and vibration issues (Ltd). It featured practical and financially viable pollution control technologies, with a particular emphasis on areas of environmental pollution mitigation. Nineteen technologies were listed dedicated to air pollution control, with an additional six targeting noise and vibration control. These technologies encompass various methods such as desulfurization, denitrification, dust removal, volatile organic compound (VOC) treatment, and deodorization. Noise and vibration control technologies predominantly focus on applications in roadways, railways, and buildings. Noteworthy examples from the catalog include fully automatic air supply reduction and exhaust concentration systems, VOC recovery membrane technology, electronic waste material, and multi-metal solid waste smelting flue gas treatment technology.

In addition, the Environmental Defense Fund's (EDF) collaborative efforts with Google in Oakland and Houston revealed significant variability in air pollution levels, highlighting previously unknown disparities. Through this collaboration, it has been revealed that air pollution levels can fluctuate by up to eightfold within a single city block. Residing in areas characterized by heightened pollution levels elevates the risk of heart attacks in the elderly by 40%, similar to the impact of a history of smoking. Satellite data indicates that approximately 24 million more US residents, twice the previous estimate, inhabit regions with unhealthy air quality. To address this challenge, the EDF and partners such as Google Earth Outreach have embarked on pilot studies utilizing novel, cost-effective air quality sensors to map pollution at hyperlocal levels, from Oakland to London. These next-generation sensors can be deployed on various platforms, including vehicles and stationary

Table 1 Technological innovations for air pollution control

Technology	Key features	Key strengths and applicability	Role and significance in air pollution control	References
Renewable energy sources	Utilizes solar, wind, and hydropower to generate electricity	Reduces pollution from traditional power plants, promotes sustainability, decreases reliance on fossil fuels	Shifting toward renewable energy sources diminishes air pollution associated with energy production, mitigating environmental and health hazards	Haines et al. [63]
Electric vehicles (EVs)	Produces significantly less air pollution compared to traditional gasoline or diesel cars	Decreases emissions as the electricity grid becomes cleaner, offers potential for long-term pollution reduction	Adoption of EVs contributes to reducing emissions from transportation, crucial for mitigating urban air pollution and improving public health	Pan et al. [64]
Air quality sensors	Measures levels of pollutants in the air	Identifies areas with high pollution levels, allows for timely intervention to reduce pollution levels, empowers individuals to make informed decisions	Air quality sensors provide real-time data on pollution levels, aiding in the implementation of targeted interventions and raising public awareness about air quality issues	Kelly et al. [65]
Air quality maps	Provides a visual representation of pollution levels in different areas	Facilitates identification of high-pollution areas, enables informed decision-making by authorities and individuals, supports pollution reduction efforts	Air quality maps offer a comprehensive view of pollution distribution, assisting policymakers and the public in addressing pollution hotspots and implementing effective control measures	Thunis et al. [66]
Artificial intelligence (AI)	Enhances accuracy in measuring air pollution	Provides detailed insights into pollution dynamics, enables hyperlocal pollution monitoring, supports decision-making for pollution management	AI-driven models improve the understanding of air pollution patterns, aiding policymakers in implementing targeted interventions and optimizing resource allocation for pollution control	Bibri et al. [67]

(continued)

Table 1 (continued)

Technology	Key features	Key strengths and applicability	Role and significance in air pollution control	References
Wearable air quality sensors	Lightweight sensors worn by individuals to monitor personal exposure to air pollution	Democratizes access to air quality data, empowers citizens to advocate for environmental justice, supports local pollution monitoring initiatives	Wearable sensors offer individuals real-time feedback on personal exposure to pollution, fostering public engagement in air quality advocacy and facilitating community-led pollution monitoring efforts	Dove et al. [68]
Pariyayantra filtration units on buses	Filters installed on buses to capture dust particles from the surrounding environment	Minimizes contribution of vehicular emissions to air pollution, requires no power to operate	Filtration units on buses mitigate vehicular emissions, improving air quality in urban areas and reducing health risks associated with exposure to pollution from transportation	Tran et al. [69]
WAYU air purification units at traffic intersections	Installed at major traffic intersections to purify air in the immediate vicinity	Reduces the impact of vehicular emissions on air quality, acts as a localized air purifier	WAYU units mitigate traffic-related pollution at congested intersections, enhancing air quality and safeguarding public health in urban areas prone to high levels of vehicular emissions	Gulia et al. [70]
Ionization technology	Neutralizes pollutants through ionization processes	Enhances air quality in target areas, provides a potential solution for pollution reduction	Ionization technology offers a novel approach to air pollution control, potentially complementing existing mitigation strategies and contributing to improved air quality in urban environments	Molina et al. [71]
Medium/large-scale smog towers	Substantial air purifiers targeting a reduction of particulate matter and pollutants	Effective in reducing pollution on a broader scale	Smog towers serve as significant interventions in combating air pollution, offering a large-scale solution for reducing particulate matter and pollutants in densely populated urban areas	Tiwary and Williams [72]

(continued)

Table 1 (continued)

Technology	Key features	Key strengths and applicability	Role and significance in air pollution control	References
Retrofitting emission control devices	Installation of emission control devices in older vehicles	Reduces emissions from vehicles adhering to older emission standards	Retrofitting emission control devices in older vehicles addresses a significant source of pollution, aligning with efforts to improve air quality and mitigate health risks associated with vehicular emissions	Huang et al. [73]
Indigenous photonic system	Developed for real-time remote monitoring of air quality parameters	Enhances accuracy and accessibility of air quality data, supports informed decision-making in pollution management	Indigenous photonic systems provide reliable data on air quality, facilitating evidence-based decision-making and enabling proactive measures to address pollution and safeguard public health	Kapri et al. [74]
Electric vehicle (EV) autonomous technology	Focuses on autonomous navigation in electric vehicles	Optimizes driving patterns to reduce emissions, contributes to traffic congestion reduction, lowers greenhouse gas emissions	Integration of autonomous technology in EVs offers a sustainable solution for reducing emissions and enhancing urban air quality, promoting a shift toward cleaner and smarter transportation systems	Pan et al. [75]
Corning air purification system	Utilizes ceramic honeycomb filters to trap particulate matter from polluted air	Reduces or maintains particulate matter levels below WHO guidelines, suitable for indoor and outdoor air treatment systems, contributes to cleaner environments	Corning air purification system effectively reduces particulate matter levels in polluted air, contributing to cleaner and healthier environments, thus addressing the adverse effects of air pollution	Chen et al. [76]

networks, enabling block-by-block insights into air quality dynamics (Environmental Defense Fund).

Similarly, the Breathe London project, initiated in January 2019 by the Mayor of London, EDF, C40, and numerous collaborators, seeks to measure and map Londoners' daily exposure to air pollution using an extensive network of advanced sensors deployed throughout the city. By visualizing pollution data in this manner, various stakeholders can advocate for cleaner air and make informed urban development decisions. The success of projects like Breathe London underlines the potential to replicate similar initiatives globally, facilitating the generation of detailed street-by-street maps of air pollution for cities worldwide. Ongoing efforts in Houston and London involve deploying mobile and stationary sensors to comprehensively monitor pollution levels and assess the effectiveness of lower-cost sensor networks in providing hyperlocal air quality data to the public [83].

The steps taken by the Ministry of Ecology and Environment in China regarding advanced technologies to combat air pollution mark a step in addressing environmental challenges. With a focus on practical and financially viable pollution control methods, the catalog showcases a range of technologies dedicated to air pollution control, as well as noise and vibration mitigation. Collaborative efforts between organizations like the EDF and Google have introduced significant variability in air pollution levels. Utilizing novel, cost-effective air quality sensors, pilot studies in cities like Oakland and London are providing valuable insights into hyperlocal pollution dynamics, empowering national and regional bodies to advocate for cleaner air and informed urban development decisions. The success of initiatives like the Breathe London project highlights the potential for similar endeavors worldwide, offering hope for more technologically empowered sustainable urban environments on a global scale.

5 Regional Partnerships and Collaborations

Regional partnerships and collaborations stand as pillars in the global battle against air pollution, an insidious threat affecting 99% of the world's population [84]. Beyond geographical boundaries, this crisis penetrates our health, climate, and economic prosperity, demanding concerted efforts across sectors for sustainable change.

One cornerstone of international cooperation is the UNECE Convention on Long-Range Transboundary Air Pollution, which includes 51 parties and pioneers a legal framework to combat cross-border pollution [85]. Through its protocols, emissions of pollutants like sulfur dioxide have plummeted by over 90% in Europe since the early 1990s, resulting in the prevention of 600,000 premature deaths annually and an additional year of life expectancy.

Beyond continental collaborations, specific regions have forged alliances tailored to their unique challenges. In Southeast Asia, the ASEAN Agreement on Transboundary Haze Pollution tackles land and forest fires, while the UN ESCAP Regional Action

Programme on Air Pollution addresses rapid urbanization and industrialization in the Asia-Pacific [86]. These initiatives are exemplars of tailored, context-specific approaches necessary for effective regional cooperation.

Africa has witnessed the emergence of landmark agreements like the Lusaka Agreement and the Nairobi Agreement, signaling a concerted effort to harmonize legislation, standards, and monitoring practices across the continent [87, 88]. In West Asia, the push for a Regional Air Quality Network reflects a growing recognition of the need for coordinated action to combat air pollution, particularly exacerbated by natural emissions like sand and dust storms (UN). Latin America and the Caribbean are also making strides through a partnership between the UNEP and the EDF [89]. With a focus on increasing awareness, enhancing management practices, and strengthening regional cooperation, this collaboration aims to deliver impactful solutions to improve air quality and protect public health across the region.

The dialogue on air quality management (AQM) in the Indo-Gangetic Plain and Himalayan Foothills (IGP-HF) aims to foster collaboration among the four nations sharing the IGP-HF airshed. This platform provides an opportunity for these countries to address their respective AQM challenges, identify commonalities, exchange knowledge, and explore avenues for partnership and financing to improve monitoring, decision-making, and abatement actions [90]. Cross-country cooperation and knowledge exchange are essential to address these challenges effectively. ICIMOD's Regional Programme on Atmosphere aims to enhance understanding of air pollution dynamics and promote mitigation measures in the Hindu Kush Himalayan region. Similarly, the World Bank's initiatives focus on AQM projects and policy reforms in the IGP-HF countries [90].

The establishment of the North-East Asia Clean Air Partnership (NEACAP) emphasizes the global concern over air pollution. Within the framework of the North-East Asian Subregional Programme for Environmental Cooperation (NEASPEC), NEACAP serves as a platform for dialogue and cooperation among North-East Asian countries to tackle air pollution issues. In September 2023, the ESCAP East and North-East Asia Office, in collaboration with the National Air Emission Inventory and Research Center of the Republic of Korea, organized a symposium on air pollution in Seoul. This symposium brought together government officials, research experts, and specialists to discuss air quality management strategies and policy implications [41].

Collaborative efforts between the UNEP and Japan signify a new alliance aimed at combating air pollution in the Asia-Pacific region. This partnership is particularly significant considering that approximately 70% of the estimated seven million premature deaths caused by air pollution annually occur within this region [91].

Known as the Asia Pacific Clean Air Partnership, this initiative is a joint endeavor between the UNEP Regional Office for Asia and the Pacific and Japan's Ministry of Environment [91]. Its primary objective is to consolidate various regional initiatives to provide comprehensive policy recommendations for addressing air pollution across the region. By facilitating coordinated action, this partnership endeavors to assist regional countries in effectively managing air pollution, thereby yielding numerous benefits, including improved public health, enhanced food and water security, and contributions to climate change mitigation efforts. This is of importance as the Asia-Pacific region is home to some of the world's most polluted cities, with particulate matter less than $PM_{2.5}$ exceeding safe limits by fourfold in certain areas.

Air pollution poses a significant challenge for countries like India and Pakistan, where cities frequently rank among the most polluted globally. Despite historical hostilities, both countries face similar health risks due to air pollution, with millions of premature deaths recorded annually. Efforts to address air pollution have been initiated, with cities like New Delhi implementing measures to mitigate pollution levels. However, challenges persist, particularly in cities like Lahore, where air pollution remains a pressing issue. Identifying the sources of air pollution is crucial, with biomass burning, industrial emissions, and transportation identified as major contributors. Collaboration between India and Pakistan is essential to address transboundary air pollution and mitigate its adverse effects on public health and the environment [92].

The suspension of industrial and social activities during the COVID-19 crisis offered a glimpse of improved air quality in cities like New Delhi. This presents an opportunity for both countries to collaborate and implement sustainable measures to maintain improved air quality post-crisis. Effective air pollution governance requires collaborative efforts between municipal and national governments. Establishing joint scientific commissions can facilitate data-driven decision-making and policy development to mitigate regional air pollution [92].

International efforts, such as the designation of September 7 as the International Day of Clean Air for Blue Skies, emphasize the urgency of addressing air pollution on a global scale [93]. Collaborative actions at regional and international levels are crucial to combat air pollution and protect public health and the environment.

Regional partnerships and collaborations are essential in addressing the pervasive issue of air pollution. These alliances facilitate the exchange of knowledge, resources, and best practices, enabling countries to effectively manage air quality and mitigate the adverse effects of pollution on public health and the environment. From the Indo-Gangetic Plain to the Asia-Pacific region, concerted efforts are underway to tackle air pollution through coordinated action and innovative solutions [94]. As the world addresses the challenges of climate change and environmental degradation, regional cooperation remains pivotal in the global effort to ensure clean air for all.

6 The Role of Multilateral Agreements to Mitigating Air Pollution

In the area of global environmental governance, Multilateral Environmental Agreements (MEAs) emerge as vital instruments forged through international collaboration, aimed at challenging the environmental dilemmas of our era, such as climate change, the erosion of biodiversity, and the challenges posed by pollution and waste [95]. These agreements, born out of a collective realization that environmental issues transcend national boundaries requiring a unified global response, are crafted and adopted by Member States within the UN framework, setting the stage for a collaborative approach to environmental stewardship.

Among these agreements, the Paris Climate Agreement stands out as a foundation of the United Nations Framework Convention on Climate Change (UNFCCC), emblematic of the international community's resolve to address climate-related challenges [96]. The UNEP, as the premier environmental authority within the United Nations, utilizes its unique position and convening power to lead the development, negotiation, and enactment of a comprehensive suite of international environmental conventions and treaties, currently encompassing 15 MEAs.

These agreements enable nations to collectively tackle environmental crises that threaten the global community, emphasizing the critical need for sustainable consumption and production (SCP) practices [97]. While no MEA exclusively addresses SCP, this concept acts as an essential link across existing agreements, enhancing their effectiveness in combating environmental issues. This is notably recognized in instruments such as the Paris Agreement and within frameworks addressing biodiversity and chemical and waste management post-2020.

The foundational principle of SCP is further reinforced by the 10-Year Framework of Programs and SDG 12, under the 2030 Agenda for Sustainable Development, which collectively underpin the multilateral cooperation essential for fostering sustainable consumption and production patterns.

This emphasis on SCP within MEAs is exemplified by:

- The Paris Climate Agreement (2015), under the UNFCCC, which underscores the significance of sustainable consumption and production patterns, particularly by developed countries, in mitigating climate change [98].
- The Aichi Biodiversity Target 4 (2011–2020) within the Convention on Biological Diversity (CBD), which aims for governments, businesses, and stakeholders at all levels to adopt sustainable production and consumption plans that keep natural resource use within ecological limits [99].
- The Basel Convention (1989), focusing on minimizing the generation of hazardous wastes [100].

Acknowledging the pivotal role of SCP in addressing the environmental crises of climate, biodiversity, and pollution, the 10-Year Framework of Programs on SCP (2012), established at the United Nations Conference on Sustainable Development, calls for a common vision that supports the attainment of MEA targets and goals.

The interconnection between achieving SCP and MEA targets is evidenced by the One Planet network's annual reporting, which illustrates how SCP tools and solutions facilitate MEA implementation. According to SDG 12.1.1 reports by Member States, a significant proportion of SCP policies contribute to the goals of climate action, biodiversity, and ecosystem conservation, and the sustainable management of marine resources, with a notable focus on circular economy practices and waste reduction, including the mitigation of plastic pollution (United Nations/ Multilateral body).

To mainstream SCP across international environmental frameworks effectively, MEAs present an unprecedented opportunity to incorporate resource efficiency, circular economy principles, and waste reduction strategies into national policies. By embedding these concepts into Nationally Determined Contributions (NDCs), for instance, countries can elevate their ambitions within the framework of the Paris Agreement. Moreover, integrating circularity and waste prevention into national strategies for chemicals and waste management can significantly advance the achievement of SDG 12's targets on sustainable consumption and production patterns [101].

The 10-Year Framework of Programs serves as a collaborative platform to enhance implementation through the dissemination of knowledge, policies, and tools; leveraging the initiatives of the One Planet network programs; and fostering engagement with other essential alliances, initiatives, and platforms. This concerted effort to embed SCP within the fabric of international environmental agreements highlights a collective commitment to transitioning toward more sustainable and resilient global consumption and production systems (One Planet Network).

MEAs and the achievement of the SDGs are marked by a transformation envisioned in the 2015 United Nations General Assembly's Resolution 70/1, which laid the groundwork for the 2030 Agenda for Sustainable Development [102]. This foundational document, targeting the SDGs, did not delineate the frequency of the assessments essential for monitoring progress toward these goals. Within this context, the UNEP has been instrumental in advancing the Strategic Plan for Biodiversity 2011–2020 and the Aichi biodiversity targets, thereby prompting various sectors toward the collective goal of sustainable development.

Multilateral environmental agreements serve as the backbone of international legal frameworks addressing specific environmental challenges, thereby augmenting national and regional efforts from a global perspective. These agreements, often referred to as "soft laws," encompass a range of thematic areas, including biodiversity conservation and chemical management, exemplified by conventions such as the Convention on Biological Diversity and the Stockholm Convention on Persistent Organic Pollutants, among others. Their strategic alignment with the development agenda has been increasingly recognized, illustrating a concerted effort to integrate sustainable development objectives with environmental governance [103].

The linkage between the SDGs and multilateral environmental agreements is evident across various goals and targets, highlighting the importance of coherent and synergistic approaches to global environmental management. For instance, Goals 12 (responsible consumption and production) and 17 (partnerships for the goals) resonate across all dimensions of sustainability, while specific goals address critical issues such as chemical management, climate action, and biodiversity conservation [104].

The adoption of the Strategic Plan for Biodiversity for the period 2011–2020 and the Aichi Biodiversity Targets reflects a global consensus on the need for setting actionable targets and indicators, facilitating national-level prioritization and implementation strategies. The role of multilateral environmental agreements in steering global, regional, and national initiatives toward the 2030 Agenda's objectives has become crucial. These agreements embody a collective commitment to address environmental issues through coordinated action, enhancing governance structures, fostering partnerships, and promoting equitable and sustainable development practices [99].

As the international community moves forward, the integration of biodiversity-related multilateral environmental agreements with the SDGs offers a promising pathway to achieving conservation and development goals in tandem. The focus on biodiversity and ecosystems, particularly through Goals 14 and 15, highlights the critical importance of maintaining the planet's ecological balance for the well-being of all. The collaborative efforts of parties to these agreements, alongside the strategic guidance provided by the Strategic Plan for Biodiversity, pave the way for a more effective approach to sustainable development, reflecting a deepened understanding of the interconnectedness of environmental and development priorities [105, 106].

An illustration of the necessity for cross-border environmental cooperation is the battle against air pollution, identified as the most significant environmental health hazard leading to seven million deaths annually across the globe. The resolution adopted at UNEA-6 on enhancing regional cooperation for global air quality improvement pays homage to the achievements of the 1979 Convention on Long-range Transboundary Air Pollution (Air Convention) and its Protocols over the past 45 years. This unique treaty has been pivotal in fostering regional collaboration toward cleaner air, setting emission reduction targets for critical pollutants, and delivering substantial benefits for climate, ecosystems, and public health [107].

The sustainable management of transboundary water resources represents another critical area where international cooperation is indispensable. A UNEA-6 resolution advocating for the implementation of integrated water resources management, including through transboundary cooperation, underscores this point. The 1992 Convention on the Protection and Use of Transboundary Watercourses and International Lakes (Water Convention) has been instrumental in facilitating cooperation on transboundary rivers, lakes, and groundwater, leading to significant improvements in water quality and biodiversity. The convention's role in driving pollution reduction in major water basins through collaborative efforts was highlighted during a UNEA side event on Transboundary Ecosystem Restoration [108].

As MEAs continue to evolve as tools for bolstering countries' resilience and sustainable development trajectories, the integration of environmental considerations into development planning through treaties like the 1991 Convention on Environmental Impact Assessment in a Transboundary Context (Espoo Convention) and its 2003 Protocol on Strategic Environmental Assessment stands as a testament to the power of proactive environmental governance [85].

The acknowledgment of environmental protection as a universal human right reinforces the imperative for inclusive environmental action. The 1998 Convention on Access to Information, Public Participation in Decision-making, and Access to Justice in Environmental Matters (Aarhus Convention) empower the public and NGOs to actively participate in environmental decision-making and hold governments accountable, thus promoting both transparency and accountability in environmental governance [109].

MEAs stand as pillars of international cooperation and environmental governance, orchestrating a unified response to the issues of climate change, biodiversity loss, and pollution [110]. Through the strategic implementation of these agreements within the framework of the UN, nations worldwide are brought together to address environmental crises with a sense of shared purpose and collective action [111]. The integration of SCP principles across MEAs, particularly highlighted in landmark agreements such as the Paris Climate Agreement and the Convention on Biological Diversity, highlights the critical nexus between environmental stewardship and sustainable development [112]. This concerted effort not only facilitates the achievement of the SDGs but also exemplifies a commitment to preserving the planet's ecological balance and ensuring the well-being of future generations [113]. As the international community advances, the synergy between MEAs and the SDGs leads a path forward, promising a sustainable and resilient future where environmental governance and sustainable development go hand in hand, embodying the aspiration for coexistence between humanity and nature.

7 Challenges and Opportunities

In the discourse on air quality and pollution, significant steps have been taken to adhere to national standards established by the Environmental Protection Agency (EPA) since 1971, reflecting advancements grounded in evolving scientific insights. The manifestation of air pollution has become less prevalent and pronounced since the 1970s, signaling a positive shift in environmental health efforts. Despite these improvements, air pollution remains a subtle yet potent threat to both public health and environmental integrity, even at minimal levels. At present, numerous locales across the USA report pollution metrics that surpass the established air quality benchmarks for at least one of the key pollutants. While there has been a decrease in the prevalence of particulate matter and ozone pollution, their levels continue to pose health risks in various regions. These pollutants, stemming from a variety of sources, can traverse extensive distances, affecting areas beyond their points of origin.

Table 2 delineates the landscape of air quality management, highlighting the challenges that have evolved alongside societal progress and the corresponding opportunities for advancement [124]. It highlights the need for ongoing refinement in air quality standards to reflect current scientific understanding, the critical need for targeted mitigation strategies to address health impacts, and the importance of identifying and addressing pollution sources through technological and policy innovations. It emphasizes the role of economic incentives, public health infrastructure, and international collaboration in overcoming these challenges.

Extensive research corroborates the adverse health consequences of prolonged and short-term exposure to $PM_{2.5}$, including premature mortality and cardiovascular

Table 2 Challenges and opportunities in air quality management

Aspect	Challenges	Opportunities	References
Air quality standards	Persistent exceedance of air quality benchmarks despite improvements since the 1970s	Refinement of air quality standards by the EPA in response to new scientific evidence, fostering a healthier environment	[114]
Health impacts	Adverse effects from $PM_{2.5}$ and ozone exposure include premature mortality, cardiovascular complications, and respiratory challenges	Enhanced public health through targeted actions to reduce $PM_{2.5}$ and ozone levels, leading to decreased healthcare utilization and mortality	[115]
Source identification	Difficulty in curbing emissions from a wide range of sources, including industrial activities, vehicles, and household heating	Targeted interventions at major pollution sources like metallurgy, electricity production, and residential heating to efficiently reduce pollutant levels	[116]
Technological advancements	The slow pace of adopting pollution abatement technologies across sectors	Implementation of modern equipment and pollution control devices across key sectors to minimize emissions effectively	[117]
Policy and regulation	Lengthy processes for environmental permitting and limited scope of current policies on pollutants	Streamlining permit processes and expanding policy coverage to include a broader range of pollutants, encouraging compliance and investment in pollution control	[118]
Economic incentives	Insufficient motivation for enterprises and households to invest in environmental protection and energy efficiency	Development of financial instruments and incentives to promote investments in cleaner technologies and practices at both the industrial and household levels	[119]
Public health infrastructure	Lack of capacity for health impact assessments and a reliance on international studies for local policymaking	Strengthening technical expertise and conducting localized studies to better understand and address the health impacts of air pollution within specific contexts	[120]
Climate change	Exacerbation of air pollution challenges due to climate change, including increased risks of wildfires and temperature-related pollution peaks	Addressing climate change through emissions reductions, renewable energy adoption, and public transportation investments to mitigate both air pollution and climate impacts	[121]
International collaboration	The global divide in air quality monitoring and management capabilities, particularly in low- and middle-income countries (LMICs)	Global efforts to improve air quality monitoring and management in LMICs through technology transfer, shared best practices, and international cooperation	[122]

(continued)

Table 2 (continued)

Aspect	Challenges	Opportunities	References
Data and knowledge gap	Limited air quality data and understanding of pollution sources and impacts in many regions	Utilization of low-cost sensors and satellite data to enhance air quality monitoring and data availability, informing more effective pollution reduction strategies	[123]

complications, which manifest through heightened hospital and emergency department visits for cardiovascular incidents and respiratory ailments [125]. Ozone exposure is similarly linked to respiratory challenges and enduring lung damage, with elevated ozone concentrations associated with increased healthcare utilization and mortality [126]. Environmental degradation and visibility impairment are further consequences of these pollutants [127].

Sulfur dioxide and nitrogen dioxide, pollutants arising predominantly from fossil fuel combustion, are under examination for their peak unhealthy levels [128]. The EPA collaborates with various stakeholders to pinpoint and address the occurrences of these peaks, given their respiratory health implications and environmental impact [129]. Additionally, the near elimination of airborne lead pollution, a significant health hazard prior to regulatory interventions under the Clean Air Act, marks a pivotal achievement, with the exception of proximities to certain industrial entities [130]. The attainment of carbon monoxide standards nationwide underscores the efficacy of vehicular emissions regulations.

The EPA's endeavors extend to refining air quality standards in response to emergent scientific evidence, exemplifying an adaptive and proactive approach to environmental stewardship. Through collaboration with state and rural entities, the EPA fosters the implementation of regulatory measures aimed at mitigating the emission of common pollutants, thereby contributing to the realization of air quality objectives [131]. This partnership model, exemplified by the Clean Air Act, facilitates a tailored approach to achieving ambient air quality standards, with a focus on technological innovation and policy guidance [132].

The ongoing battle against air pollution encompasses efforts to address climate change, prompted by the detrimental impact of carbon dioxide and other greenhouse gases on global climate patterns and oceanic chemistry. The recognition of these challenges by authoritative bodies highlights the urgency of addressing climate change as a critical public and environmental health issue [133]. Initiatives to curtail greenhouse gas emissions reflect a commitment to mitigating climate change and its ramifications, with regulatory measures targeting vehicular emissions, industrial activities, and methane emissions, among others [134].

The discourse on air pollution encompasses the realm of toxic pollutants, with a concerted effort to diminish the release of hazardous substances into the atmosphere. Regulatory frameworks and technological standards have significantly curtailed emissions of air toxicants, though elevated risks persist in urban, industrial, and transportation-heavy areas [135]. The comprehensive management of air quality also encompasses measures to safeguard the stratospheric ozone layer, essential for protecting life on Earth from harmful ultraviolet radiation. The phased elimination of ozone-depleting substances, guided by international agreements and domestic regulations, exemplifies a global commitment to environmental preservation and public health protection [136].

Addressing air pollution demands a comprehensive understanding of its sources, particularly the significant contributions from industrial activities emitting hazardous air pollutants (HAPs) and VOC [137]. The challenge lies in executing strict regulations for HAP and VOC reduction in industrial contexts globally. This effort is critical for mitigating the adverse effects of air pollution emanating from these sectors. The EPA, despite its ongoing efforts to curtail pollution, highlights four principal obstacles in air pollution abatement (Fig. 2).

The path forward is uncertain regarding the management of air pollution levels. Notably, the expansion of industrial activities parallels human population growth, posing significant regulatory and abatement challenges. The scarcity of effective pollution mitigation systems across global industrial facilities further complicates efforts to enhance air quality [138]. The lack of accountability among companies disregarding eco-friendly practices highlights an area where regulatory bodies must intervene to enforce environmental compliance [139].

In regions experiencing rapid industrialization, such as China and India, there is a dual challenge of diminishing air pollution and greenhouse gas emissions [140]. This scenario demands an analysis of technological solutions and their impact on air quality, public health, and greenhouse gas emissions [141]. For instance, while coal gasification offers clean burning options that reduce air pollutants, it substantially increases carbon dioxide emissions [142]. Conversely, the integration of renewable energy sources not only curtails emissions of air pollutants and greenhouse gases but also confers benefits for climate, health, and air quality [143].

Electrification through renewable sources stands as a pivotal strategy for carbon mitigation, offering substantial reductions in air pollutants and greenhouse gas emissions compared to traditional fossil fuel-based energy production. The deployment of renewable energy, particularly in China, embodies a significant opportunity for environmental and public health improvements. The extent of these benefits varies based on the renewable energy sources adopted and their integration into the existing power grid infrastructure [144].

The comprehensive management of air quality extends beyond immediate environmental concerns, encompassing the broader implications for public health, climate change, and societal well-being. Future research should prioritize indoor air quality, the interplay between air pollution and climate change, and the specific health impacts on vulnerable populations, including children. Additionally, advancing our understanding of the efficacy of air pollution mitigation strategies, the role of renewable energy in improving air quality, and the development of resilient and sustainable urban infrastructures will be crucial. Collaborative efforts among governments, industries, and communities are essential to implement effective policies and ensure sustainable living.

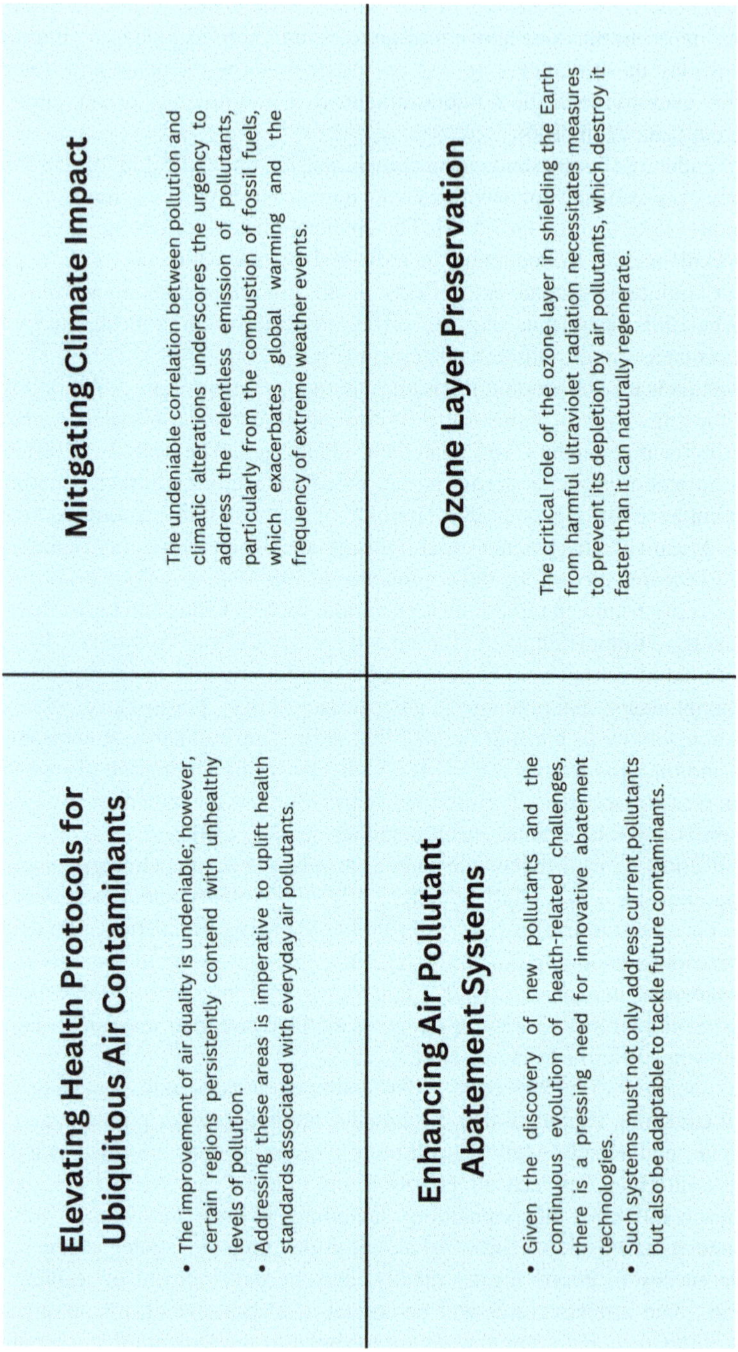

Fig. 2 Key strategies for improving air quality and mitigating climate impact

The examination of indoor air quality (IAQ) is a critical aspect of environmental health, influenced by a myriad of factors, including the socioeconomic status of households, ambient air quality, the origins of air pollutants (combustion or non-combustion), the structure and design of buildings, and the activities conducted within these spaces [145]. The components of indoor air pollution are shaped by these elements and include not only the well-documented criteria pollutants like particulate matter and inorganic gases (nitrogen dioxide, carbon monoxide) but also the significant presence of biological contaminants and volatile organic compounds.

Despite the global shift toward spending more time indoors – whether in residences, workplaces, educational institutions, or vehicles – the understanding of the full impact of indoor air pollution on health remains incomplete. Studies suggest that in the USA, adults spend upward of 87% of their time indoors, a figure that is thought to closely mirror global trends. This indoor exposure is not confined to homes but extends to schools where children are similarly at risk from pollutants that breach safe levels.

The move toward green building designs, while beneficial in aspects like cost and maintenance, does not inherently assure superior IAQ. Innovations in natural ventilation within green buildings, however, have shown promise in enhancing indoor air quality. Meanwhile, the reliance on traditional fuels for cooking and heating, particularly in lower-income regions, significantly contributes to indoor air pollution, exacerbating health risks and also influencing outdoor air quality.

The COVID-19 pandemic has further complicated the landscape of indoor air quality, with the increased use of solid fuels in lower-income countries amplifying indoor pollution despite a simultaneous decrease in ambient pollution levels. Efforts by governments and NGOs are essential in addressing these challenges, alongside addressing the IAQ issues prevalent in more affluent countries due to activities such as cooking, heating, and the use of incense [145].

In Australia, for example, measures have been introduced to phase out gas heating and cooking in favor of electric solutions in new homes by the end of 2024, highlighting the importance of context-specific strategies to improve IAQ [145]. Urbanization and the migration of rural populations to cities have also brought about significant shifts in household energy use, impacting both indoor and outdoor air pollution levels.

Addressing indoor smoking and the emissions from e-cigarettes remains a significant challenge, contributing to both second- and third-hand smoke exposure [146]. Furthermore, non-combustion sources, including personal care products and the off-gassing from furniture, add to the indoor pollution burden, raising concerns over emerging contaminants like liquid crystal monomers found in various indoor environments [147]. Addressing these emerging contaminants requires a comprehensive approach that combines clean energy transitions, improved ventilation, and targeted control strategies for specific pollutant sources.

The relationship between climate change and air pollution introduces additional complexities, with the former expected to exacerbate air quality issues through changes in meteorological patterns that affect pollutant formation and dispersion [148]. For instance, increased temperatures can heighten the risk of wildfires, which in turn boost particulate pollution levels across vast distances [149]. The interplay between air pollutants and greenhouse gases further complicates the climate scenario, impacting both environmental and human health [121].

The adoption of EVs presents an opportunity to mitigate traffic-related air pollution, although concerns remain regarding the potential risks associated with EV battery fires and non-exhaust emissions [150]. Investment in public transportation and active travel infrastructure has been identified as an effective strategy for reducing air pollution and associated health risks, particularly in urban settings [151].

Research into air pollution's impact on children's health highlights the need for targeted interventions to mitigate exposure, given their heightened vulnerability due to developing physiological and metabolic systems [152]. The evidence points to the detrimental effects of air pollution on children's respiratory and cognitive development, underscoring the importance of improving indoor air quality in homes and schools [153].

Focusing on primary pollution sources offers a pathway to significantly reduce air pollution levels [154]. Recommended actions encompass the adoption of cutting-edge technology in metal production processes, the installation of pollution control devices and transition to cleaner fuels in energy production, and the implementation of dust collectors in asphalt production [155]. For residential heating, the introduction of modern, efficient stoves could markedly improve air quality [156].

Policy reforms should broaden to cover a wider range of pollutants, including PM, and streamline the process for obtaining environmental permits to motivate investment in pollution control measures [157]. Economic incentives, possibly facilitated through collaborations with the financial sector, could spur investments in environmental upgrades both for industries and at the household level [158].

In the realm of public health, enhancing the capacity for health impact assessments and bolstering technical expertise on the nexus between environmental and health issues are imperative [159]. This would enable local health professionals to rely less on international studies and more on data reflective of country-specific context.

Addressing air pollution in LMICs requires a multilayered approach that extends beyond only identifying pollution levels to implementing viable solutions [123]. Essential steps include providing guidance on the use of air quality sensors, developing strategies for their deployment, and ensuring their operation in environments with intermittent power and data connectivity.

There is a need to enhance satellite-based remote sensing capabilities, improve data management practices, and invest in the professional development of air quality staff within LMICs. Establishing durable institutions that can manage and disseminate air quality information is crucial for building public support and driving policy changes [123].

Looking ahead, the global trajectory of air pollution and its health implications necessitates a deeper understanding and more robust interventions to mitigate its impact [160]. Despite improvements in air quality in high-income countries, low- and middle-income regions continue to face significant challenges [161]. The interconnections between air pollution, climate change, and health outcomes call for comprehensive strategies that address the root causes and adopt an integrated approach to improve air quality and protect public health.

8 Conclusion

In concluding our exploration of international policies, partnerships, and agreements aimed at addressing air pollution, we reaffirm the necessity of a cohesive and concerted international response to combat this pervasive environmental challenge. The complexity and transboundary nature of air pollution demand robust collaborative frameworks and innovative governance that can adapt to the diverse socioeconomic and environmental landscapes across the globe.

This chapter has enlisted how structured international efforts such as the Paris Climate Agreement and regional partnerships like the ASEAN Agreement on Transboundary Haze Pollution serve as cornerstones in the global fight against air pollution. These agreements exemplify how strategic cooperation can lead to substantial reductions in air pollution, thereby mitigating its impact on public health and climate change.

However, as we delve into the effectiveness of these frameworks, it becomes apparent that significant challenges persist. The disparity in capabilities between developed and developing nations, the inconsistency in policy implementation, and the ongoing struggle against emerging pollution sources highlight the need for strengthened international governance and enhanced resource allocation.

To advance global air quality management, it is imperative that future efforts not only reinforce existing treaties but also innovate new solutions that address both current and emerging air quality challenges. This involves integrating air quality management with other global sustainability goals, enhancing real-time monitoring and reporting systems, and promoting public and private investments in pollution control technologies.

Moreover, the landscape of global industry and urban development calls for adaptive policies that can respond to the evolving nature of air pollution sources. Encouraging the adoption of clean technologies, fostering sustainable urban planning, and enhancing public awareness and participation are crucial steps toward effective air quality management.

International cooperation must also extend to sharing knowledge, technologies, and strategies to combat air pollution. By fostering an inclusive international community that supports all nations – especially those with fewer resources – we enhance our collective ability to manage air pollution effectively.

In the pursuit of cleaner air, the role of international agreements becomes increasingly vital. These frameworks not only facilitate substantial environmental health improvements but also foster economic and social benefits globally. By continuing to build on the foundations laid by these international efforts, we can aspire to a future where air quality management is a pivotal aspect of global environmental governance, ensuring health and sustainability for all populations.

References

1. Isaksen ISA, Granier C, Myhre G et al (2009) Atmospheric composition change: Climate–Chemistry interactions. Atmos Environ 43:5138–5192. https://doi.org/10.1016/j.atmosenv.2009.08.003
2. Jiang K, Xing R, Luo Z et al (2024) Pollutant emissions from biomass burning: a review on emission characteristics, environmental impacts, and research perspectives. Particuology 85: 296–309. https://doi.org/10.1016/j.partic.2023.07.012
3. Chen T-M, Kuschner WG, Gokhale J, Shofer S (2007) Outdoor air pollution: nitrogen dioxide, sulfur dioxide, and carbon monoxide health effects. Am J Med Sci 333:249–256. https://doi.org/10.1097/MAJ.0b013e31803b900f
4. Bălă G-P, Râjnoveanu R-M, Tudorache E et al (2021) Air pollution exposure – the (in)visible risk factor for respiratory diseases. Environ Sci Pollut Res 28:19615–19628. https://doi.org/10.1007/s11356-021-13208-x
5. Guan W-J, Zheng X-Y, Chung KF, Zhong N-S (2016) Impact of air pollution on the burden of chronic respiratory diseases in China: time for urgent action. Lancet 388:1939–1951. https://doi.org/10.1016/S0140-6736(16)31597-5
6. Mir KA, Purohit P, Cail S, Kim S (2022) Co-benefits of air pollution control and climate change mitigation strategies in Pakistan. Environ Sci Policy 133:31–43. https://doi.org/10.1016/j.envsci.2022.03.008
7. Whitmee S, Haines A, Beyrer C et al (2015) Safeguarding human health in the Anthropocene epoch: report of The Rockefeller Foundation–Lancet Commission on planetary health. Lancet 386:1973–2028. https://doi.org/10.1016/S0140-6736(15)60901-1
8. Yatoo AM, Hamid B, Sheikh TA et al (2024) Global perspective of municipal solid waste and landfill leachate: generation, composition, eco-toxicity, and sustainable management strategies. Environ Sci Pollut Res 31:23363–23392. https://doi.org/10.1007/s11356-024-32669-4
9. Jacobs DE, Kelly T, Sobolewski J (2007) Linking public health, housing, and indoor environmental policy: successes and challenges at local and federal agencies in the United States. Environ Health Perspect 115:976–982. https://doi.org/10.1289/ehp.8990
10. Patz JA, Frumkin H, Holloway T et al (2014) Climate change: challenges and opportunities for Global Health. JAMA 312:1565. https://doi.org/10.1001/jama.2014.13186
11. Anwar MN, Shabbir M, Tahir E et al (2021) Emerging challenges of air pollution and particulate matter in China, India, and Pakistan and mitigating solutions. J Hazard Mater 416:125851. https://doi.org/10.1016/j.jhazmat.2021.125851
12. D'Amato G, Vitale C, De Martino A et al (2015) Effects on asthma and respiratory allergy of climate change and air pollution. Multidiscip Respir Med 10:39. https://doi.org/10.1186/s40248-015-0036-x
13. Wu X, Nethery RC, Sabath MB et al (2020) Air pollution and COVID-19 mortality in the United States: strengths and limitations of an ecological regression analysis. Sci Adv 6: eabd4049. https://doi.org/10.1126/sciadv.abd4049
14. Shetty SS, Deepthi D, Harshitha S et al (2023) Environmental pollutants and their effects on human health. Heliyon 9:e19496. https://doi.org/10.1016/j.heliyon.2023.e19496
15. Bolan S, Padhye LP, Jasemizad T et al (2024) Impacts of climate change on the fate of contaminants through extreme weather events. Sci Total Environ 909:168388. https://doi.org/10.1016/j.scitotenv.2023.168388
16. Owusu PA, Asumadu-Sarkodie S (2016) A review of renewable energy sources, sustainability issues and climate change mitigation. Cogent Eng 3:1167990. https://doi.org/10.1080/23311916.2016.1167990
17. Schraufnagel DE, Balmes JR, De Matteis S et al (2019) Health benefits of air pollution reduction. Ann ATS 16:1478–1487. https://doi.org/10.1513/AnnalsATS.201907-538CME
18. Chowdhury S, Marginean I, Chaudhary E et al (2024) Impact of the changing climate on air pollution, heat stress and human health. In: Health and environmental effects of ambient air pollution. Elsevier, pp 331–359

19. Ravichandran V, Archer J-M, Aiyar L et al (2024) Utilization of MD EJSCREEN to assess health outcomes in Baltimore, Maryland. Environ Justice:env.2022.0039. https://doi.org/10.1089/env.2022.0039
20. Scotford E, Misonne D, Lewis A (2023) Guide on ambient air quality legislation-air pollution series
21. Nanda VP, Pring GR (2013) Chapter ten International air pollution. In: International environmental law and policy for the 21st century. Brill Nijhoff, pp 331–425
22. Kum EG (2020) What can the World Health Organization (WHO), and the 1979 Geneva convention (CLRTAP) do, under the rules of public international law (PIL), to curb air pollution which has amplified the death toll rate of the Coronavirus-19 in the world? Available at SSRN 3635016
23. Šega IB, Čačković M, Špirić Z, Franić Z, Juretić AA, Kosanović C, Zovko VV, Godec R, Davila S, Kraljević L, Hercog P (2018) "AIR PROTECTION 2017" 10th Croatian scientific and professional conference with international participation, 3-7 October 2017, Primošten, Croatia. Arh Hig Rada Toksikol 69:A25–A28
24. Huang J, Shen J, Miao L, Zhang W (2021) The effects of emission trading scheme on industrial output and air pollution emissions under city heterogeneity in China. J Clean Prod 315:128260
25. Ohtonen K, Savolahti M, Anttila P, Vainio-Mattila B, Liljaniemi P (2023) First update of the national air pollution control programme 2030
26. Fritz Carrapatoso A (2008) Environmental aspects in free trade agreements in the Asia-Pacific region. Asia Eur J 6(2):229–243
27. Han C, Xu R, Ye T et al (2022) Mortality burden due to long-term exposure to ambient PM2.5 above the new WHO air quality guideline based on 296 cities in China. Environ Int 166:107331
28. Lelieveld J, Evans JS, Fnais M et al (2015) The contribution of outdoor air pollution sources to premature mortality on a global scale. Nature 525:367–371
29. Åström S, Källmark L, Yaramenka K, Grennfelt P (2021) European and central Asian actions on air quality: a regional summary of emission trends, policies, and programs to reduce air pollution
30. World Health Organization (2017) Progress report on the regional plan of action for implementation of the roadmap for an enhanced global response to the adverse health effects of air pollution. World Health Organization. Regional Office for the Eastern Mediterranean
31. Gouldson A, Sudmant A, Khreis H, Papargyropoulou E (2018) The economic and social benefits of low-carbon cities: a systematic review of the evidence
32. Holzer K (2023) Tackling black carbon in road transport: towards global car emission standards. In: Reducing emissions of short-lived climate pollutants. Brill Nijhoff, pp 332–363
33. de Mello Santos VH, Campos TLR, Espuny M, de Oliveira OJ (2022) Towards a green industry through cleaner production development. Environ Sci Pollut Res:1–22
34. World Health Organization (2016) Burning opportunity: clean household energy for health, sustainable development, and wellbeing of women and children
35. Premakumara DGJ, Menikpura S, Singh RK et al (2018) Reduction of greenhouse gases (GHGs) and short-lived climate pollutants (SLCPs) from municipal solid waste management (MSWM) in The Philippines: rapid review and assessment. Waste Manag 80:397–405
36. Bhuvaneshwari S, Hettiarachchi H, Meegoda JN (2019) Crop residue burning in India: policy challenges and potential solutions. Int J Environ Res Public Health 16:832
37. Milner J, Turner G, Ibbetson A et al (2023) Impact on mortality of pathways to net zero greenhouse gas emissions in England and Wales: a multisectoral modelling study. Lancet Planetary Health 7:e128–e136
38. Miller B, Hurley J (2003) Life table methods for quantitative impact assessments in chronic mortality. J Epidemiol Community Health 57:200–206
39. Castle JL, Hendry DF (2021) Can the UK achieve net-zero greenhouse gas emissions by 2050?

40. Koelemeijer R, Eerens H, Van Velze K et al (2012) Cobenefits of climate and air pollution regulations. The context of the European Commission Roadmap for moving to a low carbon economy in 2050
41. Miller J (2020) Evaluation of the north-east Asian subregional programme for environmental cooperation (NEASPEC)
42. World Health Organization (2023) Informing policy and progress. https://www.who.int/teams/environment-climate-change-and-health/air-quality-energy-and-health/policy-progress. Accessed 30 Mar 2024
43. Nundy S, Ghosh A, Mesloub A et al (2021) Impact of COVID-19 pandemic on socio-economic, energy-environment and transport sector globally and sustainable development goal (SDG). J Clean Prod 312:127705. https://doi.org/10.1016/j.jclepro.2021.127705
44. Dechezleprêtre A, Rivers N, Stadler B (n.d.) The economic costs of air pollution: evidence from Europe Economics
45. UNECE (2023) Air pollution and economic development. https://unece.org/air-pollution-and-economic-development. Accessed 30 Mar 2024
46. Greenpeace Southeast Asia (2020) Toxic air: the price of fossil fuels (full report). https://www.greenpeace.org/southeastasia/publication/3603/toxic-air-the-price-of-fossil-fuels-full-report/. Accessed 30 Mar 2024
47. Statista (2020) Infographic: the economic burden of air pollution. In: Statista daily data. https://www.statista.com/chart/20804/costs-of-air-pollution-from-fossil-fuels. Accessed 30 Mar 2024
48. World Economic Forum (2020) This is the global economic cost of air pollution. In: World economic forum. https://www.weforum.org/agenda/2020/02/the-economic-burden-of-air-pollution/. Accessed 30 Mar 2024
49. Zawudie AB, Daka DW, Dejene T, Meskerem SE (n.d.) Economic burden of diabetic mellitus among patients on follow-up care in Hospitals of southwest Shewa zone, Central Ethiopia. BMC Health Serv Res:22. https://doi.org/10.1186/s12913-022-08819-0
50. Yi X, Liu S (2021) Impact of environmental factors on pulmonary tuberculosis in multi-levels industrial upgrading area of China. Environ Res 195:110768. https://doi.org/10.1016/j.envres.2021.110768
51. Nie D, Shen F, Wang J et al (2021) Changes of air quality and its associated health and economic burden in 31 provincial capital cities in China during COVID-19 pandemic. Atmos Res 249:105328. https://doi.org/10.1016/j.atmosres.2020.105328
52. Kaur R, Pandey P (2021) Air pollution, climate change, and human health in Indian cities: a brief review. Front Sustain Cities 3
53. Dong D, Xu B, Shen N, He Q (2021) The adverse impact of air pollution on China's economic growth. Sustainability 13:9056. https://doi.org/10.3390/su13169056
54. Su S, Qamruzzaman M, Karim S (2023) Charting a sustainable future: the impact of economic policy, environmental taxation, innovation, and natural resources on clean energy consumption. Sustain For 15:13585. https://doi.org/10.3390/su151813585
55. Ross K, Chmiel JF, Ferkol T (2012) The impact of the clean air act. J Pediatr 161:781–786. https://doi.org/10.1016/j.jpeds.2012.06.064
56. Kim JB, Prunicki M, Haddad F et al (2020) Cumulative lifetime burden of cardiovascular disease from early exposure to air pollution. JAHA 9:e014944. https://doi.org/10.1161/JAHA.119.014944
57. Palacio LC, Pachajoa DC, Echeverri-Londoño CA et al (2023) Air pollution and cardiac diseases: a review of experimental studies. Dose-Response 21:15593258231212793. https://doi.org/10.1177/15593258231212793
58. Liu H, Hu T, Wang M (2021) Impact of air pollution on residents' medical expenses: a study based on the survey data of 122 cities in China. Front Public Health 9:743087. https://doi.org/10.3389/fpubh.2021.743087

59. Alshammari KH, Alshammari AF (2023) Green innovation and its effects on innovation climate and environmental sustainability: the moderating influence of Green abilities and strategies. Sustain For 15:15898. https://doi.org/10.3390/su152215898
60. OECD (2016) The economic consequences of outdoor air pollution. https://www.oecd.org/environment/indicators-modelling-outlooks/Policy-Highlights-Economic-consequences-of-outdoor-air-pollution-web.pdf
61. Van Der Wall EE (2015) Air pollution: 6.6 million premature deaths in 2050! Neth Heart J 23: 557–558. https://doi.org/10.1007/s12471-015-0763-9
62. Lavietes M (2021) What are the economic impacts of air pollution? In: World Economic Forum. https://www.clarity.io/blog/a-deep-dive-economic-impacts-of-air-pollution. Accessed 30 Mar 2024
63. Haines A, Smith KR, Anderson D, Epstein PR, McMichael AJ, Roberts I, Wilkinson P, Woodcock J, Woods J (2007) Policies for accelerating access to clean energy, improving health, advancing development, and mitigating climate change. Lancet 370(9594):1264–1281
64. Pan S, Yu W, Fulton LM, Jung J, Choi Y, Gao HO (2023) Impacts of the large-scale use of passenger electric vehicles on public health in 30 US metropolitan areas. Renew Sustain Energy Rev 173:113100
65. Kelly FJ, Fuller GW, Walton HA, Fussell JC (2012) Monitoring air pollution: use of early warning systems for public health. Respirology 17(1):7–19
66. Thunis P, Miranda A, Baldasano JM, Blond N, Douros J, Graff A, Janssen S, Juda-Rezler K, Karvosenoja N, Maffeis G, Martilli A (2016) Overview of current regional and local scale air quality modelling practices: assessment and planning tools in the EU. Environ Sci Policy 65: 13–21
67. Bibri SE, Krogstie J, Kaboli A, Alahi A (2024) Smarter eco-cities and their leading-edge artificial intelligence of things solutions for environmental sustainability: a comprehensive systematic review. Environ Sci Ecotechnol 19:100330
68. Dove G, Chen S, Fries D, Johnson V, Mydlarz C, Bello JP, Nov O (2022) From environmental monitoring to mitigation action: considerations, challenges, and opportunities for HCI. Proc ACM Hum-Comput Interact 6(CSCW2):1–31
69. Tran PT, Adam MG, Tham KW, Schiavon S, Pantelic J, Linden PF, Sofianopoulou E, Sekhar SC, Cheong DKW, Balasubramanian R (2021) Assessment and mitigation of personal exposure to particulate air pollution in cities: an exploratory study. Sustain Cities Soc 72:103052
70. Gulia S, Kaur S, Mendiratta S, Tiwari R, Goyal SK, Gargava P, Kumar R (2022) Performance evaluation of air pollution control device at traffic intersections in Delhi. Int J Environ Sci Technol 19(2):785–796
71. Molina LT, Velasco E, Retama A, Zavala M (2019) Experience from integrated air quality management in the Mexico City metropolitan area and Singapore. Atmos 10(9):512
72. Tiwary A, Williams I (2018) Air pollution: measurement, modelling and mitigation. CRC Press
73. Huang Y, Lee CK, Yam YS, Mok WC, Zhou JL, Zhuang Y, Surawski NC, Organ B, Chan EF (2022) Rapid detection of high-emitting vehicles by on-road remote sensing technology improves urban air quality. Sci Adv 8(5)
74. Kapri RK, Sharma P, Dubey PK (2021) Indigenous design and development of gated photon counter for low-rate photon regime. Mapan 36(1):59–66
75. Pan S, Fulton LM, Roy A, Jung J, Choi Y, Gao HO (2021) Shared use of electric autonomous vehicles: air quality and health impacts of future mobility in the United States. Renew Sustain Energy Rev 149:111380
76. Chen F, Wang M, Pu Z (2022) The impact of technological innovation on air pollution: firm-level evidence from China. Technol Forecast Soc Chang 177:121521
77. Chien F, Sadiq M, Nawaz MA et al (2021) A step toward reducing air pollution in top Asian economies: the role of green energy, eco-innovation, and environmental taxes. J Environ Manage 297:113420. https://doi.org/10.1016/j.jenvman.2021.113420

78. Clean Air Fund (2023) AI and wearables: innovative tech for fighting air pollution. In: Clean air fund. https://www.cleanairfund.org/news-item/ai-wearables-innovative-tech/. Accessed 30 Mar 2024
79. Feng S, Zhang R, Li G (2022) Environmental decentralization, digital finance and green technology innovation. Struct Chang Econ Dyn 61:70–83. https://doi.org/10.1016/j.strueco.2022.02.008
80. Global T (2020) Techniques of air pollution control – thermax. https://www.thermaxglobal.com/techniques-of-air-pollution-control/. Accessed 30 Mar 2024
81. González-Martín J, Kraakman NJR, Pérez C et al (2021) A state–of–the-art review on indoor air pollution and strategies for indoor air pollution control. Chemosphere 262:128376. https://doi.org/10.1016/j.chemosphere.2020.128376
82. Lu X, Zhang S, Xing J et al (2020) Progress of air pollution control in China and its challenges and opportunities in the ecological civilization era. Engineering 6:1423–1431. https://doi.org/10.1016/j.eng.2020.03.014
83. C40 (2019) Breathe London. In: C40 Knowledge Hub. https://www.c40knowledgehub.org/s/article/Breathe-London?language=en_US. Accessed 30 Mar 2024
84. Verrier RL, Mittleman MA, Stone PH (2002) Air pollution: an insidious and pervasive component of cardiac risk. Circulation 106:890–892. https://doi.org/10.1161/01.CIR.0000027434.34445.23
85. UN (1979) United Nations Treaty Collection. In: United Nations Treaties. https://treaties.un.org/Pages/ViewDetails.aspx?src=TREATY&mtdsg_no=XXVII-1&chapter=27&clang=_en. Accessed 30 Mar 2024
86. ESCAP (2022) Regional consultation for air pollution in Asia and the Pacific: South-East Asia. In: ESCAP. https://www.unescap.org/event-series/dialogue-2021/sea. Accessed 30 Mar 2024
87. Evaborhene NA, Oga JO, Nneli OV, Mburu S (2023) The WHO pandemic treaty: where are we on our scepticism? BMJ Glob Health 8:e012636. https://doi.org/10.1136/bmjgh-2023-012636
88. Journal of International Wildlife Law & Policy (2008) Lusaka agreement on co-operative enforcement operations directed at illegal trade in wild Fauna and Flora. J Int Wildl Law Policy 1:155–168. https://doi.org/10.1080/13880299809353889
89. Environmental Defense Fund (2022) New partnership to improve air quality, advance Global Health and climate goals in Latin American and Caribbean Countries. In: EDF. https://www.edf.org/media/new-partnership-improve-air-quality-advance-global-health-and-climate-goals-latin-american. Accessed 30 Mar 2024
90. Rabi Qazizada R, Heger M (2022) Air quality management in the Indo-Gangetic Plain and Himalayan foothills. In: ICIMOD. https://www.icimod.org/event/air-quality-management-in-the-indo-gangetic-plain-and-himalayan-foothills/. Accessed 30 Mar 2024
91. Brittlebank W (2014) UNEP and Japan announce partnership to battle air pollution. In: Climate action. https://www.climateaction.org/news/unep_and_japan_announce_partnership_to_battle_air_pollution/. Accessed 30 Mar 2024
92. Mehmood K, Saifullah IM et al (2020) Pakistan and India collaboration to improve regional air quality has never been more promising. Integr Environ Assess Manag 16:549–551. https://doi.org/10.1002/ieam.4292
93. Song J, Yoo S, Lim JY, Ko Y (2020) Strategic communication for establishing collectivity for UN international day of clean air for blue skies. Asian J Innov Policy 9:95–105. https://doi.org/10.7545/AJIP.2020.9.1.095
94. Waglé S, Wignaraja K (eds) (2022) The great upheaval: resetting development policy and institutions for the decade of action in Asia and the Pacific. Cambridge University Press, New York
95. One Planet Network (2021) Addressing the triple crisis. In: One Planet network. https://www.oneplanetnetwork.org/SDG-12/multilateral-environmental-agreements. Accessed 31 Mar 2024

96. Garcia MLCT (2021) The United Nations Framework Convention on Climate Change: orchestration as a governance mode. Doutorado Direto em Relações Internacionais, Universidade de São Paulo
97. Environment UN (2017) Sustainable consumption and production policies. In: UNEP – UN environment programme. http://www.unep.org/explore-topics/resource-efficiency/what-we-do/sustainable-consumption-and-production-policies. Accessed 31 Mar 2024
98. Kemp L (2018) A systems critique of the 2015 Paris agreement on climate. In: Hossain M, Hales R, Sarker T (eds) Pathways to a sustainable economy. Springer, Cham, pp 25–41
99. Carr H, Abas M, Boutahar L et al (2020) The Aichi biodiversity targets: achievements for marine conservation and priorities beyond 2020. PeerJ 8:e9743. https://doi.org/10.7717/peerj.9743
100. Kummer K (1992) The international regulation of transboundary traffic in hazardous wastes: the 1989 Basel convention. ICLQ 41:530–562. https://doi.org/10.1093/iclqaj/41.3.530
101. Hotta Y, Tasaki T, Koide R (2021) Expansion of policy domain of sustainable consumption and production (SCP): challenges and opportunities for policy design. Sustainability 13:6763. https://doi.org/10.3390/su13126763
102. United Nations/Multilateral Body (2015) A/RES/70/1 Transforming our world: the 2030 Agenda for Sustainable Development
103. Chenet H, Zamarioli L, Kretschmer B, Narvaez R (2019) From transformational climate finance to transforming the financial system for climate. https://doi.org/10.13140/RG.2.2.17998.00327
104. Coscieme L, Mortensen LF, Donohue I (2021) Enhance environmental policy coherence to meet the sustainable development goals. J Clean Prod 296:126502. https://doi.org/10.1016/j.jclepro.2021.126502
105. Ekardt F, Günther P, Hagemann K et al (2023) Legally binding and ambitious biodiversity protection under the CBD, the global biodiversity framework, and human rights law. Environ Sci Eur 35:80. https://doi.org/10.1186/s12302-023-00786-5
106. Obrecht A, Pham-Truffert M, Spehn E, et al (2021) Achieving the SDGs with biodiversity. [object object]
107. Climate & Clean Air Coalition (2024) Unpacking the newly adopted UNEA-6 resolution on promoting regional cooperation on air pollution to improve air quality globally. In: CCA coaliation. https://www.ccacoalition.org/news/unpacking-newly-adopted-unea-6-resolution-promoting-regional-cooperation-air-pollution-improve-air-quality-globally. Accessed 31 Mar 2024
108. UN (2023) Transboundary water management cooperation crucial for sustainable development, peace, security, speakers stress at conference's fourth interactive dialogue | meetings coverage and press releases. In: United Nations. https://press.un.org/en/2023/envdev2056.doc.htm. Accessed 31 Mar 2024
109. Meunier-Rubel V (2022) Interstitial law-making in public international law: a study of environmental impact assessments. Brill | Nijhoff
110. Holley C, Gunningham N, Shearing C (2013) The new environmental governance. Routledge
111. Bodin Ö (2017) Collaborative environmental governance: achieving collective action in social-ecological systems. Science 357:eaan1114. https://doi.org/10.1126/science.aan1114
112. Agyeman J (ed) (2003) Just sustainabilities: development in an unequal world. MIT Press, Cambridge
113. Agbedahin AV (2019) Sustainable development, education for sustainable development, and the 2030 agenda for sustainable development: emergence, efficacy, eminence, and future. Sustain Dev 27:669–680. https://doi.org/10.1002/sd.1931
114. Chow JC, Watson JG, Feldman HJ et al (2007) Will the circle be unbroken: a history of the U.S. National Ambient Air Quality Standards. J Air Waste Manage Assoc 57:1151–1163. https://doi.org/10.3155/1047-3289.57.10.1151

115. Basith S, Manavalan B, Shin TH et al (2022) The impact of fine particulate matter 2.5 on the cardiovascular system: a review of the invisible killer. Nanomaterials 12:2656. https://doi.org/10.3390/nano12152656
116. Anjum MS, Ali SM, Imad-ud-din M et al (2021) An emerged challenge of air pollution and ever-increasing particulate matter in Pakistan; a critical review. J Hazard Mater 402:123943. https://doi.org/10.1016/j.jhazmat.2020.123943
117. Tetteh EK, Amankwa MO, Yeboah C, Amankwa MO (2021) Emerging carbon abatement technologies to mitigate energy-carbon footprint – a review. Cleaner Mater 2:100020. https://doi.org/10.1016/j.clema.2021.100020
118. Jonidi Jafari A, Charkhloo E, Pasalari H (2021) Urban air pollution control policies and strategies: a systematic review. J Environ Health Sci Eng 19:1911–1940. https://doi.org/10.1007/s40201-021-00744-4
119. Mushafiq M, Arisar MMK, Tariq H, Czapp S (2023) Energy efficiency and economic policy: comprehensive theoretical, empirical, and policy review. Energies 16:2381. https://doi.org/10.3390/en16052381
120. Jabakhanji SB, Arnold SR, Aunan K et al (2022) Public health measures to address the impact of climate change on population health – proceedings from a stakeholder workshop. IJERPH 19:13665. https://doi.org/10.3390/ijerph192013665
121. Ofremu GO, Raimi BY, Yusuf SO et al (2024) Exploring the relationship between climate change, air pollutants and human health: impacts, adaptation, and mitigation strategies. Green Energy Resources, p 100074. https://doi.org/10.1016/j.gerr.2024.100074
122. Cromar KR, Duncan BN, Bartonova A et al (2019) Air pollution monitoring for health research and patient care. An official American Thoracic Society Workshop Report. Annals ATS 16:1207–1214. https://doi.org/10.1513/AnnalsATS.201906-477ST
123. Pinder RW, Klopp JM, Kleiman G et al (2019) Opportunities and challenges for filling the air quality data gap in low- and middle-income countries. Atmos Environ 215:116794. https://doi.org/10.1016/j.atmosenv.2019.06.032
124. Sokhi RS, Moussiopoulos N, Baklanov A et al (2021) Advances in air quality research – current and emerging challenges
125. De Bont J, Jaganathan S, Dahlquist M et al (2022) Ambient air pollution and cardiovascular diseases: an umbrella review of systematic reviews and meta-analyses. J Intern Med 291:779–800. https://doi.org/10.1111/joim.13467
126. Eckhardt CM, Wu H (2021) Environmental exposures and lung aging: molecular mechanisms and implications for improving respiratory health. Curr Envir Health Rpt 8:281–293. https://doi.org/10.1007/s40572-021-00328-2
127. Alharbi OML, Basheer AA, Khattab RA, Ali I (2018) Health and environmental effects of persistent organic pollutants. J Mol Liq 263:442–453. https://doi.org/10.1016/j.molliq.2018.05.029
128. Saxena P, Sonwani S (2019) Criteria air pollutants and their impact on environmental health. Springer, Singapore
129. D'Evelyn SM, Jung J, Alvarado E et al (2022) Wildfire, smoke exposure, human health, and environmental justice need to be integrated into forest restoration and management. Curr Environ Health Rep 9:366–385. https://doi.org/10.1007/s40572-022-00355-7
130. Sullivan M, Green D (2020) Toward eliminating children's lead exposure: a comparison of policies and their outcomes in three lead producing and using countries. Environ Res Lett 15:103008. https://doi.org/10.1088/1748-9326/abb55e
131. Kandpal V, Jaswal A, Santibanez Gonzalez EDR, Agarwal N (2024) Sustainable energy transition, circular economy, and ESG practices. In: Sustainable energy transition. Springer Nature Switzerland, Cham, pp 1–51
132. LaCount MD, Haeuber RA, Macy TR, Murray BA (2021) Reducing power sector emissions under the 1990 clean air act amendments: a retrospective on 30 years of program development and implementation. Atmos Environ 245:118012. https://doi.org/10.1016/j.atmosenv.2020.118012

133. Cifuentes-Faura J (2022) European Union policies and their role in combating climate change over the years. Air Qual Atmos Health 15:1333–1340. https://doi.org/10.1007/s11869-022-01156-5
134. Fonseca LM, Domingues JP, Dima AM (2020) Mapping the sustainable development goals relationships. Sustainability 12:3359. https://doi.org/10.3390/su12083359
135. Gowrisankaran G, Langer A, Zhang W (2022) Policy uncertainty in the market for coal electricity: the case of air toxics standards. National Bureau of Economic Research, Cambridge
136. Harris PG (ed) (2022) Routledge handbook of global environmental politics.2nd edn. Routledge, Taylor & Francis Group, London, New York
137. Nakhjiri A, Kakroodi AA (2024) Air pollution in industrial clusters: a comprehensive analysis and prediction using multi-source data. Eco Inform 80:102504. https://doi.org/10.1016/j.ecoinf.2024.102504
138. Rentschler J, Leonova N (2023) Global air pollution exposure and poverty. Nat Commun 14:4432. https://doi.org/10.1038/s41467-023-39797-4
139. Dragomir VD (2020) Theoretical aspects of environmental strategy. In: Corporate environmental strategy. Springer, Cham, pp 1–31
140. Wang Q, Su M, Li R (2018) Toward to economic growth without emission growth: the role of urbanization and industrialization in China and India. J Clean Prod 205:499–511. https://doi.org/10.1016/j.jclepro.2018.09.034
141. West JJ, Smith SJ, Silva RA et al (2013) Co-benefits of mitigating global greenhouse gas emissions for future air quality and human health. Nat Clim Change 3:885–889. https://doi.org/10.1038/nclimate2009
142. Imran M, Kumar D, Kumar N et al (2014) Environmental concerns of underground coal gasification. Renew Sustain Energy Rev 31:600–610. https://doi.org/10.1016/j.rser.2013.12.024
143. Vennemo H, Aunan K (2018) Air pollution and greenhouse gas emissions in China: an unsustainable situation in search of a solution. In: Brinkmann R, Garren SJ (eds) The Palgrave handbook of sustainability. Springer, Cham, pp 139–157
144. Gielen D, Boshell F, Saygin D et al (2019) The role of renewable energy in the global energy transformation. Energ Strat Rev 24:38–50. https://doi.org/10.1016/j.esr.2019.01.006
145. Vilcins D, Christofferson RC, Yoon J-H et al (2024) Updates in air pollution: current research and future challenges. Ann Glob Health 90:9. https://doi.org/10.5334/aogh.4363
146. Cui T, Lu R, Liu C et al (2024) Characteristics of second-hand exposure to aerosols from e-cigarettes: a literature review since 2010. Sci Total Environ 926:171829. https://doi.org/10.1016/j.scitotenv.2024.171829
147. Tagesse M, Deti M, Dadi D et al (2021) Non-combustible source indoor air pollutants concentration in beauty salons and associated self-reported health problems among the beauty salon workers. RMHP 14:1363–1372. https://doi.org/10.2147/RMHP.S293723
148. Chang J-H, Lee Y-L, Chang L-T et al (2023) Climate change, air quality, and respiratory health: a focus on particle deposition in the lungs. Ann Med 55:2264881. https://doi.org/10.1080/07853890.2023.2264881
149. Jaffe DA, O'Neill SM, Larkin NK et al (2020) Wildfire and prescribed burning impacts on air quality in the United States. J Air Waste Manage Assoc 70:583–615. https://doi.org/10.1080/10962247.2020.1749731
150. Schnell JL, Peters DR, Wong DC et al (2021) Potential for electric vehicle adoption to mitigate extreme air quality events in China. Earth's Future 9:e2020EF001788. https://doi.org/10.1029/2020EF001788
151. Bi S, Hu J, Shao L et al (2024) Can public transportation development improve urban air quality? Evidence from China. Urban Clim 54:101825. https://doi.org/10.1016/j.uclim.2024.101825
152. Johnson NM, Hoffmann AR, Behlen JC et al (2021) Air pollution and children's health – a review of adverse effects associated with prenatal exposure from fine to ultrafine particulate matter. Environ Health Prev Med 26:72. https://doi.org/10.1186/s12199-021-00995-5

153. Sadrizadeh S, Yao R, Yuan F et al (2022) Indoor air quality and health in schools: a critical review for developing the roadmap for the future school environment. J Building Eng 57: 104908. https://doi.org/10.1016/j.jobe.2022.104908
154. Kelly FJ, Fussell JC (2015) Air pollution and public health: emerging hazards and improved understanding of risk. Environ Geochem Health 37:631–649. https://doi.org/10.1007/s10653-015-9720-1
155. Chen J, Dan H, Ding Y et al (2021) New innovations in pavement materials and engineering: a review on pavement engineering research 2021. J Traffic Transp Eng 8:815–999. https://doi.org/10.1016/j.jtte.2021.10.001
156. Jones AP (1999) Indoor air quality and health. Atmos Environ 33:4535–4564. https://doi.org/10.1016/S1352-2310(99)00272-1
157. Zhang Z (2008) Asian energy and environmental policy: promoting growth while preserving the environment. Energy Policy 36:3905–3924. https://doi.org/10.1016/j.enpol.2008.07.015
158. Machiba T (2011) Eco-innovation for enabling resource efficiency and green growth: development of an analytical framework and preliminary analysis of industry and policy practices. In: Bleischwitz R, Welfens PJJ, Zhang Z (eds) International economics of resource efficiency. Physica-Verlag HD, Heidelberg, pp 371–394
159. Baum FE, Bégin M, Houweling TAJ, Taylor S (2009) Changes not for the fainthearted: reorienting health care systems toward health equity through action on the social determinants of health. Am J Public Health 99:1967–1974. https://doi.org/10.2105/AJPH.2008.154856
160. Halfon N, Hochstein M (2002) Life course health development: an integrated framework for developing health, policy, and research. Milbank Q 80:433–479. https://doi.org/10.1111/1468-0009.00019
161. Meghji J, Mortimer K, Agusti A et al (2021) Improving lung health in low-income and middle-income countries: from challenges to solutions. Lancet 397:928–940. https://doi.org/10.1016/S0140-6736(21)00458-X

Part IV
Conclusion

Call to Action for a Sustainable Air Pollution-Free World

Wisdom Ebiye Sawyer, Ligeiaziba Sylva, Ayobami O. Aigberua, and Muhammad Akram

Contents

W. E. Sawyer (✉)
Department of Community Medicine, Faculty of Clinical Sciences, Niger Delta University, Wilberforce Island, Bayelsa, Nigeria

L. Sylva
Department of Mathematics, Faculty of Science, Bayelsa Medical University, Yenagoa, Bayelsa, Nigeria

A. O. Aigberua
Quality Assurance Department, Infra Pipe Solutions Ltd, Saskatoon, SK, Canada

M. Akram
Department of Eastern Medicine, Government College University Faisalabad, Faisalabad, Punjab, Pakistan

Matthew Chidozie Ogwu and Sylvester Izah (eds.), *Sustainable Strategies for Air Pollution Mitigation: Development, Economics and Technologies*,
Hdb Env Chem (2024) 133: 551–572, DOI 10.1007/698_2024_1123,

Published online: 12 September 2024

Abstract Air pollution, caused by vehicular emissions, industrial pollutants, natural disasters, and domestic fuel consumption, remains a global concern. Major pollutants including PM2.5, PM10, ozone, sulfur dioxide, nitrogen oxides, dioxins, PCBs, trace metals (e.g., lead, mercury), PAHs, volatile and semivolatile organic compounds, and others contribute to respiratory, cardiovascular, and neurological diseases as well as cancers and endocrine disruptions. This chapter is a call to action to urgently address the impacts of air pollution and air pollutants. While international efforts have reduced air pollution-related mortality and disability-adjusted life years (DALYs), disparities persist, particularly in low- and low-middle sociodemographic index (SDI) regions. Public health strategies, including regulatory measures such as air quality monitoring and emission standards, community engagement through public awareness campaigns, educational programs, economic interventions like pollution taxes and clean technology subsidies are crucial. Achieving sustainable clean air requires technology innovation, stringent regulations, community involvement, financial incentives, and international cooperation. These efforts are essential to control air pollution effectively and advance toward a sustainable, pollution-free society globally.

Keywords Air pollution, Call to action, Clean air, Human health impacts, Sustainability

Abbreviations

AAP	Ambient air pollution
AQI	Air quality index
BHT	Butylated hydroxytoluene
CH4	Methane
CO	Carbon monoxide
CO2	Carbon dioxide
COPD	Chronic obstructive pulmonary disease
DALYs	Disability-adjusted life years
EPA	Environmental Protection Agency
EV	Electric vehicle
GBD	Global disease burden
HAP	Household air pollution
LEZs	Low emission zones
N2O	Nitrous oxide
NCDs	Noncommunicable diseases
NH3	Ammonia
NO2	Nitrogen dioxide
NOx	Nitrogen oxides
O3	Ozone
PAHs	Polycyclic aromatic hydrocarbons

PCBs	Polychlorinated biphenyls
PM	Particulate matter
PM2.5	Fine particulate matter
SDI	Sociodemographic index
SO2	Sulfur dioxide
SOx	Oxides of sulfur
SVOCs	Semivolatile organic compounds
VOCs	Volatile organic compounds

1 Introduction

Air pollution is reported to be one of the main environmental risk factors that seriously endangers public health worldwide [1]. It puts over 90% of the world's population at risk and is linked to an ever-increasing variety of diseases that are dangerous to people's health, such as respiratory infections, heart disease, stroke, and lung cancer [2–6]. The situation is particularly terrible in urban areas, where industrial emissions, vehicle exhaust, and construction dust are prevalent. Children, the elderly, and those with preexisting health conditions are especially vulnerable to the detrimental effects of poor air quality. Efforts to monitor and control air pollution to safeguard public health and reduce the incidence of air pollution-related diseases are critical in these regions.

An estimated seven million premature deaths globally are attributed to air pollution, involving both ambient and household air pollution (HAP) [3, 4]. The majority of these deaths (more than 90%) occur in low- and middle-income countries, where the prevalence of air pollution-related illness is significantly higher in comparison to high-income nations [1, 3, 7]. In low- and middle-income countries, factors such as reliance on biomass fuels for cooking and heating, inadequate industrial regulations, and rapid urbanization contribute to the high levels of air pollution [8]. Additionally, the lack of access to quality healthcare exacerbates the health impacts of pollution.

More specifically, the projected 4.2 million yearly fatalities associated with ambient air pollution (AAP) result from a rapid rise in morbidity and mortality [3, 9]. This alarming trend highlights the urgent need to meticulously monitor pollution statistics and implement best practices to achieve a new balance between sustainability, human productivity, and well-being. Key detrimental ambient air pollutants impacting human health include ozone (O3), sulfur dioxide (SO2), nitrogen oxides (NO and NO2), and particulate matter (PM). These pollutants are linked to various health issues such as respiratory and cardiovascular diseases, which contribute significantly to the global burden of disease [10].

Household air pollution (HAP) is also a critical concern responsible for approximately 3.8 million deaths annually [11–13]. Although its prevalence is declining, HAP remains a significant health threat, particularly in low- and middle-income countries. Around 2.7–3 billion people, predominantly in developing nations, still rely on unsustainable and harmful fuels for cooking, heating, and lighting [14–16]. This reliance on inefficient energy sources not only perpetuates health risks but also poses a serious danger to environmental well-being.

Both ambient air pollution (AAP) and household air pollution (HAP) exacerbate preexisting medical disorders and significantly raise the risk of serious health consequences. The most well-known and well-researched health impacts of air pollution are respiratory disorders. However, there is also substantial evidence linking air pollution to adverse effects on the heart, brain, reproductive and neurological systems, and increased cancer risk [3]. Studies have shown that pollutants can trigger cardiovascular diseases, leading to heart attacks and strokes, whilst contributing to neurodegenerative diseases such as Alzheimer's and Parkinson's [17, 18]. Additionally, air pollution has been linked to adverse pregnancy outcomes, including low birth weight and preterm births.

Noncommunicable diseases (NCDs) constitute the majority of the global disease burden (GBD), and air pollution is one of the main risk factors for these conditions. In 2016, among individuals aged 30–69, air pollution was a contributing factor in 16% of all NCD deaths [2, 3, 6]. Chronic diseases such as asthma, chronic obstructive pulmonary disease (COPD), and cardiovascular diseases are significantly aggravated by exposure to air pollutants. These diseases not only reduce life expectancy but also diminish the quality of life, leading to increased healthcare costs and economic burden on affected communities.

An estimated seven million premature deaths occur each year due to diseases linked to air pollution, including heart disease, lung cancer, stroke, acute respiratory infections (ARIs), and COPD [19–21]. According to the WHO [4], 37% of outdoor air pollution-related premature deaths were due to ischemic heart disease and stroke; 18% and 23% of deaths were due to chronic obstructive pulmonary disease and acute lower respiratory infections, respectively; and 11% of deaths were due to cancer within the respiratory tract.

This chapter centers on air pollution and its consequences, emphasizing the imperative to collectively mitigate its impacts. It highlights the critical requirements for clean air as well as air quality management strategies and policies to mitigate the health impacts of air pollution. The implementation of regulatory measures, advancement of technological innovations, and promotion of public health interventions are essential steps to curbing millions of premature deaths and enhancing overall public health.

2 Air Pollution: Sources and Classes

Air pollution originates from a variety of sources, including the burning of fossil fuels, industrial processes, agricultural activities, waste management, and natural events like wildfires and volcanic eruptions (Fig. 1). These sources release different

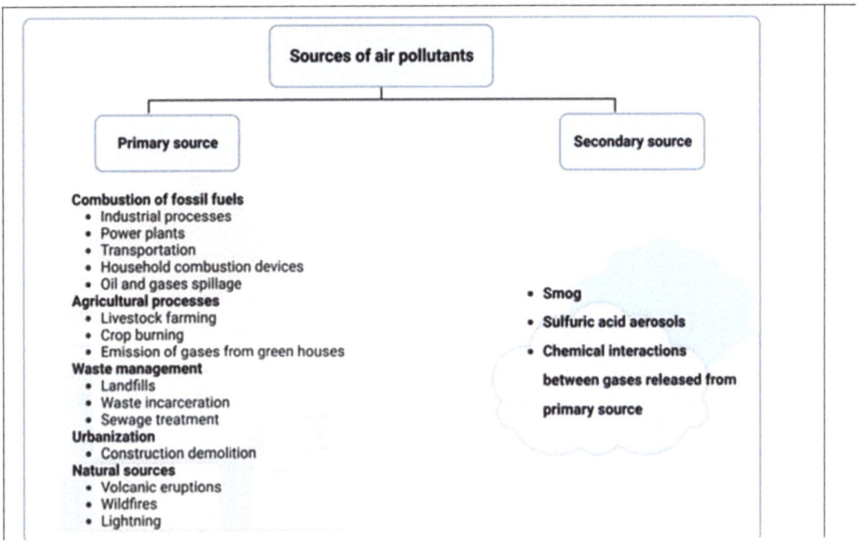

Fig. 1 Sources of air pollutants. Source: Omolaoye et al. [22]. https://creativecommons.org/licenses/by/4.0/)

classes of pollutants into the atmosphere, such as gaseous pollutants (e.g., oxides of sulfur, nitrogen, and ozone), particulate matter (PM2.5 and PM10), trace metals (e.g., lead, cadmium, and arsenic), polycyclic aromatic hydrocarbons, organic compounds (e.g., dioxins and PCBs), volatile organic compounds (e.g., toluene and hexane), and semivolatile organic compounds (e.g., butylated hydroxytoluene and diethyl phthalate) (Fig. 2).

2.1 Source of Air Pollution

Air pollution can be broadly categorized into two types: primary and secondary pollution (Fig. 1). Each type has distinct sources and mechanisms of formation, which contribute to the complex nature of air quality issues worldwide. Primary pollutants are those directly emitted into the atmosphere from identifiable sources involving various human activities and natural processes.

Energy Production and Electricity Generation One of the major sources of primary air pollutants is the burning of fossil fuels for energy production and electricity generation. This includes coal-fired power plants, natural gas facilities, and oil refineries. The combustion of these fuels releases a variety of harmful substances, such as sulfur dioxide (SO2), nitrogen oxides (NOx), carbon monoxide (CO), and particulate matter (PM). These emissions contribute significantly to air pollution and pose serious health risks to nearby populations [23].

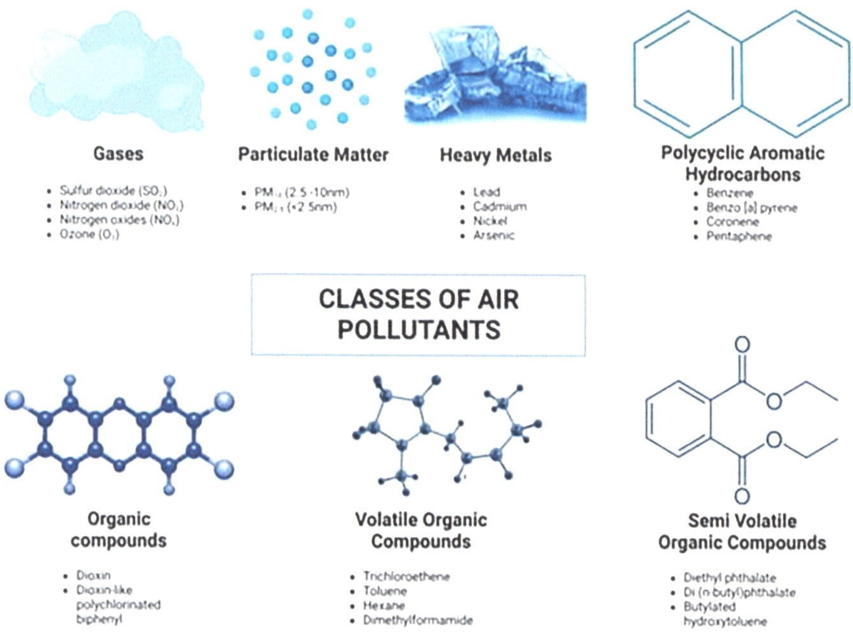

Fig. 2 Sources and classes of air pollutants. Source: Omolaoye et al. [22]. https://creativecommons.org/licenses/by/4.0/)

Transportation The transportation sector, including vehicles such as cars, trucks, buses, and airplanes, is another significant source of primary pollutants. The burning of gasoline and diesel fuels releases pollutants such as nitrogen oxides (NOx), carbon monoxide (CO), and volatile organic compounds (VOCs). Additionally, wear and tear from tires and brake pads contribute to particulate matter (PM) emissions. Urban areas with high traffic volumes often experience elevated levels of these pollutants, leading to poor air quality and health problems for residents.

Agricultural Operations Agricultural activities also contribute to primary air pollution. The use of fertilizers and pesticides, soil tilling, and the management of livestock produce pollutants such as ammonia (NH3), methane (CH4), and nitrous oxide (N2O). These emissions can degrade air quality, particularly in rural areas where agricultural practices are intensive. Furthermore, dust from plowing and harvesting can increase levels of particulate matter in the air.

Waste Management Waste management practices, including burning solid waste in landfills and incinerators, release a complex mix of pollutants into the atmosphere. These pollutants include carbon dioxide (CO2), methane (CH4), and a variety of toxic substances such as dioxins and furans. Open waste burning is particularly harmful [24], releasing large quantities of particulate matter and other hazardous compounds that can affect local and regional air quality.

Manufacturing and Construction Industrial processes, such as the production of metals and cement, release pollutants during manufacturing. These activities emit substances like sulfur dioxide (SO2), nitrogen oxides (NOx), and particulate matter (PM) into the atmosphere that are harmful to both the environment and humans. Construction and demolition projects also generate dust and other particulates, further contributing to air pollution. The disruption of soil and materials during these activities can release both coarse and fine particles into the air [23].

Natural Sources Natural events also contribute to primary air pollution. Volcanic eruptions release large amounts of gases, particulate matter, and aerosols into the atmosphere [25, 26]. Similarly, uncontrolled fire can emit a significant quantity of smoke, consisting of carbon dioxide (CO2), carbon monoxide (CO), and fine particulate matter (PM2.5). These natural sources can cause temporary but severe air quality issues, particularly in regions prone to such events.

Secondary pollutants are not directly emitted but form through chemical reactions between primary pollutants in the atmosphere. These reactions often involve sunlight and other environmental factors. Some of these secondary air pollutants include ground-level ozone, acid rain, and smog.

Ground-Level Ozone (O3) Ground-level ozone is a key secondary pollutant formed when sunlight interacts with volatile organic compounds (VOCs) and nitrogen oxides (NOx) [27, 28]. This photochemical reaction creates ozone (O3) at ground level, which is a major component of smog. Unlike the protective ozone layer in the upper atmosphere, ground-level ozone is harmful to human health, causing respiratory problems and other health issues [29, 30].

Acid Rain Acid rain is formed when nitrogen dioxide (NO2) and sulfur dioxide (SO2) react with water vapor in the atmosphere [31]. Nitrogen dioxide reacts to form nitric acid (HNO3), while sulfur dioxide reacts to produce sulfuric acid (H2SO4) aerosols. These acidic compounds can fall to the ground as precipitation, damaging ecosystems, buildings, and water bodies [32]. Acid rain also contributes to soil acidification, negatively impacting plant and aquatic life [32].

Smog Smog, particularly photochemical smog, forms when nitrogen oxides (NOx) and volatile organic compounds (VOCs) undergo complex chemical reactions in the presence of sunlight [33–35]. This type of smog is prevalent in urban areas with high levels of vehicular and industrial emissions. It can reduce visibility and cause a variety of health problems, including respiratory and cardiovascular issues.

2.2 *Classes of Air Pollutants*

Air pollutants are diverse and can be categorized into several classes based on their chemical composition and physical state. These classes include gases, particulate matter, trace metals, polycyclic aromatic hydrocarbons, organic compounds, volatile organic compounds, and semivolatile organic compounds (Fig. 2).

Gases Gaseous pollutants are a major class of air contaminants and include oxides of sulfur (SOx), nitrogen (NOx), and ozone (O3). Sulfur oxides, primarily sulfur dioxide (SO2), are released from burning fossil fuels such as coal and oil. Nitrogen oxides, including nitrogen dioxide (NO2), are emitted from vehicle engines, power plants, and industrial processes. Ozone (O3), a secondary pollutant, forms in the atmosphere through reactions involving NOx and volatile organic compounds (VOCs) in the presence of sunlight.

Particulate Matter Particulate matter (PM) refers to tiny particles suspended in the air, which can be classified based on their aerodynamic diameter [8]. The two most commonly monitored sizes are PM2.5 and PM10. PM2.5 consists of fine particles with diameters less than 2.5 μm, while PM10 includes coarse particles up to 10 μm in diameter [8]. These particles can be composed of various substances, including dust, soot, smoke, and liquid droplets, originating from both natural and anthropogenic sources.

Trace Metals Trace metals are another important class of air pollutants, involving metals such as lead, cadmium, nickel, arsenic, etc. [36, 37]. These metals can be released into the atmosphere through industrial activities, mining operations, waste incineration, and vehicle emissions. Trace metals are often found in particulate form, adhering to dust and other airborne particles.

Polycyclic Aromatic Hydrocarbons Polycyclic aromatic hydrocarbons (PAHs) are a class of organic compounds composed of multiple aromatic rings [38–40]. Common PAHs include benzene, benzo[a]pyrene, coronene, and pentaphene. These compounds are primarily generated during the incomplete combustion of organic materials such as coal, oil, gas, wood, and garbage. PAHs are of particular concern due to their potential to bind with particulate matter and persist in the environment.

Organic Compounds Organic compounds in the air can be broadly classified into various categories, including dioxins and polychlorinated biphenyls (PCBs) [41]. Dioxins are byproducts of industrial processes, particularly those involving chlorine, such as waste incineration and chemical manufacturing [42]. PCBs, although banned or restricted in many countries, persist in the environment and can still be released from old electrical equipment and other sources [43, 44].

Volatile and Semivolatile Organic Compounds Volatile organic compounds (VOCs) are a large group of carbon-based chemicals that easily evaporate at room temperature [45, 46]. Examples include toluene, hexane, and dimethylformamide. These compounds are emitted from a variety of sources, including vehicle exhaust, industrial processes, and the use of solvents and paints. Semivolatile organic compounds (SVOCs) such as butylated hydroxytoluene (BHT) and diethyl phthalate are less volatile than VOCs but can still evaporate into the air under certain conditions [47–50]. SVOCs are often found in consumer products like plastics, personal care products, and pesticides.

3 Health Impacts of Common Air Pollutants

Gaseous pollutants irritate the respiratory system and contribute to chronic respiratory diseases and mortality (Fig. 3; [52–54]). PM2.5 and PM10 can lead to respiratory infections, lung cancer, and cardiovascular diseases, exacerbating asthma and COPD and increasing hospital admissions and mortality rates [8, 51, 55–57].

Trace metals like lead, cadmium, and arsenic, emitted into the atmosphere from various sources, including industrial processes, pose severe health risks. Lead exposure impacts nearly every organ system, particularly affecting the nervous system. Children exposed to lead may experience developmental delays, learning difficulties, and reduced IQ. Cadmium exposure is associated with kidney damage, bone loss, and lung cancer, while inhalation can lead to emphysema and other lung diseases. Arsenic, a carcinogen, causes skin lesions, cardiovascular issues, neurotoxicity, and diabetes whilst increasing the risk of lung cancer and respiratory diseases (Fig. 4; [58]).

Polycyclic aromatic hydrocarbons (PAHs) are organic compounds with multiple aromatic rings that can bind to particulate matter, increasing their environmental persistence and inhalation risk. Known for their carcinogenic properties, PAHs like benzo[a]pyrene can lead to lung, skin, and bladder cancers. They also cause respiratory issues, immune suppression, and reproductive problems [59–61]. Chronic exposure to PAHs can induce DNA damage and mutations, heightening cancer risks. Dioxins and polychlorinated biphenyls (PCBs), persistent environmental pollutants,

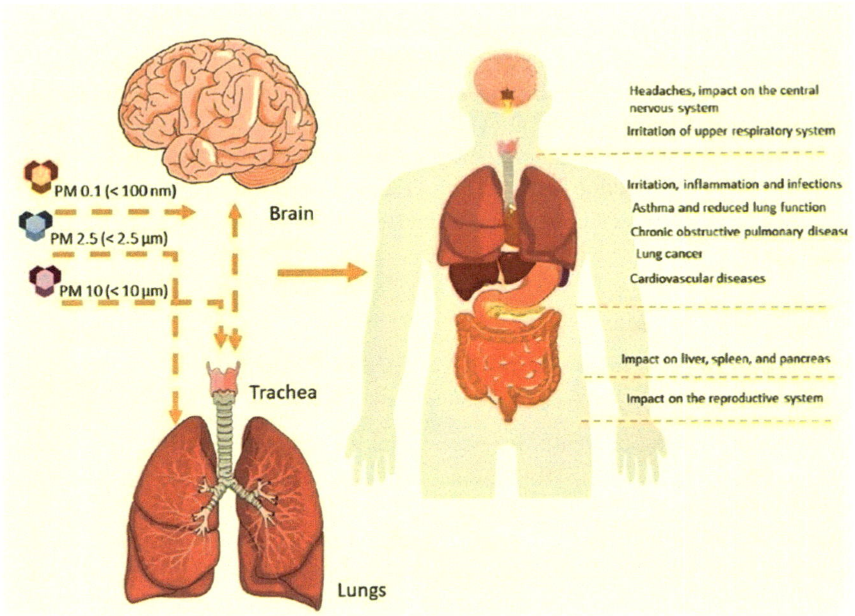

Fig. 3 Health impact of air pollution particularly particulates. Source: Zorena et al. [51]. https://creativecommons.org/licenses/by/4.0/)

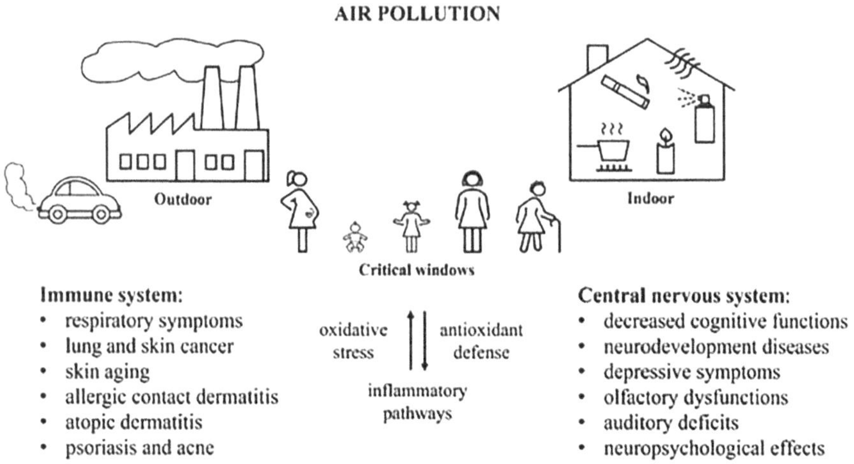

Fig. 4 Impacts of air pollutants on immune and nervous system [58]. https://creativecommons.org/licenses/by/4.0/)

accumulate in the food chain. Dioxins, byproducts of industrial processes, are highly toxic, impacting reproduction, development, immune function, and hormone regulation [62, 63] and causing cancer. PCBs lead to skin conditions like chloracne, liver damage [64, 65], and neurotoxic effects, particularly affecting children's cognitive and motor skills. Volatile organic compounds (VOCs), such as toluene and hexane, and semivolatile organic compounds (SVOCs), like butylated hydroxytoluene (BHT) and diethyl phthalate, originate from various sources, including vehicles and consumer products. VOCs irritate the eyes, nose, and throat, damage the liver and kidneys, and affect the central nervous system [66–68]. Toluene causes headaches and cognitive impairment, while hexane induces nerve damage. SVOCs persist in the environment and bioaccumulate, posing long-term health risks. Diethyl phthalate disrupts endocrine function, impacting reproduction and development [69–71]. BHT, a food preservative, may promote tumors and harm liver and kidney functions [72, 73].

4 Growing Mortality Rates Due to Air Pollution

Air pollution remains a significant public health challenge globally, with various sociodemographic index (SDI) regions exhibiting distinct trends in disability-adjusted life years (DALYs) attributable to air pollution. Dhimal et al. [3] analyzed data on mortality and disability-adjusted life years (DALYs) attributable to air pollution from 1990 to 2019. They reported the impacts of air pollution on the overall burden of disease by sociodemographic index (SDI), gender, and type of pollution (Fig. 5). Figure 5a illustrates the DALY rate per 100,000 population across

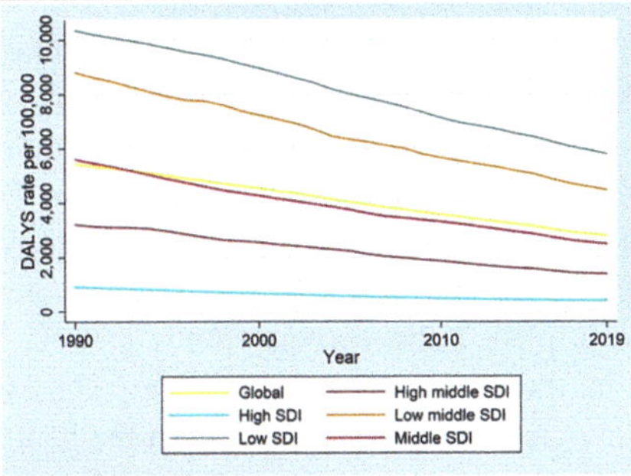

[A] DALYs attributable to air pollution from 1990 to 2019

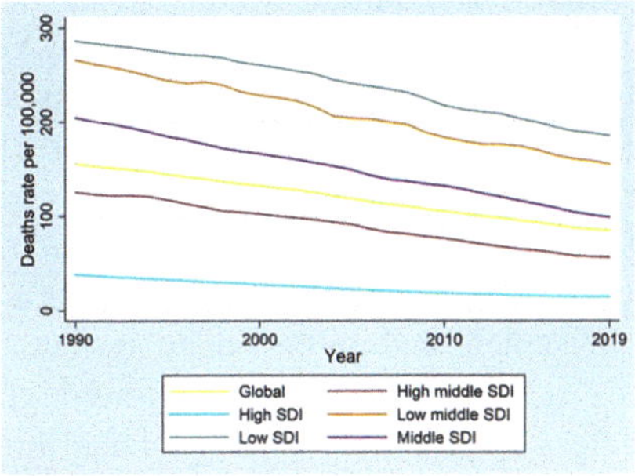

[B] Deaths attributable to air pollution from 1990 to 2019

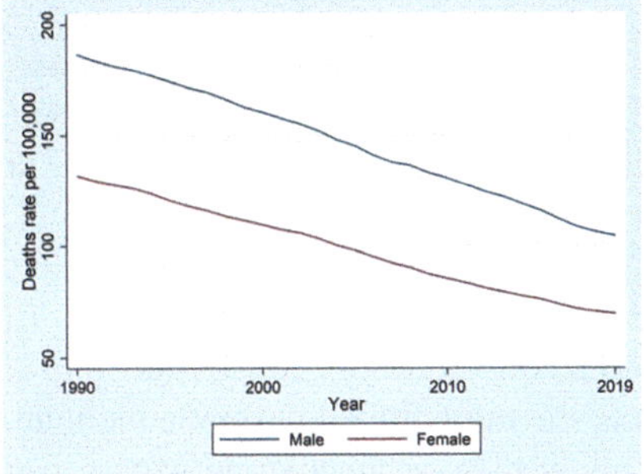

[C] Gender-wise death rate attributable to air pollution from 1990 to 2019.

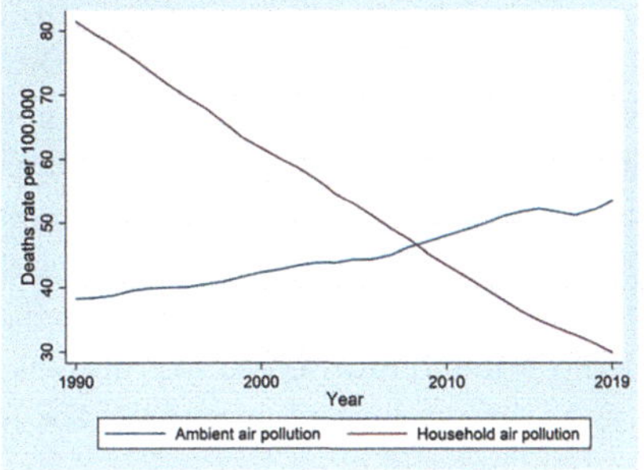

[D]. Different types of air pollution and attributable deaths.

Fig. 5 Air pollution-related DALYs and mortality (1990–2019): sociodemographic, gender-specific, and pollution-type insights [3] https://creativecommons.org/licenses/by/4.0/

different SDI regions from 1990 to 2019, highlighting a global trend of declining DALY rates attributable to air pollution over the 29 years [3]. Globally, the DALY rate attributable to air pollution has declined from 1990 to 2019, reflecting improvements in air quality and public health interventions. This trend indicates a positive shift toward reducing the burden of diseases associated with air pollution. However, the decline in DALY rates is not uniform across different sociodemographic regions, highlighting the disparities in exposure and health outcomes. In 1990, the DALY rate in the low SDI region was significantly high, indicating severe health impacts from air pollution [3]. Despite a gradual reduction over the years, the burden remains substantial, likely due to persistent exposure to high levels of pollutants and limited access to healthcare and pollution mitigation measures. Similarly, the low-middle SDI region has shown a decreasing trend in DALY rates, yet it still records high rates compared to global averages. In the middle SDI region, the DALY rate was higher than the global average in 1990, but began to decline steadily from the mid-1990s [3]. This trend suggests that middle SDI regions have made significant progress in combating air pollution, possibly through enhanced regulatory measures, industrial upgrades, and public health campaigns. The high-middle SDI region exhibited somewhat undulating trends in DALY rates between 1990 and 2008, reflecting fluctuations in air quality and possibly varying implementation of pollution control measures. However, from 2010 to 2019, the DALY rate in this region showed a more consistent decline [3]. The high SDI region has consistently experienced the lowest DALY rate attributable to air pollution from 1990 through 2019. The stability in DALY rates, particularly between 2008 and 2019, indicates that robust air quality management and public health systems are in place. The varying trends in DALY rates across different SDI regions highlight the need for tailored approaches to address air pollution. While global efforts have led to a general decline in the health impacts of air pollution, significant disparities persist, particularly in low and low-middle SDI regions. Public health policies must consider these disparities and prioritize interventions in the most affected areas.

Based on the mortality rate per 100,000 population attributable to air pollution (Fig. 5b), there was a decline in the absolute mortality rates, which varied significantly between regions [3], thereby highlighting the persistent disparities in health outcomes related to air pollution exposure. In the low SDI region, which has consistently experienced the highest death rates attributable to air pollution since 1990, the mortality rate remains significantly higher than the global average. The middle SDI and low middle SDI regions also had higher death rates compared to the global average. In the low middle SDI region, the mortality rate exhibited a unique trend: a slight increase in the late 1990s, followed by an undulating pattern from 1997 to 2015 and a modest decline thereafter [3]. Conversely, the high SDI region has consistently maintained the lowest mortality rate per 100,000 attributable to air pollution from 1990 to 2019 [3]. The steady decline in death rates in this region can be attributed to stringent environmental regulations, advanced pollution control technologies, and a higher standard of healthcare.

Based on the gender-wise death rates attributable to air pollution from 1990 to 2019 (Fig. 5c), the death rate was attributed to air pollution. In 1990, it was significantly higher for males at 185 per 100,000, whereas for females, it was 135 per 100,000 [3]. By 2019, these rates had decreased to 110 for males and 70 for females, respectively [3]. Despite the overall reduction in death rates, the higher susceptibility of males to air pollution-related mortality persists, suggesting that gender-specific factors may influence the health impacts of air pollution.

Several factors could contribute to gender disparity. Occupational exposure is a key consideration, as men are more likely to work in industries with higher exposure to air pollutants, such as construction, manufacturing, and transportation. These occupations often involve prolonged exposure to particulate matter (PM), volatile organic compounds (VOCs), and other hazardous pollutants, which can lead to chronic respiratory and cardiovascular conditions. Additionally, lifestyle factors such as higher smoking rates among men may compound the adverse health effects of air pollution, leading to a higher overall mortality rate.

Another important factor is biological differences between genders. Studies have shown that hormonal differences can influence the body's response to air pollutants [74, 75]. For instance, estrogen in females could offer some protective cardiovascular effects, potentially mitigating the impact of air pollution on heart health. On the other hand, testosterone in males may exacerbate inflammatory responses, increasing vulnerability to the harmful effects of pollutants [76, 77]. Furthermore, socioeconomic factors and healthcare access can also play a role in the observed gender differences. Women, particularly in developing countries, often have limited access to healthcare services, which can affect their ability to receive timely treatment for air pollution-related illnesses, based on the death rate per 100,000 population attributable to ambient air pollution (AAP) and household air pollution (HAP) from 1990 to 2019 (Fig. 5d) [3]. The decline can be attributed to several factors. Primarily, there has been a global shift toward cleaner cooking fuels and technologies, reducing the reliance on biomass, coal, and kerosene, which are major sources of HAP. International efforts, such as the Sustainable Development Goals (SDGs) and specific initiatives like the Global Alliance for Clean Cookstoves, could have played crucial roles in promoting cleaner alternatives. Additionally, increased awareness and improved socioeconomic conditions in many developing countries have contributed to this reduction.

5 Strategies for Achieving Sustainable Clean Air

Reducing air pollution requires a multifaceted approach that combines technological innovations, regulatory measures, community engagement, and economic interventions. The following strategies highlight various methods and provide examples of how they can be implemented to achieve sustainable clean air.

5.1 Regulatory Measures and Policy Interventions

Emission Standards and Regulations Governments can play a crucial role in reducing air pollution by setting stringent emission standards for various sectors, including industries, vehicles, and power plants. For example, the European Union's Euro 6 standards mandate significant reductions in emissions [78, 79] of nitrogen oxides (NOx) and particulate matter (PM) from cars and trucks. These regulations compel manufacturers to develop cleaner technologies and ensure that new vehicles emit fewer pollutants, thus improving overall air quality. Similarly, the Environmental Protection Agency (EPA) enforces the Clean Air Act in the United States, which sets national standards for air pollutants and regulates emissions from stationary and mobile sources [80, 81].

Air Quality Monitoring and Enforcement The implementation of robust air quality monitoring systems is essential for enforcing regulations and identifying pollution hotspots. Also, advanced monitoring networks provide real-time data on pollutant levels, enabling authorities to take timely actions. The USEPA uses the air quality index (AQI) to inform the public about daily air quality levels and issue health advisories when necessary [82–84]. This system not only raises public awareness, but also helps enforce compliance with air quality standards by identifying areas where pollution levels exceed permissible limits.

Low Emission Zones (LEZs) Cities can combat air pollution by designating low-emission zones where only vehicles meeting specific emission standards are allowed. London's ultra-low emission zone (ULEZ) is a prime example [85], charging higher fees for high-polluting vehicles while incentivizing cleaner transportation options. This measure could result in a significant reduction in NOx and PM levels, improving air quality and public health. Similar zones have been implemented in cities like Berlin and Milan, demonstrating the effectiveness of LEZs in mitigating urban air pollution [86, 87].

5.2 Technological Innovations

Clean Energy Technologies Transitioning to renewable energy sources such as wind, solar, and hydroelectric power is a fundamental strategy for reducing air pollution. By decreasing reliance on fossil fuels, these technologies lower greenhouse gas emissions and other pollutants. Denmark, for instance, has made substantial investments in wind energy [88–90], which now supplies a substantial part of the country's electricity needs. This shift not only reduces air pollution but also promotes energy security and sustainability.

Electric and Hybrid Vehicles The use and promotion of electric and hybrid vehicles can significantly reduce emissions from the transportation sector. Norway offers substantial tax incentives for electric vehicle (EV) buyers, including exemptions from purchase taxes and tolls [91, 92]. As a result, EVs make up substantial number of all new car sales in the country. This transition to cleaner vehicles has led to a notable decrease in urban air pollution from vehicular emission.

Emission Control Technologies The installation of advanced emission control technologies can significantly reduce harmful emissions from various sources. For instance, catalytic converters in vehicles convert toxic gases such as CO and NOx into less harmful substances before they are emitted [93, 94]. Industrial smokestacks equipped with scrubbers can remove pollutants like sulfur dioxide (SO2) and particulate matter from exhaust gases [95]. Additionally, the use of selective catalytic reduction (SCR) systems in diesel engines helps reduce NOx emissions by up to 90%, making a substantial impact on air quality [96].

5.3 Sustainable Urban Planning

Green Spaces and Urban Forests The integration of green spaces and urban forests into city planning can greatly improve air quality. Trees and plants absorb pollutants such as CO2, NO2, and particulate matter while producing oxygen, thus enhancing the urban environment [97, 98]. These green spaces serve as natural filters, reducing pollution levels and providing residents with healthier living conditions.

Public Transportation Systems Investing in efficient and accessible public transportation systems is essential for reducing the number of private vehicles on the road, thus lowering emissions. Curitiba, Brazil, is renowned for its bus rapid transit (BRT) system, which has significantly reduced traffic congestion and pollution [99–101]. By offering a reliable and affordable alternative to private car use, Curitiba's BRT system encourages more people to use public transport, resulting in cleaner air and a more sustainable urban environment.

Pedestrian and Bicycle Infrastructure Developing pedestrian-friendly and bicycle-friendly infrastructure encourages nonmotorized transportation, reducing vehicular emissions. Amsterdam is a leading example, with an extensive lane network and pedestrian zones promoting cycling and walking [102, 103]. This infrastructure do not only reduce air pollution but also improves public health by encouraging physical activity. Other cities, like Copenhagen and Portland, have also invested in similar infrastructure, demonstrating the effectiveness of promoting nonmotorized transport in achieving sustainable clean air.

5.4 Community Engagement and Education

Public Awareness Campaigns Educating the public about the sources and health impacts of air pollution is crucial for encouraging behavioral changes that can lead to improved air quality. This campaign could raise awareness about the adverse effects of human activities such as fireworks on air quality and health. Public awareness campaigns like this can effectively change social norms and inspire individuals to adopt practices that reduce air pollution.

Citizen Science Initiatives Engaging communities in monitoring air quality through citizen science projects empower individuals to contribute to data collection and pollution mitigation efforts. The Air Quality Egg project is a notable example, where citizens can build and use their own air quality monitoring devices [104, 105]. These devices help individuals track local air quality, fostering a sense of ownership and responsibility toward the environment. Citizen science initiatives not only provide valuable data for researchers and policymakers but also educate and mobilize the public to take action against air pollution.

School Programs and Curricula The integration of air quality education into school curricula fosters environmental stewardship from a young age. Programs like the Clean Air Campaign in Georgia, USA, educate students about air pollution and encourage actions to reduce emissions. By teaching children about the causes and consequences of air pollution, these programs could inspire future generations to prioritize air quality and engage in sustainable practices.

5.5 *Economic and Social Interventions*

Subsidies and Incentives for Clean Technologies The provision of financial incentives for adopting clean technologies encourages their widespread use [91, 92]. Subsidies and incentives can make clean technologies more affordable and attractive, accelerating their integration into mainstream use. These economic interventions could help reduce reliance on fossil fuels and lower overall emissions, contributing to cleaner air.

Pollution Taxes and Carbon Pricing The implementation of taxes on high-polluting activities and carbon pricing mechanisms internalizes the environmental costs of pollution. Sweden's carbon tax, introduced in 1991 [106], has significantly reduced greenhouse gas emissions while supporting economic growth. By putting a price on carbon, these measures can create a financial incentive for businesses and individuals to reduce their carbon footprint. Pollution taxes and carbon pricing can encourage the development and adoption of cleaner technologies, leading to a decrease in emissions and improvement in air quality [107, 108].

Support for Research and Development Investing in research and development of new technologies and pollution control methods fosters innovation and long-term solutions. The European Union's Horizon 2020 program funds research projects focused on sustainable development and air quality improvement [109–111]. By supporting scientific research and technological innovation, governments and organizations can develop more effective strategies for combating air pollution.

5.6 *International Cooperation and Agreements*

Global Initiatives and Protocols The participation in international agreements, such as the Paris Agreement, commits countries to reducing greenhouse gas emissions and combating climate change. The Montreal Protocol, aimed at phasing out ozone-depleting substances [112–114], has also contributed to air quality improvements. These global initiatives provide a framework for countries to collaborate on addressing air pollution and climate change.

6 Conclusion

Most air pollutants have detrimental health impacts on humans. Thus, there is pressing need to achieve sustainable clean air. Achieving sustainable clean air necessitates a robust approach that combines regulatory measures, technological advancements, sustainable urban planning, community engagement, economic interventions, and international cooperation. Through the implementation of stringent emission standards, the establishment of robust air quality monitoring systems, and the creation of low-emission zones, governments can significantly reduce air pollution. Technological advancements such as clean energy sources, electric vehicles, and emission control technologies further contribute to this goal. Additionally, sustainable urban planning, including the integration of green spaces, efficient public transportation, and pedestrian-friendly infrastructure, plays a crucial role in reducing emissions and improving air quality. Public awareness campaigns, citizen science initiatives, and school programs play crucial roles in educating and mobilizing individuals to take action against air pollution. Economic incentives, pollution taxes, and investments in research and development drive the adoption of cleaner technologies. Furthermore, international agreements can ensure that efforts to improve air quality are coordinated and effective. Through the implementation of these strategies and fostering a culture of environmental stewardship, air pollution can be significantly reduced, leading to improved public health and environmental quality for future generations.

References

1. Cohen AJ, Brauer M, Burnett R, Anderson HR, Frostad J, Estep K, Balakrishnan K, Brunekreef B, Dandona L, Dandona R et al (2017) Estimates and 25-year trends of the global burden of disease attributable to ambient air pollution: an analysis of data from the global burden of diseases study 2015. Lancet 389:1907–1918
2. World Health Organization (2018) Noncommunicable diseases country profiles 2018. file:///C:/Users/hp/Downloads/9789241514620-eng-1.pdf
3. Dhimal M, Chirico F, Bista B, Sharma S, Chalise B, Dhimal ML, Ilesanmi OS, Trucillo P, Sofia D (2021) Impact of air pollution on global burden of disease in 2019. Processes 9(10):1719
4. World Health Organization (2022) Ambient (outdoor) air pollution. Available online: https://www.who.int/news-room/fact-sheets/detail/ambient-(outdoor)-air-quality-and-health
5. Shaddick G, Thomas ML, Mudu P, Ruggeri G, Gumy S (2020) Half the world's population are exposed to increasing air pollution. NPJ Clim Atmos Sci 3:1–5
6. World Health Organization (2018) Health benefits far outweigh the costs of meeting climate change goals. Available online: https://www.who.int/news/item/05-12-2018-health-benefits-far-outweigh-the-costs-of-meeting-climate-change-goals
7. Landrigan PJ, Fuller R, Acosta NJR, Adeyi O, Arnold R, Basu NN, Baldé AB, Bertollini R, Bose-O'Reilly S, Boufford JI et al (2018) The lancet commission on pollution and health. Lancet 391:462–512

8. Richard G, Sawyer WE, Izah SC (2021) Health implications of air quality index of fine and coarse particulates during outdoor combustion of biomass in The Niger Delta, Nigeria. Int J World Policy Dev Stud 7(2):21–26
9. World Health Organization (2019) Ambient air pollution: training for health care providers (No. WHO/CED/PHE/EPE/19.12. 14). World Health Organization
10. Ohimain EI, Izah SC, Abah SO (2013) Air quality impacts of smallholder oil palm processing in Nigeria. J Enviro Protection 4:83–98
11. Das I, Lewis JJ, Ludolph R, Bertram M, Adair-Rohani H, Jeuland M (2021) The benefits of action to reduce household air pollution (BAR-HAP) model: a new decision support tool. PLoS One 16(1):e0245729
12. Ahmed F, Hossain S, Hossain S, Fakhruddin ANM, Abdullah ATM, Chowdhury MAZ, Gan SH (2019) Impact of household air pollution on human health: source identification and systematic management approach. SN Appl Sci 1:1–19
13. Shrivastava P, Shrivastava SR (2018) exploRing the impacT of household aiR polluTion on womens healTh: a naRRaTive review. Ann SBV 7(1)
14. Bhattacharyya SC (2012) Energy access programmes and sustainable development: a critical review and analysis. Energy Sustain Dev 16(3):260–271
15. Kaygusuz K (2012) Energy for sustainable development: a case of developing countries. Renew Sustain Energy Rev 16(2):1116–1126
16. Puzzolo E, Pope D (2017) Clean fuels for cooking in developing countries. Encyclopedia of sustainable technologies, pp 289–297
17. Mallhi TH, Butt MH, Ahmad A, Misbah S, Atique S, Khan YH (2021) Air pollutants and neurological disorders: from exposure to preventive interventions. In: Environmental contaminants and neurological disorders. Springer, Cham, pp 31–47
18. Hahad O, Frenis K, Kuntic M, Daiber A, Münzel T (2021) Accelerated aging and age-related diseases (CVD and neurological) due to air pollution and traffic noise exposure. Int J Mol Sci 22(5):2419
19. Bălă GP, Râjnoveanu RM, Tudorache E, Motişan R, Oancea C (2021) Air pollution exposure—the (in) visible risk factor for respiratory diseases. Environ Sci Pollut Res 28(16): 19615–19628
20. Franchini M, Mannucci PM, Harari S, Pontoni F, Croci E (2015) The health and economic burden of air pollution. Am J Med 128(9):931–932
21. World Health Organization (2016) Ambient air pollution: a global assessment of exposure and burden of disease
22. Omolaoye TS, Skosana BT, Ferguson LM, Ramsunder Y, Ayad BM, Du Plessis SS (2024) Implications of exposure to air pollution on male reproduction: the role of oxidative stress. Antioxidants 13(1):64
23. Seiyaboh EI, Izah SC (2019) Impacts of soil pollution on air quality under Nigerian setting. J Soil Water Sci 3(1):45–53
24. Izah SC, Angaye CN, Aigberua AO, Nduka JO (2017) Uncontrolled bush burning in The Niger Delta region of Nigeria: potential causes and impacts on biodiversity. Int J Mol Ecol Conserv 7(1):1–15
25. von Glasow R, Bobrowski N, Kern C (2009) The effects of volcanic eruptions on atmospheric chemistry. Chem Geol 263(1–4):131–142
26. Self S (2006) The effects and consequences of very large explosive volcanic eruptions. Philos Trans R Soc A Math Phys Eng Sci 364(1845):2073–2097
27. Pugliese SC, Murphy JG, Geddes JA, Wang JM (2014) The impacts of precursor reduction and meteorology on ground-level ozone in the greater Toronto area. Atmos Chem Phys 14(15):8197–8207
28. Hui K, Yuan Y, Xi B, Tan W (2023) A review of the factors affecting the emission of the ozone chemical precursors VOCs and NOx from the soil. Environ Int 172:107799
29. Liu H, Liu S, Xue B, Lv Z, Meng Z, Yang X, Xue T, Yu Q, He K (2018) Ground-level ozone pollution and its health impacts in China. Atmos Environ 173:223–230

30. Soares AR, Silva C (2022) Review of ground-level ozone impact in respiratory health deterioration for the past two decades. Atmos 13(3):434
31. Shammas NK, Wang LK, Wang MHS (2020) Sources, chemistry and control of acid rain in the environment. In: Handbook of environment and waste management: acid rain and greenhouse gas pollution control, pp 1–26
32. Seiyaboh EI, Izah SC (2017) A review of impacts of gas flaring on vegetation and water resources in The Niger Delta Region of Nigeria. Int J Econ Energy Environ 2(4):48–55
33. Falkowska L (2016) Environmental characteristics of gaseous pollutants and related adverse health effects. Synergic influence of gaseous, particulate, and biological pollutants on human health. CRC Press, Boca Raton, pp 3–38
34. Gera H, Bhasin M (2023) Photochemical smog – a review. J Sci Res Teach 2(7):210–228
35. Rani B, Singh U, Chuhan AK, Sharma D, Maheshwari R (2011) Photochemical smog pollution and its mitigation measures. J Adv Sci Res 2(04):28–33
36. Izah SC, Uzoekwe SA, Aigberua AO (2021) Source, geochemical spreading and risks of trace metals in particulate matter 2.5 within a gas flaring area in Bayelsa state, Nigeria. Adv Environ Technol 7(2):101–118
37. Uzoekwe SA, Izah SC, Aigberua AO (2021) Environmental and human health risk of heavy metals in atmospheric particulate matter (PM_{10}) around gas flaring vicinity in Bayelsa state, Nigeria. Toxicol Environ Health Sci 13(4):323–335. https://doi.org/10.1007/s13530-021-00085-7
38. Boehm PD (1964) Polycyclic aromatic hydrocarbons (PAHs). In: Environmental forensics. Academic Press, pp 313–337
39. Lee BK (2010) Sources, distribution and toxicity of polyaromatic hydrocarbons (PAHs) in particulate matter. In: Air pollution. IntechOpen
40. Pampanin DM, Sydnes MO (2013) Polycyclic aromatic hydrocarbons a constituent of petroleum: presence and influence in the aquatic environment. Hydrocarbon 5:83–118
41. Kanan S, Samara F (2018) Dioxins and furans: a review from chemical and environmental perspectives. Trends Environ Anal Chem 17:1–13
42. Fiedler H, Hutzinger O, Timms CW (1990) Dioxins: sources of environmental load and human exposure. Toxicol Environ Chem 29(3):157–234
43. Ross G (2004) The public health implications of polychlorinated biphenyls (PCBs) in the environment. Ecotoxicol Environ Saf 59(3):275–291
44. Ododo MM, Wabalo BK (2019) Polychlorinated biphenyls (PCBs) and their impacts on human health: a review. J Environ Pollut Hum Health 7(2):73–77
45. Zhang X, Gao B, Creamer AE, Cao C, Li Y (2017) Adsorption of VOCs onto engineered carbon materials: a review. J Hazard Mater 338:102–123
46. Khatib M, Haick H (2022) Sensors for volatile organic compounds. ACS Nano 16(5): 7080–7115
47. Weschler CJ, Nazaroff WW (2008) Semivolatile organic compounds in indoor environments. Atmos Environ 42(40):9018–9040
48. Cecchi T (2014) Identification of representative pollutants in multiple locations of an Italian school using solid phase micro extraction technique. Build Environ 82:655–665
49. Cabanes A, Valdés FJ, Fullana A (2020) A review on VOCs from recycled plastics. Sustain Mater Technol 25:e00179
50. Parthasarathy S (2016) Understanding exposures to volatile and semivolatile organic compounds in indoor environments (Doctoral dissertation, UC Berkeley)
51. Zorena K, Jaskulak M, Michalska M, Mrugacz M, Vandenbulcke F (2022) Air pollution, oxidative stress, and the risk of development of type 1 diabetes. Antioxidants 11:1908. https://doi.org/10.3390/antiox11101908
52. Greenberg N, Carel RS, Derazne E, Bibi H, Shpriz M, Tzur D, Portnov BA (2016) Different effects of long-term exposures to SO2 and NO2 air pollutants on asthma severity in young adults. J Toxic Environ Health A 79(8):342–351

53. Khaniabadi YO, Daryanoosh M, Sicard P, Takdastan A, Hopke PK, Esmaeili S, De Marco A, Rashidi R (2018) Chronic obstructive pulmonary diseases related to outdoor PM 10, O 3, SO 2, and NO 2 in a heavily polluted megacity of Iran. Environ Sci Pollut Res 25:17726–17734
54. Patocka J, Kuca K (2014) Irritant compounds: respiratory irritant gases. Milit Med Sci Lett 83(2):73–82
55. Li T, Hu R, Chen Z, Li Q, Huang S, Zhu Z, Zhou LF (2018) Fine particulate matter (PM2. 5): the culprit for chronic lung diseases in China. Chronic Dis Transl Med 4(03):176–186
56. Li X, Liu X (2021) Effects of PM2. 5 on chronic airway diseases: a review of research progress. Atmos 12(8):1068
57. Misiukiewicz-Stepien P, Paplinska-Goryca M (2021) Biological effect of PM10 on airway epithelium-focus on obstructive lung diseases. Clin Immunol 227:108754
58. Serafini MM, Maddalon A, Iulini M, Galbiati V (2022) Air pollution: possible interaction between the immune and nervous system? Int J Environ Res Public Health 19(23):16037
59. Ravanbakhsh M, Yousefi H, Lak E, Ansari MJ, Suksatan W, Qasim QA, Asban P, Kianizadeh M, Mohammadi MJ (2023) Effect of polycyclic aromatic hydrocarbons (PAHs) on respiratory diseases and the risk factors related to cancer. Polycycl Aromat Compd 43(9): 8371–8387
60. Bukowska B, Mokra K, Michałowicz J (2022) Benzo [a] pyrene – environmental occurrence, human exposure, and mechanisms of toxicity. Int J Mol Sci 23(11):6348
61. Venkatraman G, Giribabu N, Mohan PS, Muttiah B, Govindarajan V, Alagiri M, Rahman PSA, Karsani SA (2024) Environmental impact and human health effects of polycyclic aromatic hydrocarbons and remedial strategies: a detailed review. Chemosphere:141227
62. Szajner J, Czarny-Dzialak M, Dziechciaz M, Pawlas N, Walosik A (2021) Dioxin-like compounds (DLCs) in the environment and their impact on human health. J Elem 26(2)
63. Shree P, Mohit K, Singh IK, Singh DK (2020) Dioxin–exposure routes, pathways, and human health implications. In: Dioxin. CRC Press, pp 83–112
64. Reggiani G, Bruppacher R (1985) Symptoms, signs and findings in humans exposed to PCBs and their derivatives. Environ Health Perspect 60:225–232
65. Kimbrough RD, Barnes MM, Greig JB (1973) The toxicity of polychlorinated polycyclic compounds and related chemicals. CRC Crit Rev Toxicol 2(4):445–498
66. Rouf Z, Dar IY, Javaid M, Dar MY, Jehangir A (2022) Volatile organic compounds emission from building sector and its adverse effects on human health. Ecol Health Effects Building Mater:67–86
67. Tsai WT (2019) An overview of health hazards of volatile organic compounds regulated as indoor air pollutants. Rev Environ Health 34(1):81–89
68. Niaz K, Bahadar H, Maqbool F, Abdollahi M (2015) A review of environmental and occupational exposure to xylene and its health concerns. EXCLI J 14:1167
69. Basso CG, de Araújo-Ramos AT, Martino-Andrade AJ (2022) Exposure to phthalates and female reproductive health: a literature review. Reprod Toxicol 109:61–79
70. Witorsch RJ, Thomas JA (2010) Personal care products and endocrine disruption: a critical review of the literature. Crit Rev Toxicol 40(Suppl 3):1–30
71. Lyche JL, Gutleb AC, Bergman Å, Eriksen GS, Murk AJ, Ropstad E, Saunders M, Skaare JU (2009) Reproductive and developmental toxicity of phthalates. J Toxic Environ Health, Part B 12(4):225–249
72. Malkinson AM (1983) Putative mutagens and carcinogens in foods III. Butylated hydroxytoluene (BHT). Environ Mutagen 5(3):353–362
73. Ito N, Fukushima S, Tsuda H (1985) Carcinogenicity and modification of the carcinogenic response by BHA, BHT, and other antioxidants. CRC Crit Rev Toxicol 15(2):109–150
74. Clougherty JE (2010) A growing role for gender analysis in air pollution epidemiology. Environ Health Perspect 118(2):167–176
75. Thomson EM (2019) Air pollution, stress, and allostatic load: linking systemic and central nervous system impacts. J Alzheimers Dis 69(3):597–614

76. Montano L, Donato F, Bianco PM, Lettieri G, Guglielmino A, Motta O, Bonapace IM, Piscopo M (2021) Air pollution and COVID-19: a possible dangerous synergy for male fertility. Int J Environ Res Public Health 18(13):6846
77. Dutta S, Sengupta P, Slama P, Roychoudhury S (2021) Oxidative stress, testicular inflammatory pathways, and male reproduction. Int J Mol Sci 22(18):10043
78. Rodríguez F, Bernard Y, Dornoff J, Mock P (2019) Recommendations for post-Euro 6 standards for light-duty vehicles in the European Union. Communications 49(30):847102–847129
79. Selleri T, Melas AD, Joshi A, Manara D, Perujo A, Suarez-Bertoa R (2021) An overview of lean exhaust denox aftertreatment technologies and nox emission regulations in the european union. Catalysts 11(3):404
80. National Research Council (2006) State and federal standards for mobile-source emissions. Division on Earth, Life Studies, Board on Environmental Studies, & Committee on State Practices in Setting Mobile Source Emissions Standards. National Academies Press
81. Li WW (2020) Air pollution, air quality, vehicle emissions, and environmental regulations. In: Traffic-related air pollution. Elsevier, pp 23–49
82. Mirabelli MC, Ebelt S, Damon SA (2020) Air quality index and air quality awareness among adults in the United States. Environ Res 183:109185
83. Mirabelli MC, Sarnat SE, Damon SA (2019) Air quality index and air quality awareness among adults in the United States. In: C45. Effects of the environment on pulmonary health. American Thoracic Society, p A4909
84. Brook RD, Rajagopalan S, Al-Kindi S (2024) Public health relevance of US EPA air quality index activity recommendations. JAMA Netw Open 7(4):e245292
85. Wagner R (2023) Developing the understanding of air pollutant emissions in urban areas: using new analysis and measurement approaches (Doctoral dissertation, University of York)
86. Ezeah C, Finney K, Nnajide C (2015) A critical review of the effectiveness of low emission zones (lez) as a strategy for the management of air quality in major European cities. J Multidiscip Eng Sci Technol 2(7):1860–1868
87. Bernardo V, Fageda X, Flores-Fillol R (2020) Mitigating congestion in urban areas: urban tolls versus low emission zones *apeles de E*, 65
88. Sovacool BK (2013) Energy policymaking in Denmark: implications for global energy security and sustainability. Energy Policy 61:829–839
89. Hvelplund F, Østergaard PA, Meyer NI (2017) Incentives and barriers for wind power expansion and system integration in Denmark. Energy Policy 107:573–584
90. Meyer NI (2007) Learning from wind energy policy in the EU: lessons from Denmark, Sweden and Spain. Eur Environ 17(5):347–362
91. Bjerkan KY, Nørbech TE, Nordtømme ME (2016) Incentives for promoting battery electric vehicle (BEV) adoption in Norway. Transp Res Part D: Transp Environ 43:169–180
92. Mersky AC, Sprei F, Samaras C, Qian ZS (2016) Effectiveness of incentives on electric vehicle adoption in Norway. Transp Res Part D: Transp Environ 46:56–68
93. Dey S, Mehta NS (2020) Automobile pollution control using catalysis. Resour Environ Sustain 2:100006
94. Manojkumar R, Haranethra S, Muralidharan M, Ramaprabhu A (2021) IC engine emission reduction using catalytic converter by replacing the noble catalyst and using copper oxide as the catalyst. Mater Today Proc 45:769–773
95. Mkhize PL (2020) Use of air dispersion modelling to determine the impact of gas emissions from coal-fired boilers in South African Durban basin (Doctoral dissertation)
96. Johnson TV (2014) Review of selective catalytic reduction (SCR) and related technologies for mobile applications. In: Urea-SCR technology for deNOx after treatment of diesel exhausts. New York, Springer, pp 3–31
97. Abd El Aziz NG, Mahgoub MH, Azza MMM, Farahat MM, Abouziena HF (2015) Potentiality of ornamental plants and woody trees as phytoremediators of pollutants in the air: a review. Int J ChemTech Res 8:468–482

98. Buraerah MF, Patandjengi B, Suryani S, Hamzah A, Demmalino EB (2023) The effect of vegetation in reducing air pollution in an urban environment: a review. In: IOP conference series: earth and environmental science, vol 1253. IOP Publishing, p 012105
99. Ni X (2018) Solutions to public transport challenges: the solution of urban public transportation–the development and application of BRT. Papers presented at the 2018 37th southern African transport conference 9-12 July 2018 Pretoria, South Africa. Theme "Towards a desired transport future: safe, sufficient and affordable". http://hdl.handle.net/2263/69599
100. Bonicelli E (2015) Social sustainability and mobility in Curitiba: bus rapid transit in the "green city" of Brazil (Doctoral dissertation, University of Illinois at Urbana-Champaign)
101. Gustafsson HR, Kelly EA (2016) Developing the sustainable city: Curitiba, Brazil, as a case study. In: How cities will save the world. Routledge, pp 81–96
102. Koglin T, te Brömmelstroet M, van Wee B (2021) Cycling in Copenhagen and Amsterdam. In: Buehler R, Pucher J (eds) Urban and industrial environments. MIT Press. https://doi.org/10.7551/mitpress/11963.003.0022
103. Feddes F, de Lange M, te Brömmelstroet M (2020) Hard work in paradise. The contested making of Amsterdam as a cycling city. In: The politics of cycling infrastructure. Policy Press, pp 133–156
104. Castell N, Viana M, Minguillón MC, Guerreiro C, Querol X (2013) Real-world application of new sensor technologies for air quality monitoring. ETC/ACM Technical Paper, 16, 34
105. Ramsay DB (2016) LearnAir: toward intelligent, *Personal air quality monitoring* (Doctoral dissertation, Massachusetts Institute of Technology, School of Architecture and Planning, Program in Media Arts and Sciences)
106. Shmelev SE, Speck SU (2018) Green fiscal reform in Sweden: econometric assessment of the carbon and energy taxation scheme. Renew Sustain Energy Rev 90:969–981
107. Shahzad U (2020) Environmental taxes, energy consumption, and environmental quality: theoretical survey with policy implications. Environ Sci Pollut Res 27(20):24848–24862
108. Akhtar MZ, Zaman K, Rehman FU, Nassani AA, Haffar M, Abro MMQ (2022) Evaluating pollution damage function through carbon pricing, renewable energy demand, and cleaner technologies in China: blue versus green economy. Environ Sci Pollut Res 29(17):24878–24893
109. Mazzucato M (2018) Mission-oriented research & innovation in the European Union. European Commission, 36
110. Curzio AQ, Silvani A (2020) Research and development in European Union Politics. In: Oxford research encyclopedia of politics. https://doi.org/10.1093/acrefore/9780190228637.013.1050
111. Pacheco-Torgal F (2014) Eco-efficient construction and building materials research under the EU Framework Programme Horizon 2020. Construct Build Mater 51:151–162
112. Madronich S, Bernhard GH, Neale PJ, Heikkilä A, Andersen MS, Andrady AL, Aucamp PJ, Bais AF, Banaszak AT, Barnes PJ, Neale RE (2024) Continuing benefits of the Montreal Protocol and protection of the stratospheric ozone layer for human health and the environment. Photochem Photobiol Sci:1–29
113. Canan P, Andersen SO, Reichman N, Gareau B (2015) Introduction to the special issue on ozone layer protection and climate change: the extraordinary experience of building the Montreal Protocol, lessons learned, and hopes for future climate change efforts. J Environ Stud Sci 5(2):111–121
114. Mintz JA (1991) Progress toward a healthy sky: an assessment of the London amendments to the Montreal protocol on substances that deplete the ozone layer. Yale J Int'l L 16:571

The manufacturer's authorised representative in the EU is Springer Nature Customer Service Centre GmbH, Europaplatz 3, 69115 Heidelberg, Germany. If you have any concerns regarding our products, please contact ProductSafety@springernature.com

Printed and bound by CPI Group (UK) Ltd, Croydon, CR0 4YY
01/10/2026
02225969-0019